The Master Handbook of IC Circuits

About the Author

Delton T. Horn is a professional technical writer who has written more than three dozen books on computers and electronics, including *The Master Handbook of IC Circuits, Second Edition; Electronic Alarm and Security Systems: A Technician's Guide; Basic Electronics Theory, Fourth Edition;* and several others published by TAB/McGraw-Hill. He is also the author of the book *Comedy Improvisation.* A former electronics technician and alarm systems specialist, Mr. Horn wrote some of the original software for the Tandy computers. He lives in Glendale, Arizona, with his cats Isaac and Smoky.

The Master Handbook of IC Circuits

Third Edition

Delton T. Horn

McGraw-Hill

New York San Francisco Washington, D.C. Auckland Bogotá
Caracas Lisbon London Madrid Mexico City Milan
Montreal New Delhi San Juan Singapore
Sydney Tokyo Toronto

Library of Congress Cataloging-in-Publication Data

Horn, Delton T.
The master handbook of IC circuits / Delton T. Horn. — 3rd ed.
p. cm.
Includes index.
ISBN 0-07-030562-5. — ISBN 0-07-030563-3 (pbk.)
1. Integrated circuits—Handbooks, manuals, etc. I. Title.
TK7874.H678 1997 97-7937
621.3815—dc21 CIP

McGraw-Hill

A Division of The ***McGraw-Hill*** *Companies*

1 2 3 4 5 6 7 8 9 0 FGR/FGR 9 0 2 1 0 9 8 7

ISBN 0-07-030562-5 (HC)
0-07-030563-3 (PBK)

The sponsoring editor for this book was Scott Grillo, the editing supervisor was Scott Amerman, and the production supervisor was Don Schmidt. It was set in ITC-Century Light by Jana Fisher through the services of Barry E. Brown (Broker—Editing, Design and Production).

Printed and bound by Quebecor/Fairfield.

McGraw-Hill books are available at special quantity discounts to use as premiums and sales promotions, or for use in corporate training programs. For more information, please write to the Director of Special Sales, McGraw-Hill, 11 West 19th Street, New York, NY 10011. Or contact your local bookstore.

Contents

Introduction

This newly revised edition of *The Master Handbook of IC Circuits* is a one-source reference guide containing hundreds of practical circuits or bu-circuits using dozens of different integrated circuits, both digital and linear. This book is a handy addition to any electronics bookshelf or workbench, whether you are a professional technician, or a hobbyist. It is intended as a companion volume to *The Master IC Cookbook*, Third Edition, which uses the same section categories, and provides more extensive data on hundreds of currently available ICs of many differing types.

At this writing, all of the ICs described in this volume have been confirmed as commercially available from major hobbyist suppliers. However, remember that electronics is always a rapidly changing field, and some chips may become obsolete and unavailable without warning.

This new edition has been rearranged to make information on specific types of devices easier to find. The circuits are divided into ten sections, each spotlighting a particular type of integrated circuit. Digital devices are covered in Secs. 1 through 3, and linear/analog devices are featured in Secs. 4 through 9. Section 10 deals with the special-case devices which combine both analog and digital circuitry and functions.

In each section, ICs are listed in numerical order. Any prefix numbers are generally ignored. In most cases, the prefix letters simply indicate the manufacturer. For example, National Semiconductor uses LM, Rohm uses BA and BU, Motorola often uses MC, and Signetics uses NE. In most cases, prefix letters can be safely ignored in parts selection. For example, an LM567 is exactly the same device as an NE567. The only difference is who actually manufactured that particular unit.

Suffix letters are used for various purposes. Sometimes they indicate a case style, an operating temperature rating, or an improved version of an older device. In most cases, suffix letters can also be ignored, but be aware that there are some exceptions, which will be noted in the chip description, when appropriate. When in doubt about chip compatibility, always check the relevant specification sheets, whether in this book, or those supplied by the manufacturer/supplier.

There is no way a book like this can be 100 percent comprehensive, or even fully updated. There are thousands of ICs on the market today, and a number will be introduced between the time this is written and the finished books roll of the presses. The most representative, popular, and useful devices are the primary focus in this book.

1
SECTION
74xx digital

In this section, the 74xx line of digital/logic devices will be covered. Devices range from simple logic gates, through flip-flops, counters, registers, and more exotic and complex special-purpose devices.

This standardized numbering scheme was first devised for TTL (Transistor-Transistor Logic) ICs. As various sub-families of TTL were developed, the same numbering system was used. Except for standard TTL, an identifying letter (or group of letters) is added after the "74" and before the device type number (xx) to indicate the logic sub-family. Devices of a given type number are pin-for-pin function compatible, regardless of the sub-family. (Devices of different sub-families may not be completely compatible electrically, however.)

Standard TTL:	74xx (no code letter)
Low-power TTL:	74Lxx
Schottky TTL:	74Sxx
Low-power Schottky TTL:	74LSxx
Advanced low-power Schottky TTL:	74ALSxx
FAST TTL:	74Fxx
CMOS:	74Cxx
High-speed CMOS:	74HCxx
High-speed CMOS/TTL Compatible:	74HCTxx
Advanced CMOS:	74ACxx and 74ACTxx

7400 quad 2-input NAND gate

This device consists of four 2-input NAND gates. Each gate can be used independently or in combination to form a wide variety of gating or cross-coupled bistable flip-flop circuits.

The NAND gate is a variation on the basic AND gate, delivering a LOW output only when all inputs are HIGH. If one or more of the inputs is LOW, the output will be HIGH.

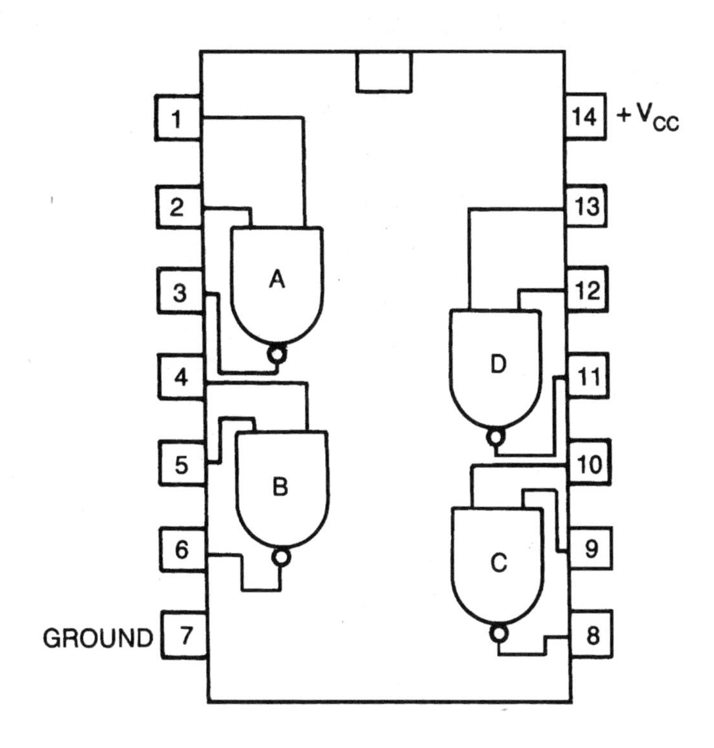

7400 Truth Table

INPUTS		OUTPUT
A	B	
0	0	1
0	1	1
1	0	1
1	1	0

Gated inverter

By using only one input for data, the other input can function as a control input. When this control input is HIGH, the gate operates as an inverter (the output has the opposite logic state as the data input). Making the control input LOW disables the device, and the output will always be HIGH, with the data input ignored.

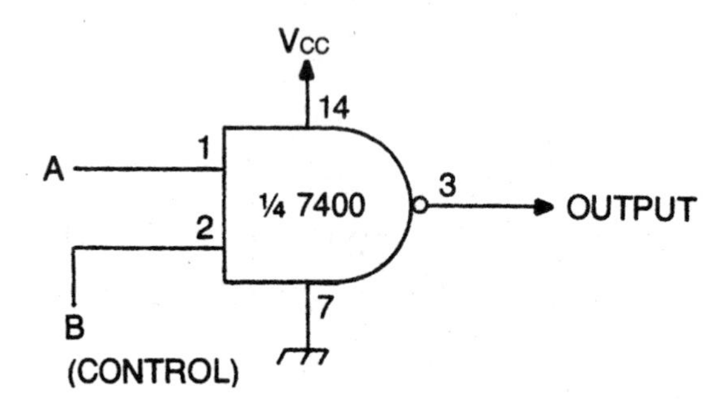

Inverter

By shorting the inputs together, a NAND gate can function as an inverter. Since the inputs are shorted, they must always be at the same logic state. Checking the truth table for the 7400, we can see that a LOW input results in a HIGH output, and a HIGH input results in a LOW output.

This trick can be very convenient in moderate to large circuits, in which there may be some unused NAND gates available. There is no need to add a dedicated inverter chip to the cost and bulk of the circuit, if just one or two such stages are required.

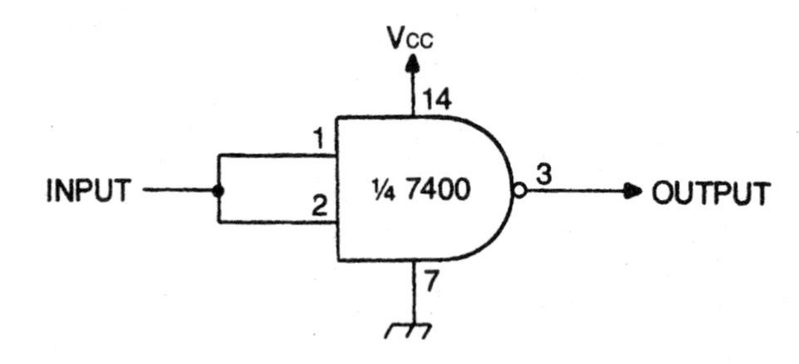

AND gate

A NAND gate is theoretically an AND gate with its output inverted. Actually NAND gates are easier and less expensive to fabricate on ICs, so they are more common. Inverting the output of a NAND gate returns us to the AND gate function. The output is HIGH if and only if both inputs are HIGH. If either or both inputs are LOW, the output will be LOW.

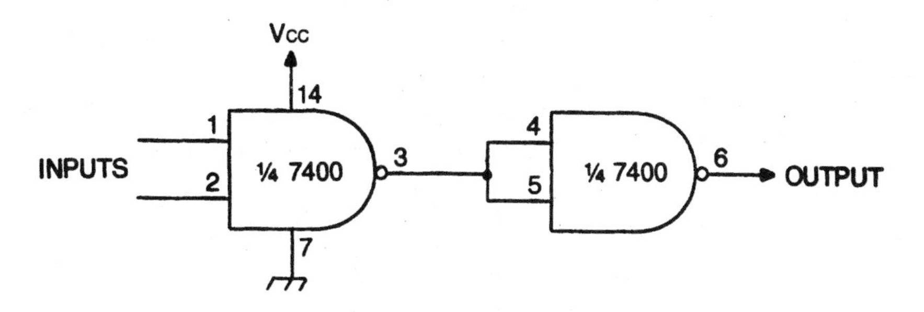

OR gate

Inverting the inputs of a NAND gate results in the OR function. The output is HIGH if either or both outputs are HIGH. The output is LOW if and only if both inputs are LOW.

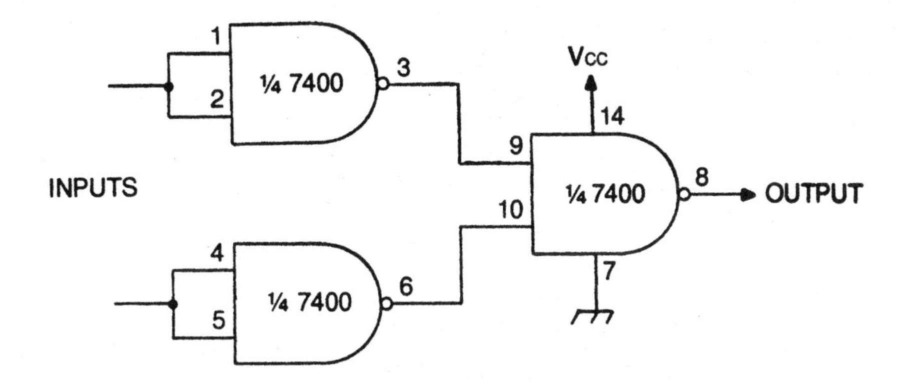

AND-OR gate

This 4-input circuit performs a more complex logic gating function. The output is HIGH if (A AND B) OR (C AND D) are HIGH.

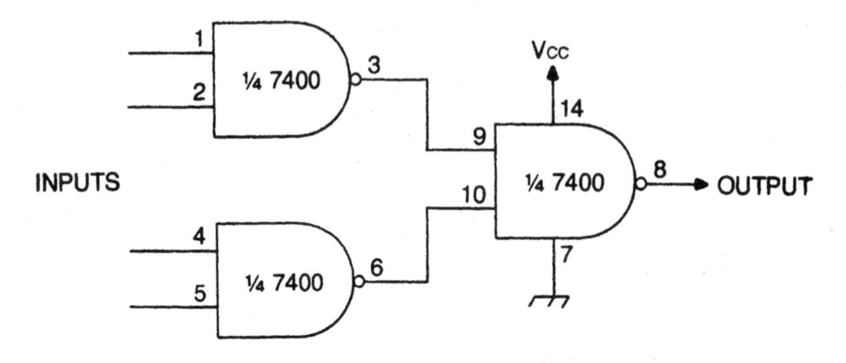

Truth Table

INPUTS				OUTPUT
A	B	C	D	
0	0	0	0	0
0	0	0	1	0
0	0	1	0	0
0	0	1	1	1
0	1	0	0	0
0	1	0	1	0
0	1	1	0	0
0	1	1	1	1
1	0	0	0	0
1	0	0	1	0
1	0	1	0	0
1	0	1	1	1
1	1	0	0	1
1	1	0	1	1
1	1	1	0	1
1	1	1	1	1

NOR gate

A NOR gate can be simulated by inverting the inputs and the output of a NAND gate. The output is HIGH if and only if neither output is HIGH. If either input (or both) goes HIGH, the output is forced LOW.

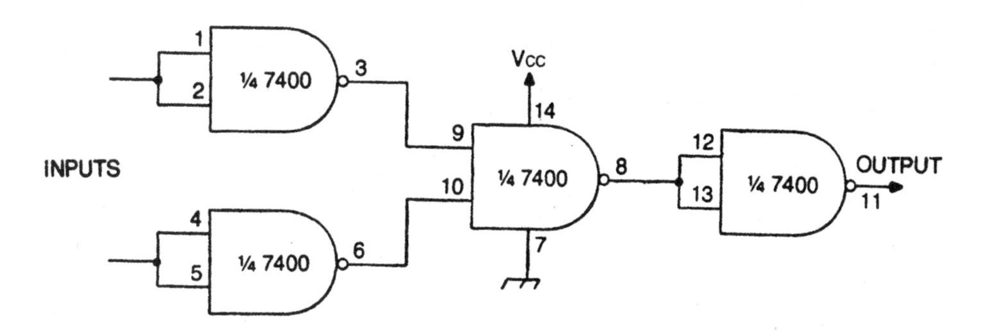

4-input NAND gate

This gating circuit expands the basic NAND function to include four inputs. It functions in the same basic way as the 2-input version. The output is LOW if and only if *all* inputs are HIGH. As long as one or more inputs are LOW, the output will be HIGH.

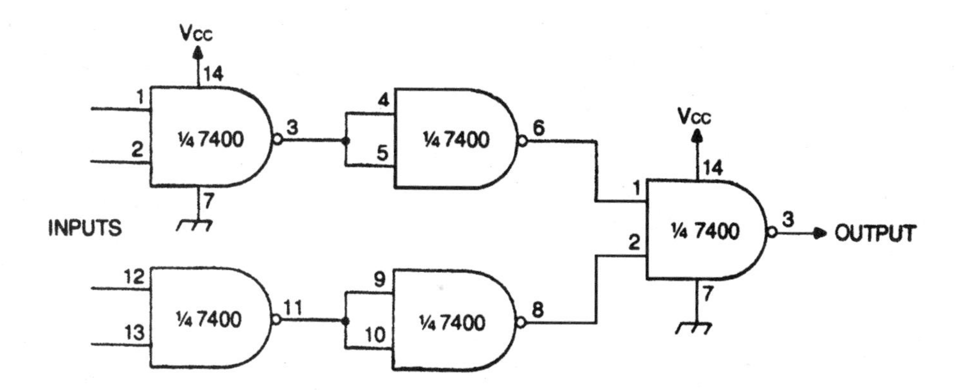

EXCLUSIVE-OR gate

The output of this gating circuit is HIGH whenever any one of the inputs is HIGH, but not both. In other words, the output is HIGH when the inputs are at opposite logic states. If the inputs are the same, the output is LOW.

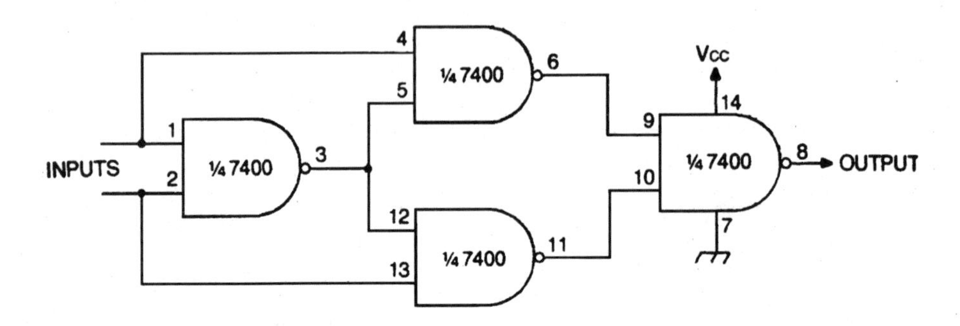

EXCLUSIVE-NOR gate

This rather unusual gating function works in exactly the opposite way as the previous circuit. The output is LOW when the inputs are at opposite logic states. If the inputs are the same, the output is HIGH.

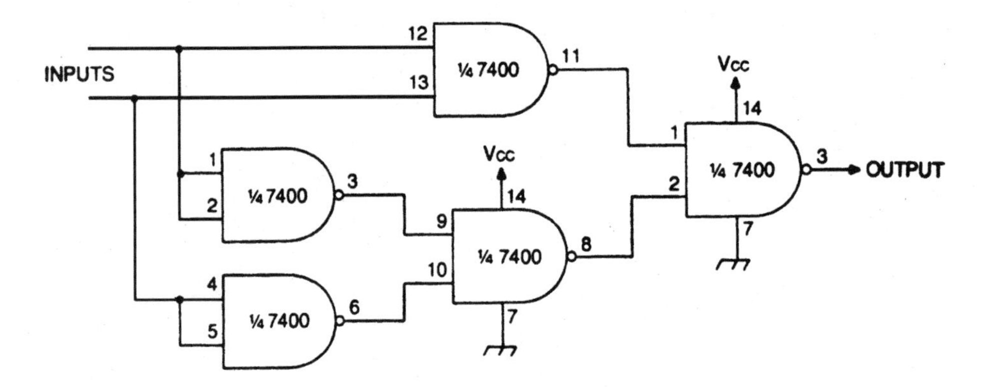

Majority logic circuit

This gating circuit permits multiple inputs to "vote" on the output state, where the majority rules. In this 3-input version, at least two inputs must be at the same logic state, and the output will be at this same "majority" logic state. The basic concept of this circuit can be expanded to include any odd number of inputs.

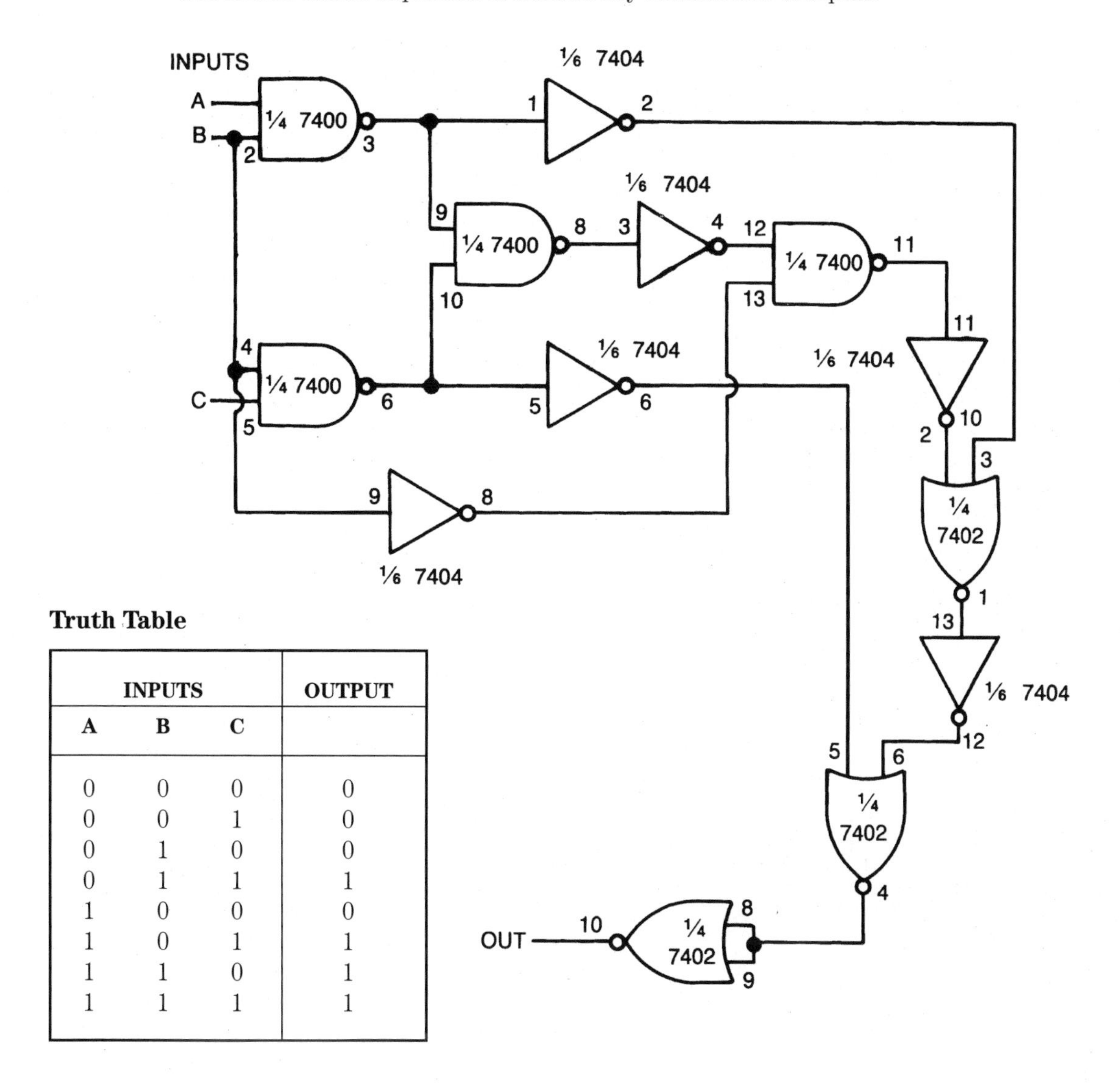

Truth Table

INPUTS			OUTPUT
A	B	C	
0	0	0	0
0	0	1	0
0	1	0	0
0	1	1	1
1	0	0	0
1	0	1	1
1	1	0	1
1	1	1	1

Half adder

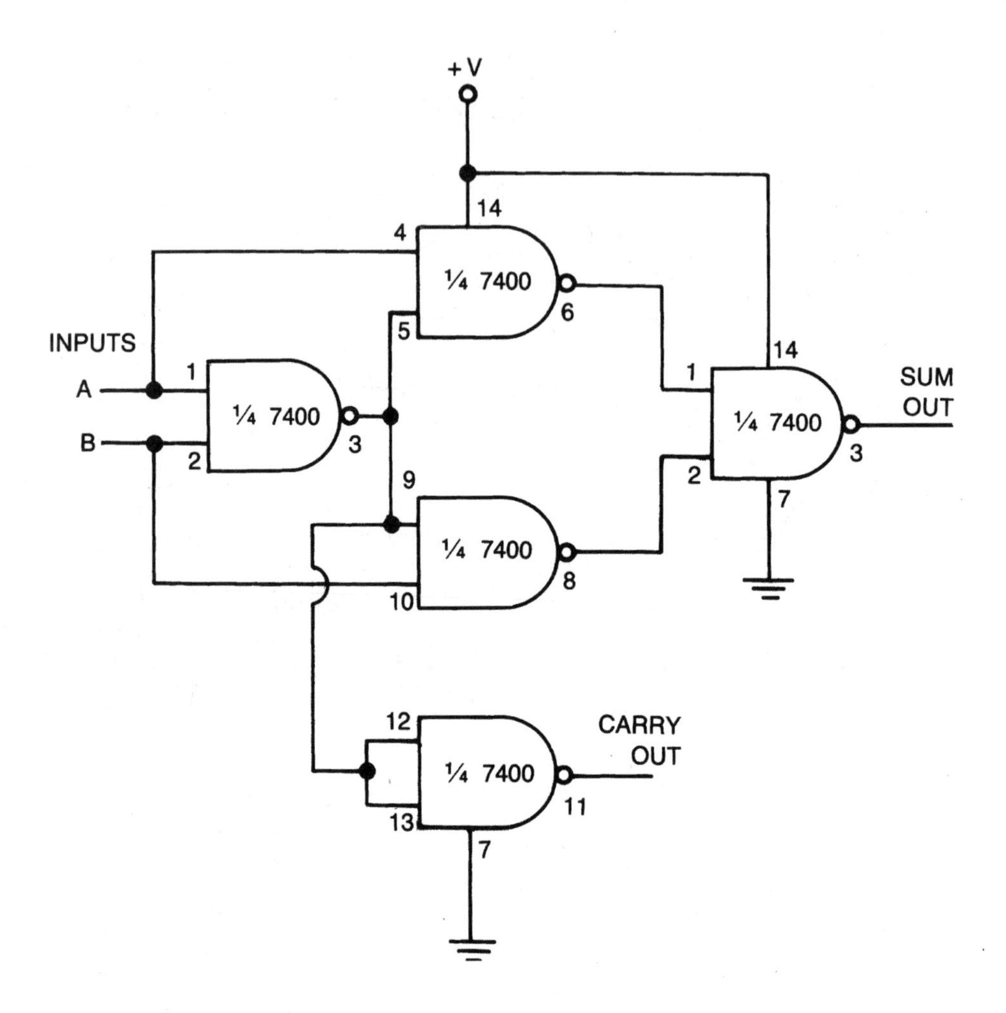
+V
14
4
¼ 7400
6
5
INPUTS
A
B
1
2
¼ 7400
3
9
¼ 7400
8
10
14
1
¼ 7400
3
2
7
SUM
OUT
12
¼ 7400
11
13
7
CARRY
OUT

RS latch

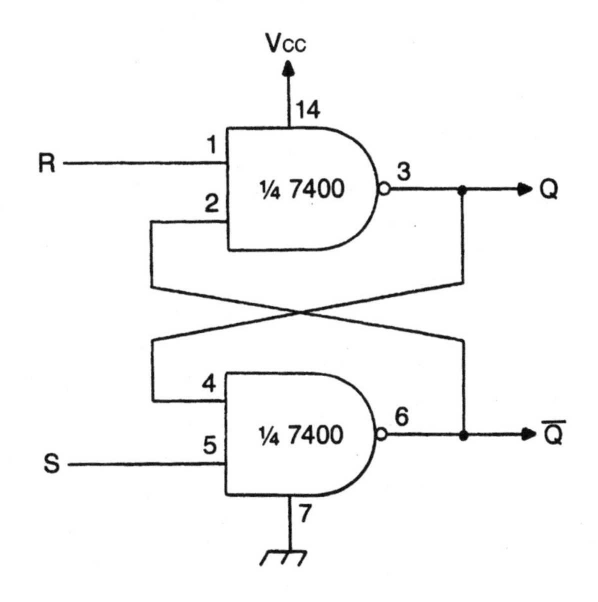

Gated RS latch

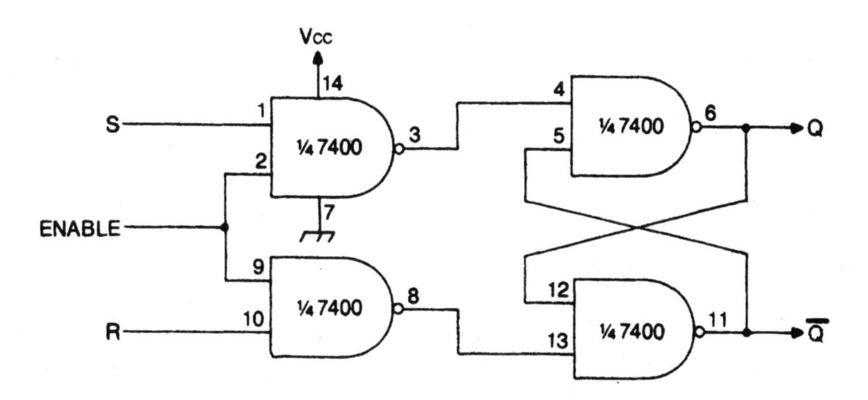

D-type flip-flop

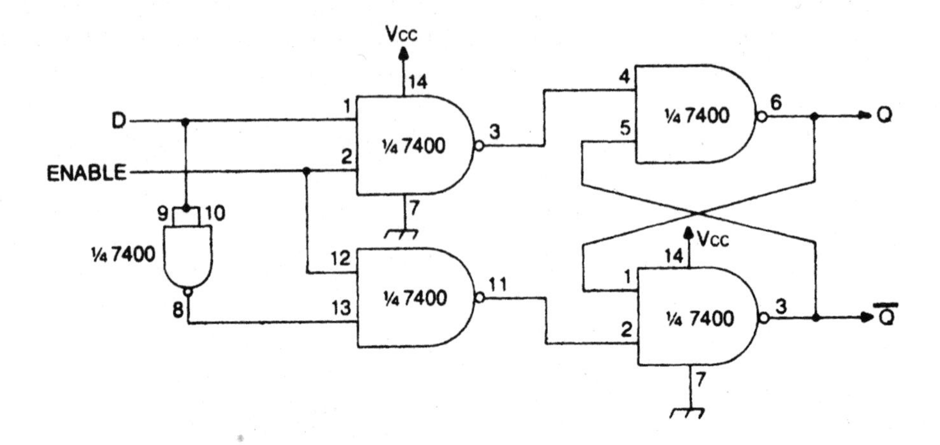

Bounceless switch

Mechanical switches often bounce a little when opened or closed. This is of little or no importance in most linear circuits, but in many digital circuits, such as counters, each bounce can appear to the circuit as a separate switch operation. A debouncing circuit such as this one ensures smooth, reliable switching action, with no data errors due to mechanical bouncing.

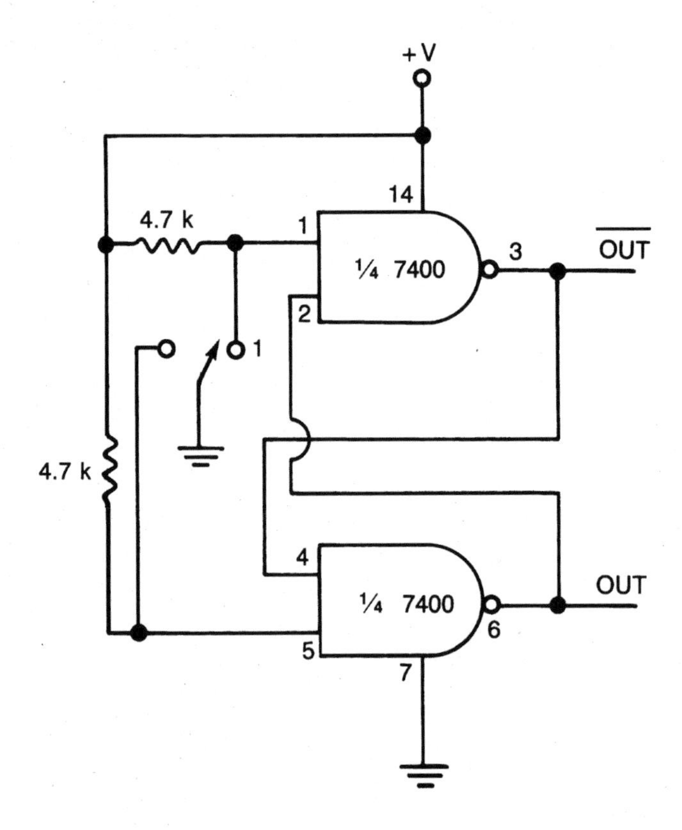

Dual LED blinker

When one LED is lit, the other will be dark. Different capacitor values will result in other flash rates. If both capacitors have an equal value, each LED will be lit for the same amount of time. Unequal capacitances will result in one LED being lit longer than the other.

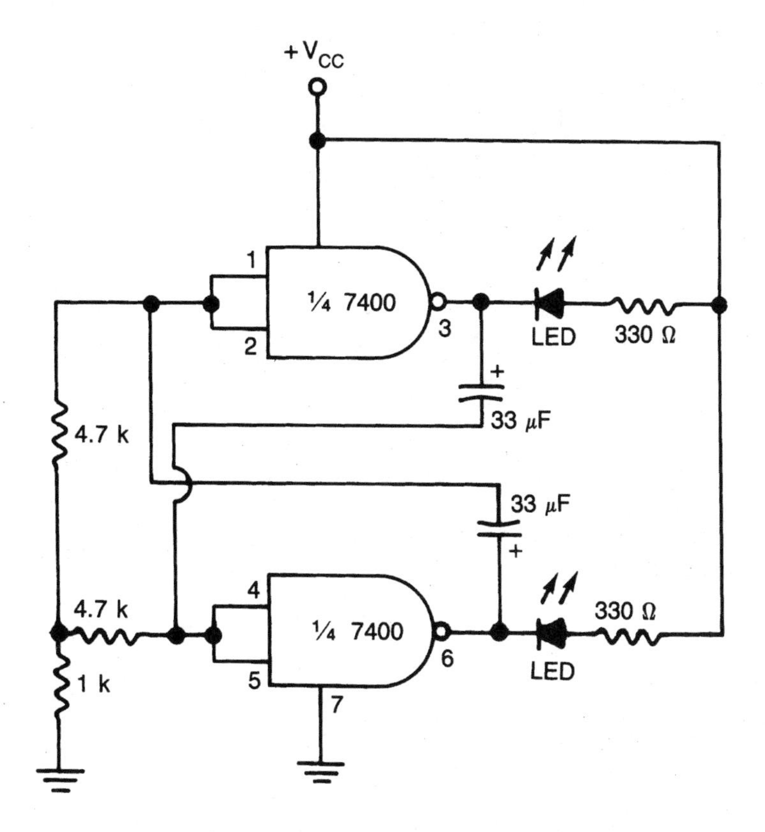

Single-value BCD decoder

A 4-bit BCD value is fed to inputs A through D. If this BCD value equals the preset value, the output will go LOW. For any other input combination, the output will be HIGH.

The desired BCD value is "programmed" by adding or removing the appropriate input inverter stages. Use an inverter if you want that bit to be a logic 0 (LOW). For a logic 1 (HIGH) bit omit the input inverter stage.

As the circuit is shown here, the preset value is BCD 1000 (decimal 8) to bring the output LOW. Any other input will produce a HIGH output. For some applications it may be desirable to invert the output of this circuit to produce a HIGH signal only when the desired BCD value appears at the inputs.

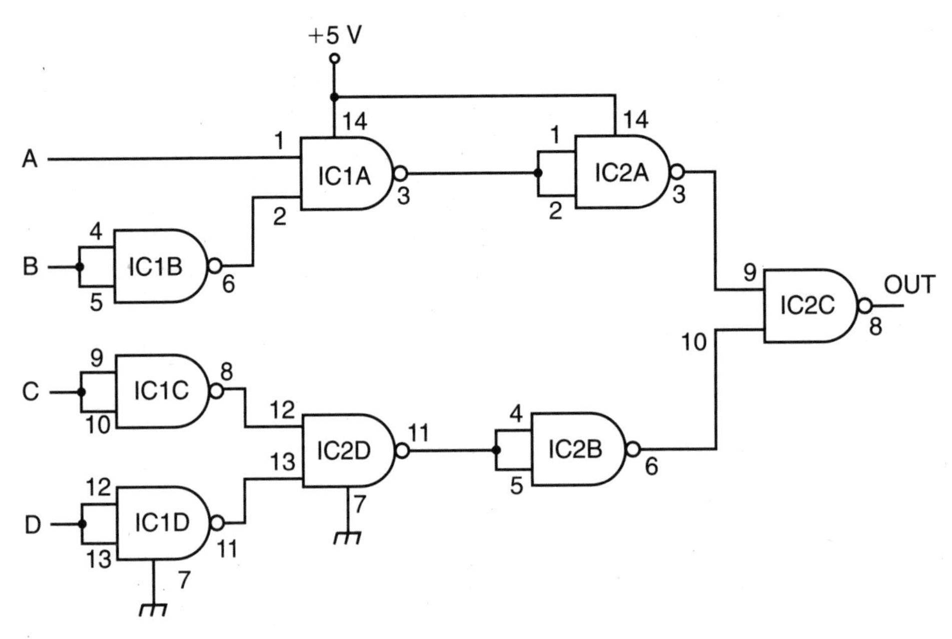

8-input NAND gate

This gating circuit expands the basic NAND function to include eight inputs. It functions in the same basic way as the 2-input version. The output is LOW if and only if *all* inputs are HIGH. As long as one or more inputs are LOW, the output will be HIGH.

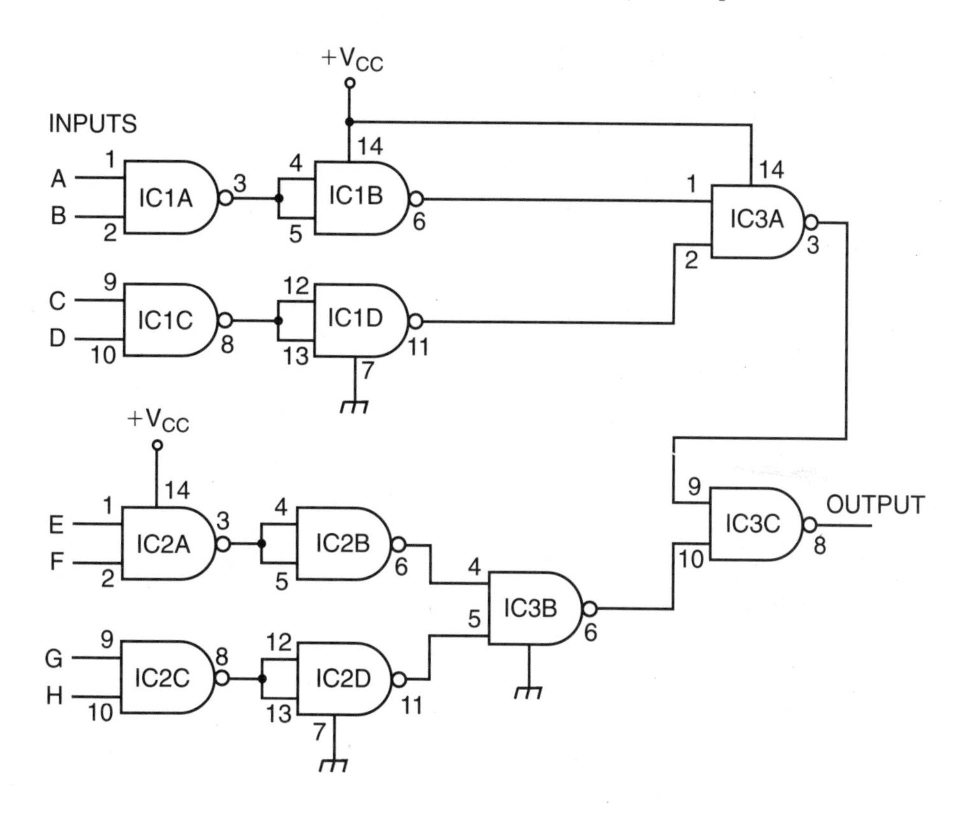

Gated NAND gate

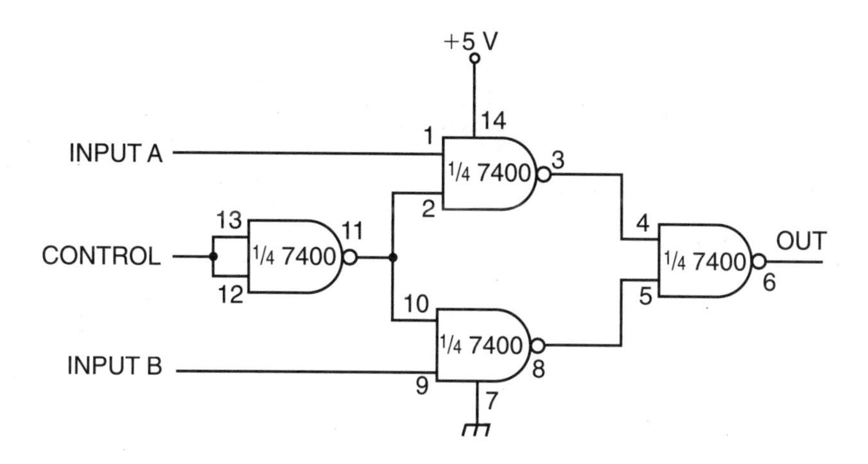

7402 quad 2-input NOR gate

This device consists of four 2-input NOR gates. The NOR gate is a variation on the basic OR gate, delivering a HIGH output if and only if *all* inputs are LOW.

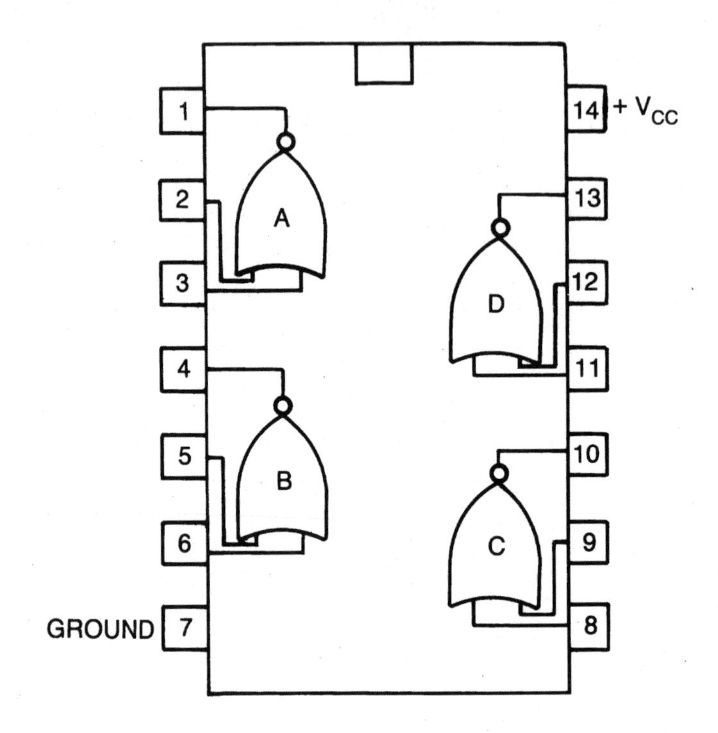

7402 Truth Table

INPUTS		OUTPUT
A	**B**	
0	0	1
0	1	0
1	0	0
1	1	0

Inverter

By shorting the inputs together, a NOR gate can function as an inverter. Since the inputs are shorted, they must always be at the same logic state. Checking the truth table for the 7400, we can see that a LOW input results in a HIGH output, and a HIGH input results in a LOW output.

This trick can be very convenient in moderate to large circuits, in which there may be some unused NAND gates available. There is no need to add a dedicated inverter chip to the cost and bulk of the circuit, if just one or two such stages are required.

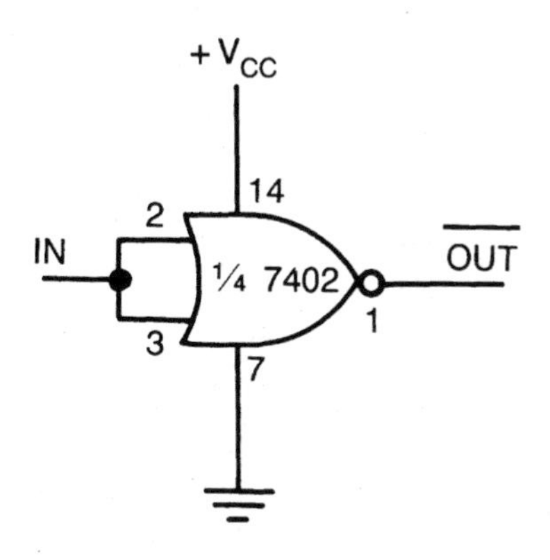

OR gate

A NOR gate is theoretically an OR gate with its output inverted. Actually, NOR gates are easier and less expensive to fabricate on ICs, so they are more common.

Inverting the output of a NOR gate returns us to the OR gate function. The output is LOW if and only if both inputs are LOW. If either or both inputs are HIGH, the output will be HIGH.

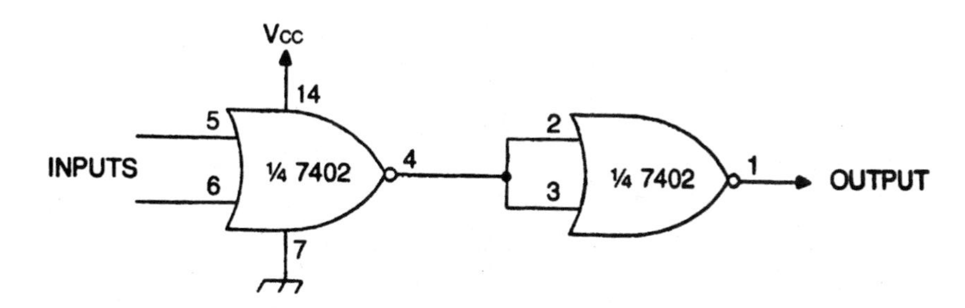

EXCLUSIVE-OR gate

The output of this gating circuit is HIGH whenever any one of the inputs is HIGH, but not both. In other words, the output is HIGH when the inputs are at the opposite logic states. If the inputs are the same, the output is LOW.

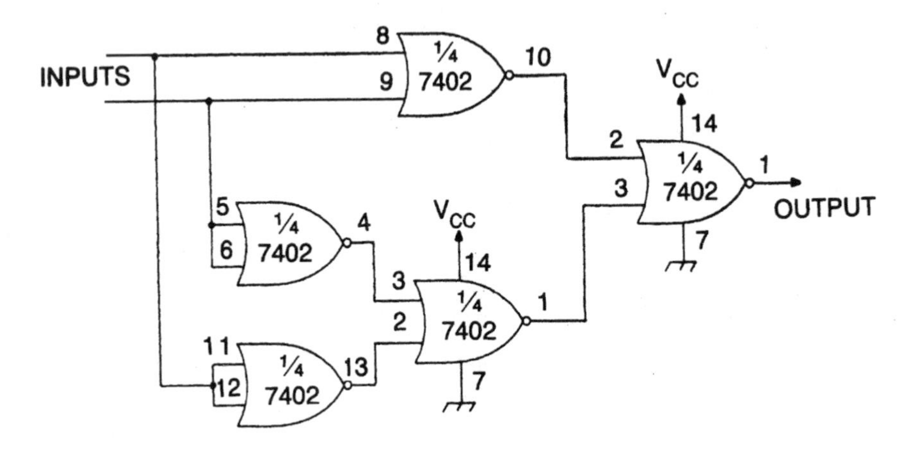

AND gate

Inverting the inputs of a NOR gate produces the AND gate function. The output is HIGH if and only if both inputs are HIGH.

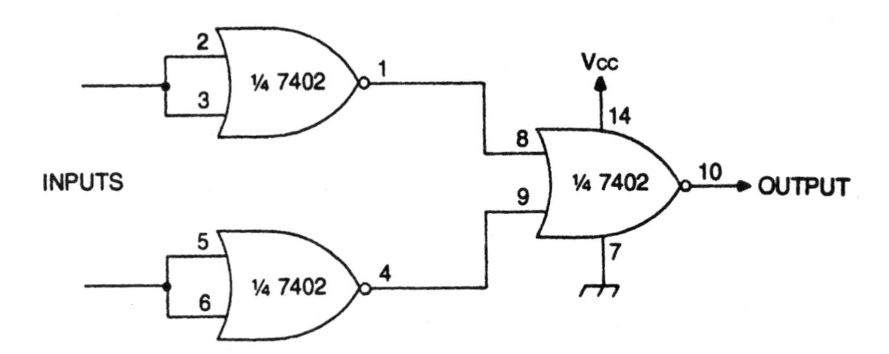

4-input NOR gate

This gating circuit expands the basic NOR function to include four inputs. It functions in the same basic way as the 2-input version. The output is HIGH if and only if *all* inputs are LOW. As long as one or more inputs are HIGH, the output will be LOW.

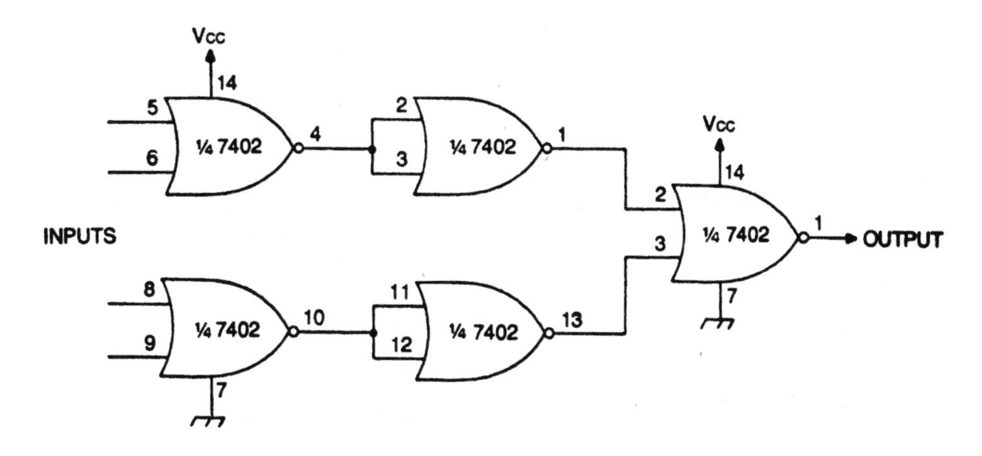

RS latch

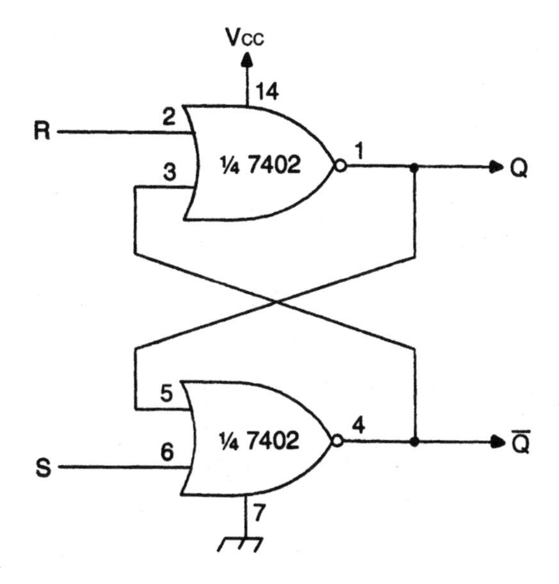

One-shot

This type of circuit is also known as a *monostable multivibrator.* The timing period is determined by the values of the resistor and the capacitor. Increasing the value of either component results in a longer timing period.

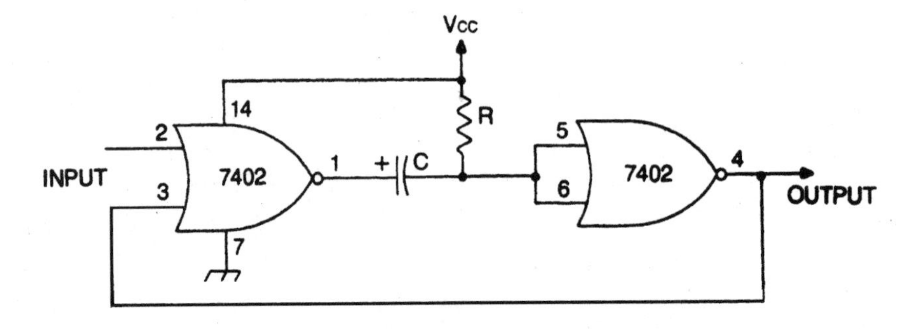

7404 HEX inverter

The 7404 contains eight single-input/single-output inverter/buffer stages. Only the power supply connections are in common. The output of an inverter always has the opposite logic state as its input. A LOW input results in a HIGH output, and vice versa.

An inverter is also sometimes called a NOT gate.

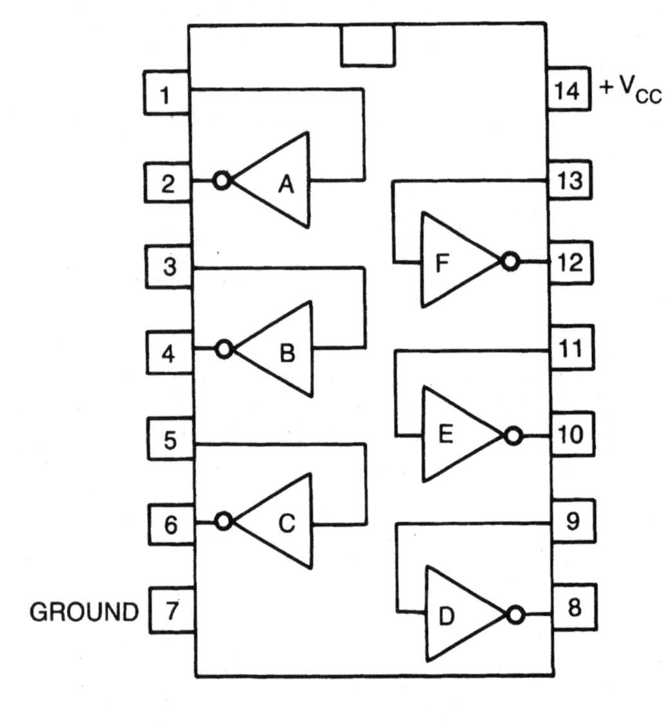

7404 Truth Table

INPUT	OUTPUT
0	1
1	0

Non-inverting buffer

Reinverting the output of an inverter creates a noninverting buffer. The output state is the same as the input state. Buffers are used to boost heavily loaded digital signals.

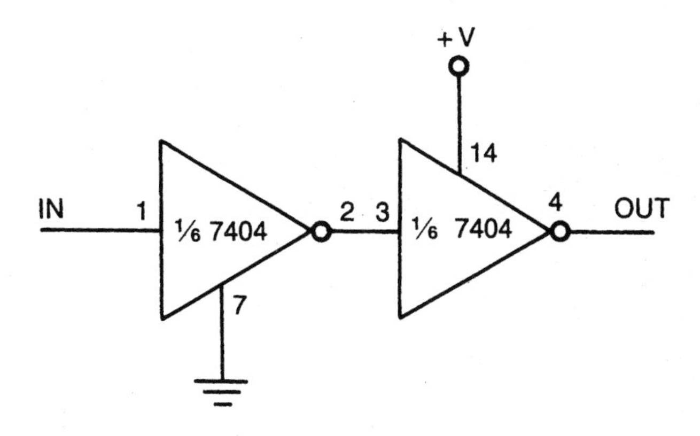

Switch debouncer

Mechanical switches often bounce a little when opened or closed. This is of little or no importance in most linear circuits, but in many digital circuits, such as counters, each bounce can appear to the circuit as a separate switch operation. A debouncing circuit such as this one ensures smooth, reliable switching action, with no data errors due to mechanical bouncing.

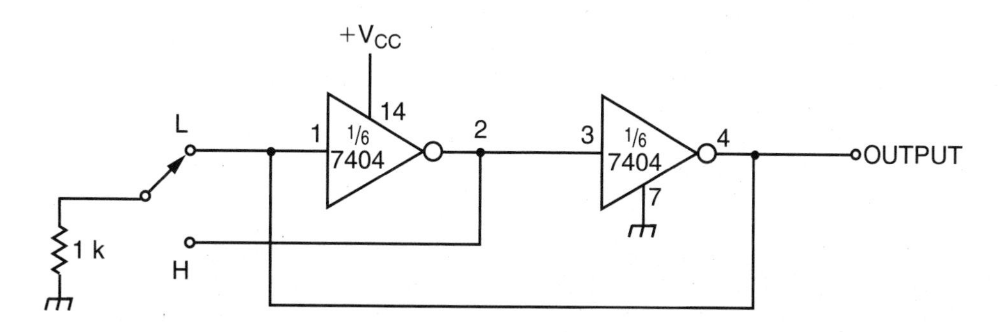

1-of-2 demultiplexer

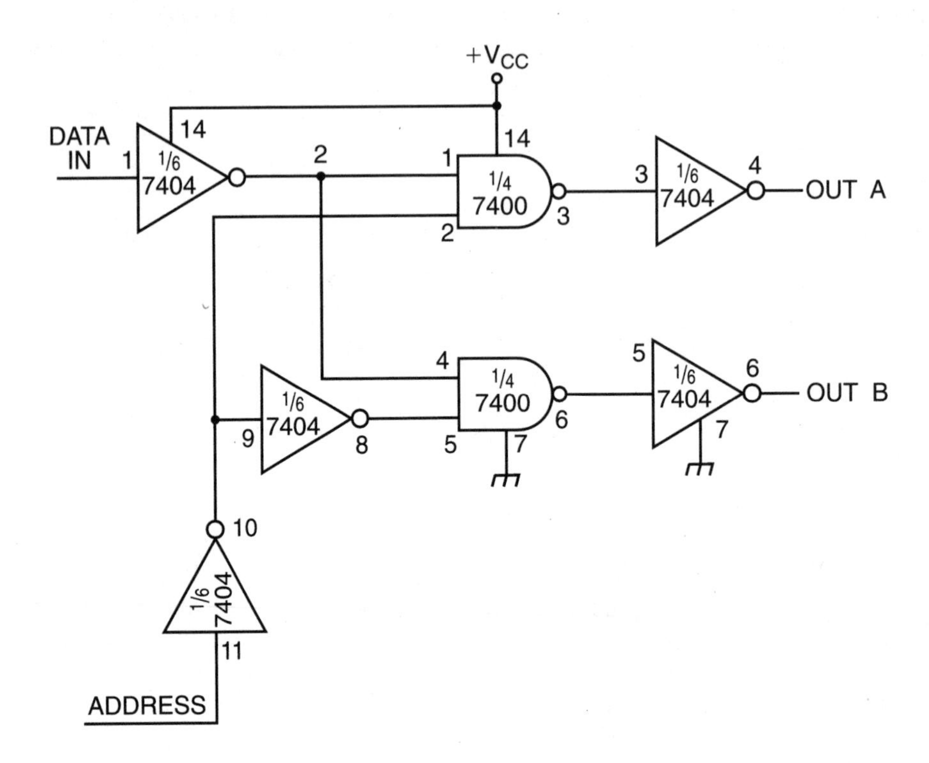

Inverting expansion buffer

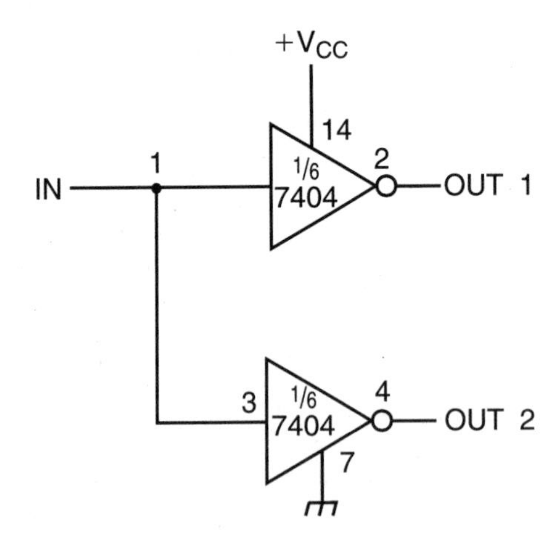

Oscillator

The output is a square wave. When one output goes HIGH, the other goes LOW, and vice versa. The frequency is set by the resistor and capacitor values. If both sets have equal values, the output will be a symmetrical square wave (50 percent duty cycle). Unequal component values will result in an asymmetrical rectangle wave with a proportional duty cycle.

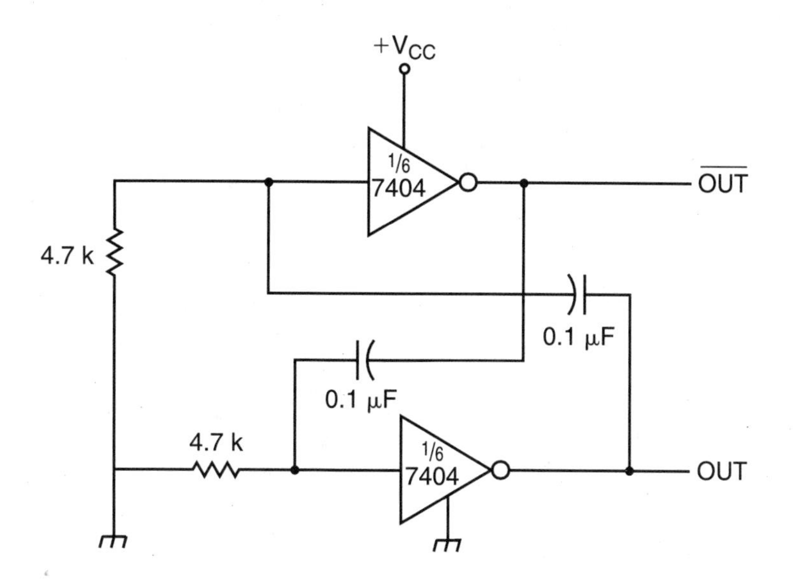

7408 quad 2-input AND gate

The 7408 contains four functionally independent 2-input/single-output AND gates. The output of an AND gate is HIGH if and only if all outputs are HIGH.

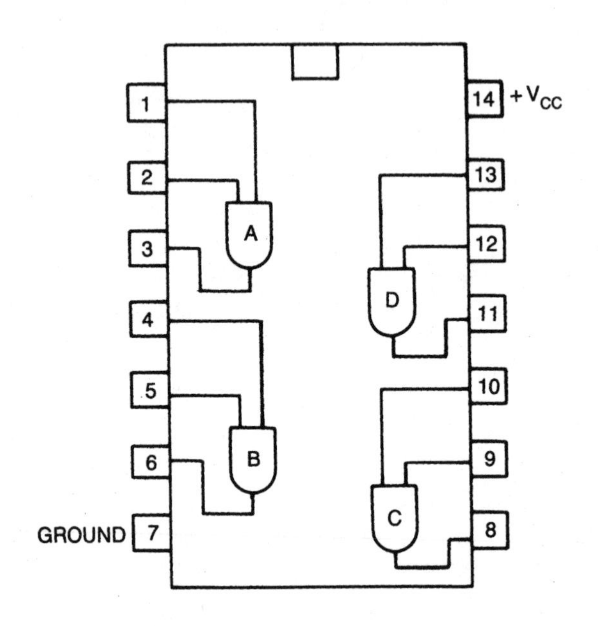

7408 Truth Table

INPUTS		OUTPUT
A	B	
0	0	0
0	1	0
1	0	0
1	1	1

Non-inverting buffer

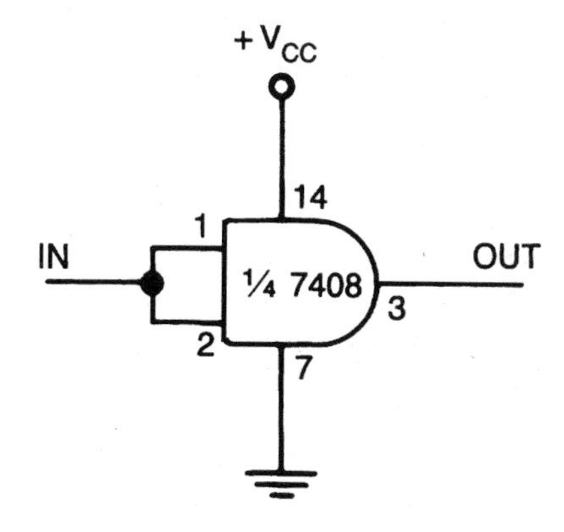

Gated buffer

By using only one input for data, the other input can function as a control input. When this control input is HIGH, the gate operates as an ordinary buffer (the output matches the logic state of the data input). Making the control input LOW disables the device, and the output will always be LOW, with the data input ignored.

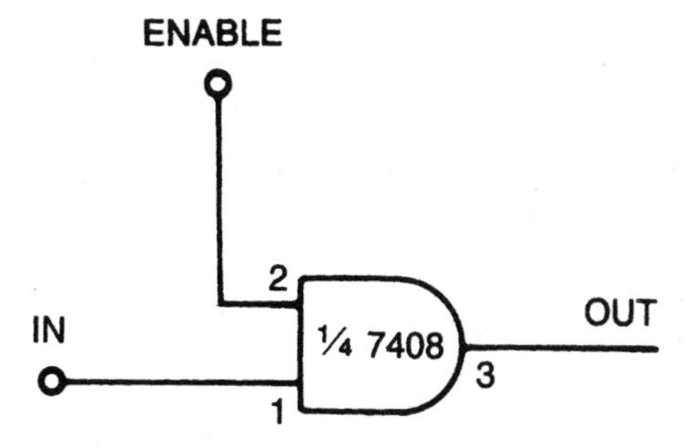

NOR gate

Inverting the inputs of an AND gate creates a NOR gate. The output is HIGH if and only if none of the inputs is HIGH. In other words, to achieve a HIGH output, all inputs must be LOW.

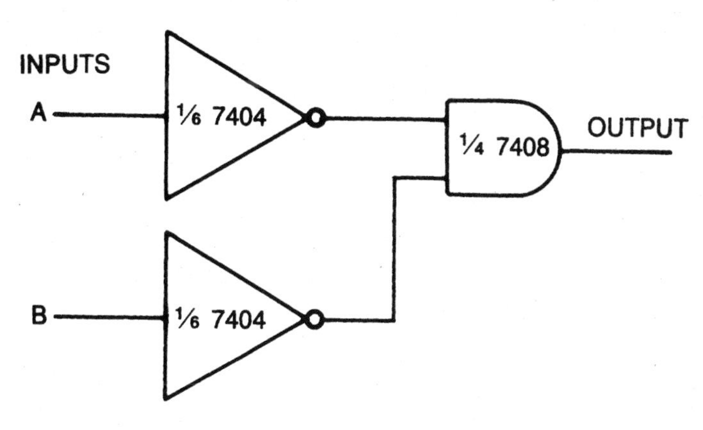

4-input AND gate

This circuit expands the basic AND gate function to include four inputs. It works in exactly the same way as a regular 2-input AND gate. The output is HIGH if and only if all inputs are HIGH. If any one (or more than one) of the inputs goes LOW, the output will be LOW.

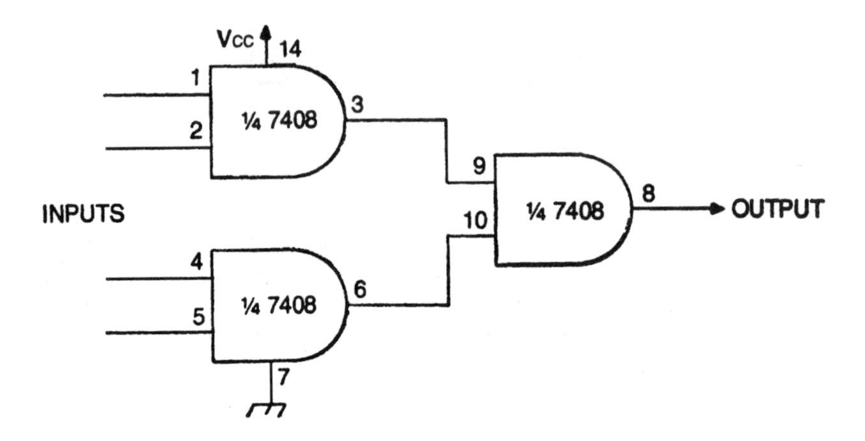

4-input NAND gate

This circuit expands the basic NAND gate function to include four inputs. It works in exactly the same way as a regular 2-input NAND gate. The output is LOW if and only if *all* inputs are HIGH. If any one (or more than one) of the inputs goes LOW, the output will be HIGH.

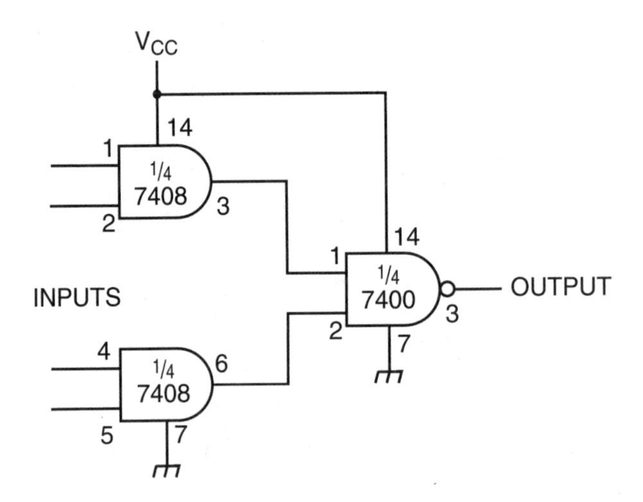

AND/OR gate

This circuit is an example of a nonstandard logic function. The output is HIGH if inputs A and B are both HIGH or if inputs C and D are both HIGH.

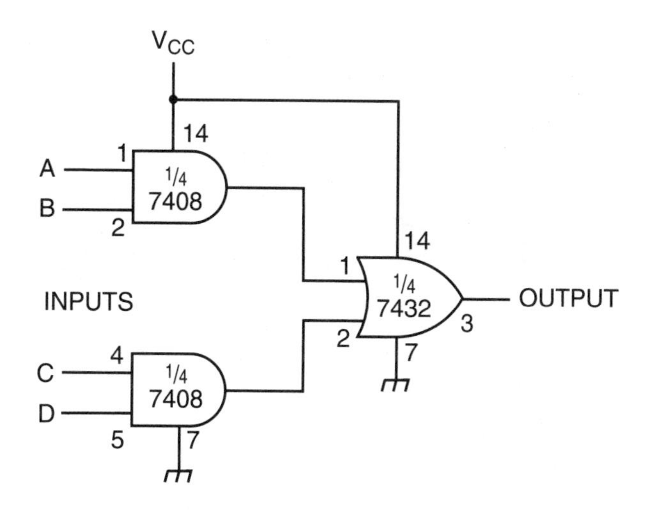

Truth Table

INPUTS				OUTPUT
A	**B**	**C**	**D**	
0	0	0	0	0
0	0	0	1	0
0	0	1	0	0
0	0	1	1	1
0	1	0	0	0
0	1	0	1	0
0	1	1	0	0
0	1	1	1	1
1	0	0	0	0
1	0	0	1	0
1	0	1	0	0
1	0	1	1	1
1	1	0	0	1
1	1	0	1	1
1	1	1	0	1
1	1	1	1	1

Two-input data selector

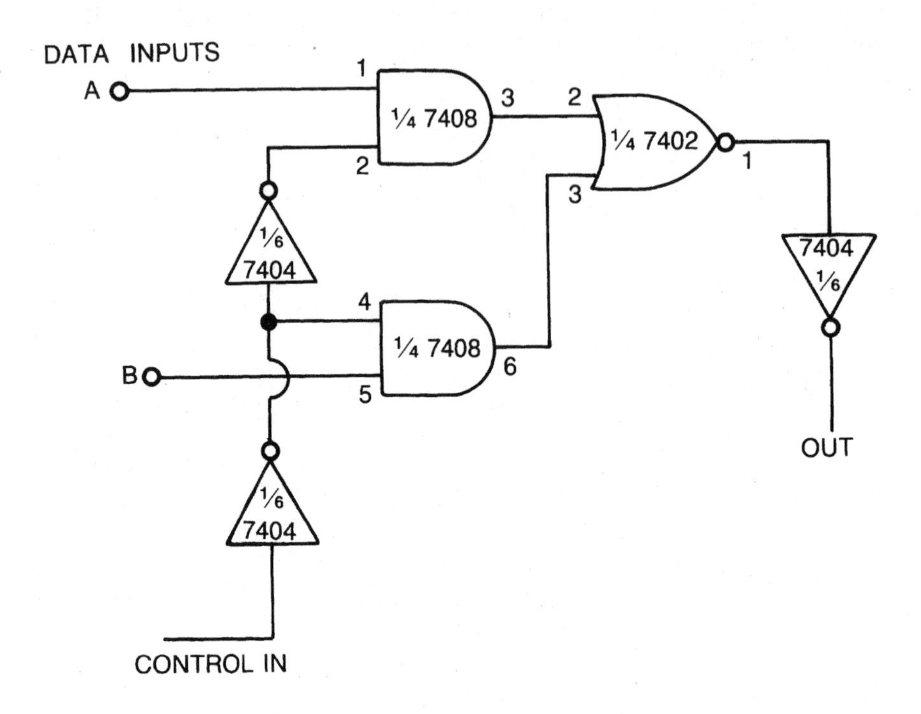

1-of-2 demultiplexer

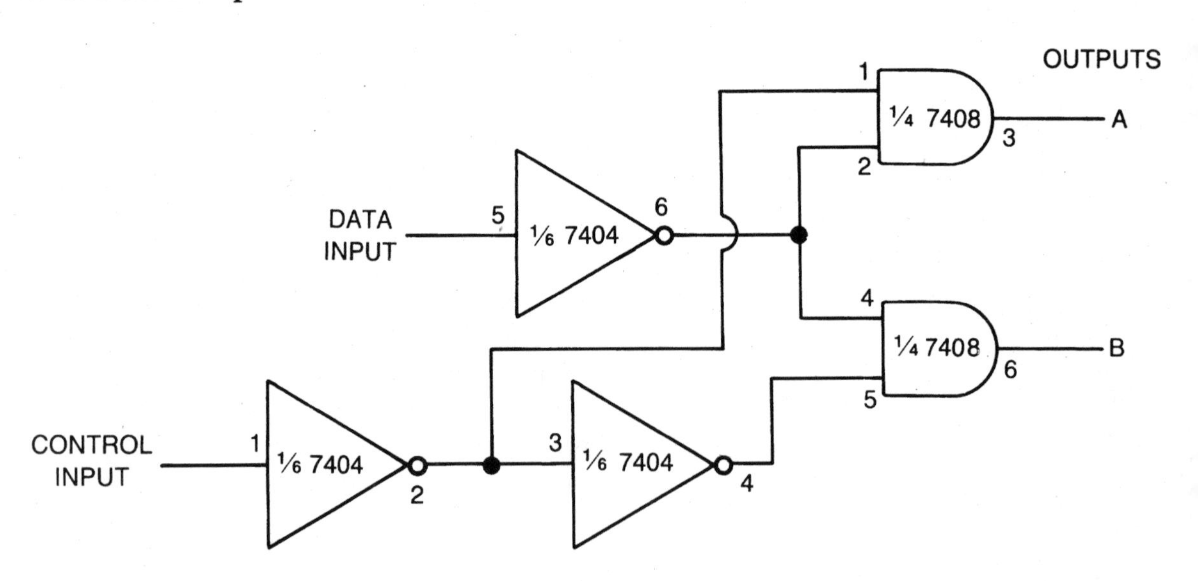

7432 quad 2-input OR gate

The 7432 contains four functionally independent 2-input/single-output OR gates. The output of an OR gate is HIGH whenever one or more of the inputs are HIGH. The output is LOW if and only if *all* inputs are LOW.

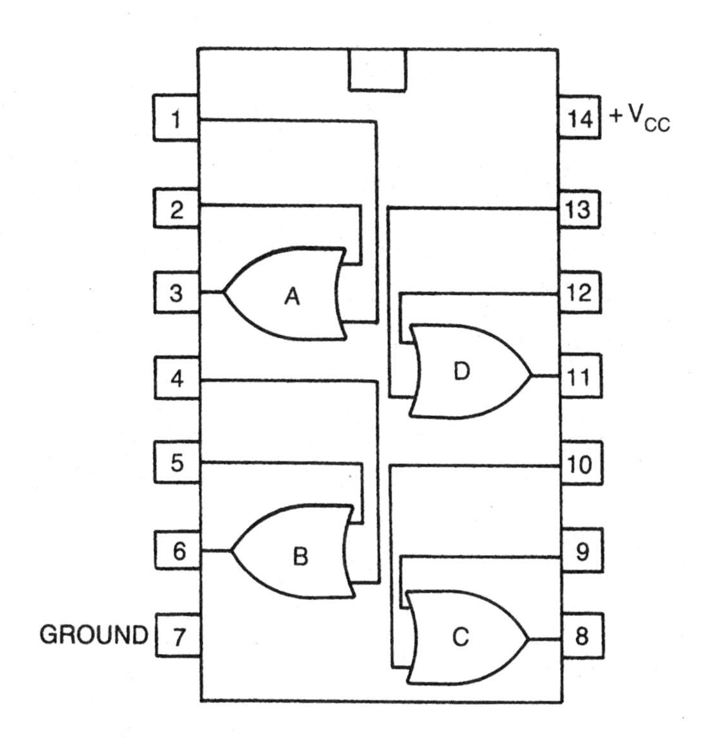

7432 Truth Table

INPUTS		OUTPUT
A	**B**	
0	0	0
0	1	1
1	0	1
1	1	1

NOR gate

Inverting the output of an OR gate creates a NOR gate. The output is HIGH if and only if neither input is HIGH. That is, all inputs must be LOW to produce a HIGH output.

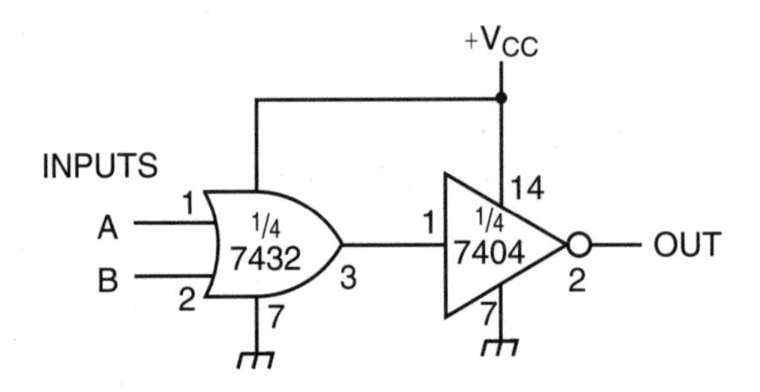

NAND gate

Inverting the inputs of an OR gate creates a NAND gate. The output is LOW if and only if *all* inputs are HIGH.

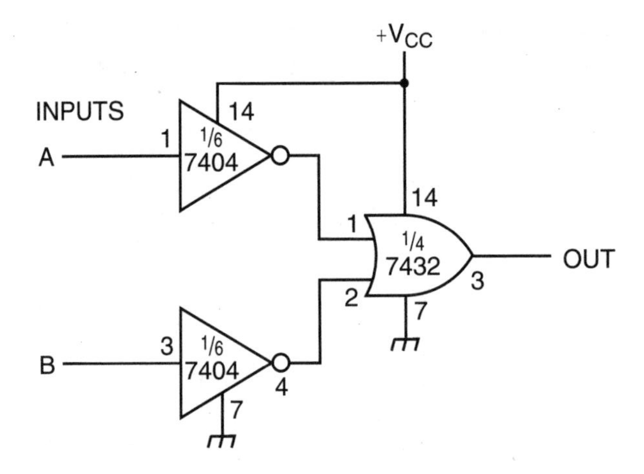

AND/OR gate

This circuit is an example of a nonstandard logic function. The output is HIGH if inputs A and B are both HIGH or if inputs C and D are both HIGH.

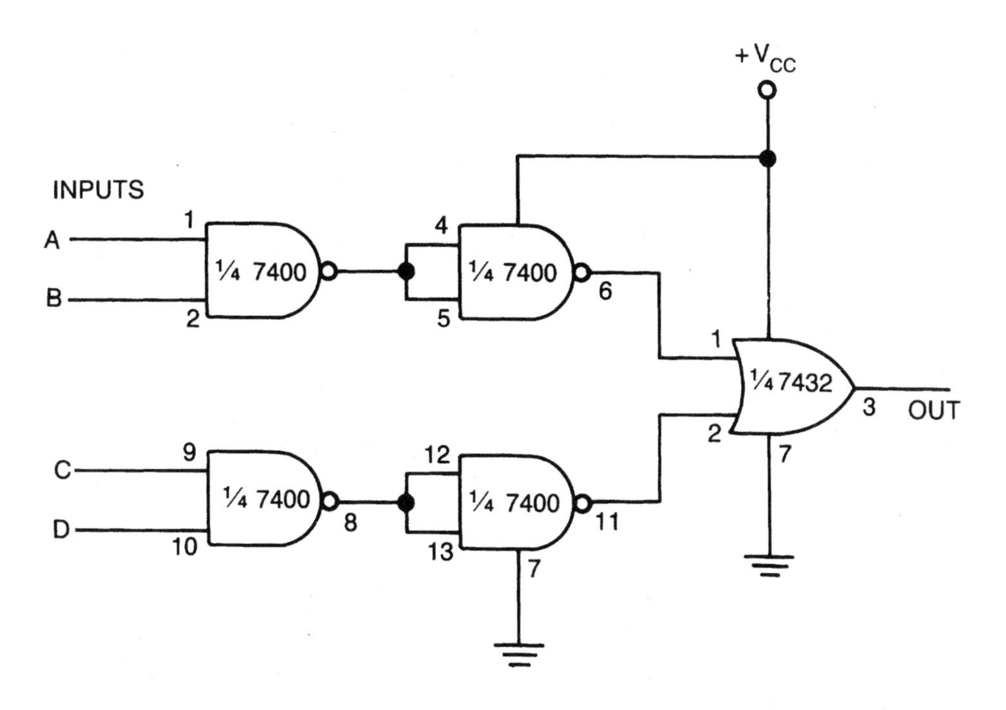

Truth Table

INPUTS				OUTPUT
A	B	C	D	
0	0	0	0	0
0	0	0	1	0
0	0	1	0	0
0	0	1	1	1
0	1	0	0	0
0	1	0	1	0
0	1	1	0	0
0	1	1	1	1
1	0	0	0	0
1	0	0	1	0
1	0	1	0	0
1	0	1	1	1
1	1	0	0	1
1	1	0	1	1
1	1	1	0	1
1	1	1	1	1

2-input data selector

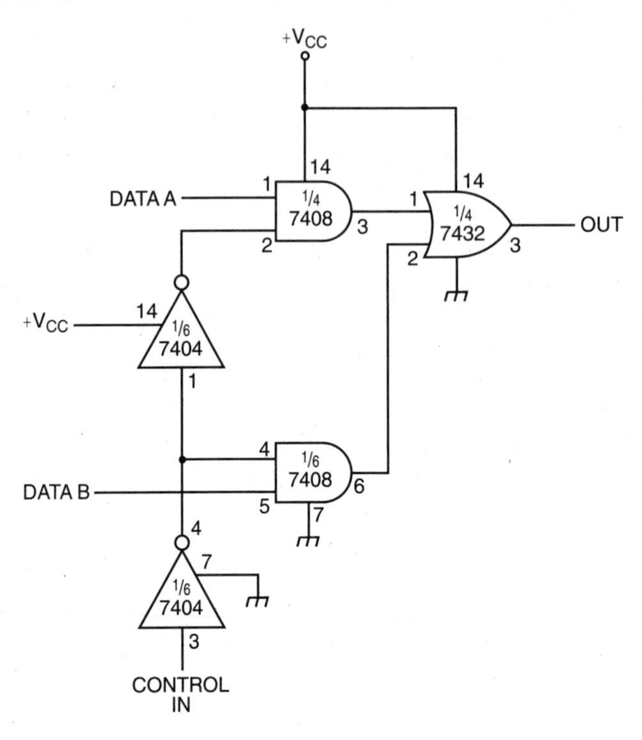

7447 BCD-to-7-segment decoder/driver

The 7447 7-segment decoder accepts a 4-bit BCD code input and provides the appropriate outputs for selection of segments in a 7-segment matrix display (LED or LCD) used for representing the decimal numerals 0 through 9. The seven (inverted) outputs (a, b, c, d, e, f, g) of the decoder select the corresponding segments in the matrix.

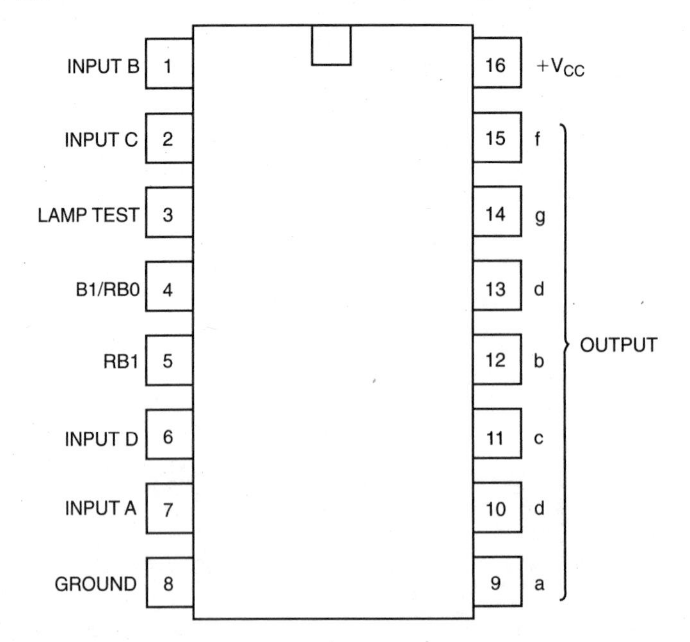

Manual BCD/7-segment demonstrator

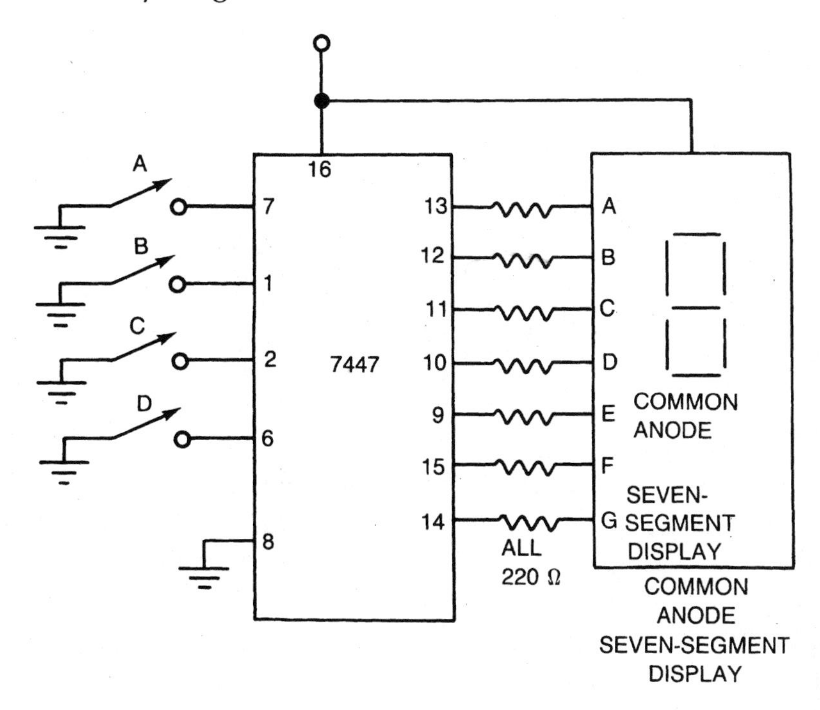

Flashing display

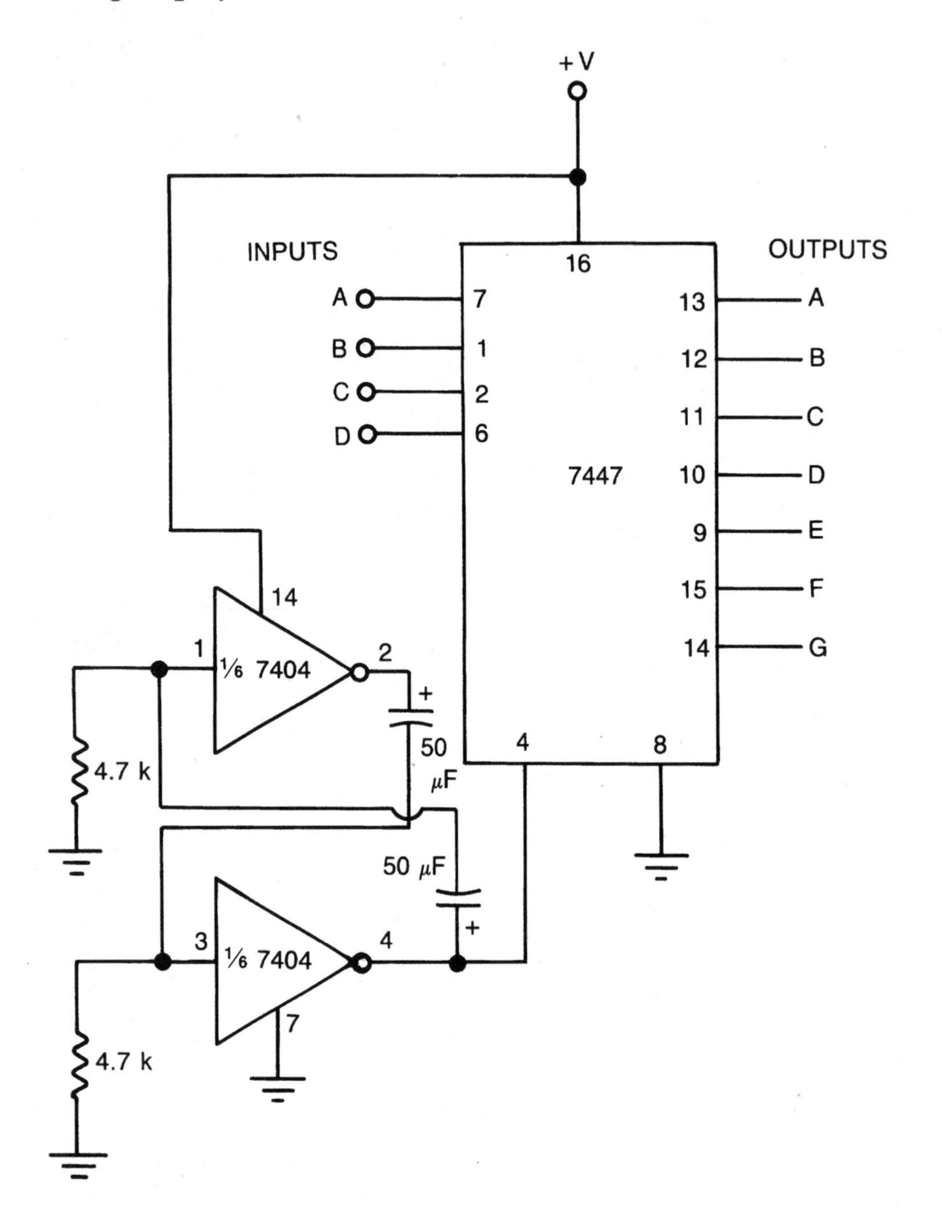

Dimmed display

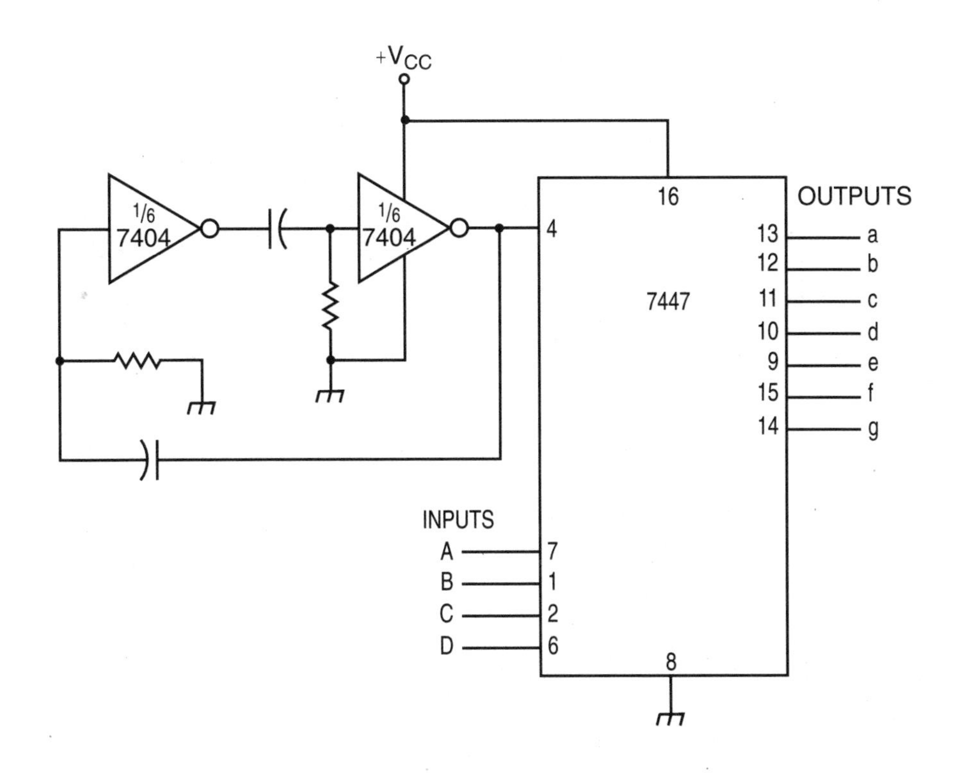

7473 dual JK flip-flop with clear

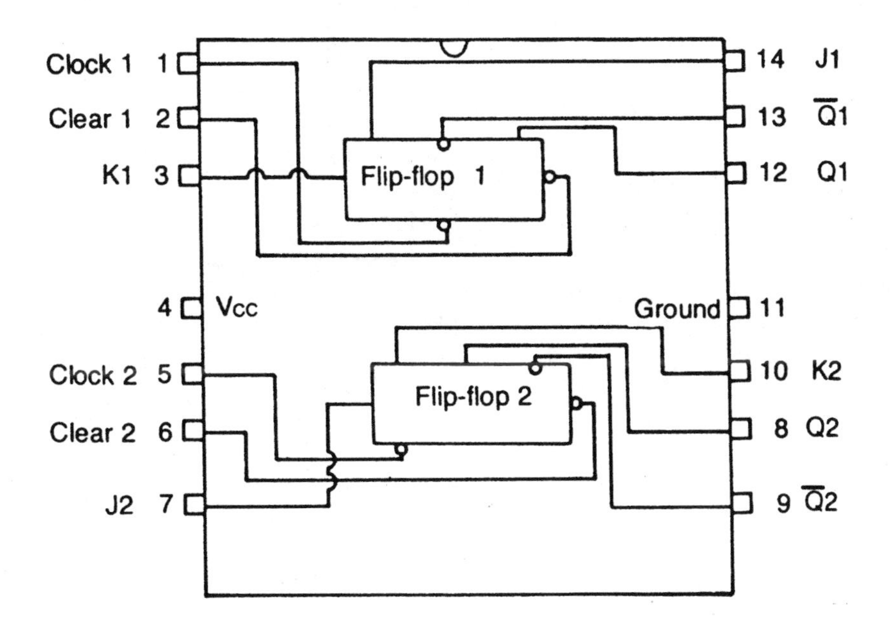

Divide by 2

The output frequency of this circuit will be one-half the input frequency.

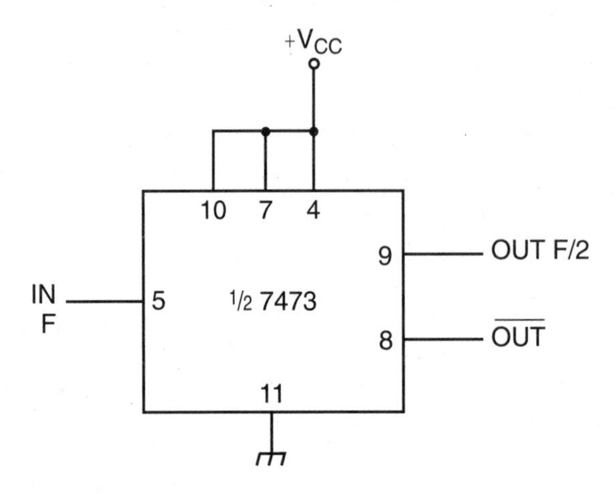

Divide by 3

The output frequency of this circuit will be one-third the input frequency.

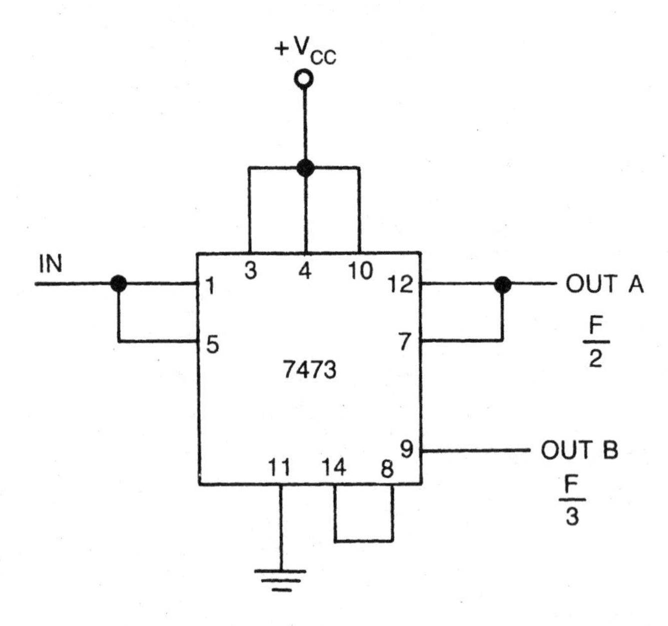

Divide by 4

The output frequency of this circuit will be one-fourth the input frequency.

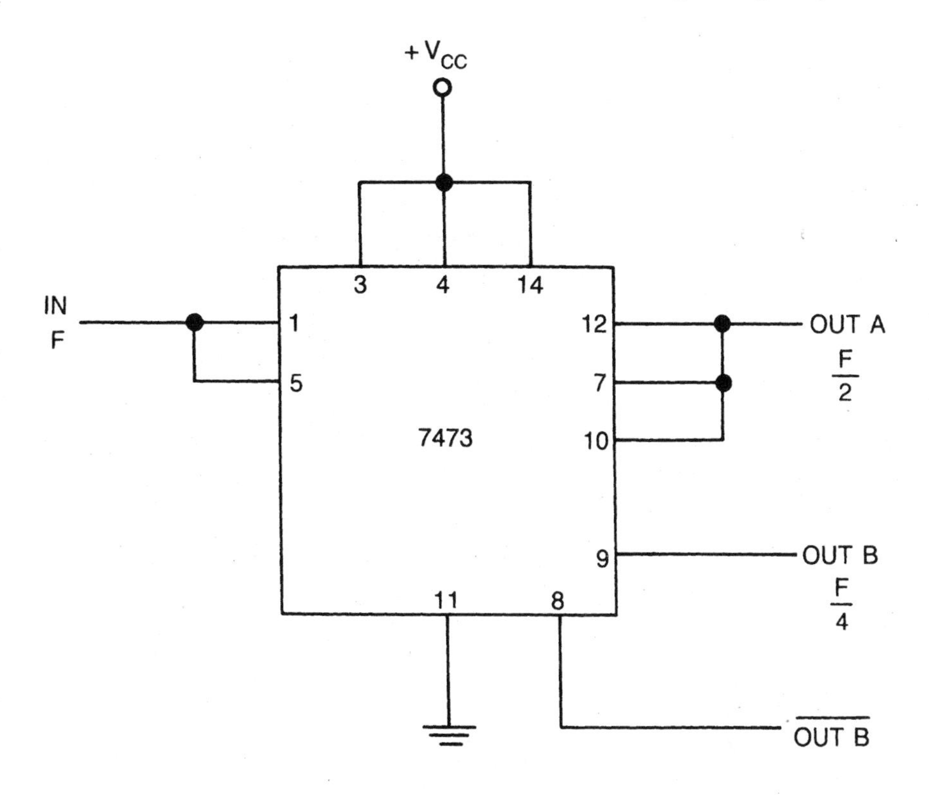

7474 dual D positive edge triggered flip-flop with preset and clear

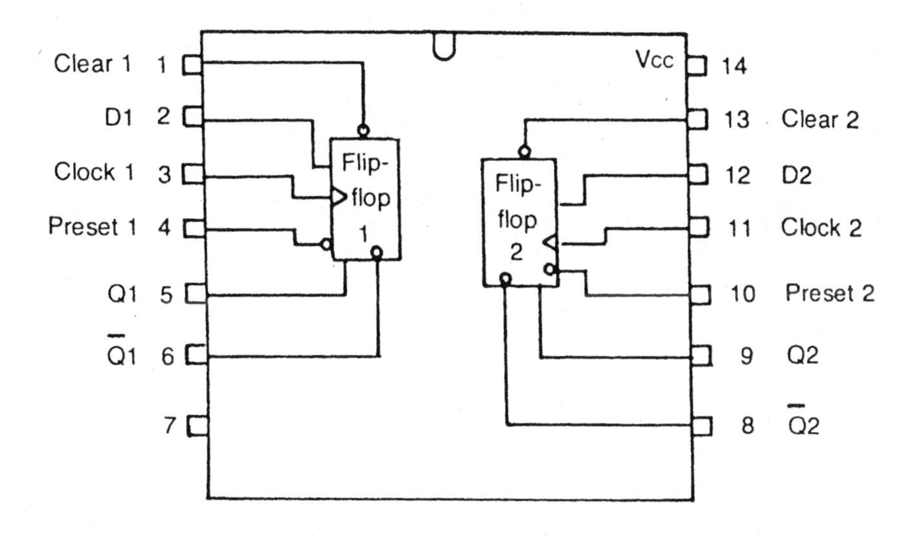

Divide by 2 counter

The output frequency of this circuit will be one-half the input frequency.

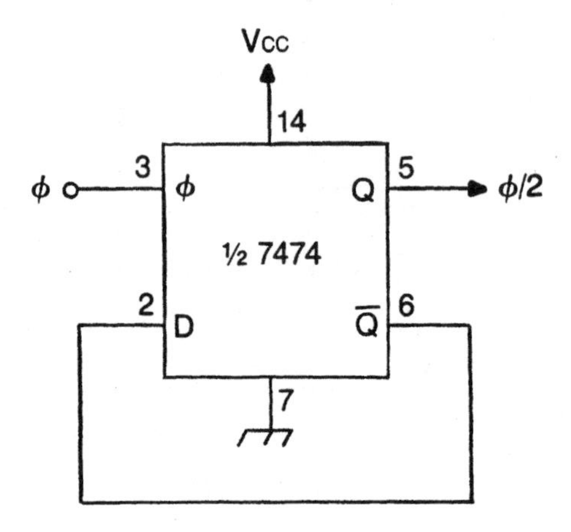

Phase detector

The output LED lights up when the two input signals are in phase. That is, both are HIGH at the same time, and both are LOW at the same time.

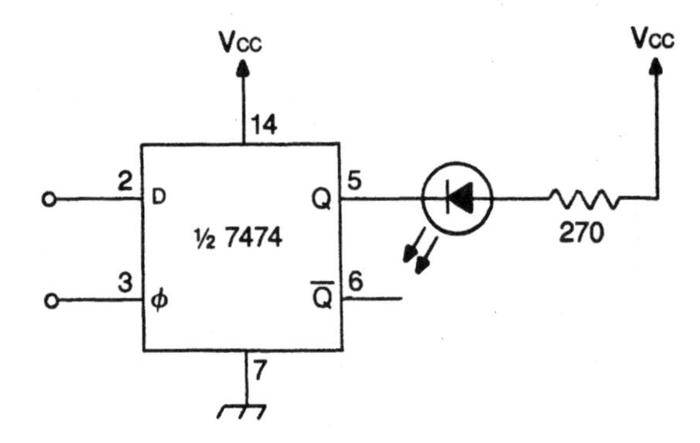

Wave shaper

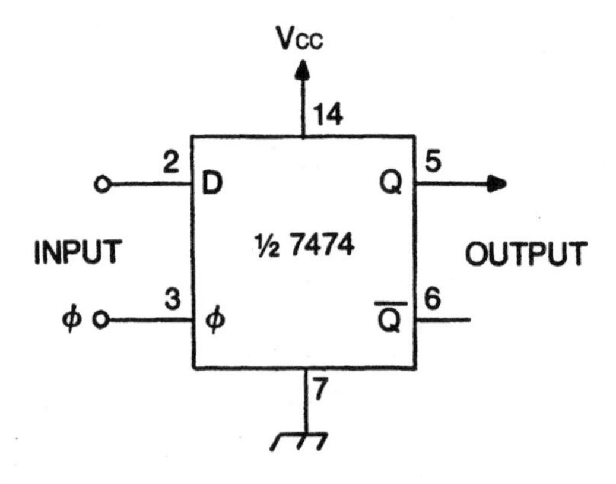

2-bit storage register

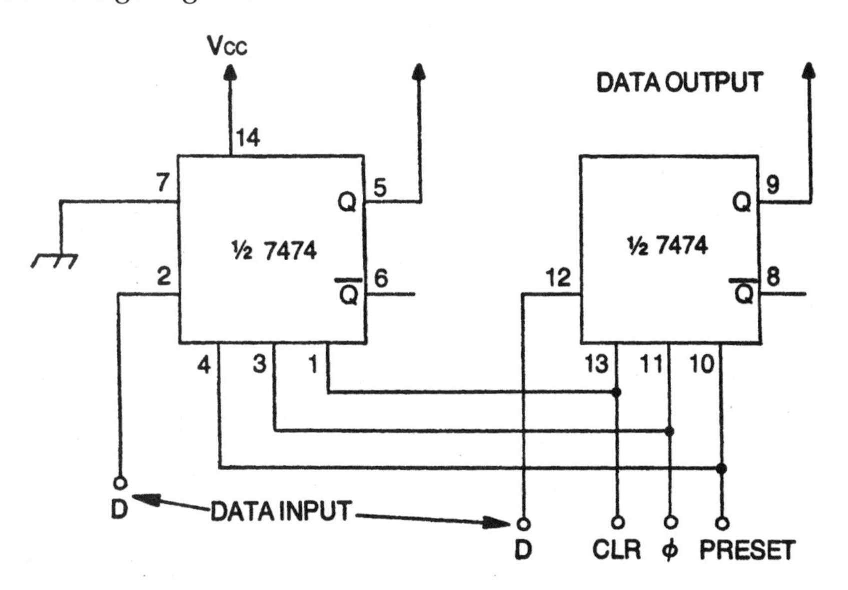

7475 quad latch

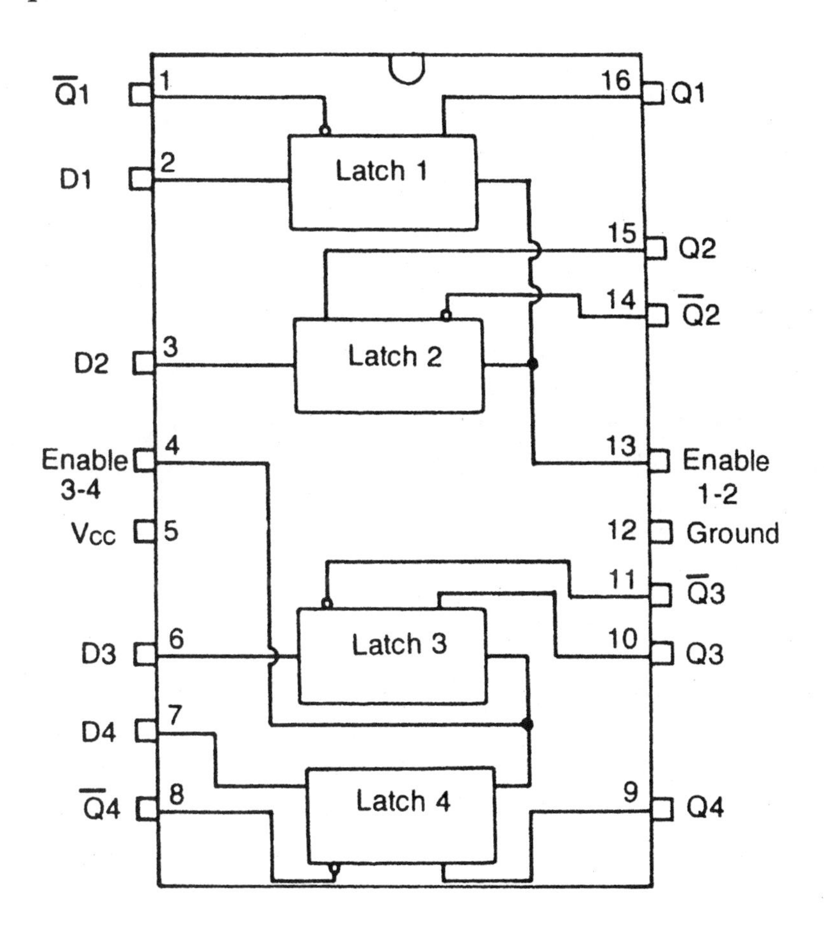

4-bit data latch

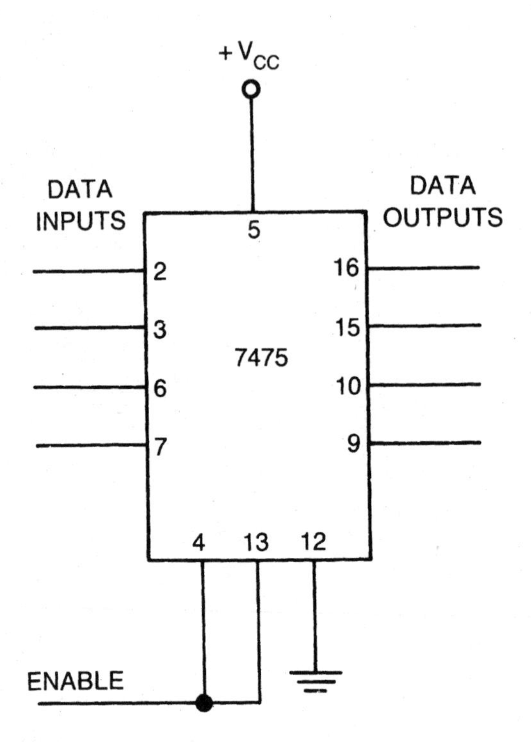

Decimal counter

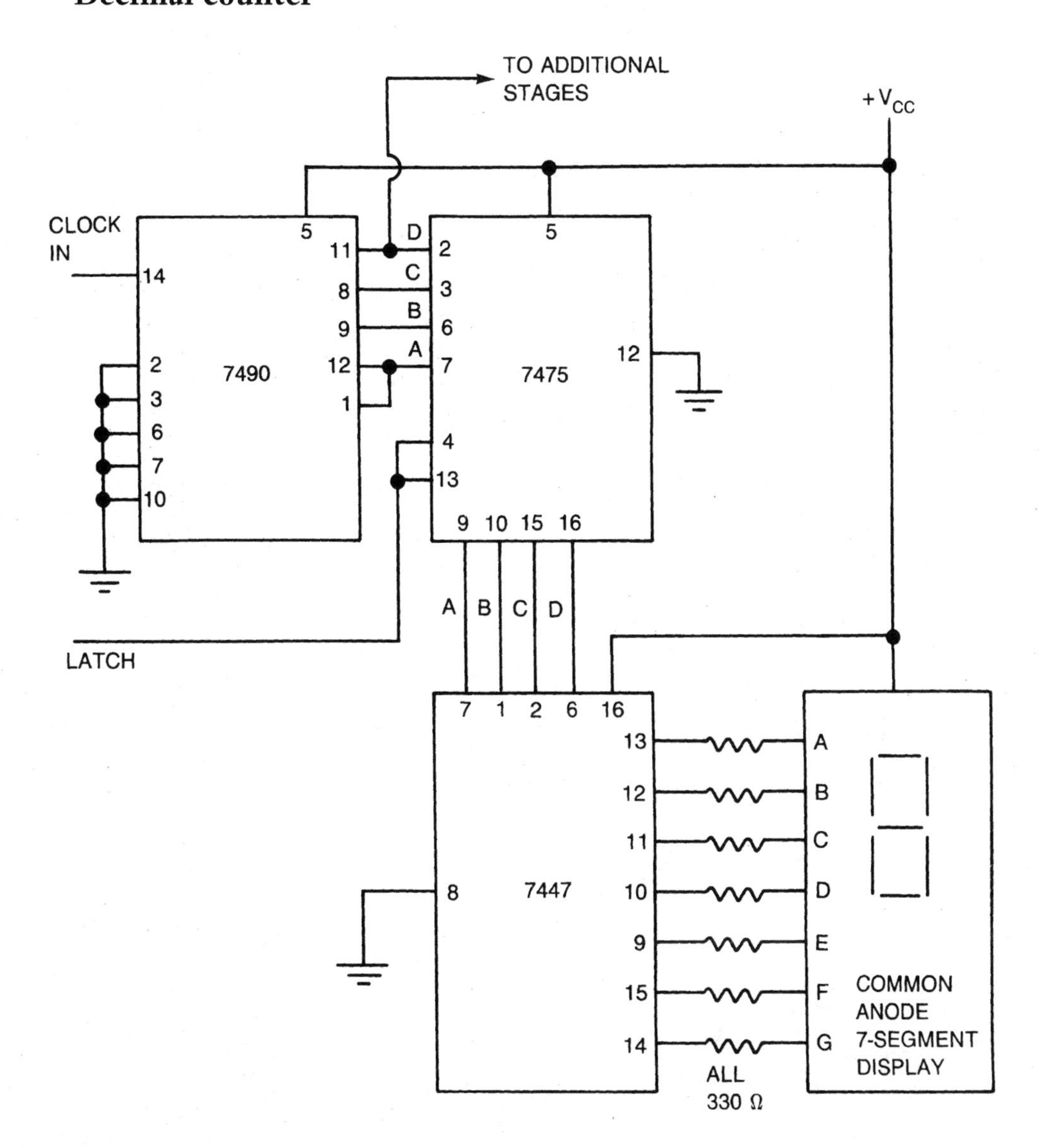

7476 dual JK flip-flop

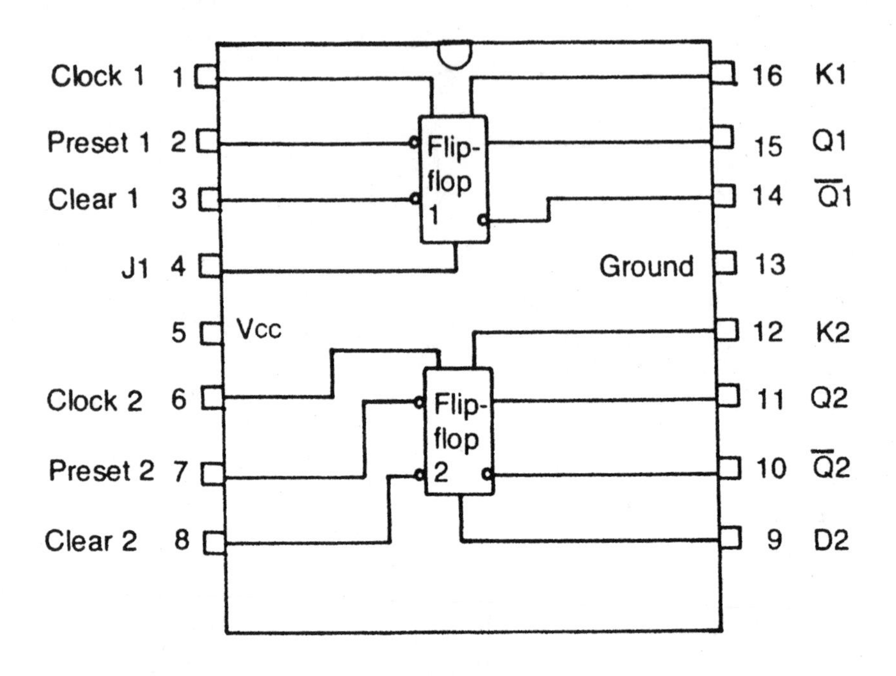

4-bit serial shift register

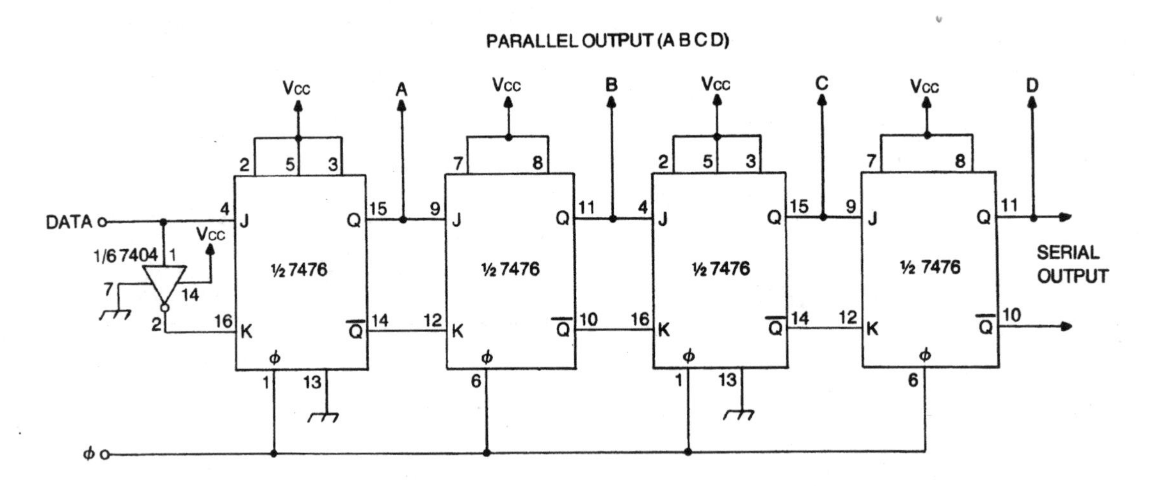

4-bit binary counter

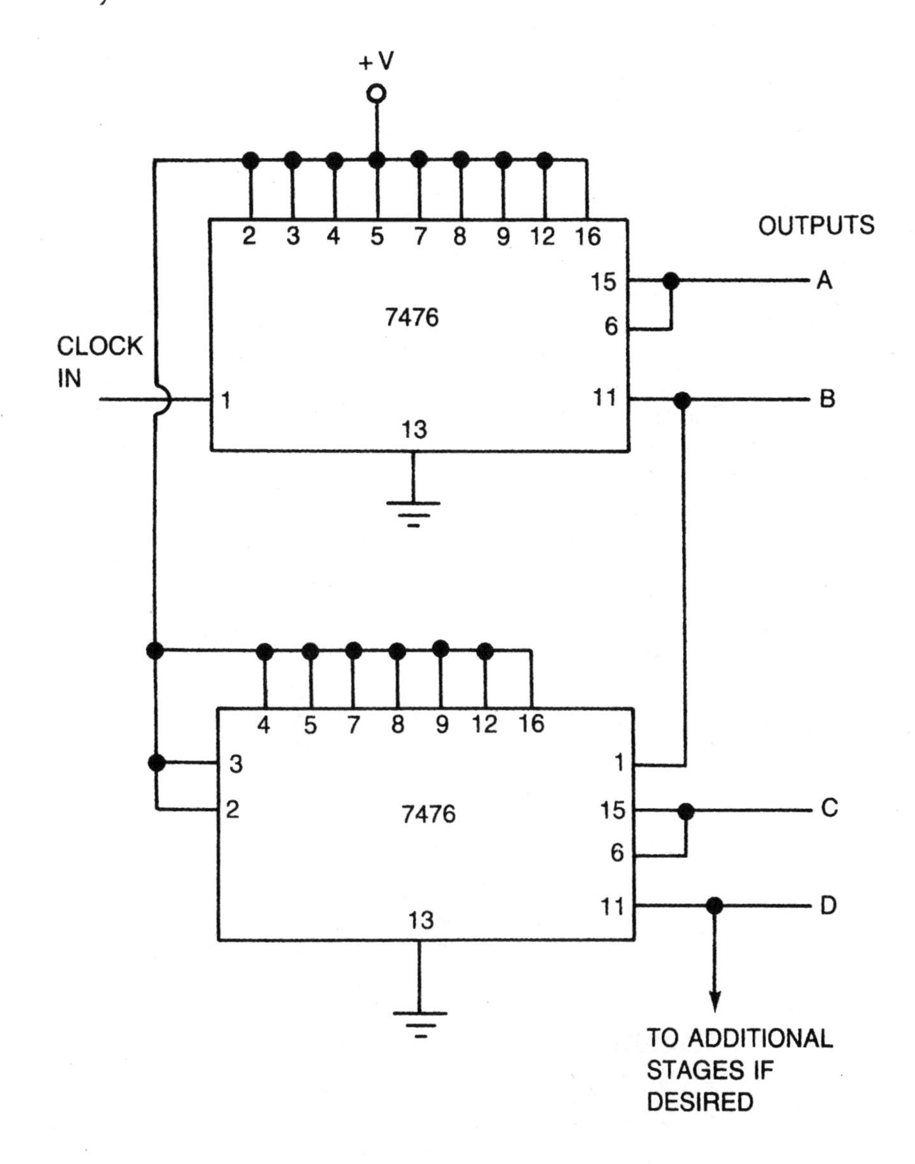

7490 decade counter

The 7490 is a 4-bit ripple-type decade counter. The device consists of four master/slave flip-flops internally connected to provide a divide-by-2 section, and a divide-by-5 section. Each section has a separate clock input to initiate stage changes of the counter on the HIGH-to-LOW clock transition. State changes of the Q outputs do not occur simultaneously, because of internal ripple delays. Therefore, decoded output signals are subject to decoding spikes, and should not be used for clocks or strobes. The Q_0 output is designed and specified to drive the rate fan-out plus the CP1 input of the device.

Because the output from the divide-by-2 section is not internally connected to the succeeding stages, the device can be operated in various counting modes. To operate as a divide-by-2, or a divide-by-5 counter, no external interconnections are required. The first flip-flop is used as a binary element for the divide-by-2 function (CP_0 as the input, and Q_0 as the output). The CP_1 input is used to obtain divide-by-5 operation at the Q_3 output.

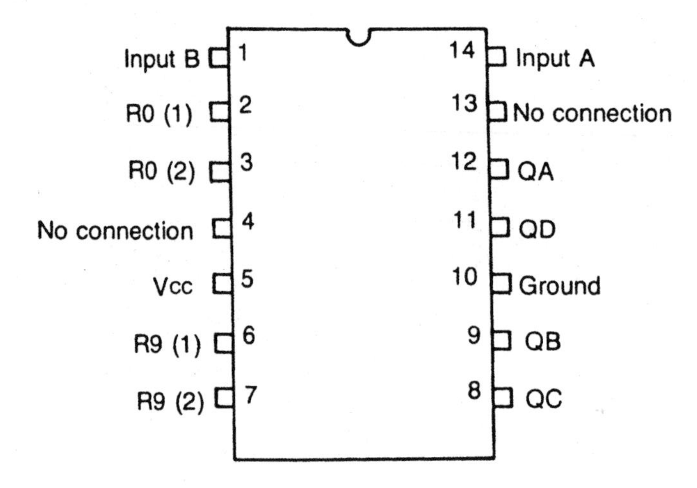

Divide-by-5 counter

To operate as a divide-by-5 counter, no external interconnections are required. The CP_1 input is used to obtain divide-by-5 operation at the Q_3 output.

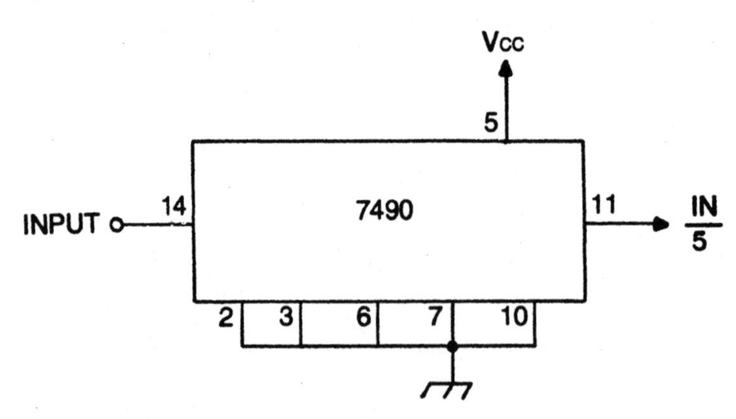

Divide-by-6 counter

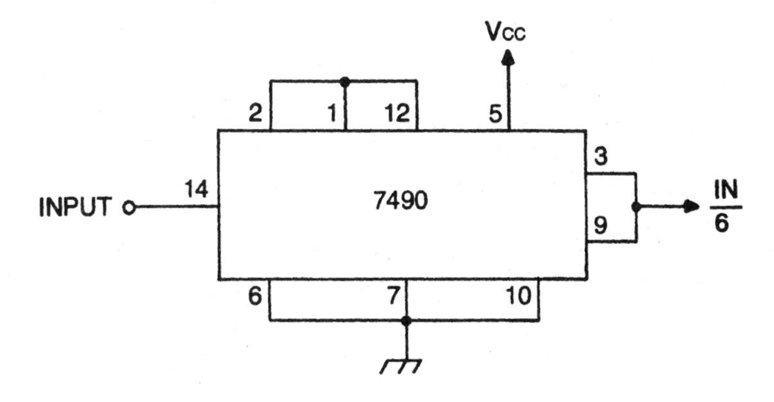

Divide-by-7 counter

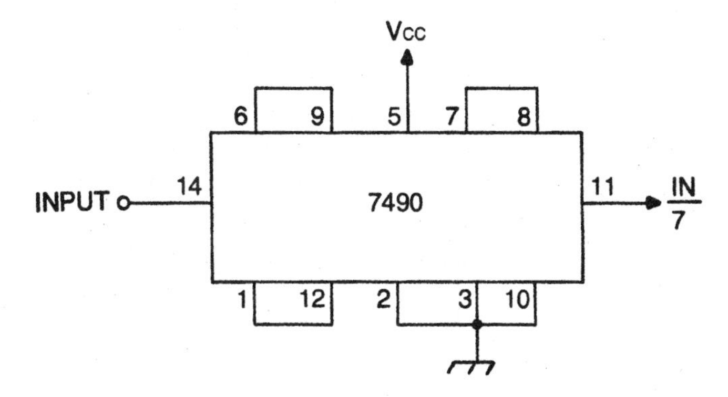

Divide-by-8 counter

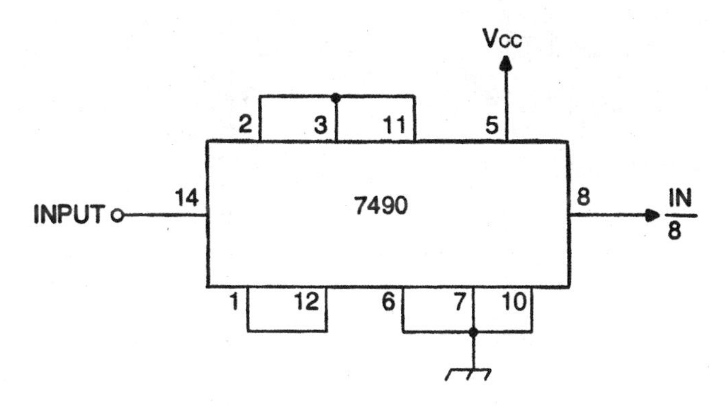

Divide-by-9 counter

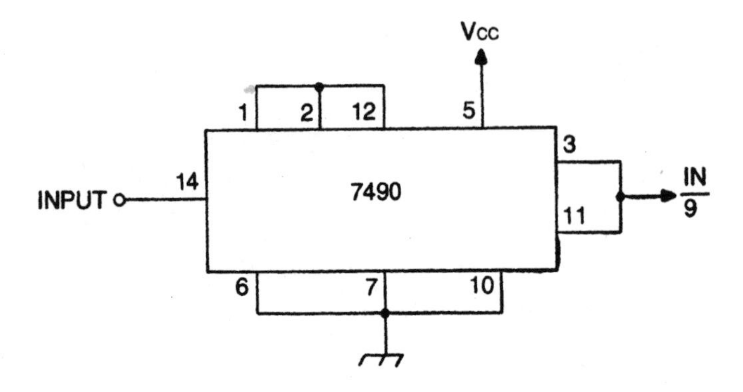

Divide-by-10 counter

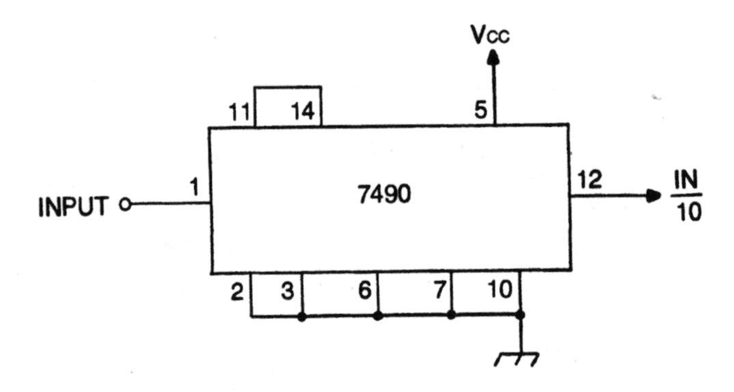

7492 divide-by-12 binary counter

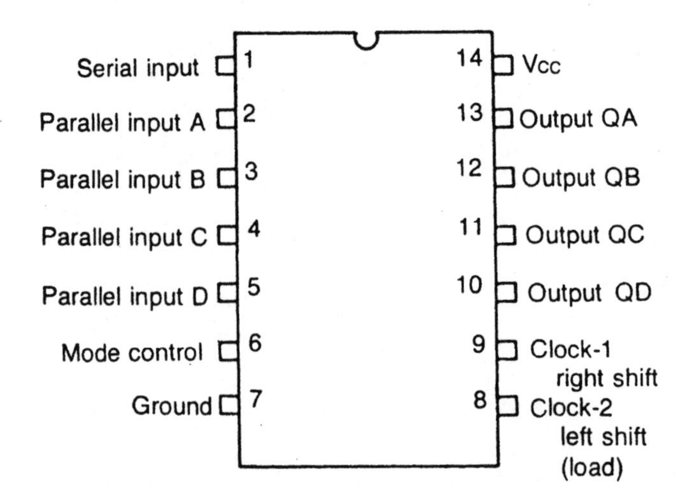

10 Hz pulse source

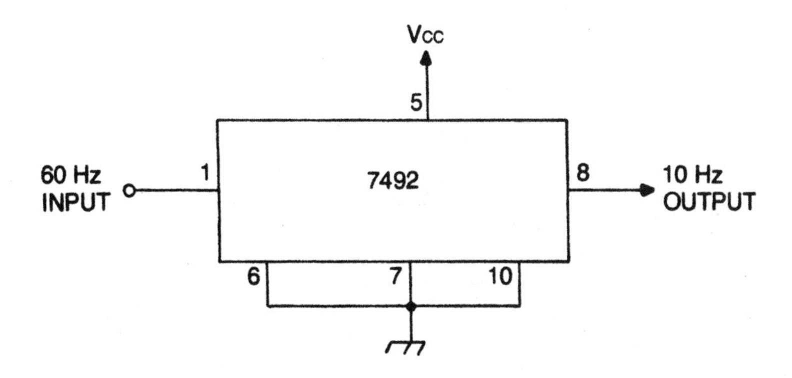

Divide-by-7 counter

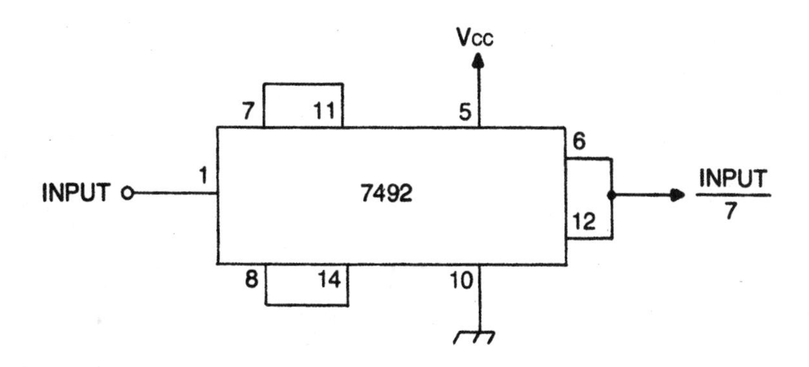

Divide-by-9 counter

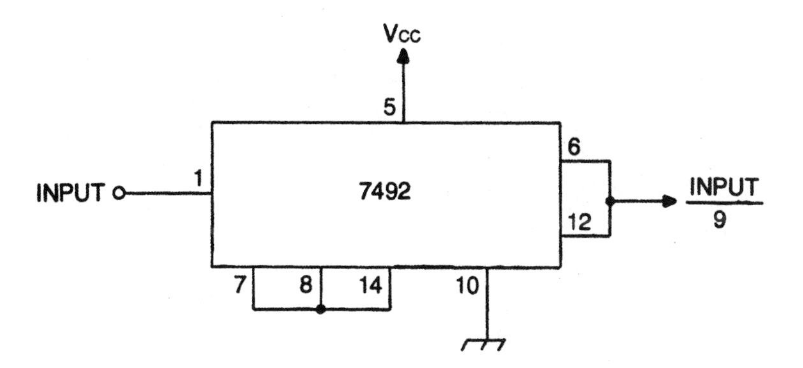

Divide-by-12 counter

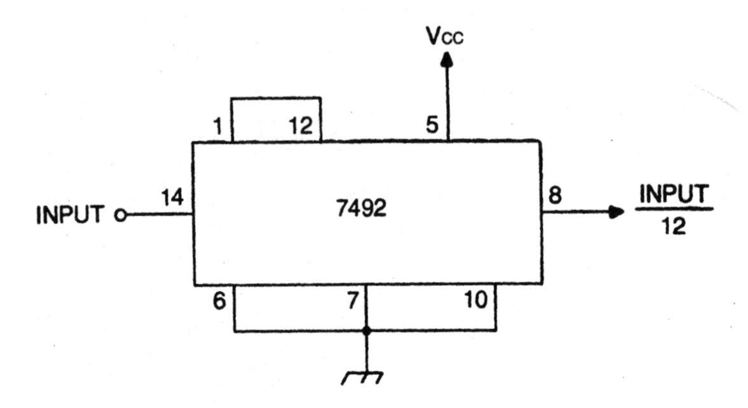

Divide-by-120 counter

The 7490 divides the input signal by 10, then the 7492 divides the result by a further 12, resulting in a total division factor of 120. Other value dividers can similarly be cascaded to achieve almost any desired division factor.

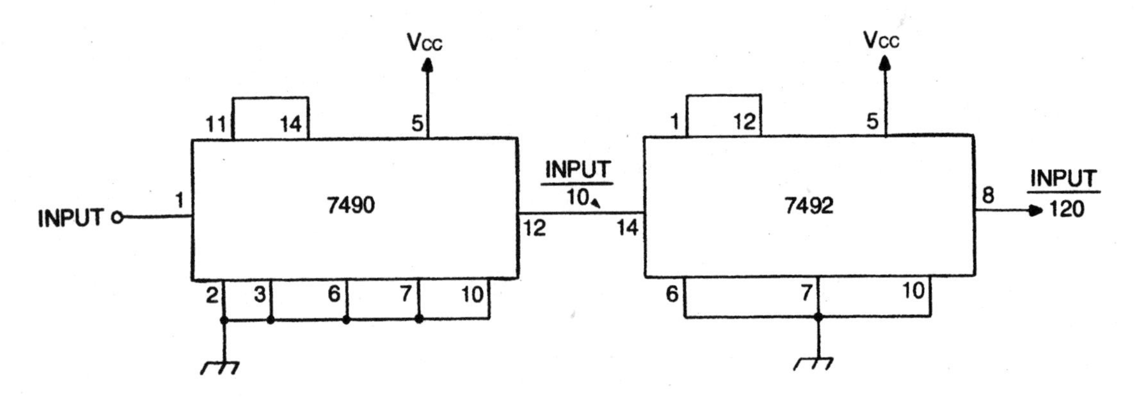

74123 dual retriggerable one-shot with clear

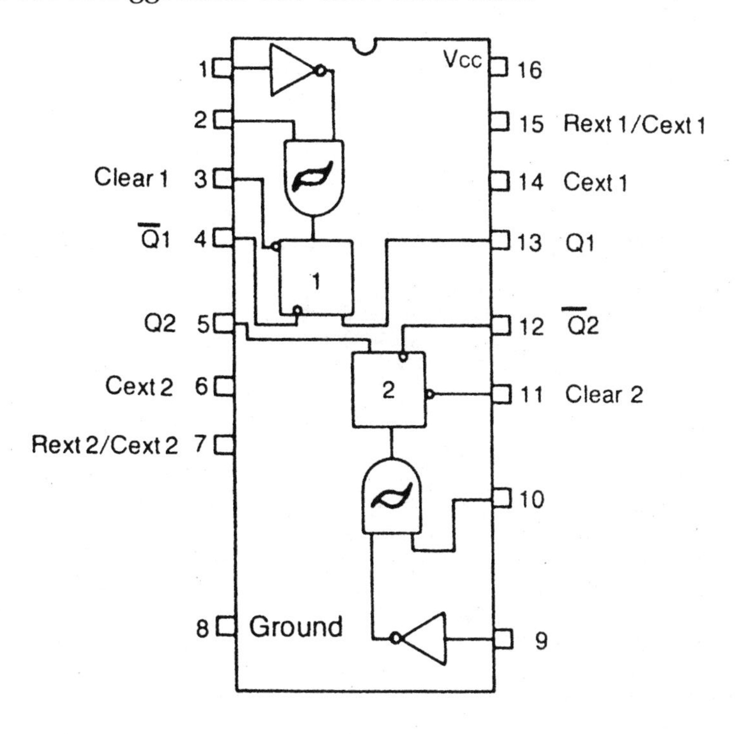

Missing pulse detector

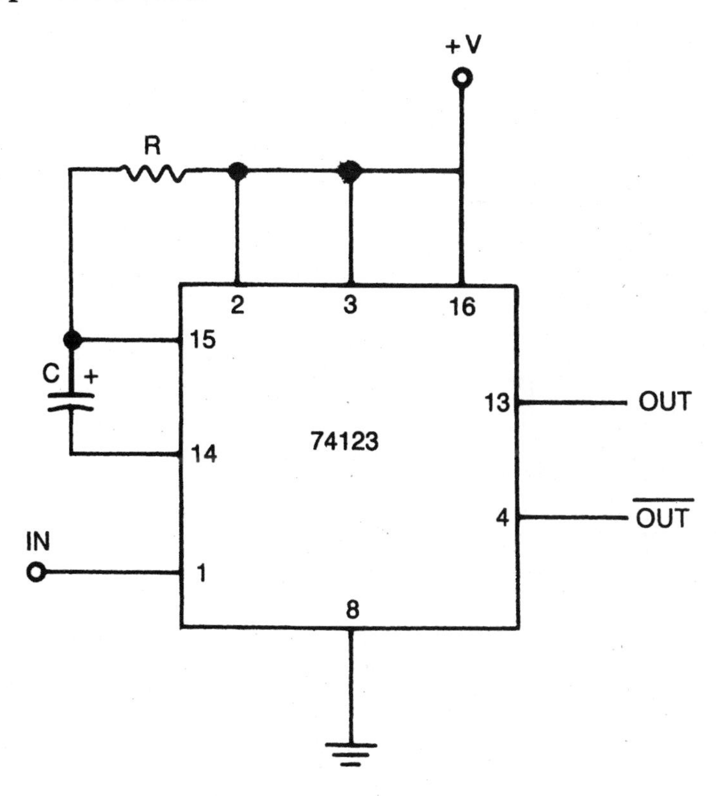

One-shot monostable multivibrator (or timer)

The timing period is determined by the values of resistor R and capacitor C. Increasing the value of either (or both) of these component values increases the timing period.

An electrical trigger signal can be substituted in place of the manual push-button switch shown here.

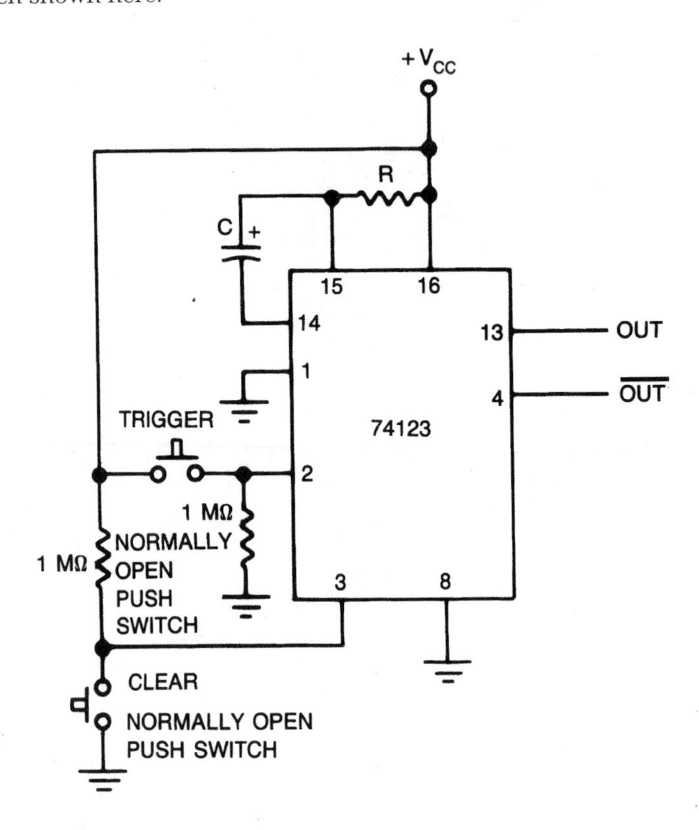

Sound-effect generator

The circuit can generate a wide variety of unusual sounds and tones, by adjusting the settings of the two potentiometers. For additional effects, you might also want to experiment with alternate capacitor values. The two capacitors in this circuit do not have to have equal values.

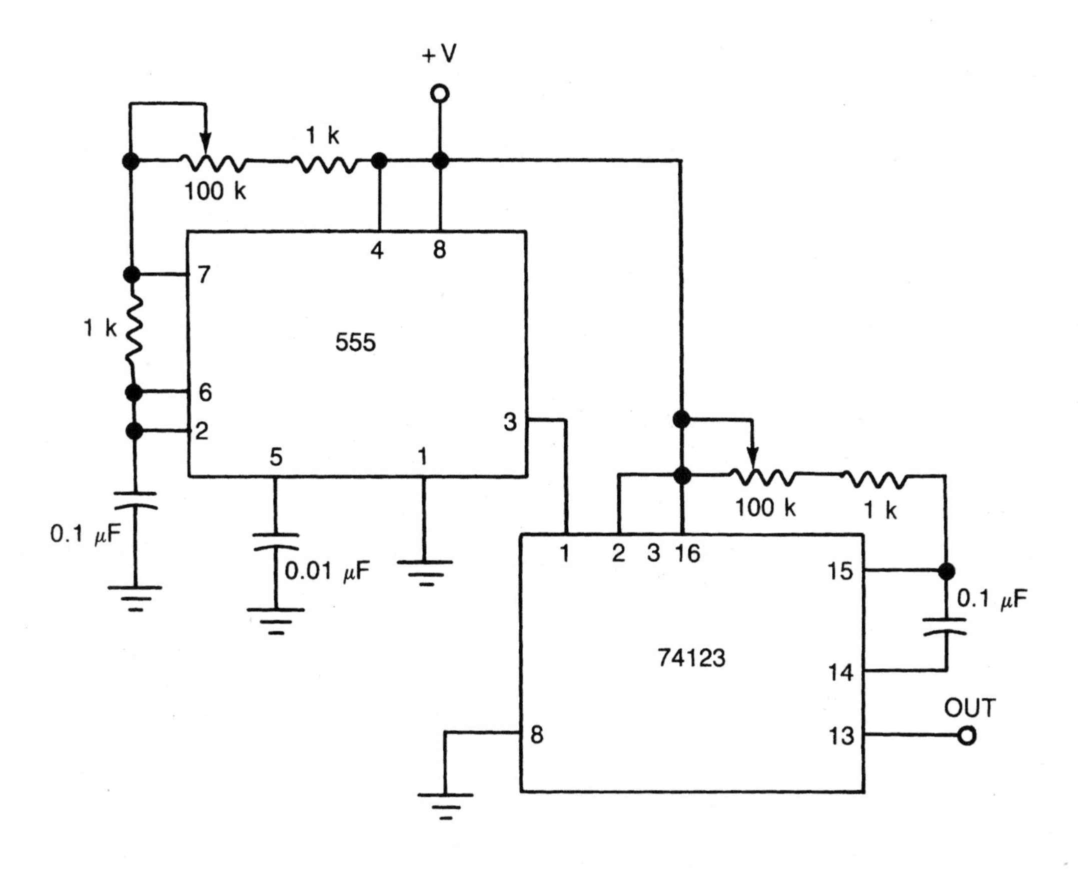

74132 quad 2-input NAND Schmitt trigger

The 74132 contains four 2-input NAND gates that operate in the same logic pattern as any other 2-input NAND gates. Unlike most standard gates, however, each gate circuit within this chip contains a 2-input Schmitt trigger, so these gates are capable of transforming slowly changing input signals into sharply defined, jitter-free output signals. In addition, they have greater noise margins than conventional NAND gates.

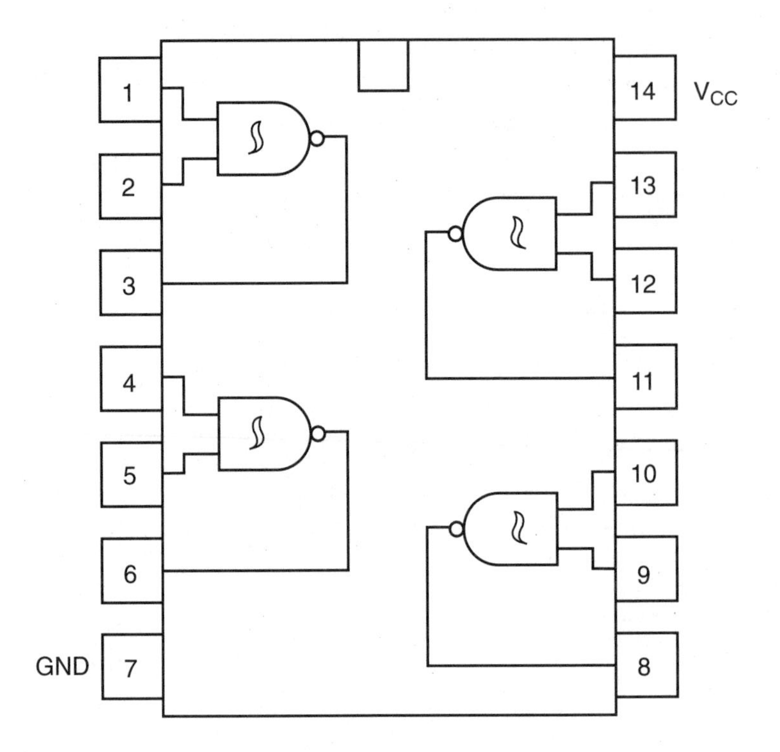

Wave-shaper and pulse restorer

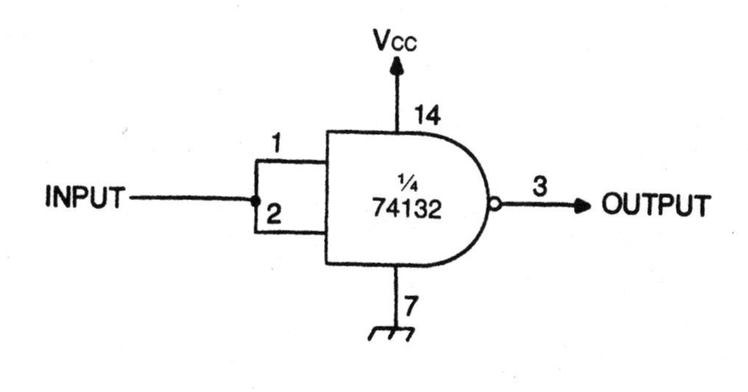

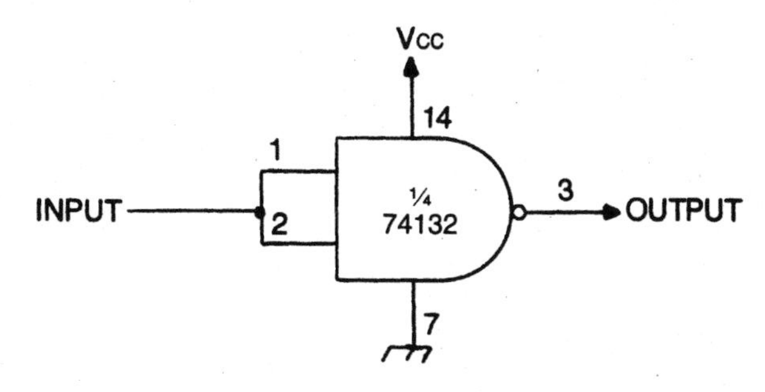

Noise eliminator

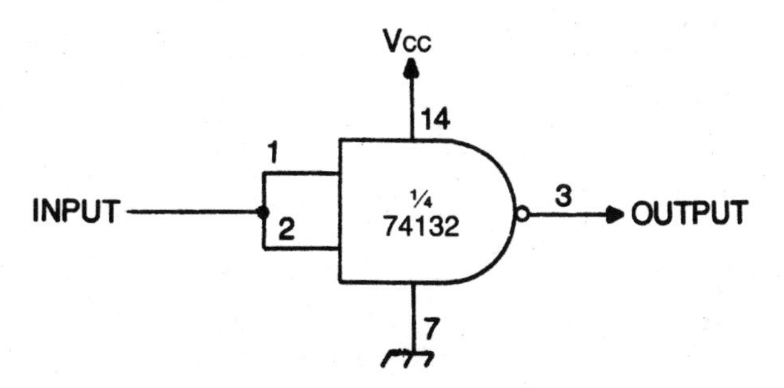

Threshold detector

The various circuits shown on pages 53 and 54 are all actually the same simple circuit. Basically, all we have here is a NAND gate with its inputs shorted together to form an inverter logic function. However, the Schmitt trigger built into the 74132's NAND gates can perform the special functions described in the circuit names. The only difference comes from the nature of the input signal, and the particular effects we are interested with. The output will be a nice, clean logic LOW or HIGH, with no noise, flutter, or other fluctuations.

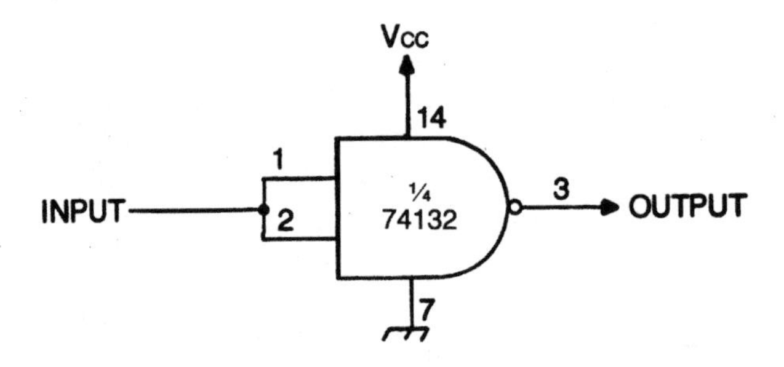

Gated pulse shaper

This circuit is similar to those shown previously, except just one input is used for the data input. The other input serves as a gate control. When this gate signal is LOW, the circuit is effectively disabled. The output will always be LOW, regardless of the data input. Making the gate input HIGH enables the device, which then performs just like the circuit(s) on pages 53 and 54.

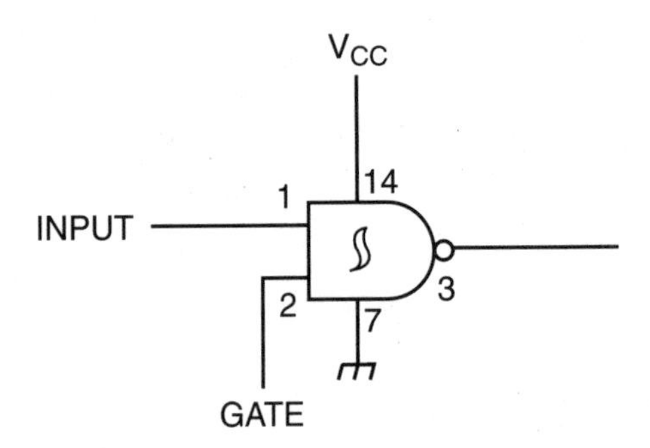

Deluxe pulse shaper/stretcher

This circuit cleans up the input signal, and produces a clean output pulse with a fixed duration, regardless of the duration of the input pulse. In some critical applications, it might be advisable to use the 7555 CMOS timer, instead of the standard 555 timer shown here. (Both of these devices are covered in later sections of this book.)

The timing period of the output pulse is set by the values of resistor R and capacitor C. Increasing the value of either (or both) of these components results in a longer output pulse. The timing formula is

$$T = 1.1RC$$

where T is the timing period in seconds, R is the resistance in megohms (1 megohm = 1,000,000 ohms), and C is the capacitance in microfarads (1 microfarad = 0.000001 farad).

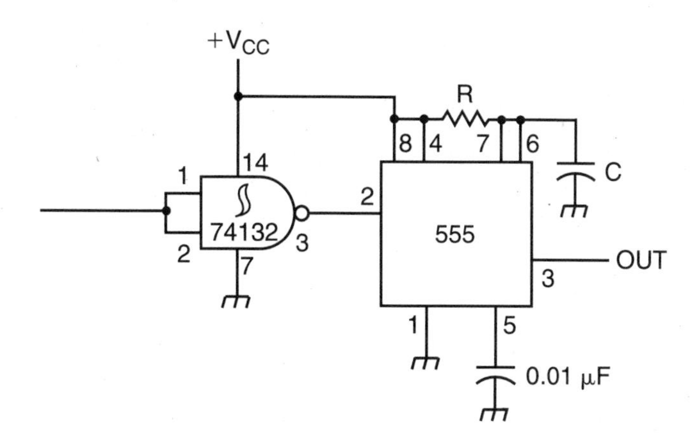

74138 1-of-8 decoder/demultiplexer

The 74138 decoder accepts three binary weighted inputs (A0, A1, and A2) and, when enabled, provides eight mutually exclusive active LOW outputs (0–7). The device features three enable inputs: two active LOW (E1 and E2), and one active HIGH (E3). Every output will be HIGH unless E1 and E2 are LOW, and E3 is HIGH. This multiple enable function allows easy parallel expansion of the device to a 1-of-32 (5 lines to 32 lines) decoder with just four 74138s and one inverter.

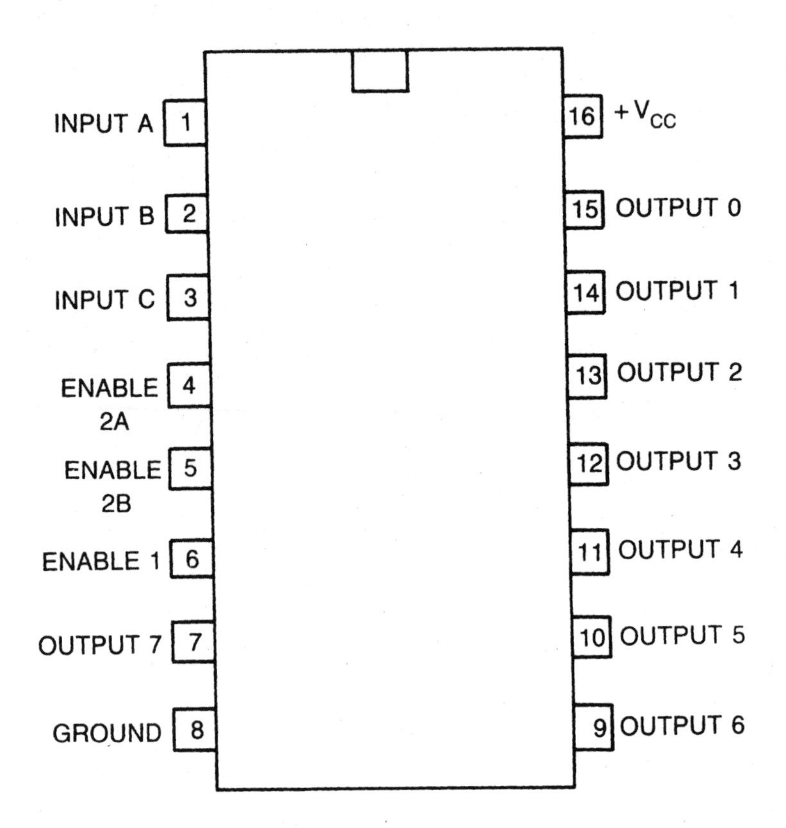

3-to-8 line decoder/demultiplexer

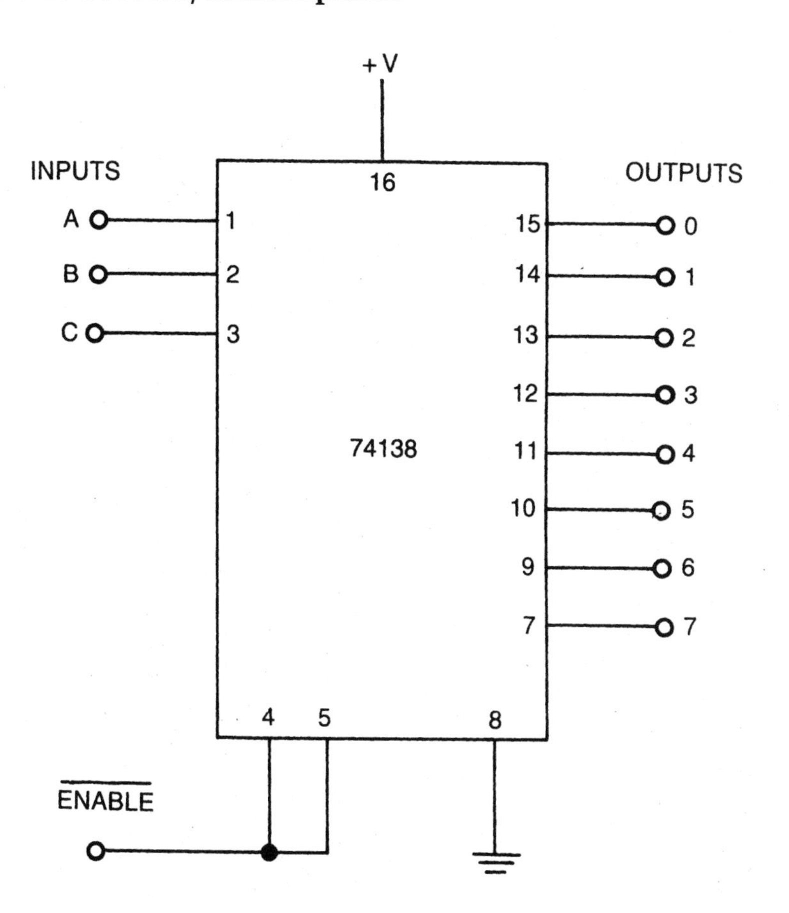

1-of-8 multiplexer

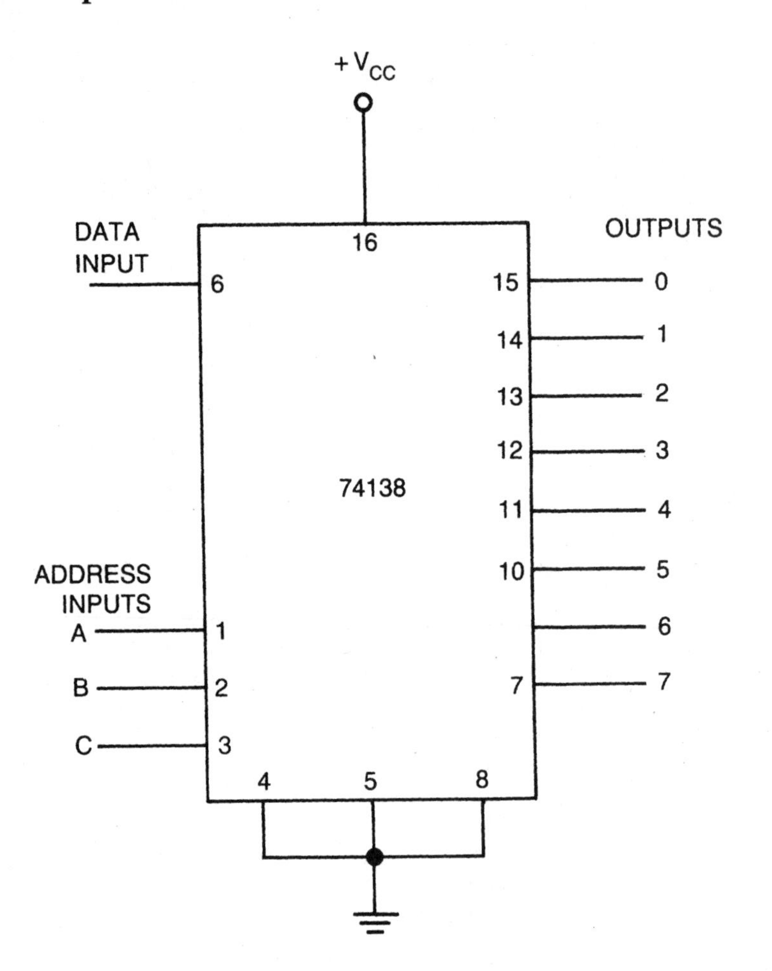

4-to-16 line decoder

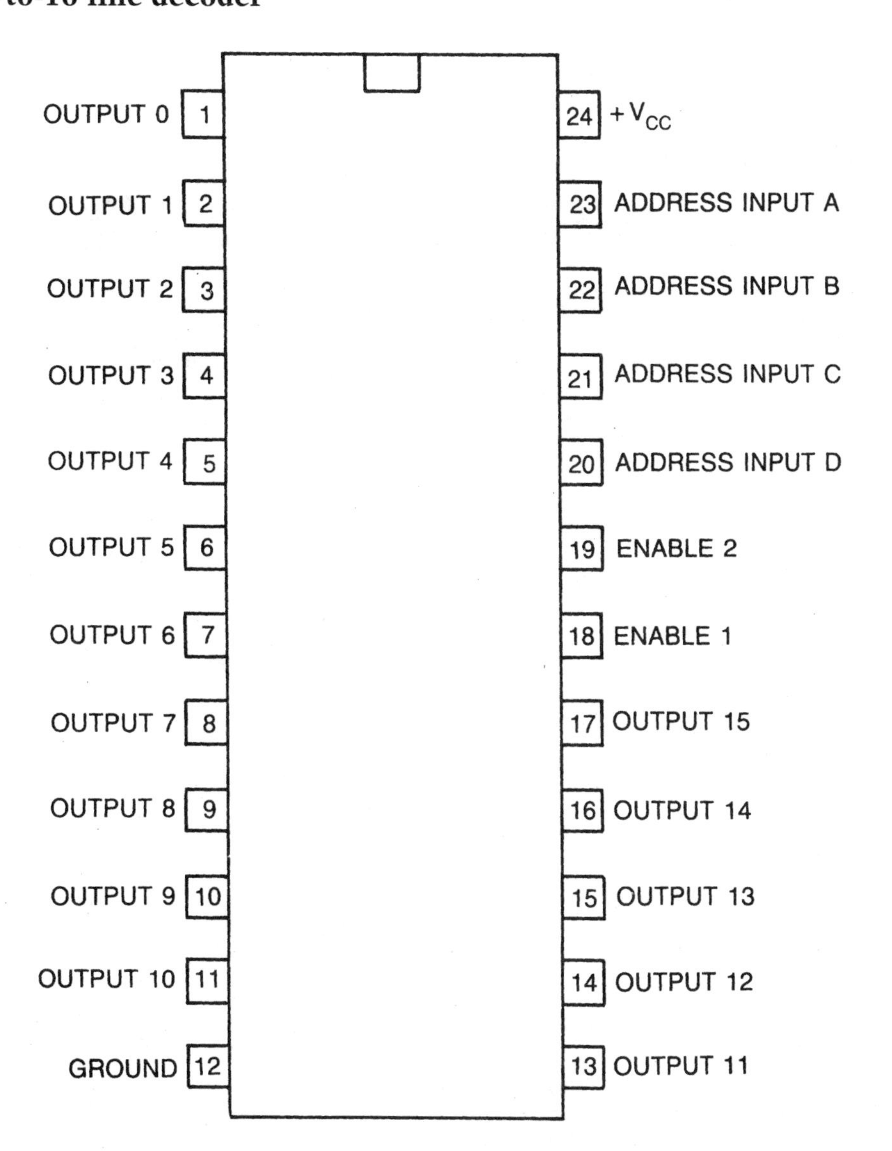

Basic 1-of-16 demultiplexer

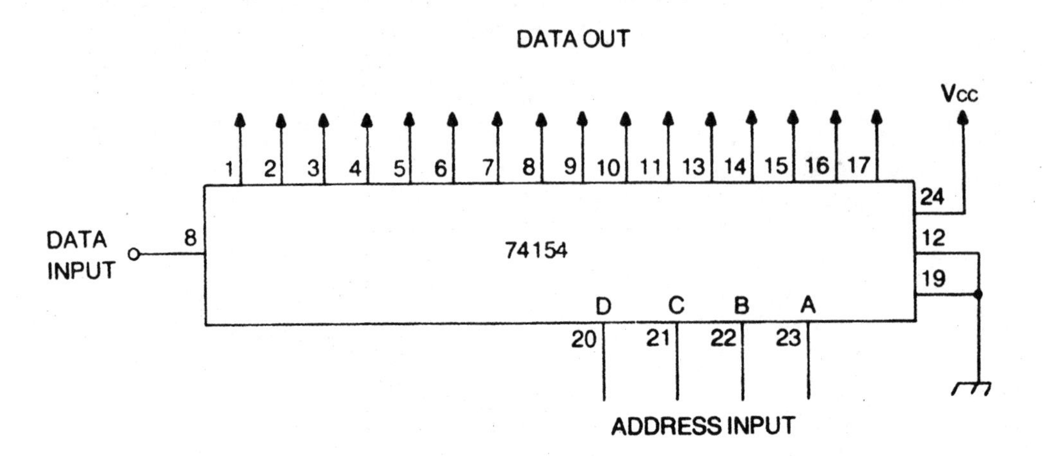

Binary-to-hexadecimal converter

This circuit accepts a 4-bit binary value (from 0000 to 1111) and translates it into the appropriate single hexadecimal (base 16) digit (from 0 to F).

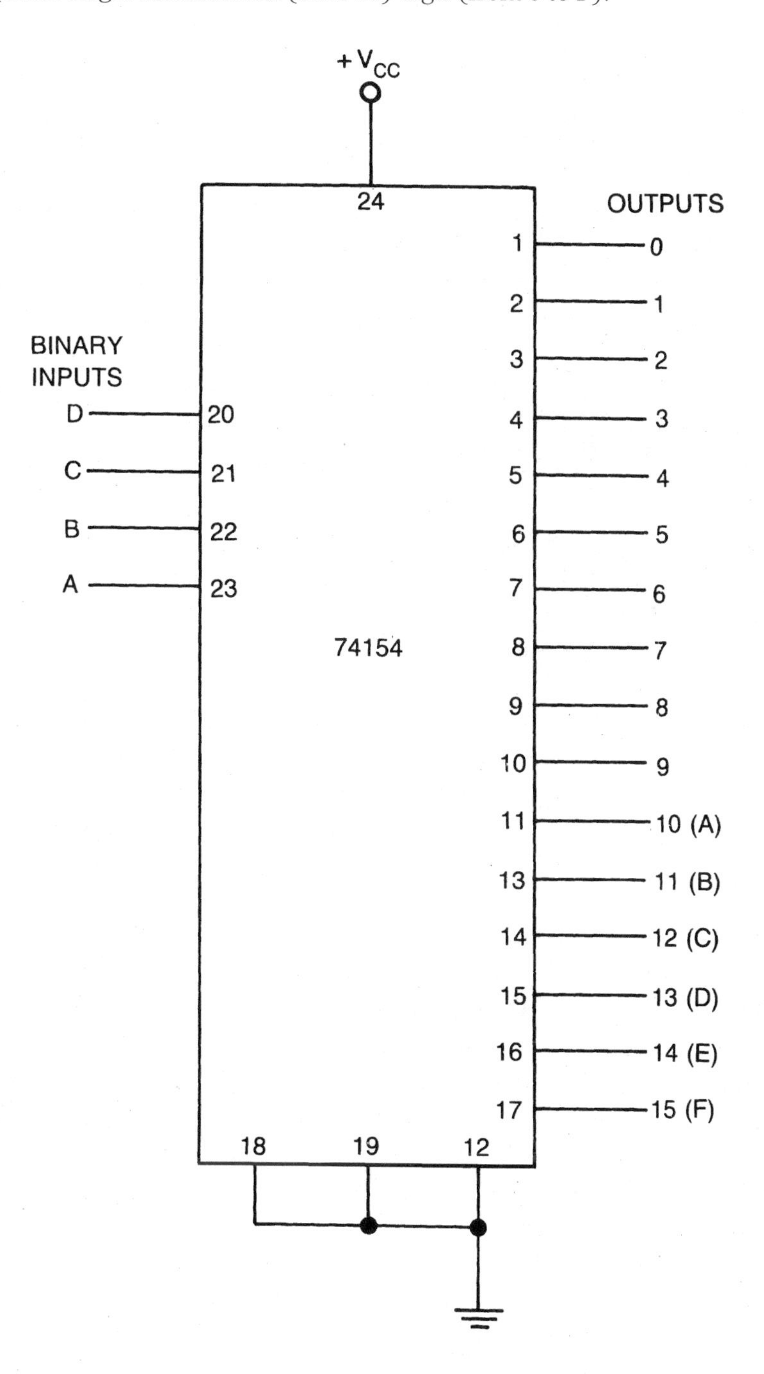

Up/down counter flasher display

This circuit is a great eye-catching display. The lighted LED appears to move back and forth. The speed of the "movement" is set by the 100K potentiometer. The speed range can be changed by substituting a different timing capacitor for the 0.22 µF unit shown here. Larger capacitances will result in slower flash times.

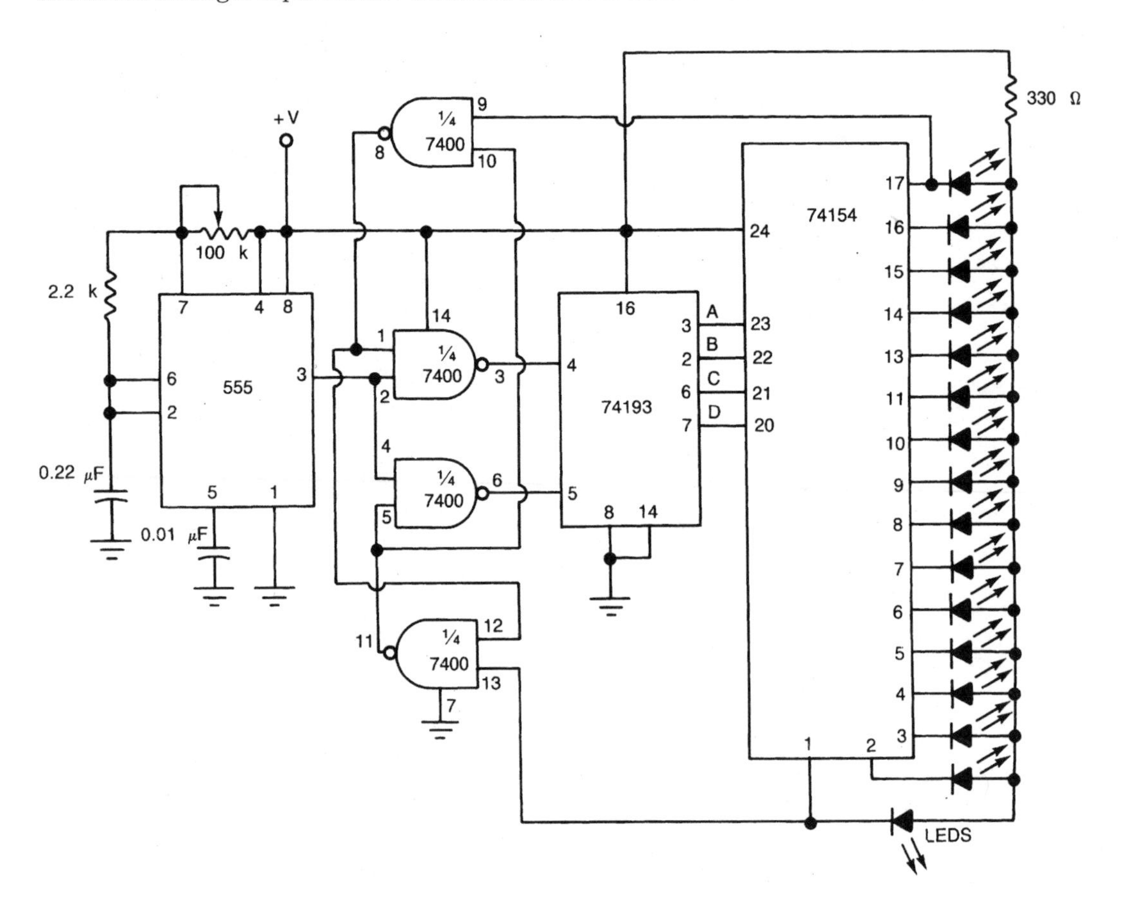

74192 BCD up/down counter

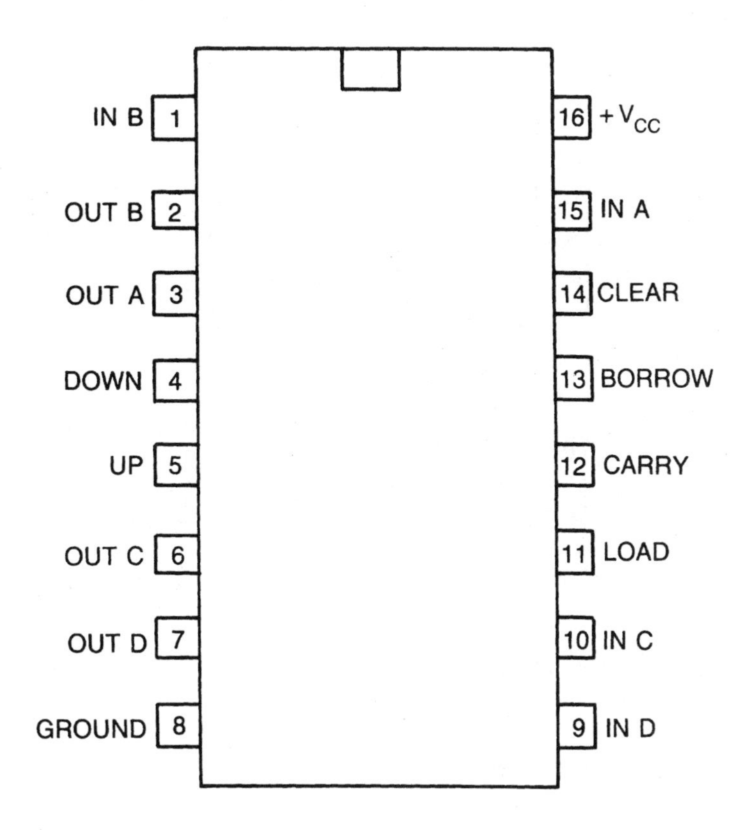

Cascaded up counter

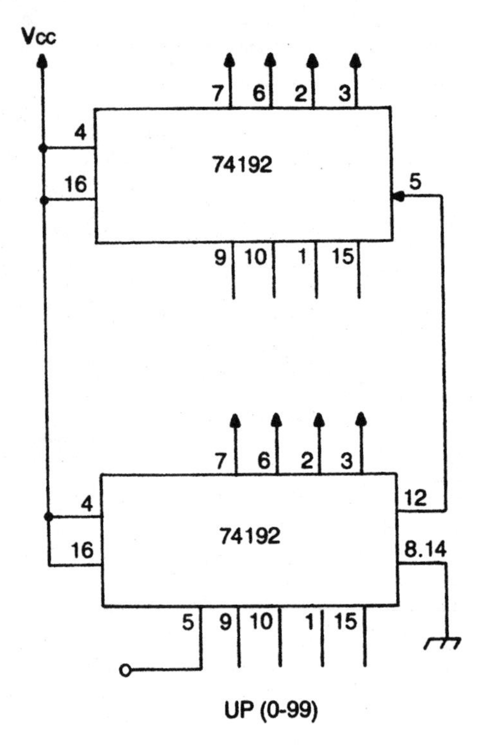

Cascaded down counter

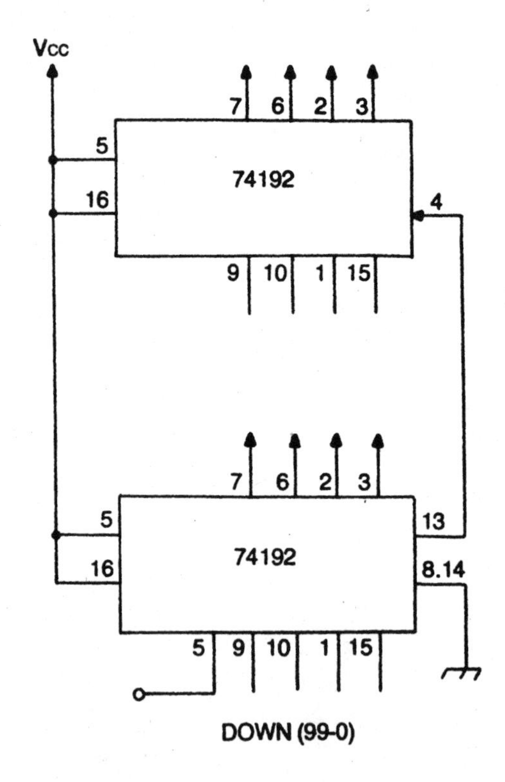

Count up to N and recycle

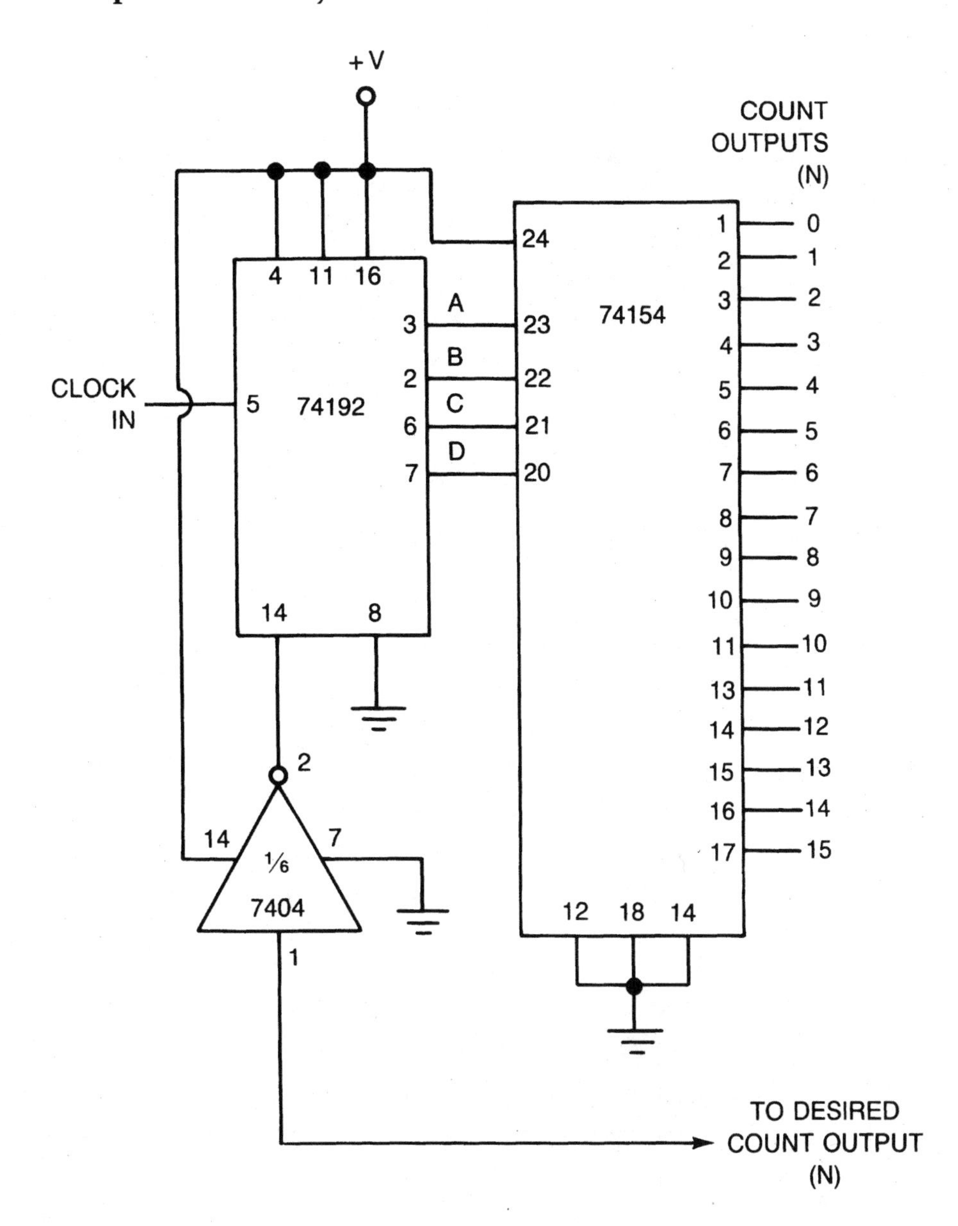

Count up to N and stop

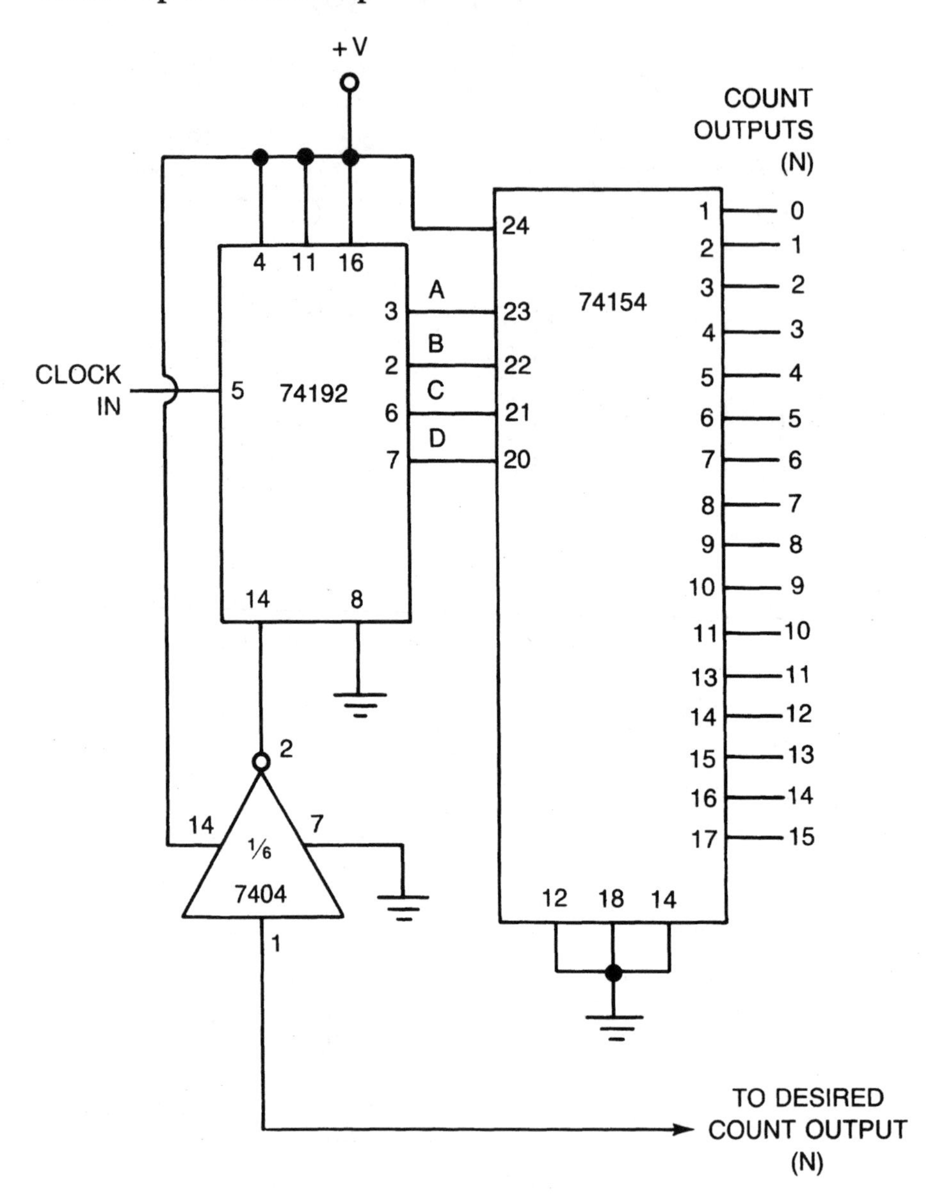

74193 4-bit up/down counter

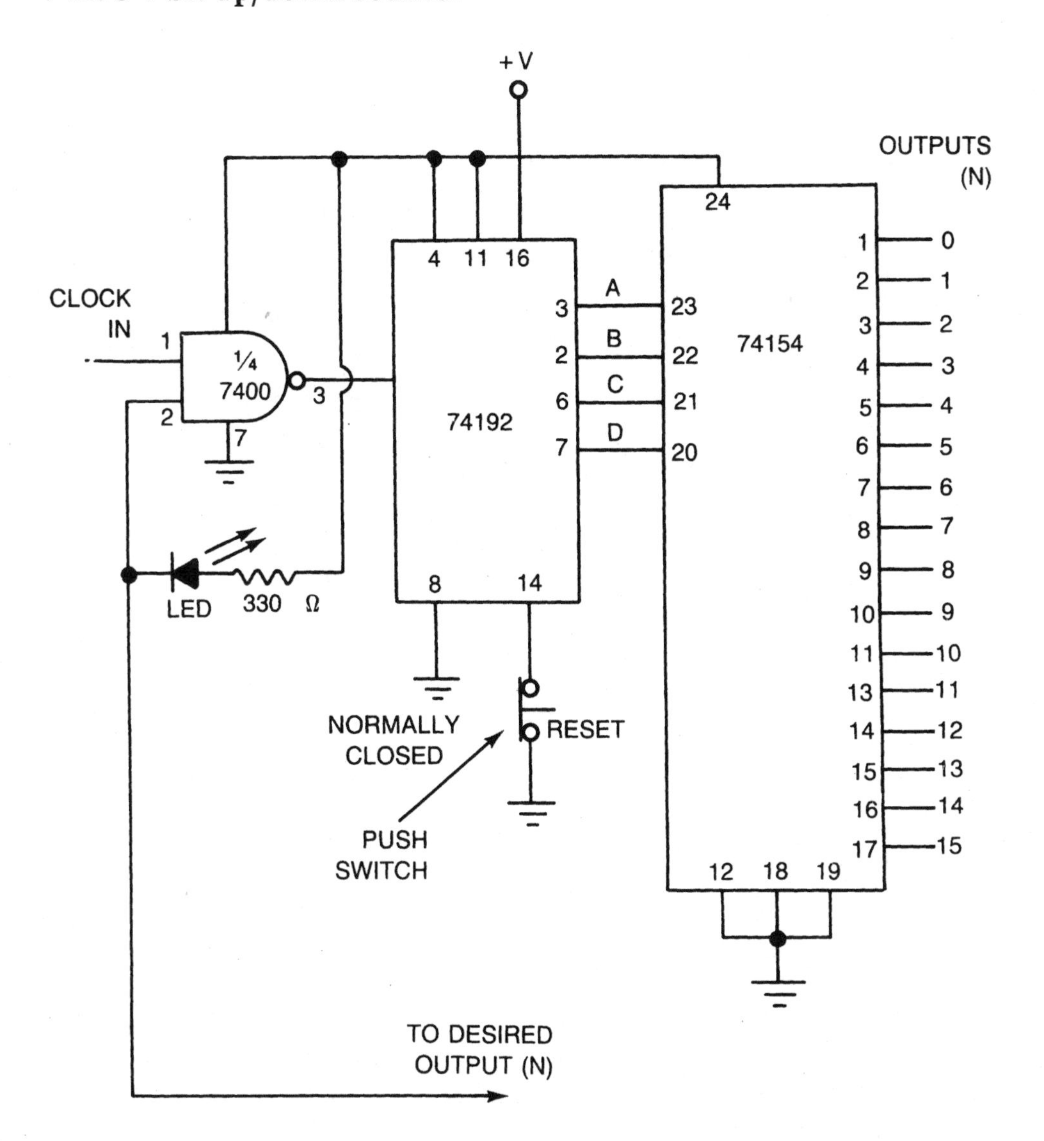

Count down from N and recycle

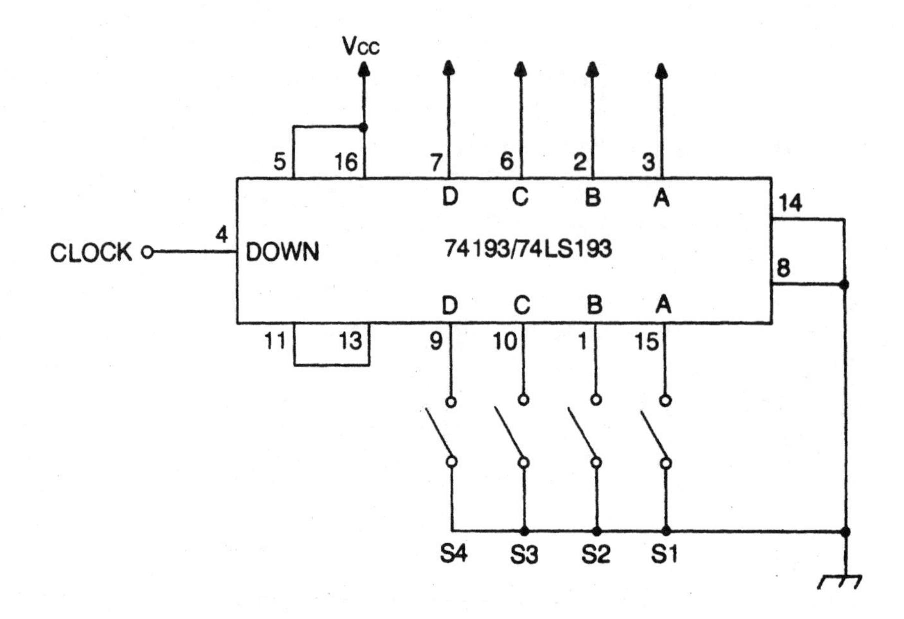

2
SECTION
4xxx digital

In this section, the 4xxx line of digital/logic devices will be covered. Devices range from simple logic gates, through flip-flops, counters, registers, and more exotic and complex special-purpose devices.

This standardized numbering scheme was created specifically for CMOS (Complementary Metal-Oxide Semiconductor) ICs, as opposed to TTL devices. CMOS chips also sometimes use the 74Cxx numbering scheme, with the same pin-outs (though not electrical compatibility) with the standard 74xx numbering scheme. There is no connection between the 4xxx and 74xx numbering schemes. Devices with similar numbers in these two series have different pin-outs, and usually completely different functions. For example, a 7406 is a hex inverter buffer/driver, but a 4006 is an 18-stage static shift register.

Many manufacturers use a "CD" prefix. For example, a CD4009 chip is the same as a 4009. You may also encounter CMOS chips numbered 14xxx. Again, there is no functional difference. A 14018 is identical to a 4018.

While CMOS devices can perform the same logic functions as TTL devices, they are not electrically compatible. A CMOS gate cannot directly drive a TTL gate, or vice versa. They can be interfaced, but extra external circuitry is required. These two logic families do not use the same voltage levels to represent the LOW and HIGH states, so confusion can result. In effect, TTL and CMOS gates don't speak the same language.

One of the biggest advantages of CMOS over TTL is much greater flexibility in the supply voltage. Most TTL devices require a tightly regulated 5 V supply voltage. CMOS gates can be operated on anything from 3 V to 15 V. Generally speaking, supply voltages of 9 V to 12 V will be the best choice.

Power dissipation in a CMOS gate varies with the operating frequency. At dc or low frequencies, almost no power at all is dissipated. At high frequencies, however, a CMOS gate can dissipate 10 mW to 15 mW, and sometimes even more.

The propagation delay of standard CMOS gates tends to be rather slow. Delays of 90 nS are not uncommon for 4xxx gates.

4001 quad 2-input NOR gate

The 4001 contains four independent NOR gates, each with two inputs. Only the power supply connections are common to all four logic elements.

A NOR gate produces a HIGH input if and only if all of its inputs are LOW. If one or more inputs goes HIGH, the output goes LOW.

4001 Truth Table

Inputs	Output
0 0	1
0 1	0
1 0	0
1 1	0

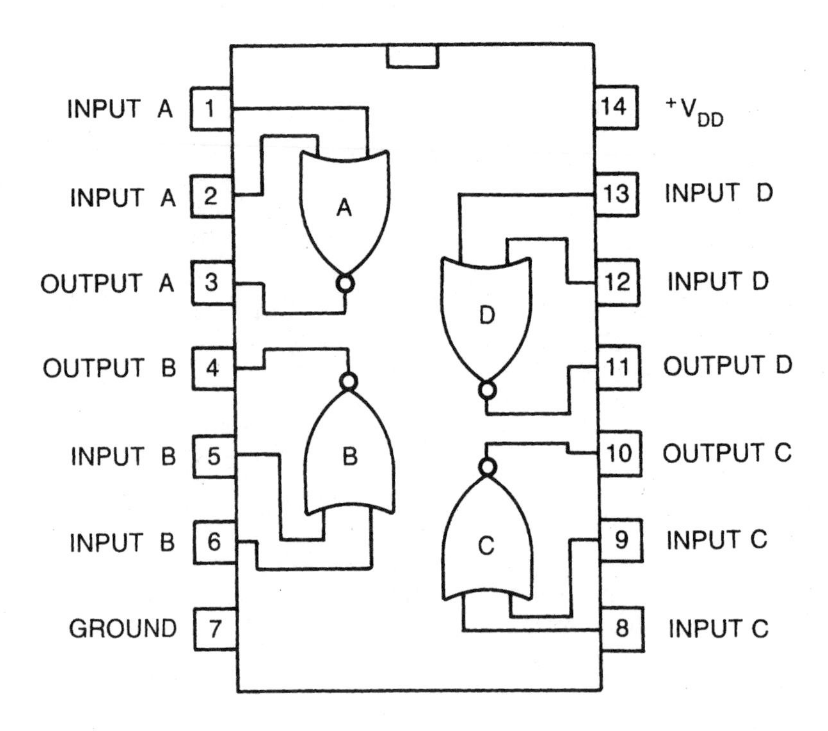

Increased fan-out NOR gate

In some large systems, you may need to drive more output devices in parallel than a single gate can handle without signal degradation and probable errors or erratic operation. The solution is to drive a pair of identical gates in parallel, essentially boosting the strength of the gate's output signal.

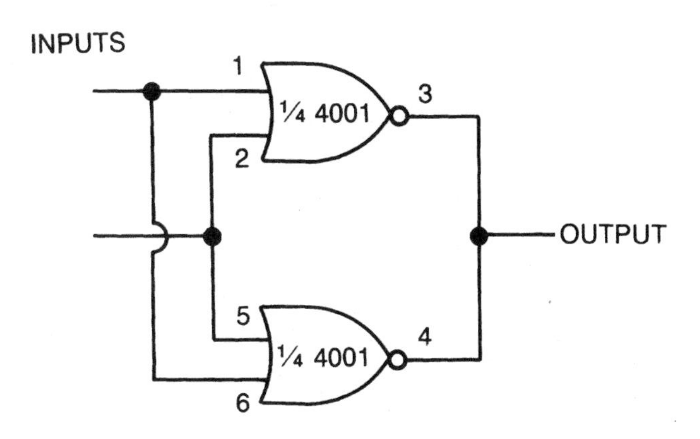

Increased fan-out inverter

In some large systems, you may need to drive more output devices in parallel than a single gate can handle without signal degradation and probable errors or erratic operation. The solution is to drive a pair of identical gates in parallel, essentially boosting the strength of the gate's output signal.

Since the inputs are shorted together, this extra-strength output gate functions as an inverter. The output is always at the opposite state as the input.

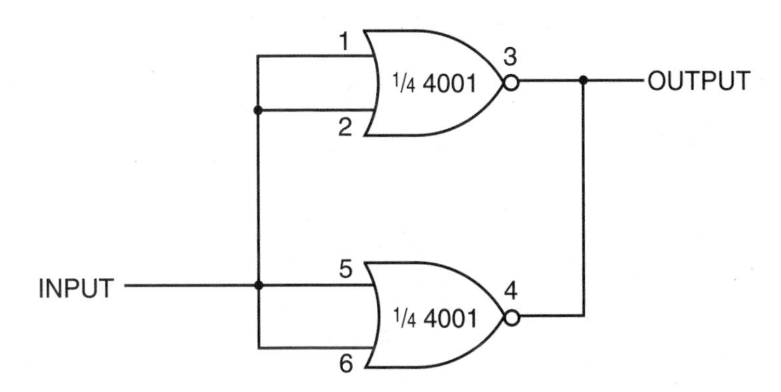

OR gate

Inverting the output of a NOR gate creates an OR gate. The output is HIGH if any one (or more) of the inputs is HIGH. The output is LOW if and only if all inputs are LOW.

OR Gate Truth Table

Inputs		Output
A	B	
0	0	0
0	1	1
1	0	1
1	1	1

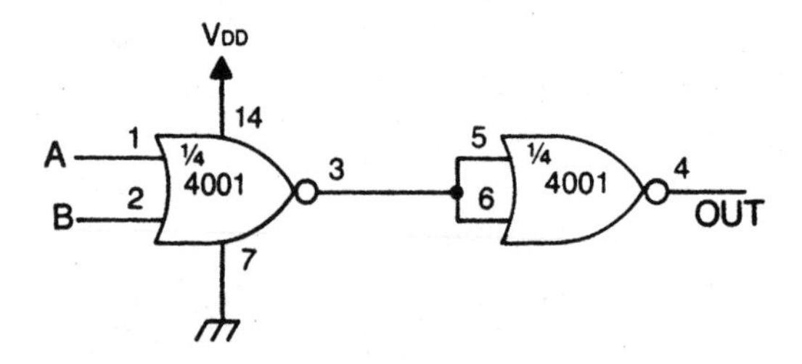

RS latch

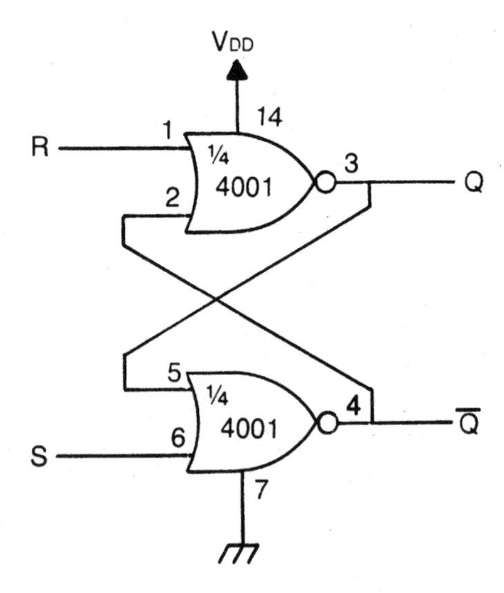

Square wave generator

The output signal's frequency can be adjusted via the 250K potentiometer. The overall frequency range can be changed by substituting a different capacitor value. Increasing the capacitance slows down the circuit's operating frequency.

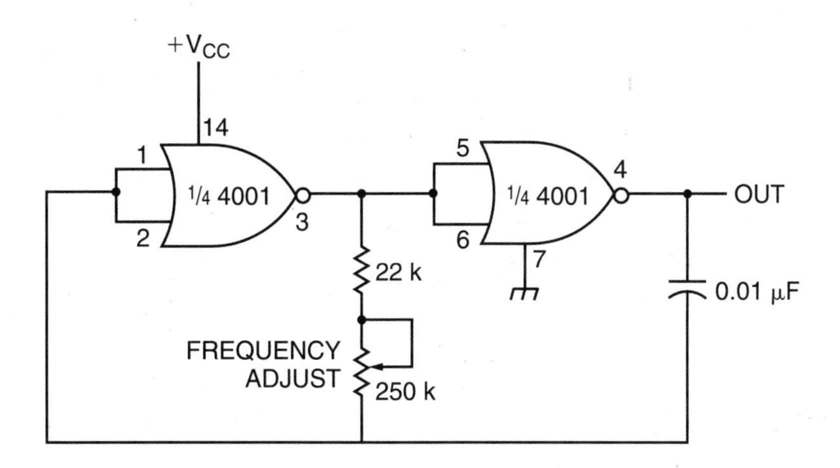

Gated square-wave generator

The output signal's frequency can be adjusted via the 250K potentiometer. The overall frequency range can be changed by substituting a different capacitor value. Increasing the capacitance slows down the circuit's operating frequency.

Unlike the circuit above, this signal generator is logic gated. The circuit is enabled and produces a square-wave output signal when the ENABLE input is LOW. When this input is brought HIGH, the signal generator is disabled, and the output is held at a continuous HIGH level.

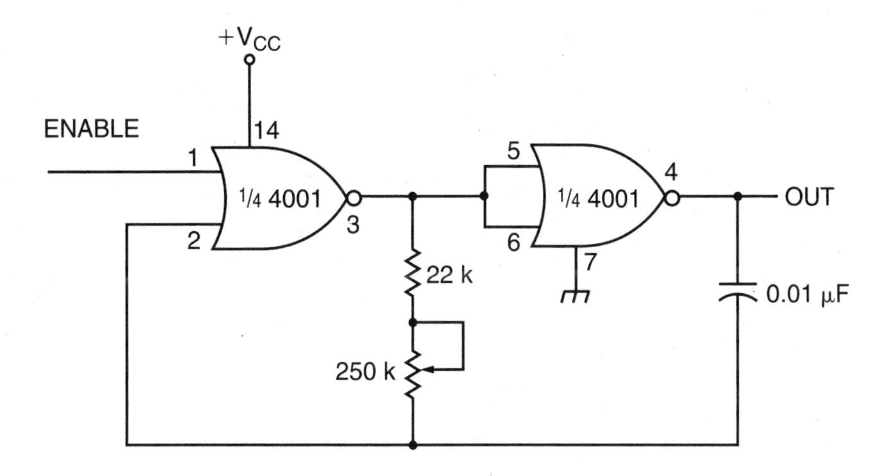

Switch debouncer

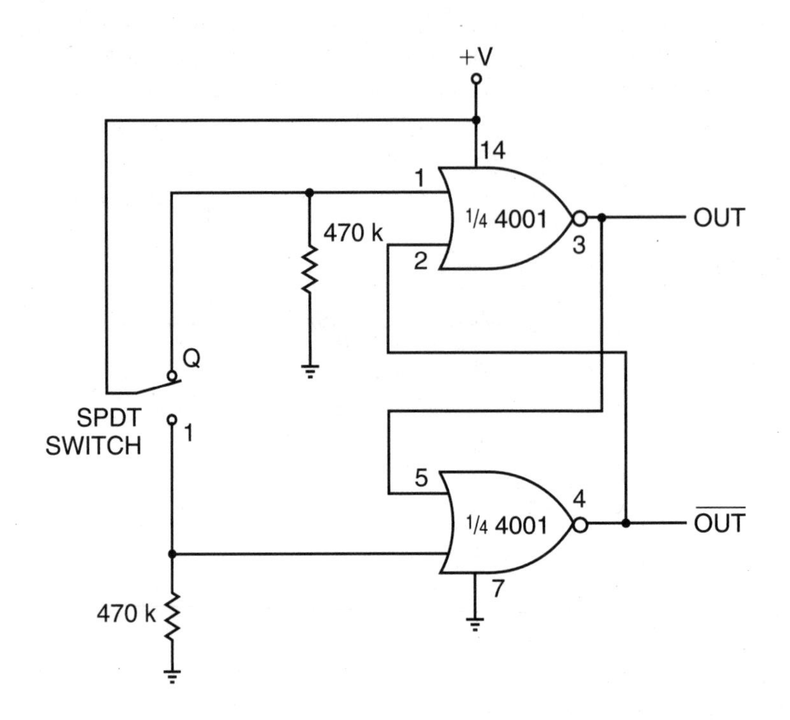

Double LED flasher

This circuit lights up two LEDs alternately. When one is lit, the other is dark, and vice versa. The flash rate can be adjusted via the 100K potentiometer. The overall flash rate range can be changed by altering the value of the capacitor. Increasing the capacitance slows down the circuit's operating frequency. If the frequency is increased above about 3 to 5 Hz or so, the eye will not be able to distinguish between the individual flashes of the LEDs. Both LEDs will appear to be continuously lit, though perhaps at a somewhat lower brightness than normal.

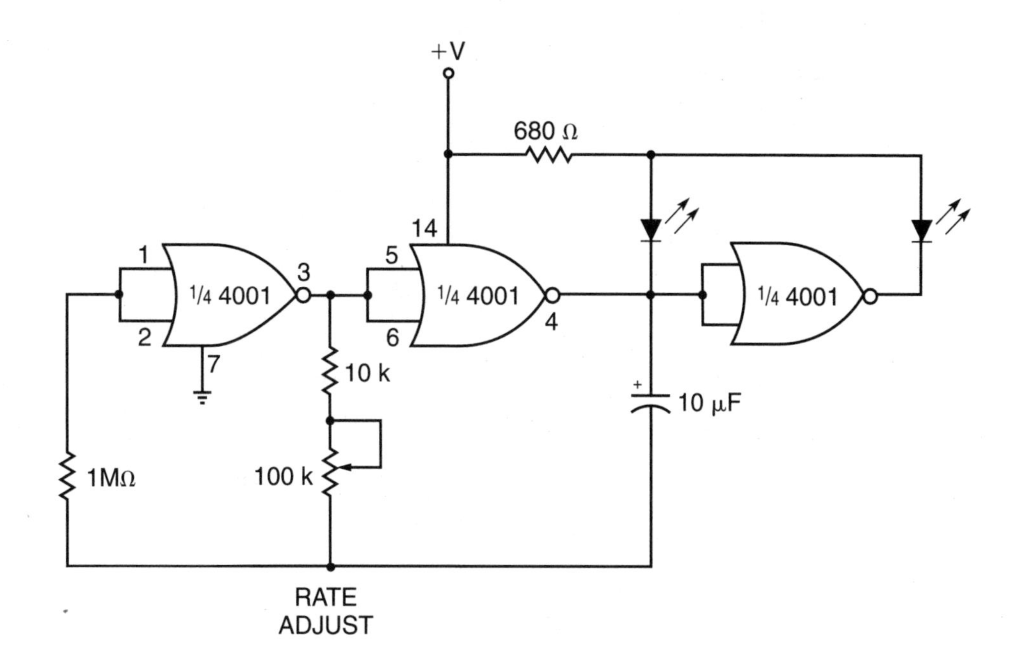

Crystal oscillator

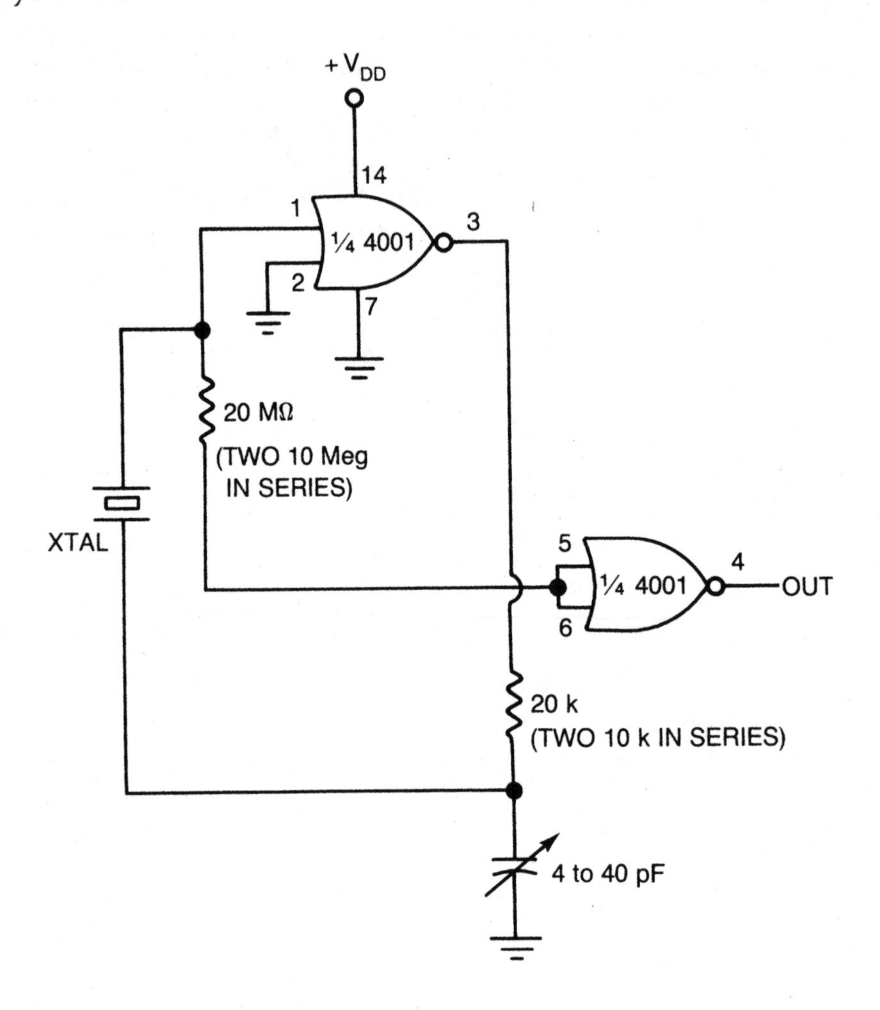

"Decision" maker

While the push-button switch is held closed, an oscillator alternately lights and extinguishes the two LEDs at a rate far too fast for the eye to follow. Both LEDs will appear to be continuously lit, although they may appear a little dimmer than normal. When the switch is released (opened) the oscillator is disabled. Whichever LED was lit last will remain on, while the other will be dark. In effect, one of the two LEDs will be randomly decided on, a little like an electronic flip of a coin.

Unlike many digital circuits with manual switches, no switch debouncing would be called for in this application. Any mechanical bouncing when the switch is released will only add to the desired randomness.

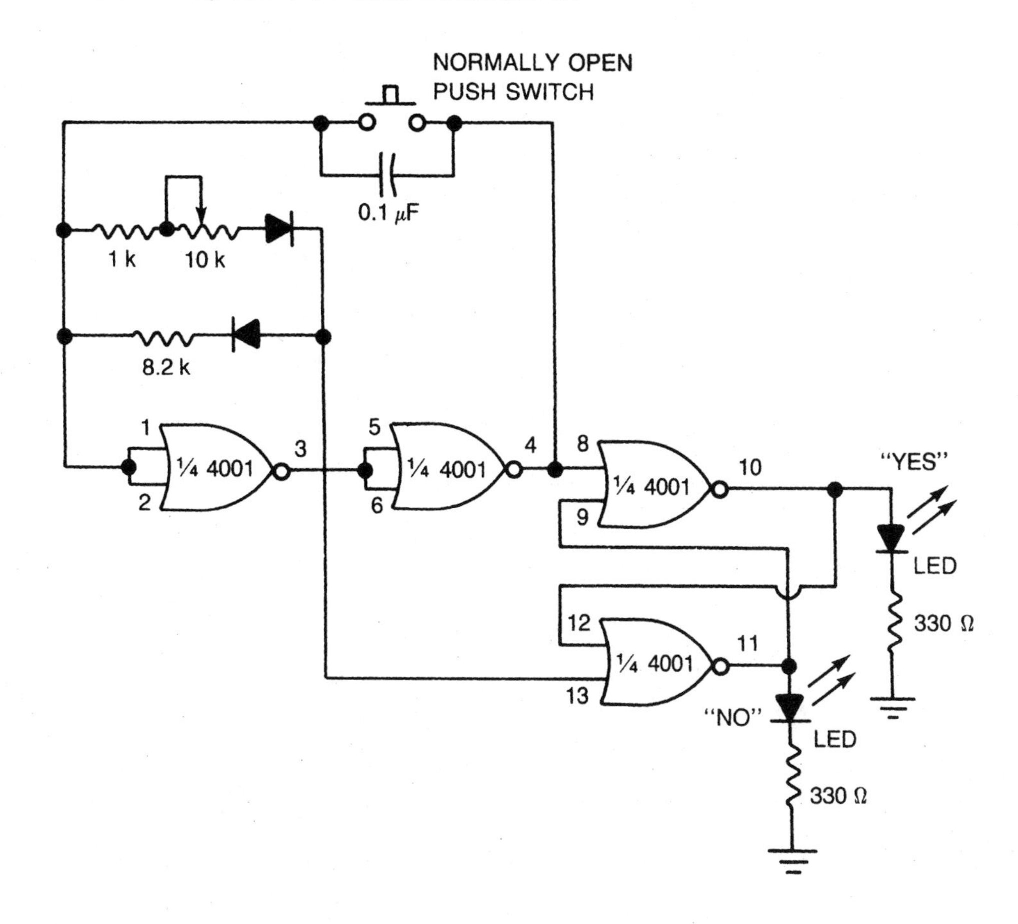

4011 quad 2-input NAND gate

The 4011 contains four NAND gates, each with two inputs. Only the power supply connections are common to both gates.

As with any NAND gate, the output is LOW if and only if all inputs are HIGH. If one or more inputs goes LOW, the output will go HIGH.

4011 Truth Table

Inputs	Output
0 0	1
0 1	1
1 0	1
1 1	0

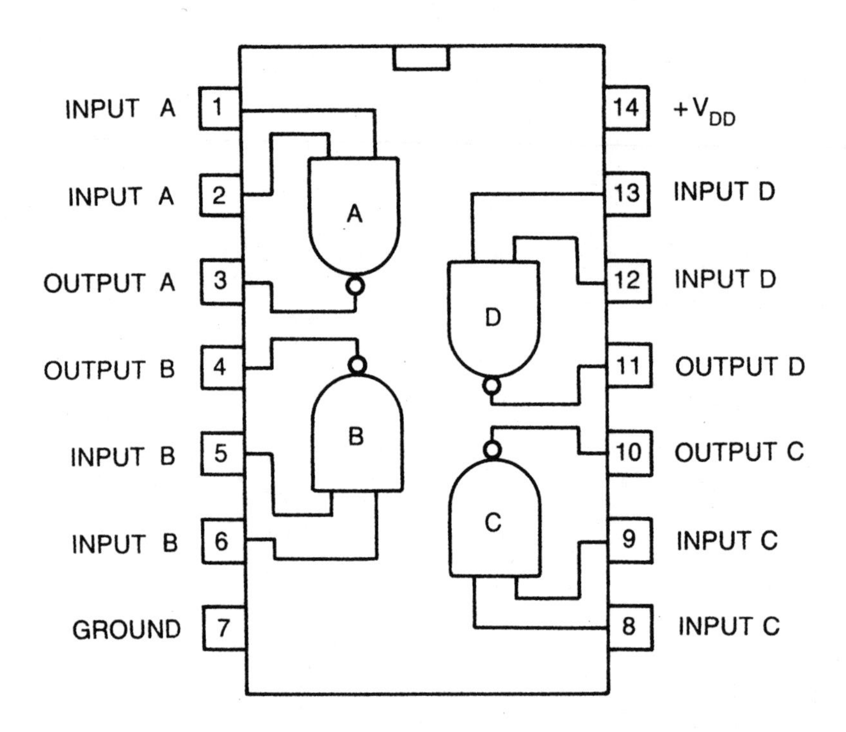

Inverter

Shorting together the inputs of a NAND gate results in an inverter. The output will always be at the opposite state as the input. A LOW input will produce a HIGH output, and vice versa.

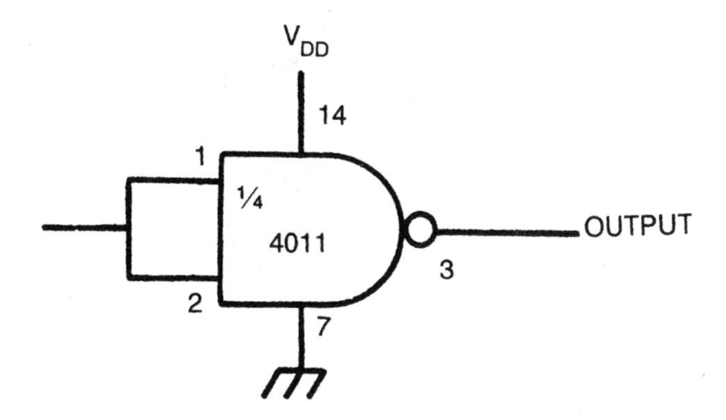

AND gate

Inverting the output of a NAND gate produces an AND gate. The output is HIGH if and only if all inputs are HIGH. Whenever any one (or more) of the inputs goes LOW, the output will also be forced LOW.

AND gate Truth Table

Inputs		Output
0	0	0
0	1	0
1	0	0
1	1	1

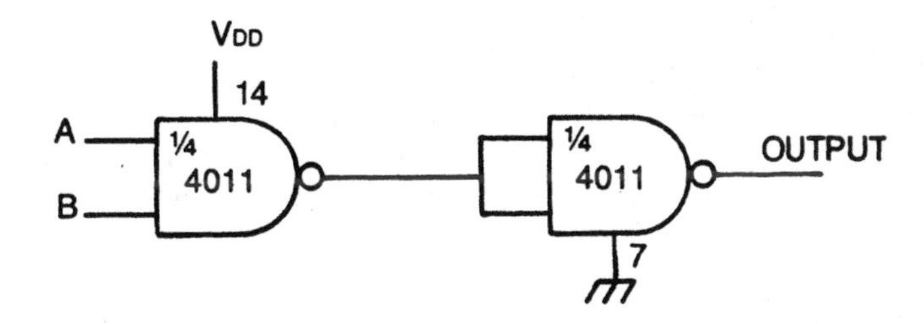

AND/OR gate

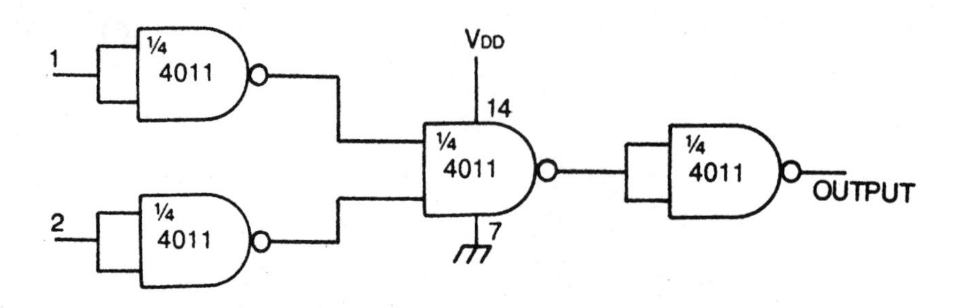

NOR gate

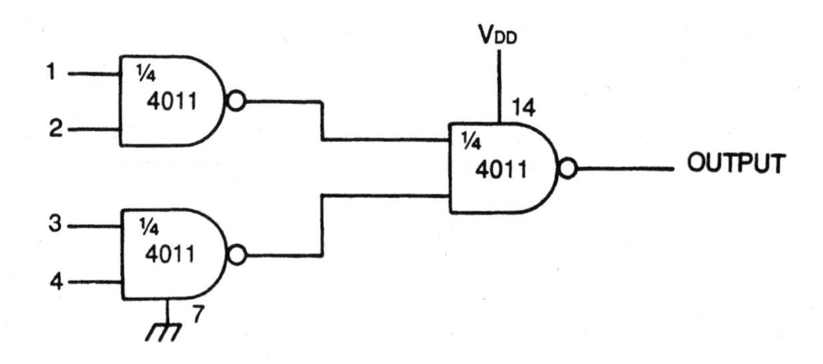

Quad input NAND gate

This circuit expands the basic NAND gate to include four inputs, instead of just two. It functions in the same way. The output is LOW if and only if all inputs are HIGH. As long as at least one (or more) of the inputs is LOW, the output of this circuit will be HIGH.

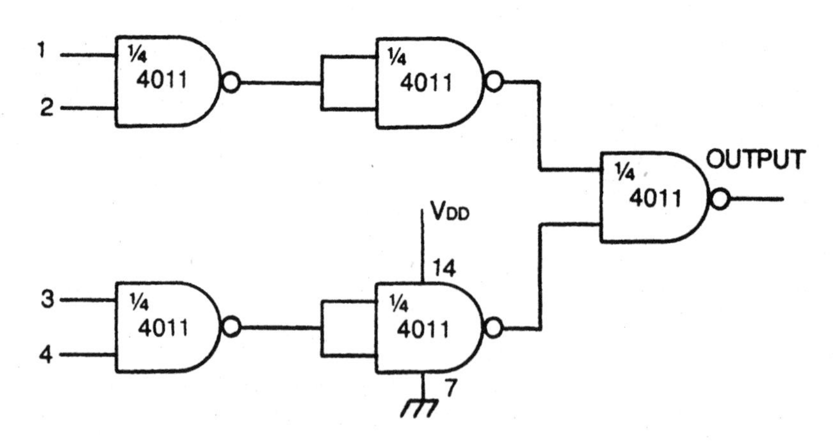

EXCLUSIVE-OR gate

An EXCLUSIVE-OR (or X-OR) gate is an important variation on the basic OR gate function. The output is HIGH if and only if input 1 is HIGH *or* if input 2 is HIGH, but *not* both. If both inputs are HIGH, or both inputs are LOW, the output will be LOW.

The EXCLUSIVE-OR gate can be used as a simple 1-bit digital comparator, or *difference detector.*

EXCLUSIVE-OR Gate Truth Table

Inputs	Output
0 0	0
0 1	1
1 0	1
1 1	0

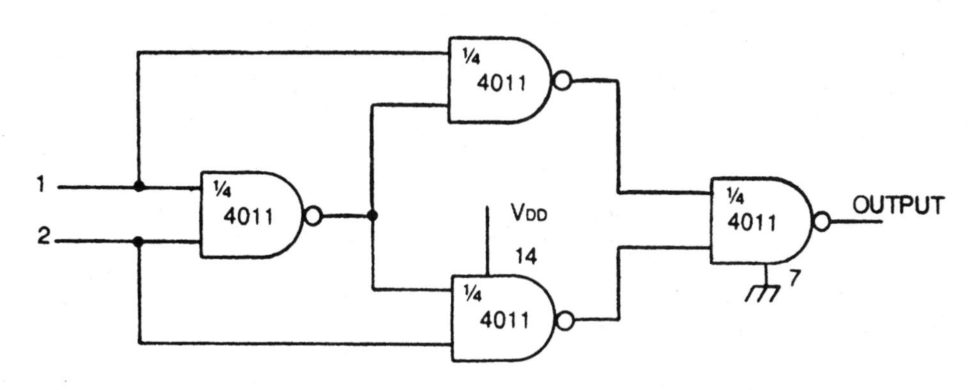

EXCLUSIVE-NOR gate

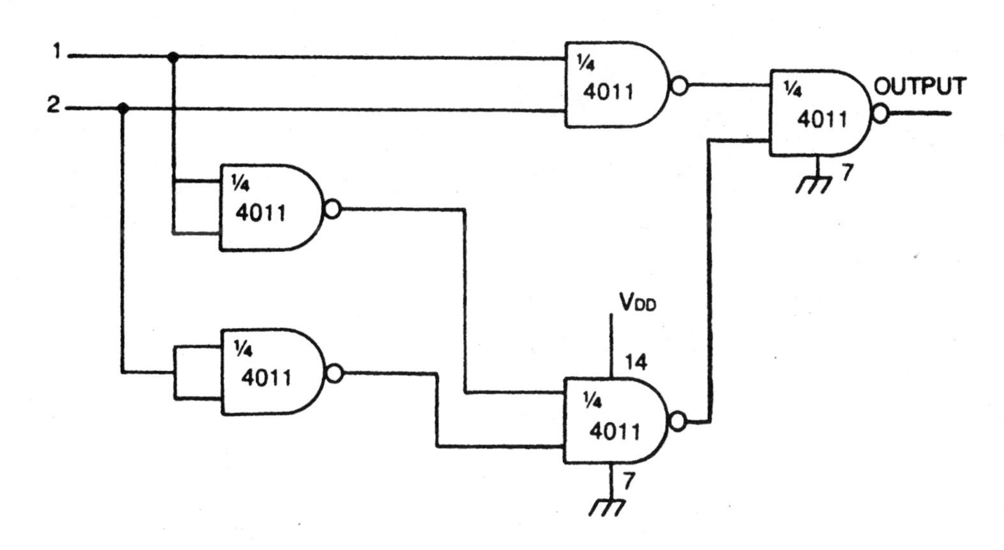

Simple logic probe

This very simple but effective piece of test equipment can be used to determine the logic state at any given point within a digital CMOS circuit. The ground clip should be connected to the ground of the circuit under test. The probe tip is touched to the specific connection point to be examined.

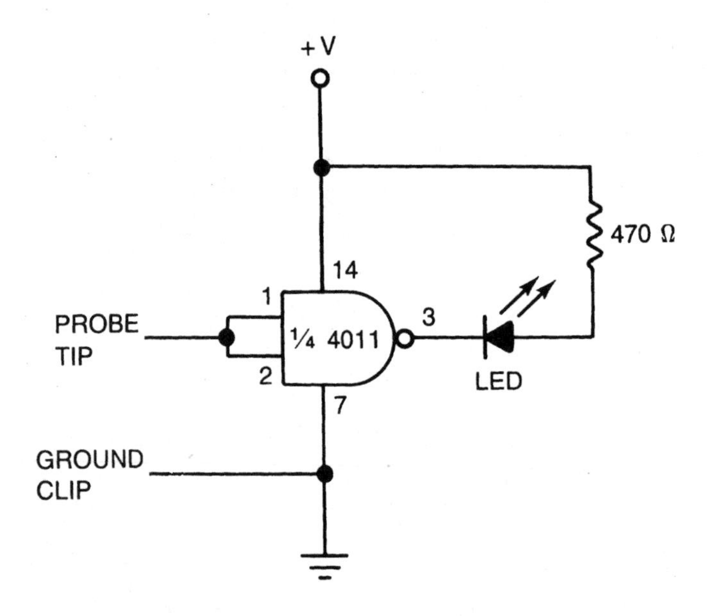

Improved logic probe

This very simple but effective piece of test equipment can be used to determine the logic state at any given point within a digital CMOS circuit. The ground clip should be connected to the ground of the circuit under test. The probe tip is touched to the specific connection point to be examined.

This improved version of the basic logic probe circuit uses two LEDs instead of just one for less ambiguity. In the one LED version, a dark LED might indicate a logic LOW, or no electrical connection, while a lit LED could indicate either a steady-state HIGH signal, or a high-speed pulse (alternating between LOW and HIGH). If only LED1 lights up, there is a HIGH signal. If only LED2 lights up, there is a LOW signal. If both LEDs light up, the connection point is carrying high-frequency pulses. If neither LED lights up, there is no electrical connection to the probe tip, or the circuit point being tested is electrically dead (no signal).

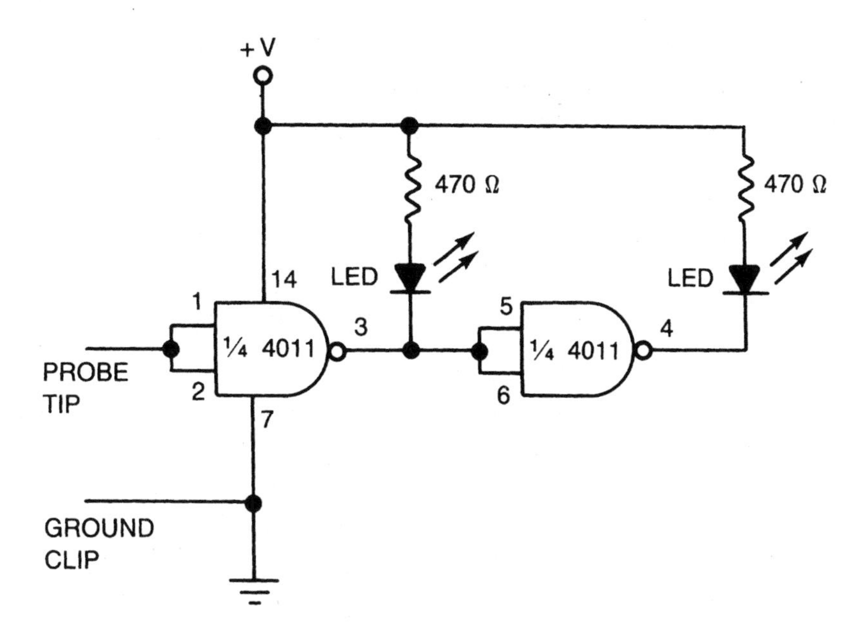

Clock pulse generator

The pulse frequency can be adjusted via the 100K potentiometer, or by substituting an alternate value for the capacitor. A smaller capacitance will produce a higher pulse frequency.

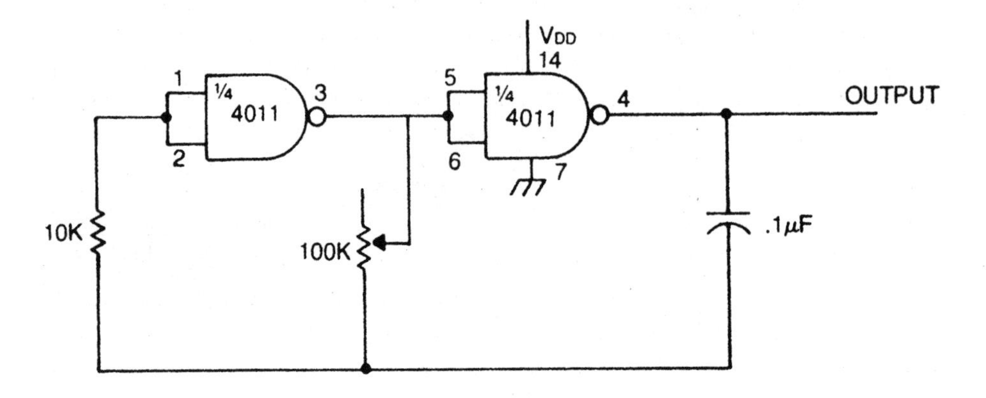

LED blinker

The flash rate of this circuit can be adjusted by changing the value of either the 100K resistor or the 5 μF capacitor. Smaller component values will result in higher frequencies. If the frequency is raised above a few hertz, the eye will not be able to distinguish between the individual flashes, and the LED will appear to be continuously lit, although perhaps at a lower than normal intensity.

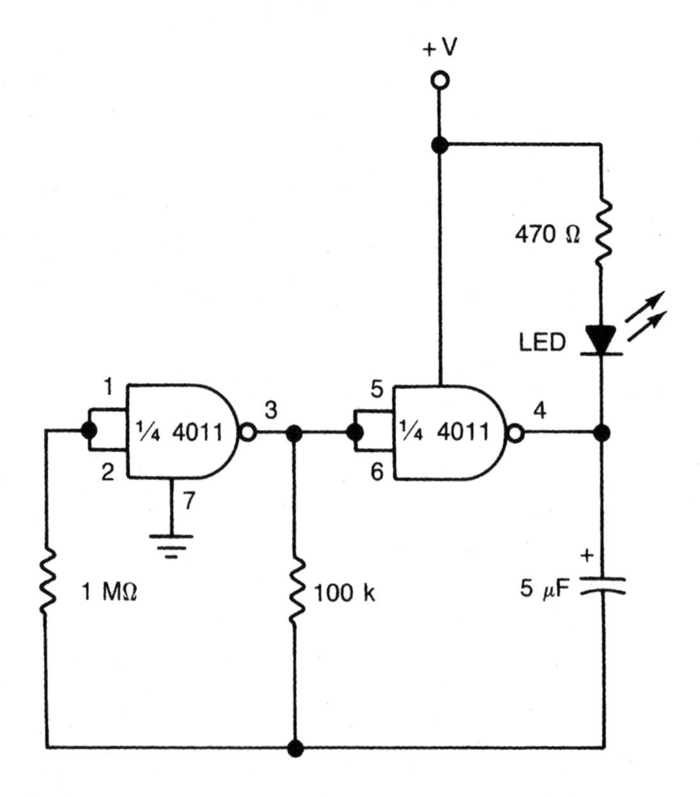

Double LED blinker

The two LEDs in this circuit flash alternately. When one is lit, the other is dark, and vice versa. The flash rate can be adjusted by changing the value of one or both of the capacitors. Smaller capacitances result in higher frequencies. If the two capacitors have equal values (as shown here), each LED will remain on for an equal amount of time per cycle. If unequal capacitances are used, one LED will stay lit longer per cycle than the other. If the frequency is raised above a few hertz, the eye will not be able to distinguish between the individual flashes, and the LEDs will both appear to be continuously lit, although perhaps at a lower than normal intensity.

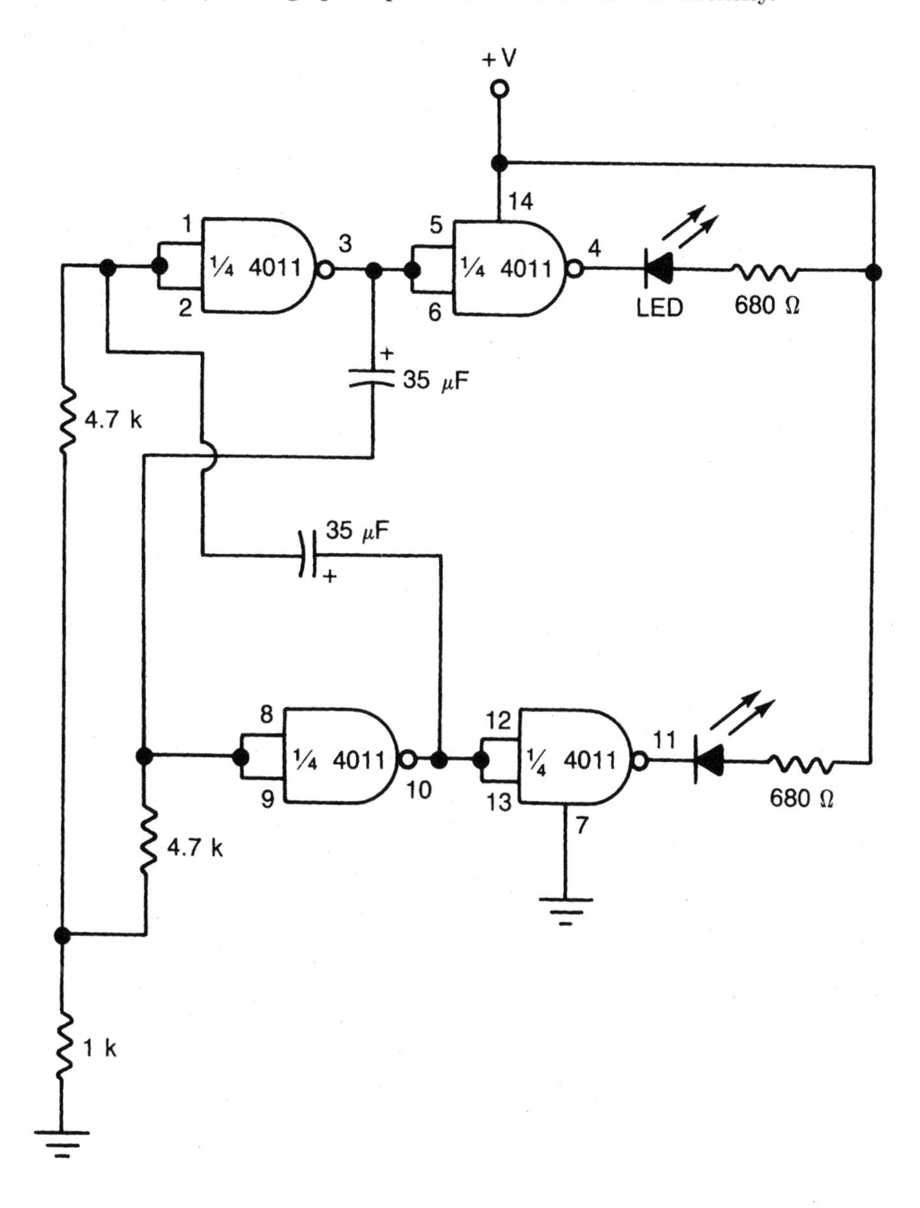

Frequency doubler

The frequency of the output signal will be twice that of the input signal (within the switching speed limits of the gates used, of course). The discrete component values in this circuit do not directly affect the output signal frequency.

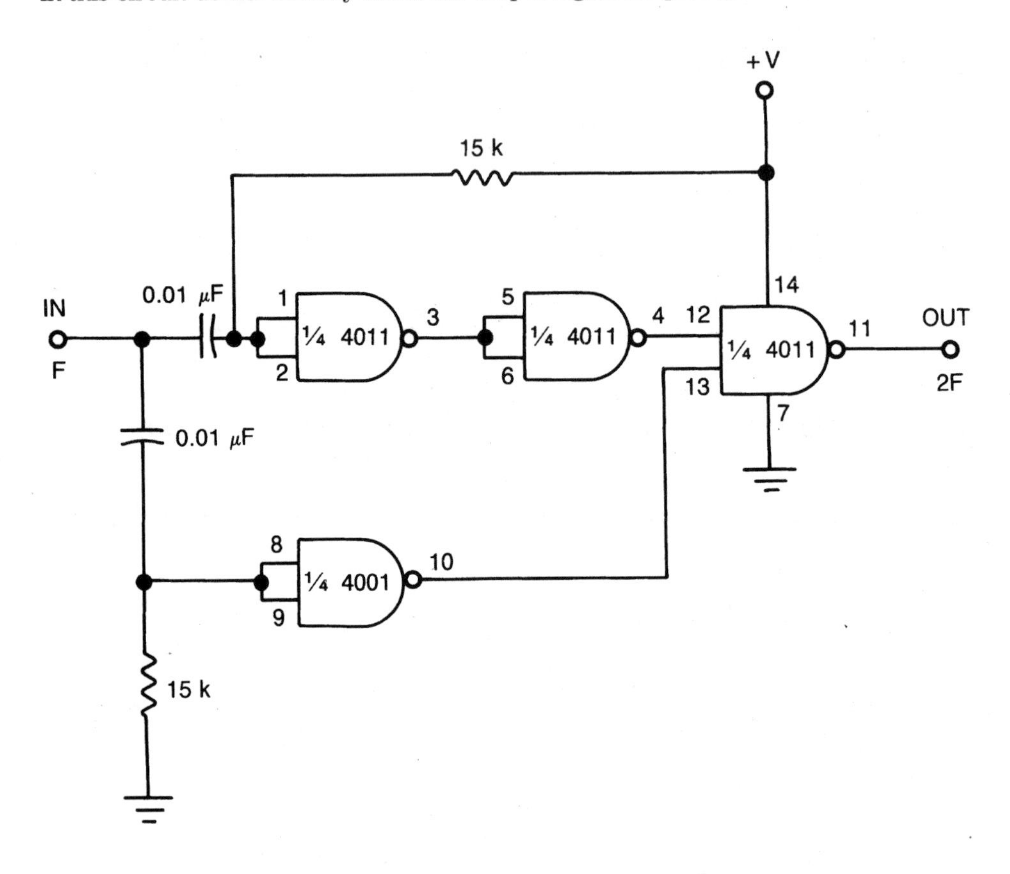

Touch switch

This electronic switch is activated by touching a fingertip across the two conductive touch pads. It is very, very important to use *battery power only* in a touch switch circuit to avoid a possibly dangerous, or even fatal, shock hazard.

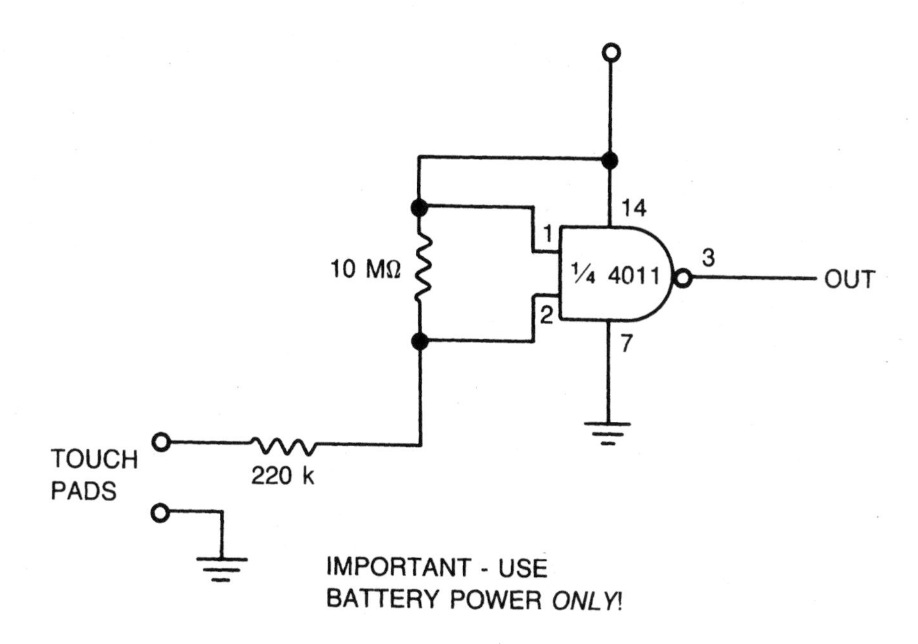

Touch switch with time delay

This electronic switch is activated by touching a fingertip across the two conductive touch pads. It is very, very important to use *battery power only* in a touch switch circuit to avoid a possibly dangerous, or even fatal, shock hazard.

When the shorting fingertip is removed from the touch pads, there will be a brief delay before the circuit shuts itself off. This time delay period is determined by the values of resistor R and capacitor C. Experiment with alternate component values.

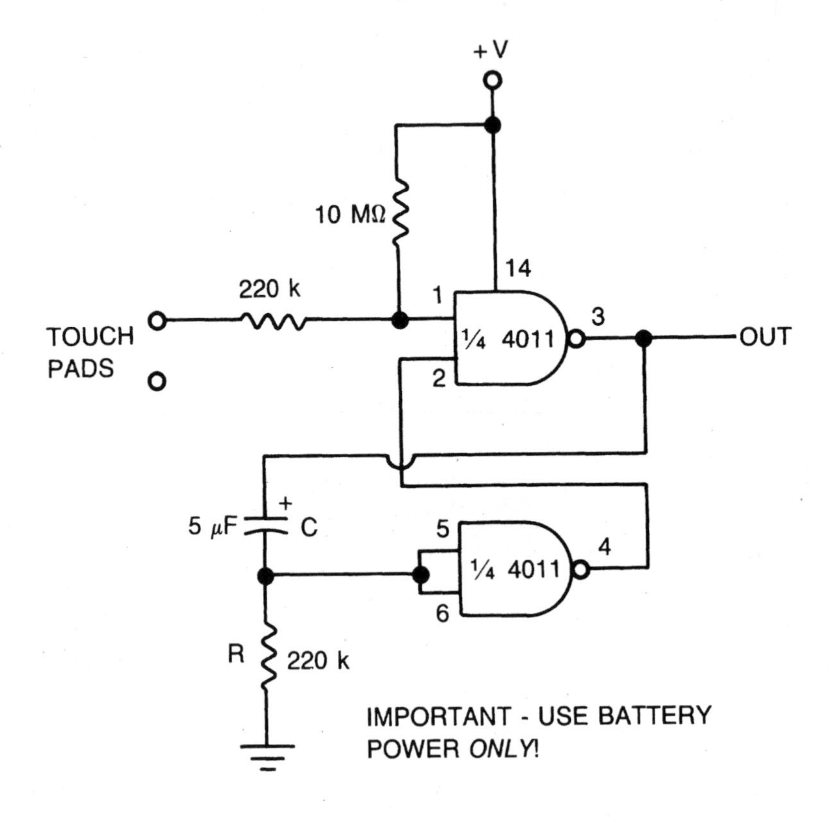

Linear amplifier

Usually digital gates cannot be used in linear applications, but this circuit actually forces a NAND gate to amplify an analog signal. The gain is determined primarily by the ratio of the two resistors in the circuit.

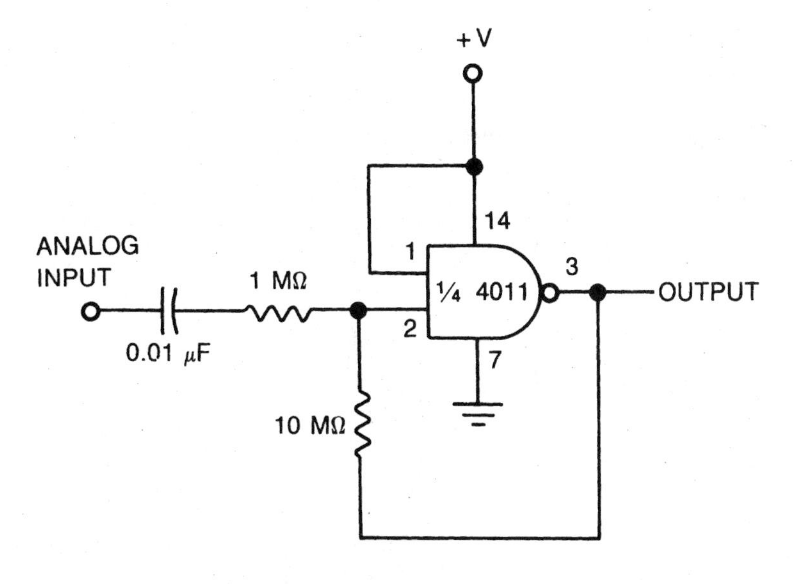

4012 dual 4-input NAND gate

The 4012 contains a pair of NAND gates, each with four inputs. Only the power supply connections are common to both gates.

As with any NAND gate, the output is LOW if and only if all inputs are HIGH. If one or more inputs goes LOW, the output will go HIGH.

4012 Truth Table

Inputs				Output
0	0	0	0	1
0	0	0	1	1
0	0	1	0	1
0	0	1	1	1
0	1	0	0	1
0	1	0	1	1
0	1	1	0	1
0	1	1	1	1
1	0	0	0	1
1	0	0	1	1
1	0	1	0	1
1	0	1	1	1
1	1	0	0	1
1	1	0	1	1
1	1	1	0	1
1	1	1	1	0

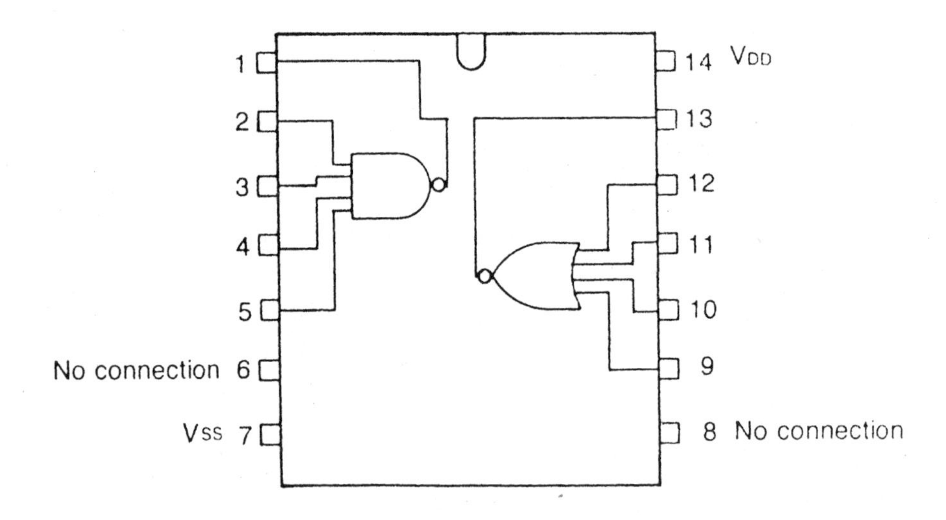

4-input NAND gate with enables

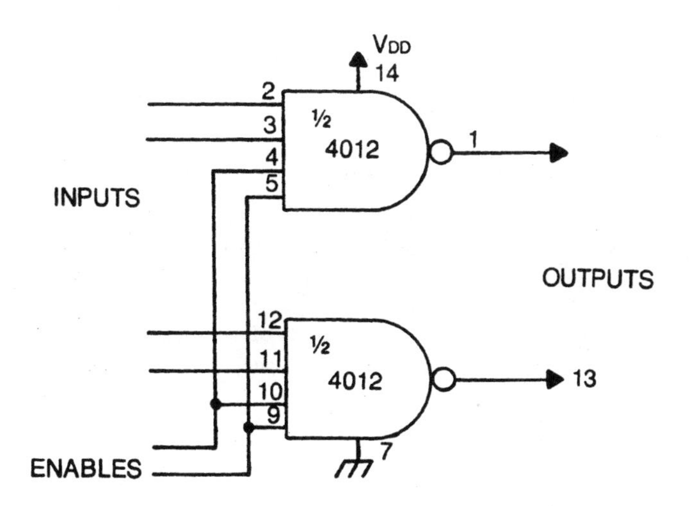

4-stage stepped wave generator

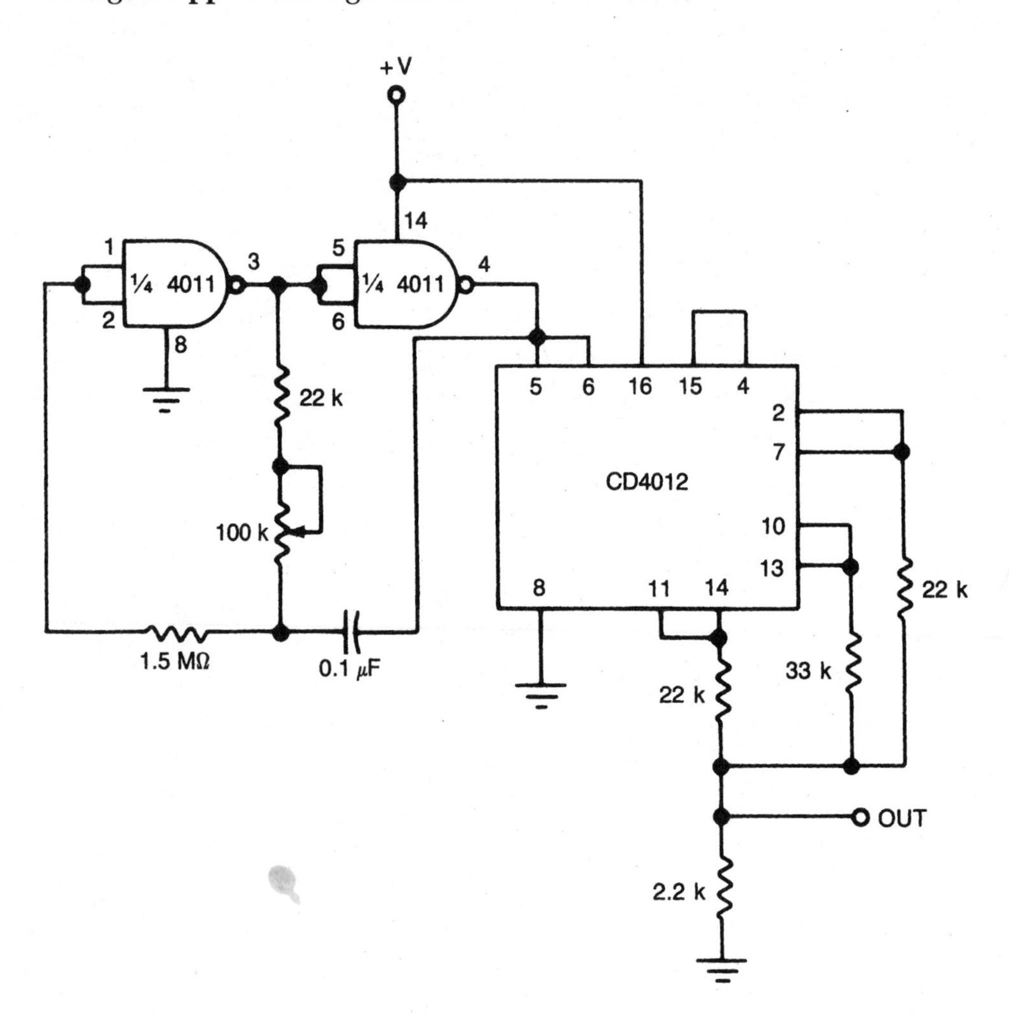

CD4013 dual D-type flip-flop

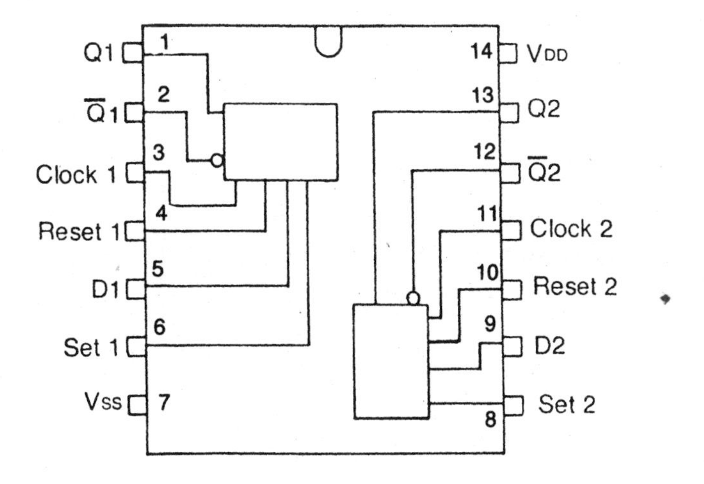

Frequency halver

The frequency of the input signal is divided by 2, so the output frequency is one-half the input frequency.

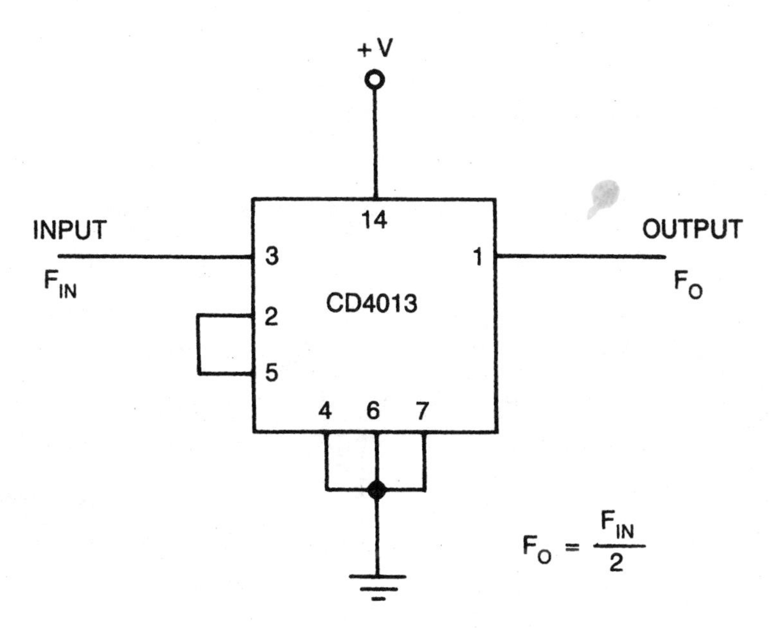

4-step sequencer

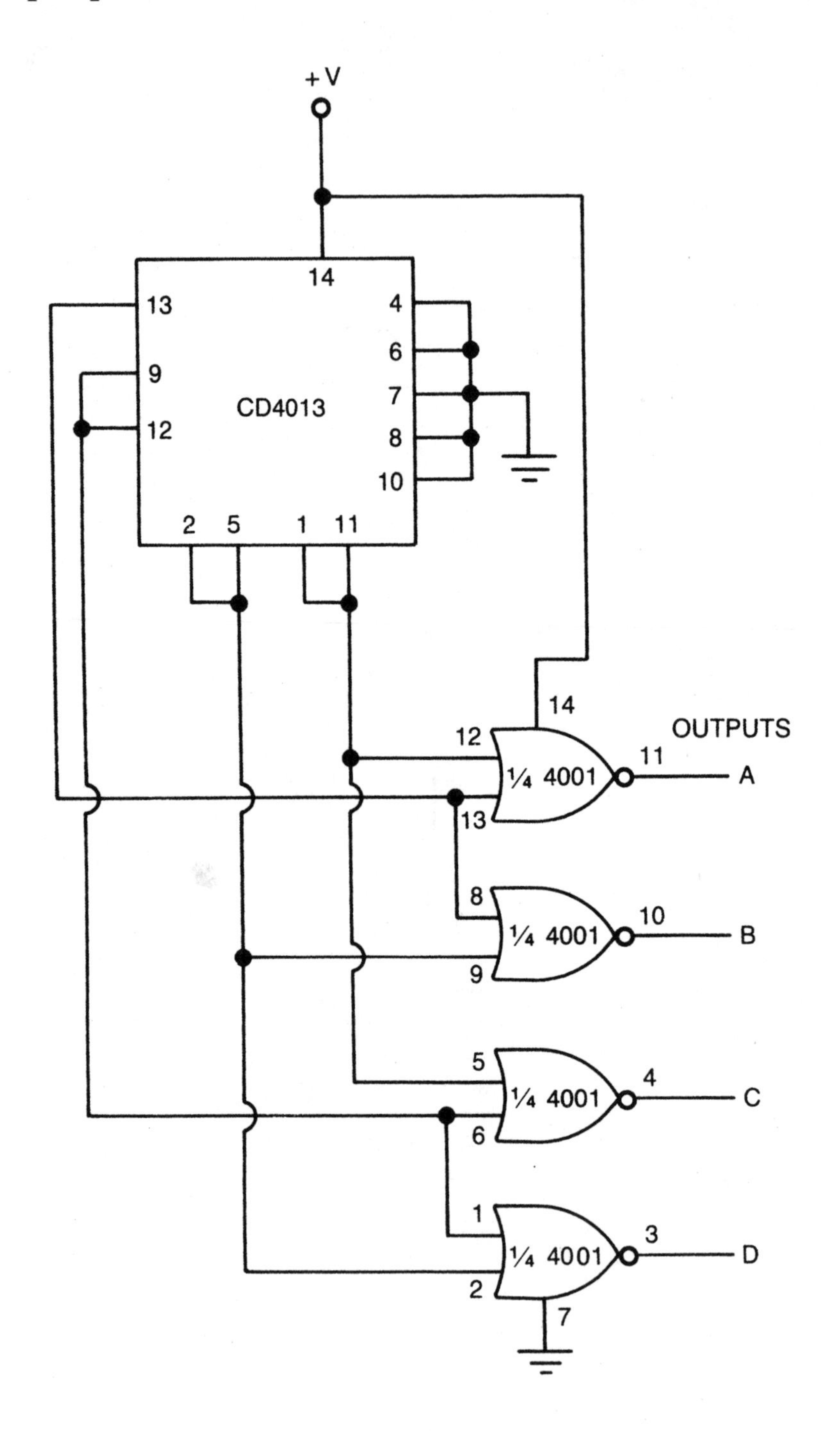

8-step binary counter

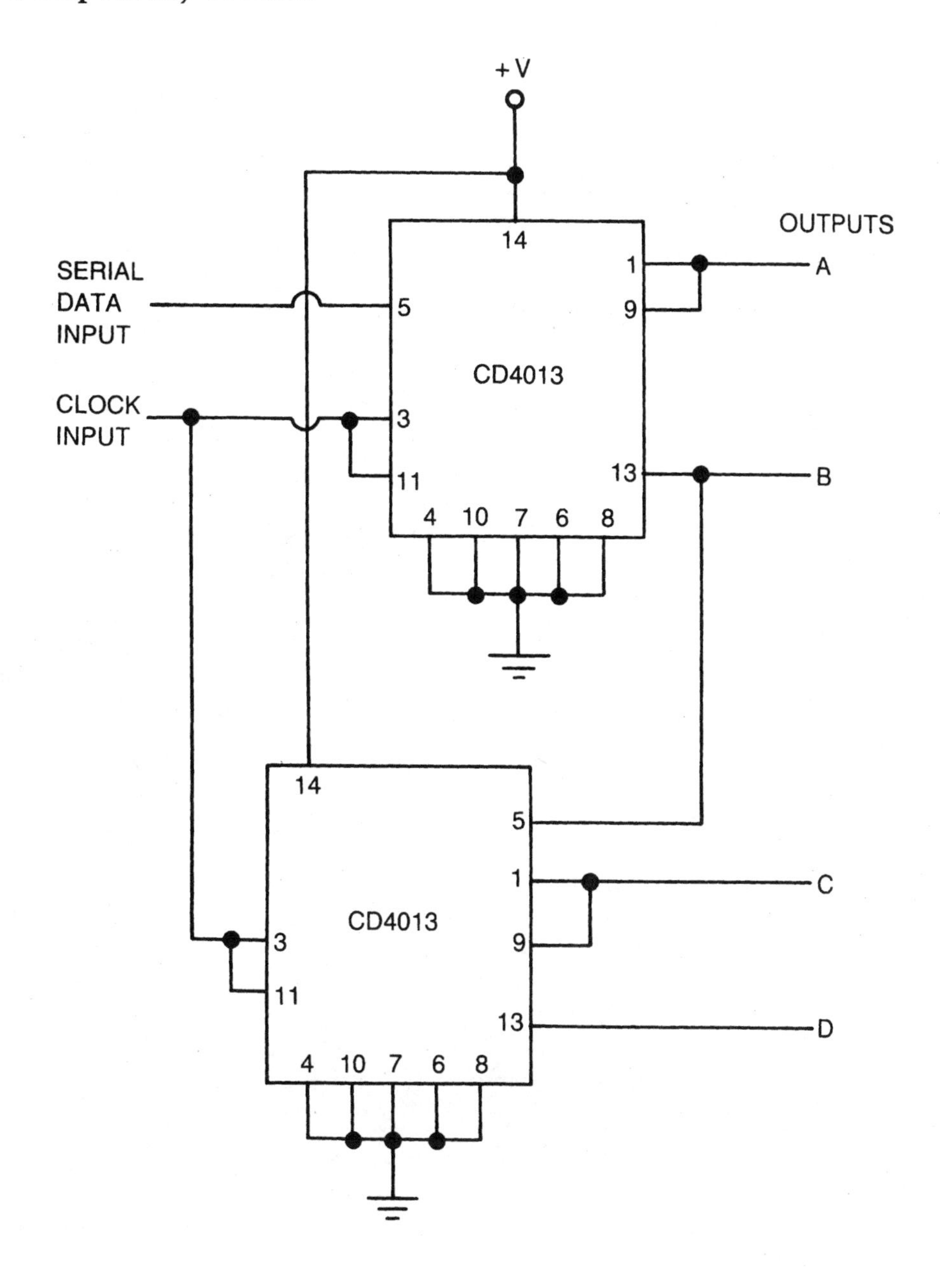

Shift register

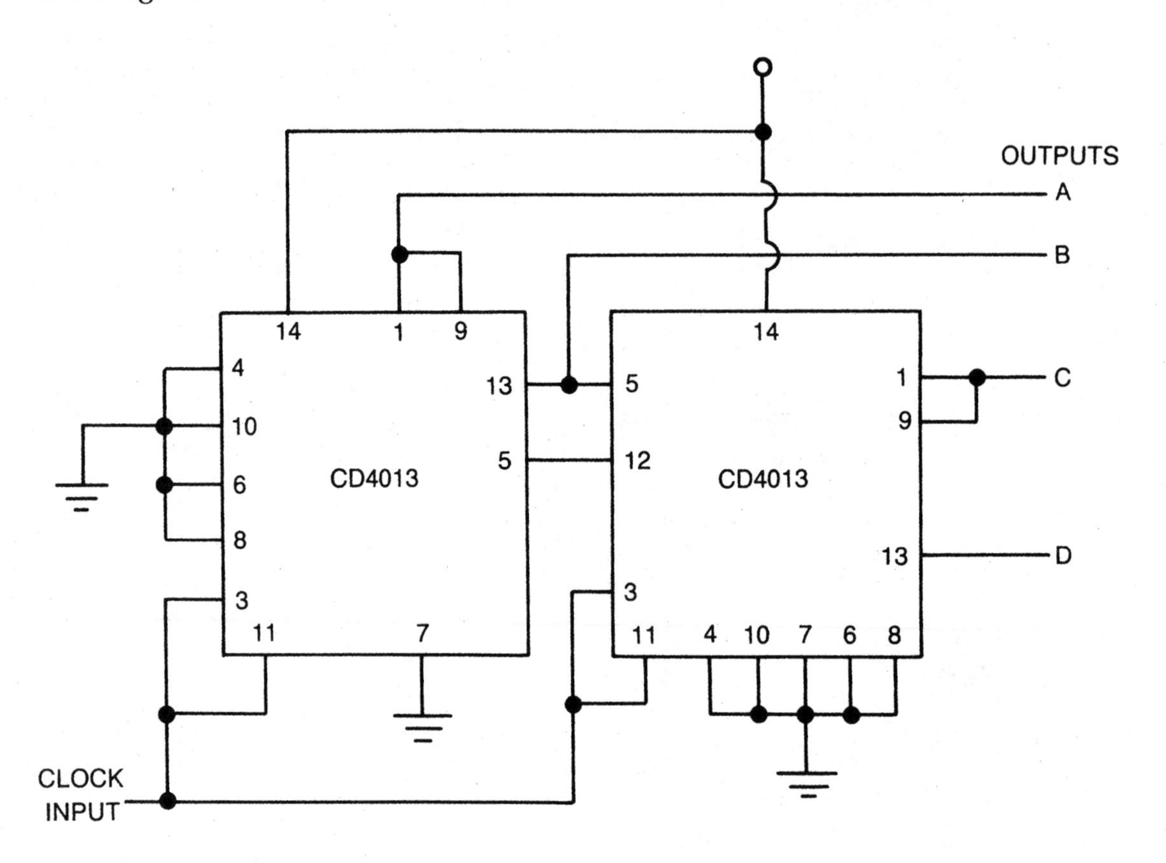

Random binary number generator

When the push-button switch is held closed, a 4-stage counter will be cycled at a very rapid rate, so all four output indicator LEDs will appear to be continuously lit (although probably a little dimmer than normal). When the switch is released (opened) the current 4-bit binary value will be held and indicated by the four LEDs.

Unlike many digital circuits with manual switches, no switch debouncing would be called for in this application. Any mechanical bouncing when the switch is released will only add to the desired randomness.

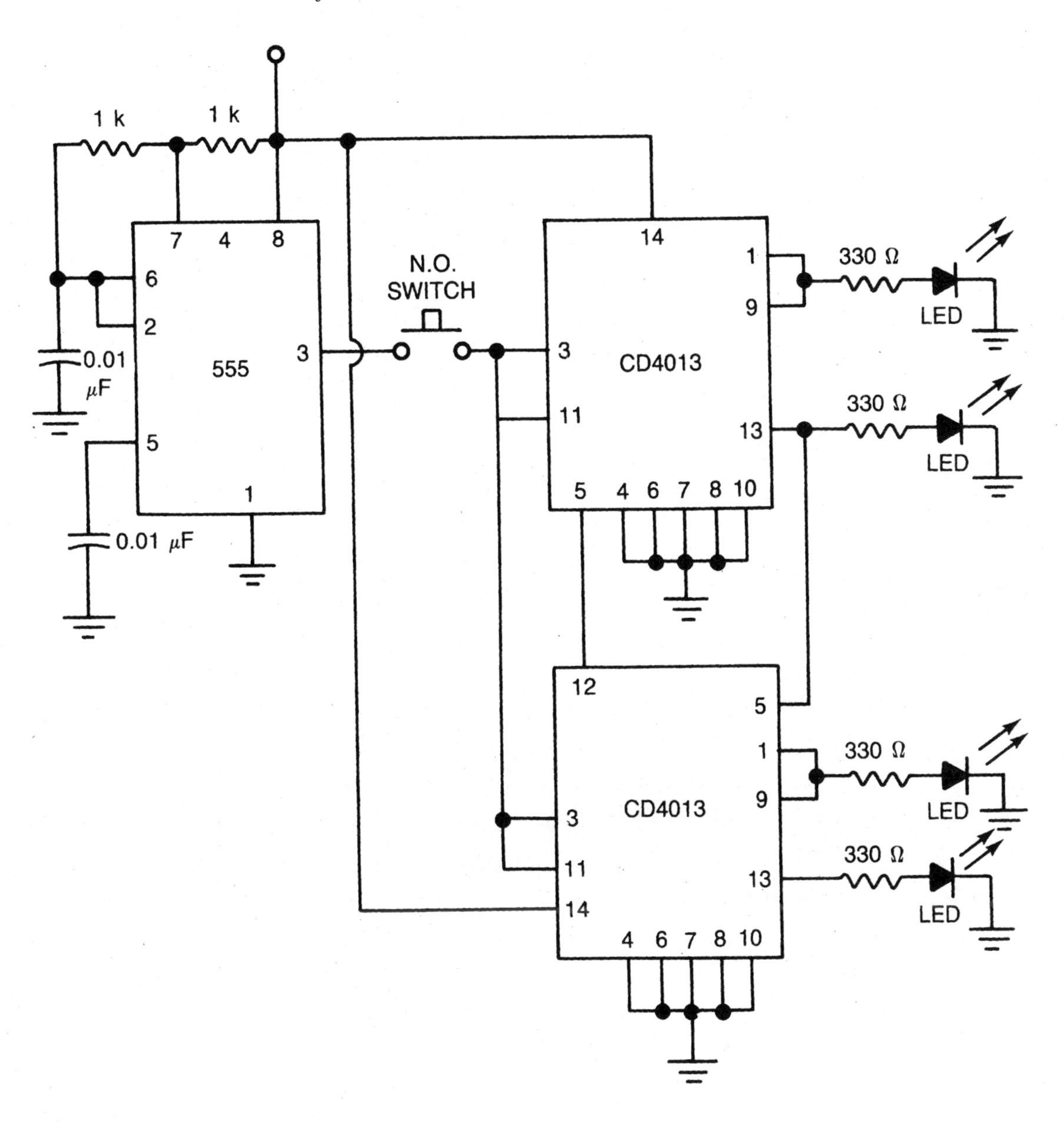

4017 decade counter/decoder

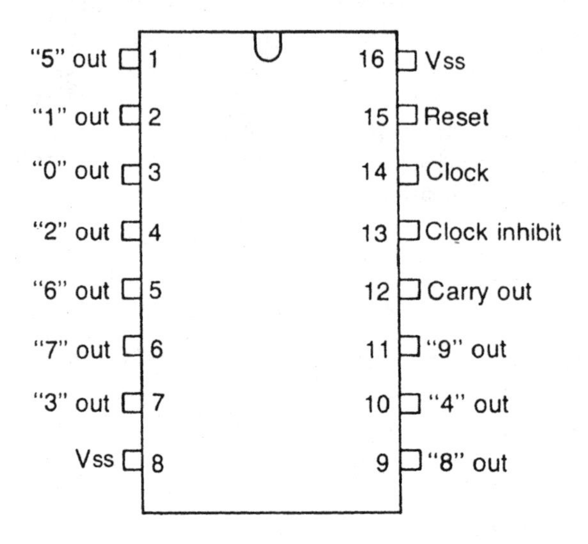

Random number generator

High-frequency clock pulses are fed to the input of this circuit. When the push-button switch is open (its normal state), no pulses can reach the counter, so nothing happens. When the switch is closed, the 4017 starts counting the input pulses from 0 to 9, and recycling. Since the clock frequency is very high, the eye can't distinguish between the flashes of the output LEDs. All ten LEDs will appear to be continuously lit at a low intensity. When the switch is released (opened), no further clock pulses can get through, so the clock stops at its last count value. Since it is impossible to predict what the final count value will be, this circuit, in effect, randomly selects a number from 0 to 9.

Unlike most digital counter circuits, there is no need at all to consider switch debouncing. Any mechanical bounces of the switch will only add to the desirable randomness of the circuit.

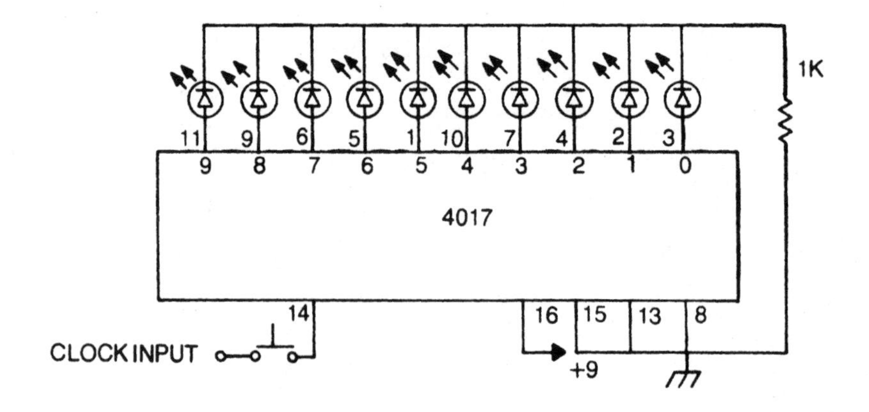

Count to 7 and halt

Other count values up to 9 can be similarly selected, simply by connecting pin No. 13 to the appropriate counter output.

When the selected count value (7 in this example) is reached, the counter stops, and any further clock pulses at the input will be ignored until the counter is reset by momentarily raising pin No. 15 to HIGH.

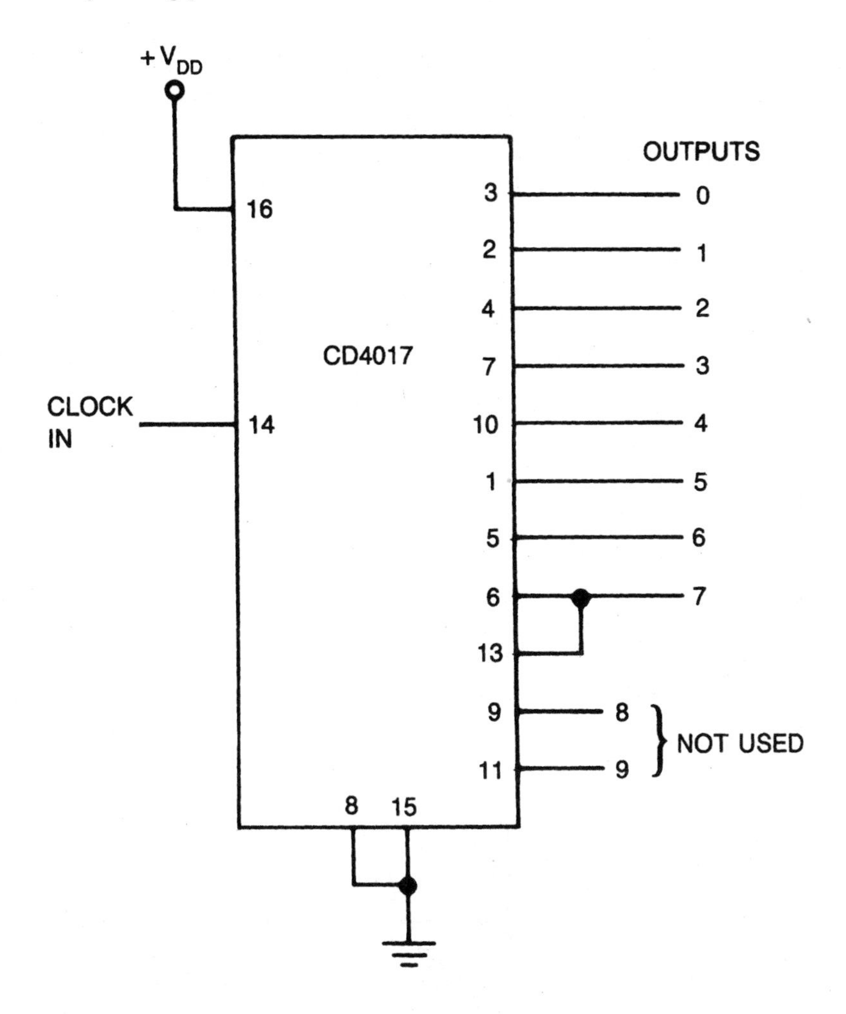

Count to 7 and recycle

Other count values up to 9 can be similarly selected, simply by connecting pin No. 15 to the appropriate counter output.

When the selected count value (7 in this example) is reached, the counter automatically jumps back to 0, and the counting cycle is repeated.

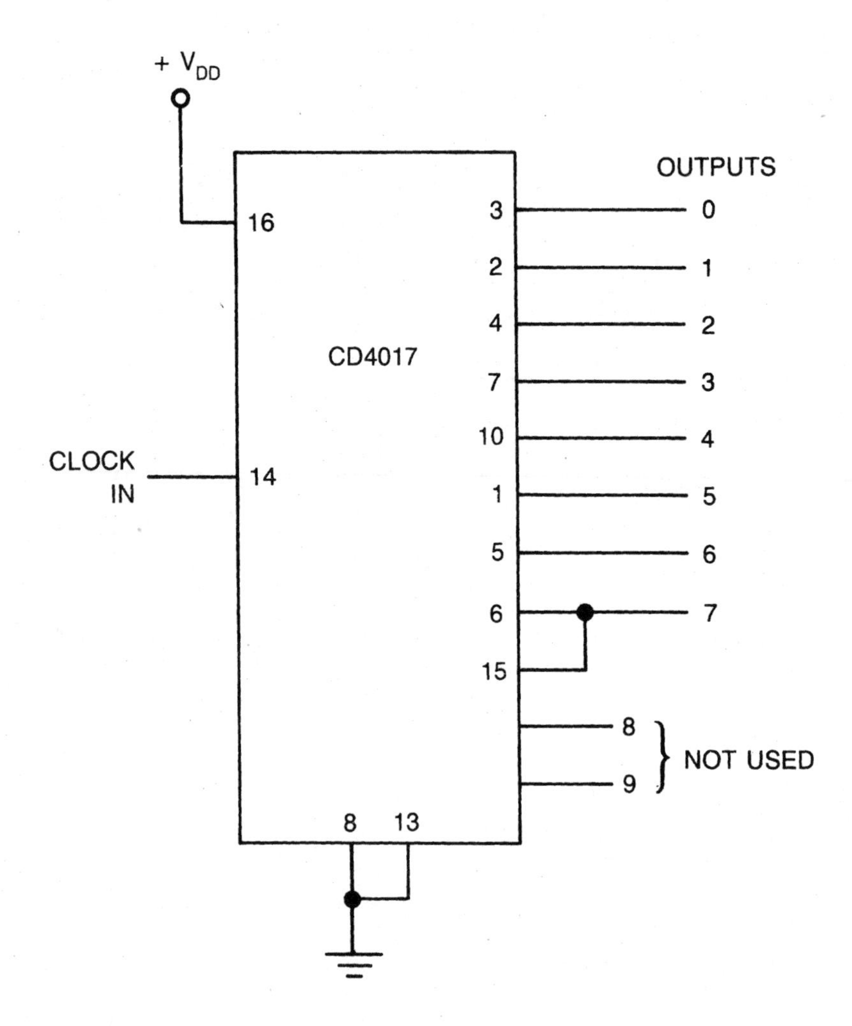

Count to 54 and recycle

A pair of 4017s can be cascaded to create count values ranging from 10 to 99. The cascaded counters work just like a single 4017.

When the selected count value (54 in this example) is reached, the counter automatically jumps back to 0, and the counting cycle is repeated.

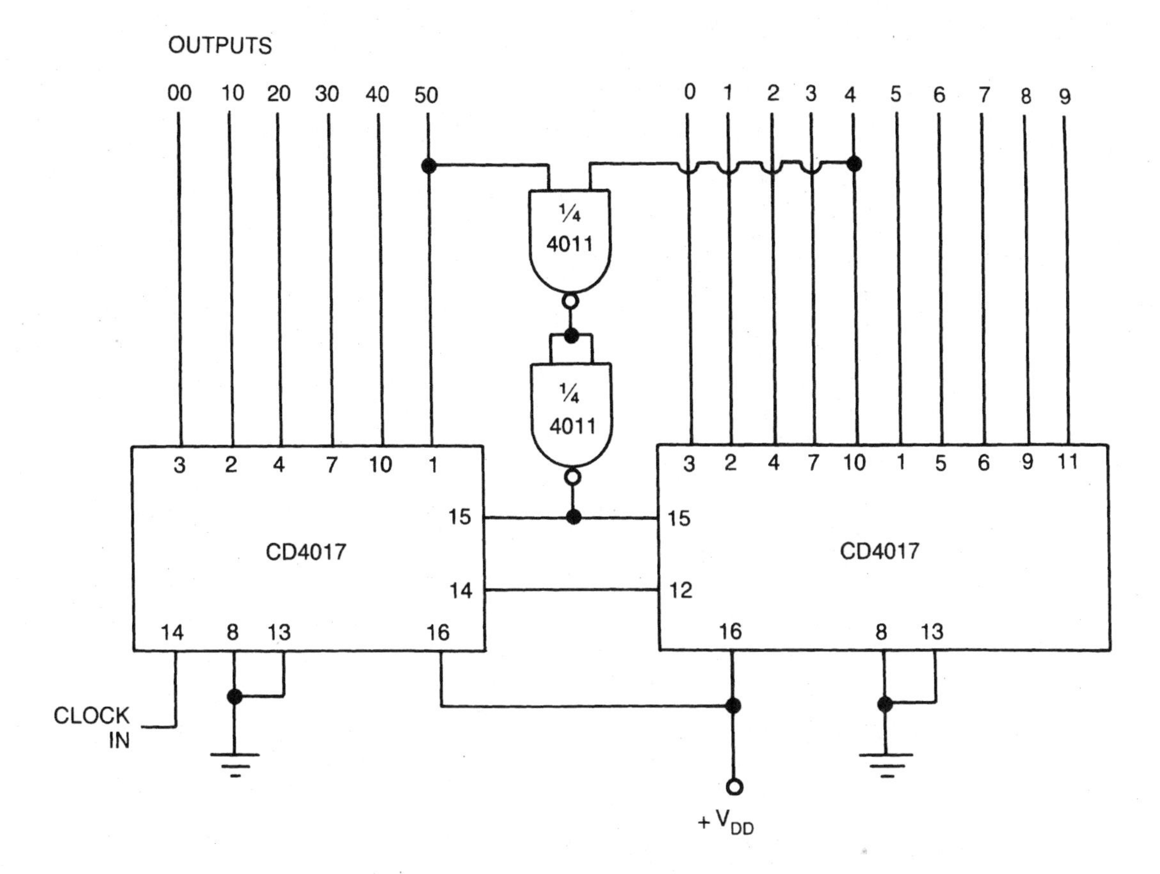

Simple 1 Hz timebase

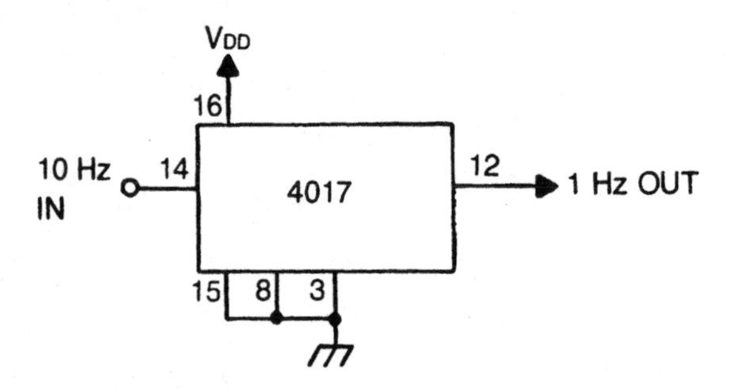

Simple 10 Hz timebase

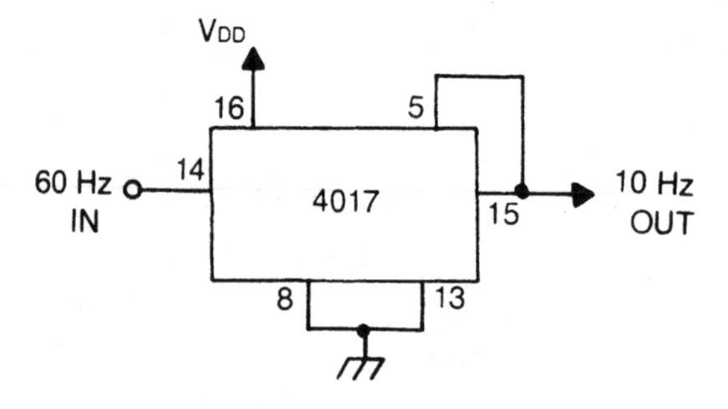

Complete 1 Hz timebase

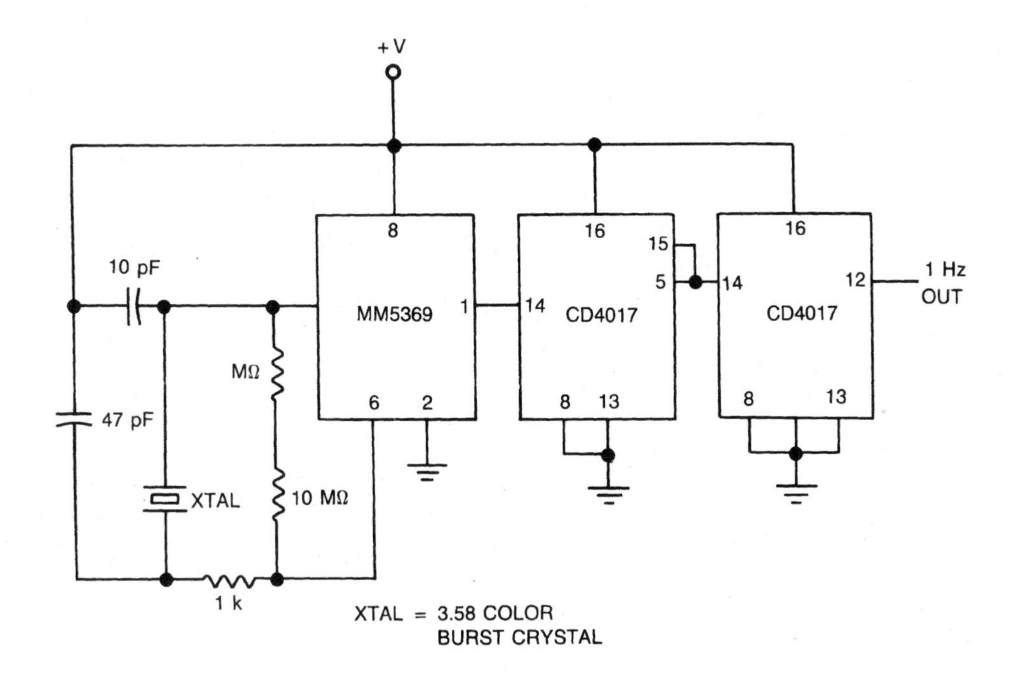

4023 triple 3-input NAND gate

The 4023 contains three independent NAND gates, each with three inputs. The output of a NAND (or Not AND) gate is LOW if and only if *all* inputs are HIGH. If one or more inputs goes LOW, the output will go HIGH.

Only the power supply connections are common to all three gates on this chip. All inputs are protected against static discharge and latching conditions.

4023 Truth Table

Inputs			Output
A	**B**	**C**	
0	0	0	1
0	0	1	1
0	1	0	1
0	1	1	1
1	0	0	1
1	0	1	1
1	1	0	1
1	1	1	0

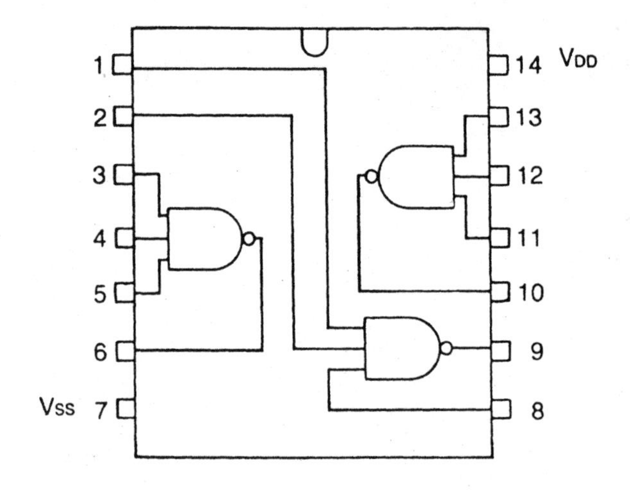

6-input OR gate

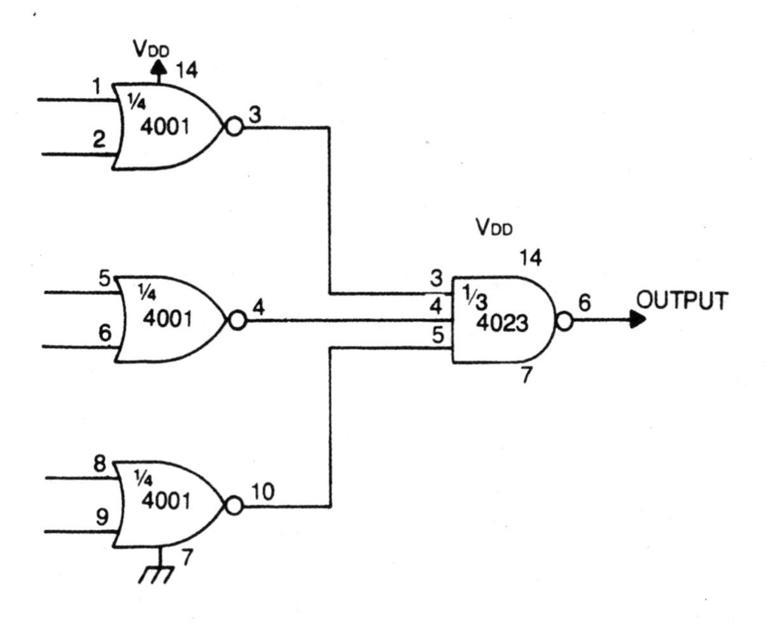

9-input NAND gate

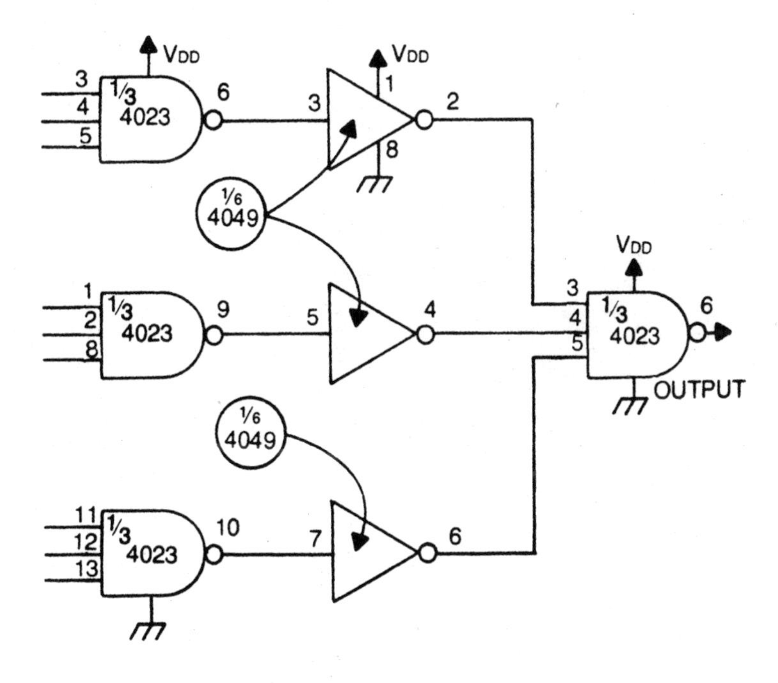

4027 dual JK flip-flop

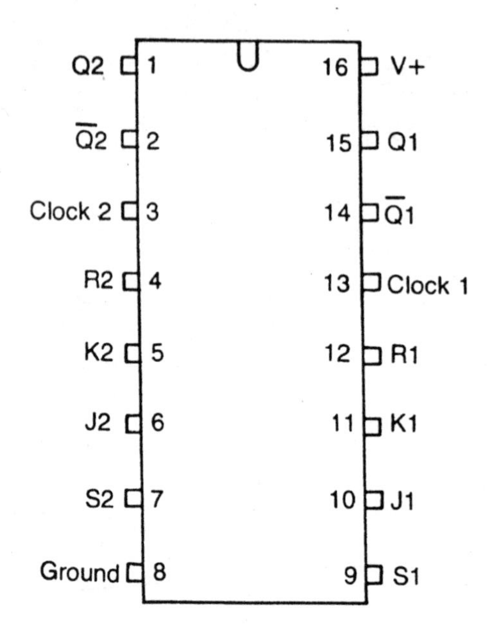

Divide-by-2 counter

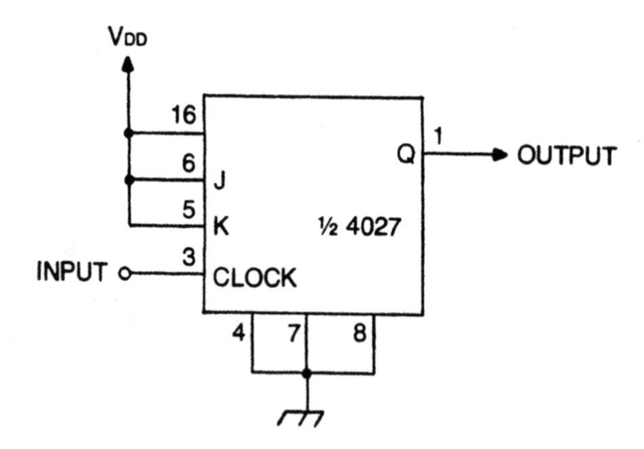

Divide-by-3 counter

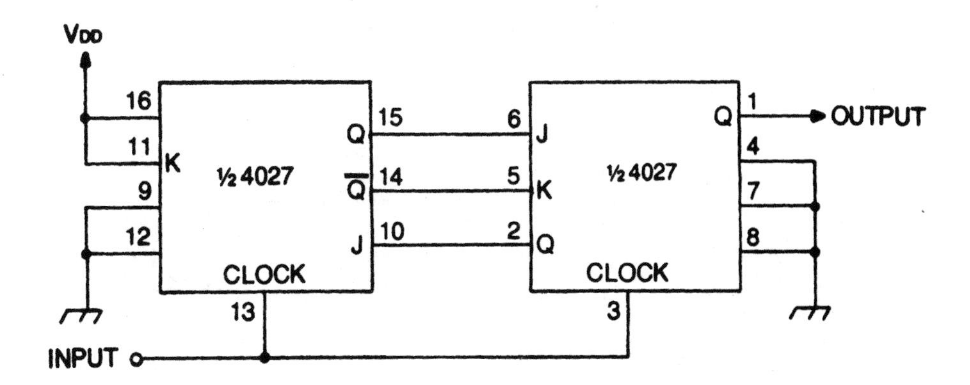

Divide-by-4 counter

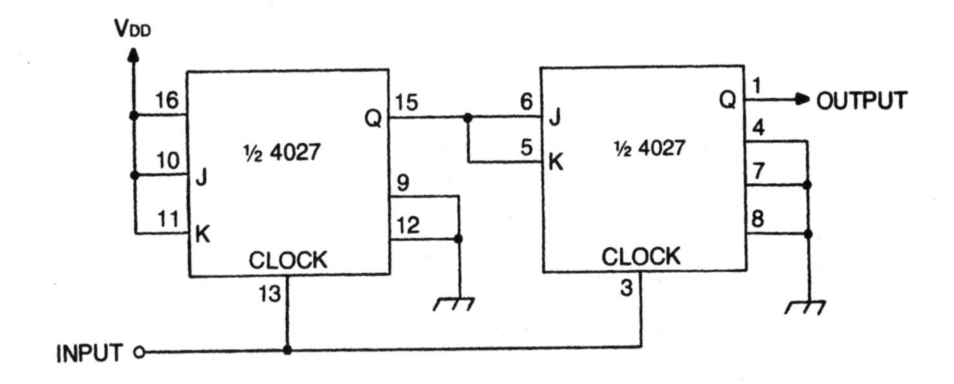

Divide-by-5 counter

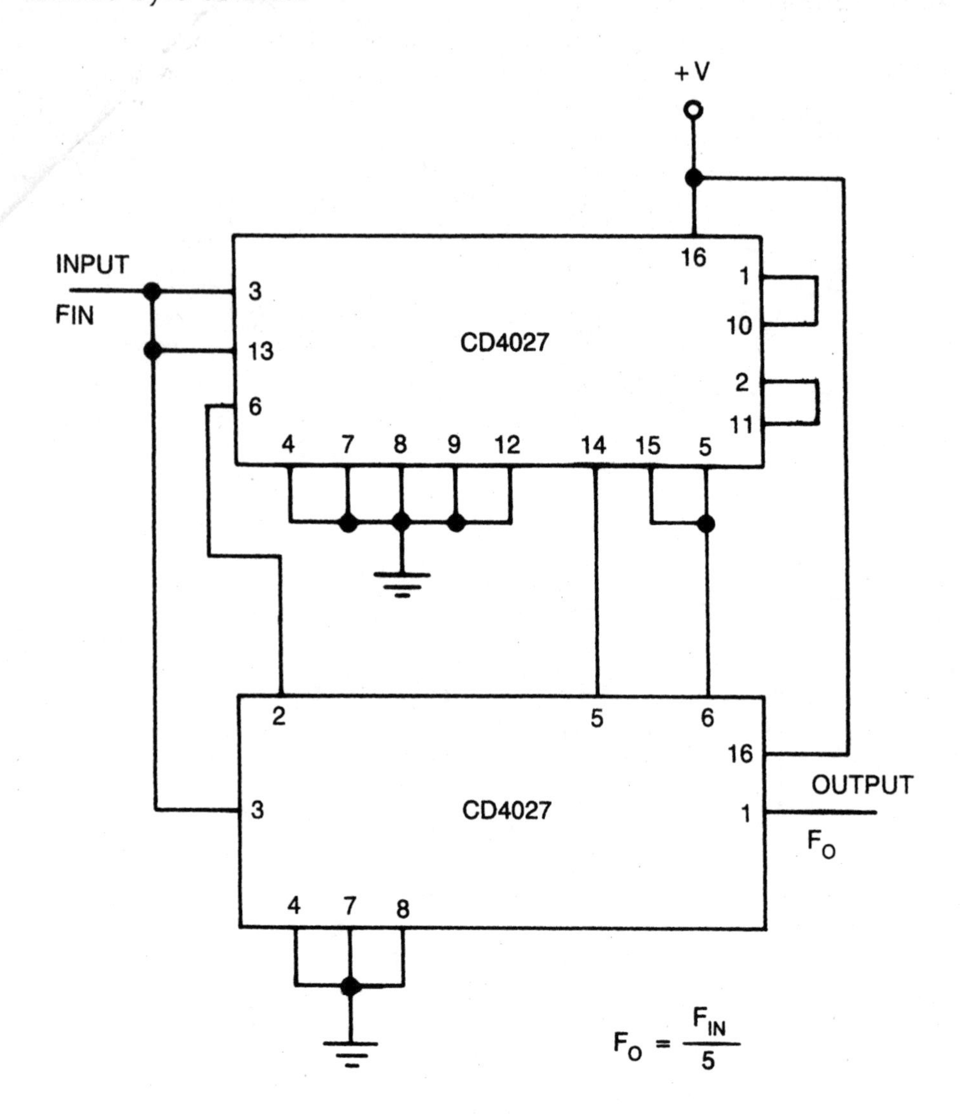

4028 BCD-to-decimal decoder

The 4028 is a BCD (Binary Coded Decimal) to decimal or binary-to-octal decoder. A BCD code applied to the four inputs (A, B, C, and D) results in a HIGH level at the selected one-of-ten decimal decoded outputs (0–9). Similarly, a 3-bit binary code applied to inputs A, B, and C is decoded in octal at outputs 0 through 7. A HIGH signal at the D input inhibits octal decoding and causes outputs 0 through 7 to go LOW.

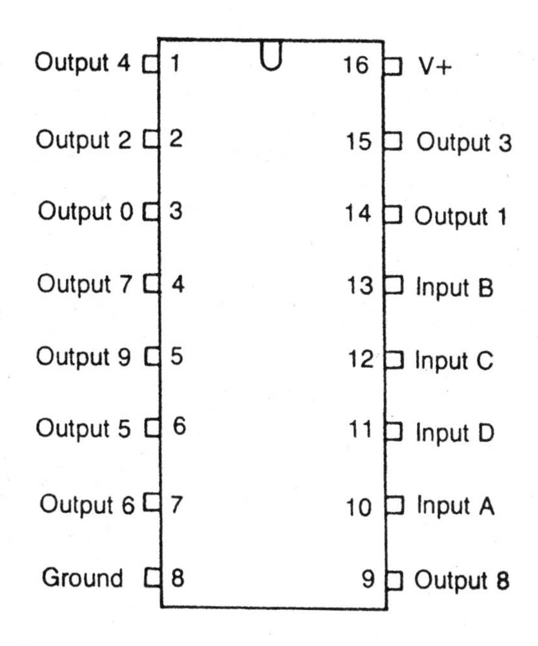

1-of-8 decoder

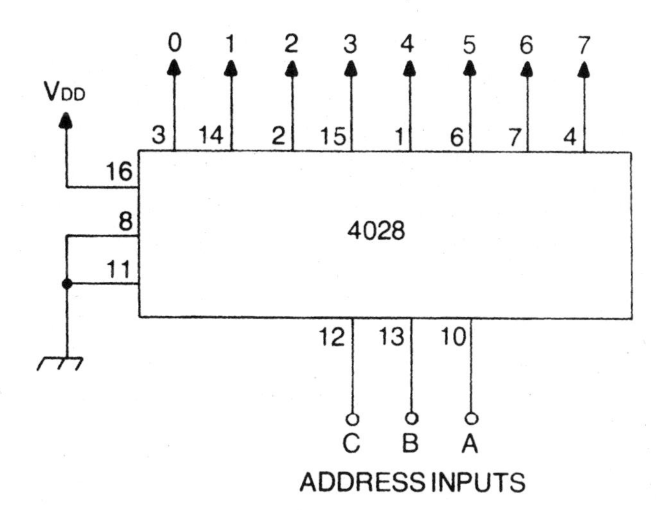

4046 micropower phase-locked loop

The 4046 micropower phase-locked loop (PLL) consists of a low-power, linear voltage-controlled oscillator (VCO), a source follower, a zener diode, and two phase comparators. The two phase comparators have a common signal input and a common comparator input. The signal input can be directly coupled for a large voltage signal, or capacitively coupled to the self-biasing amplifier at the signal input for a small voltage signal.

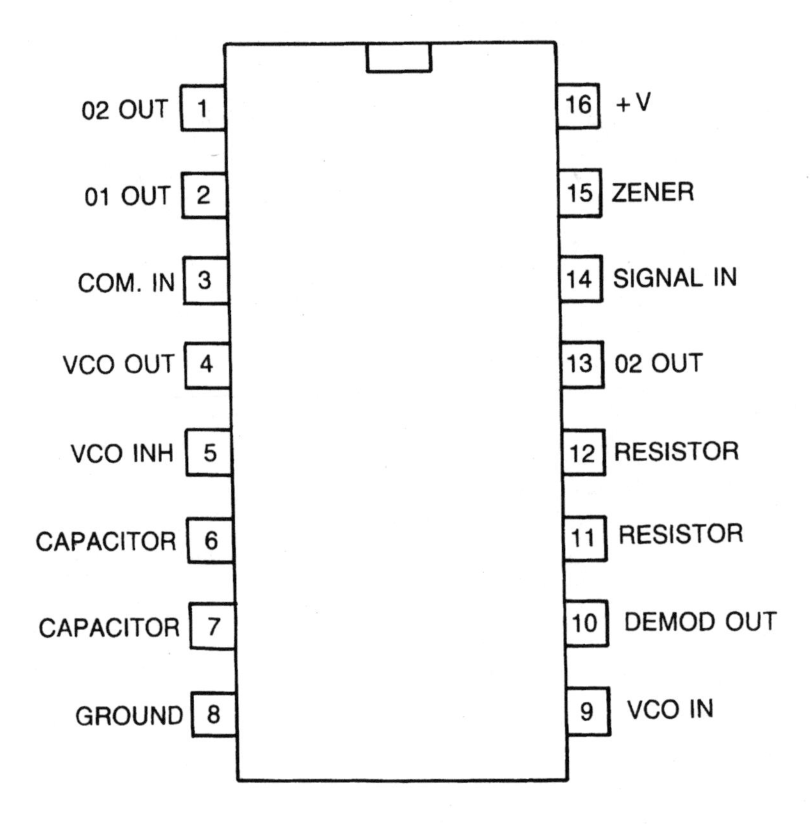

Basic phase-locked loop circuit

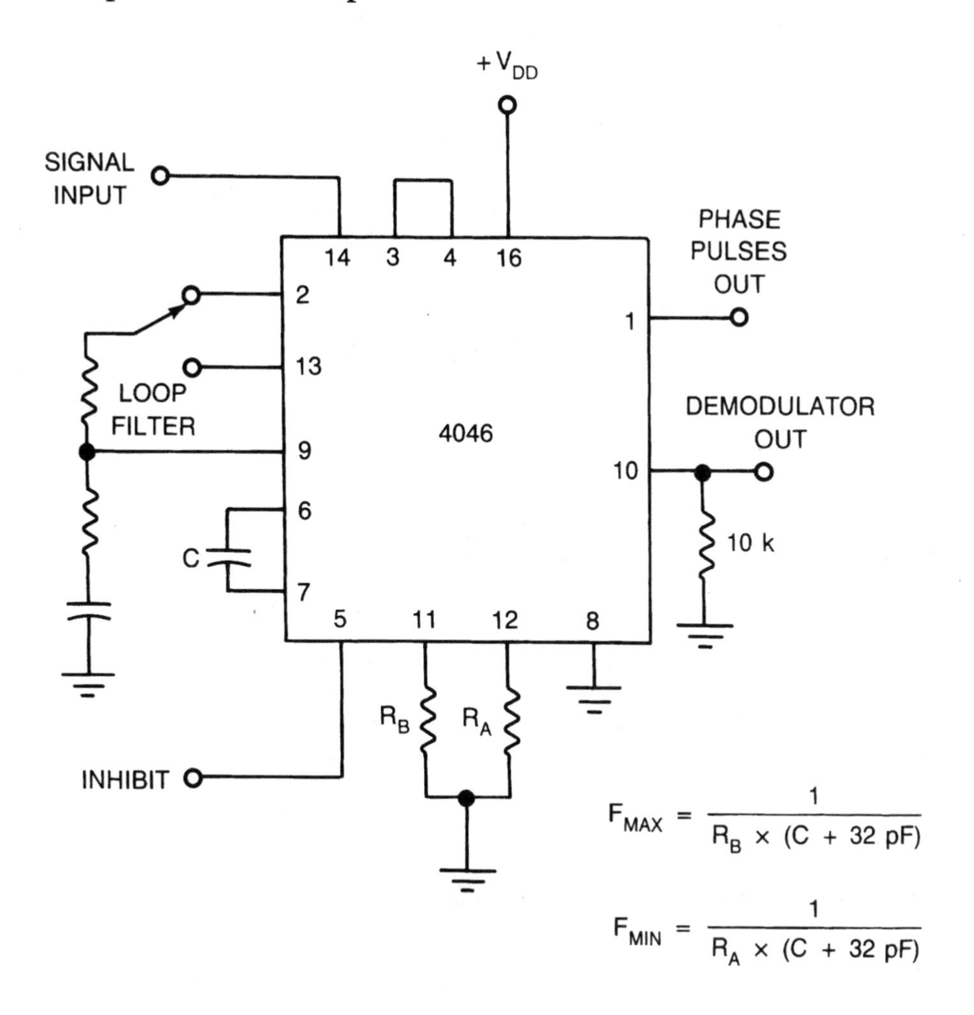

VCO

The output frequency of this voltage-controlled oscillator is proportional to the input voltage. The frequency range can be altered by changing the external resistor and capacitor values.

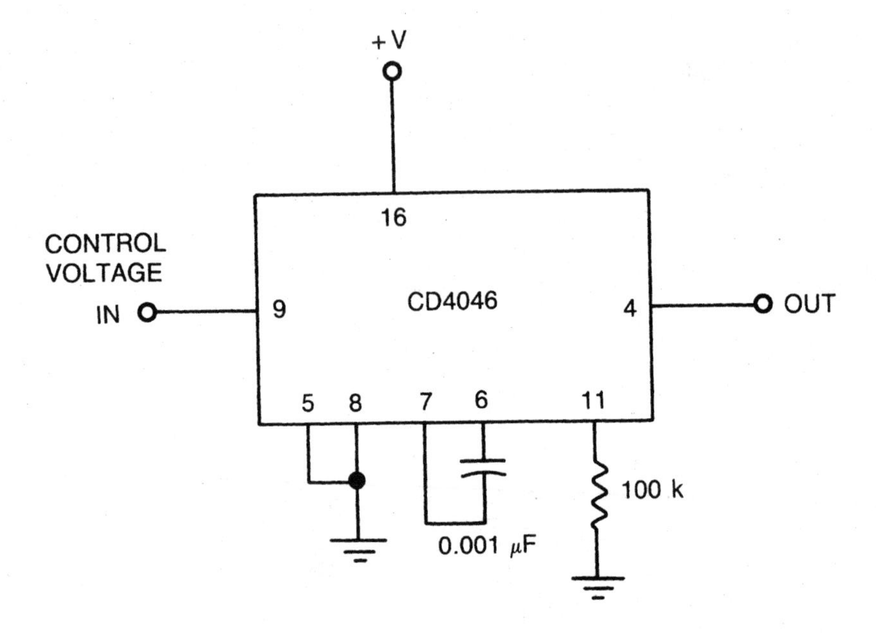

Tone burst generator

This rather unusual tone generator circuit puts out bursts or "pockets" of tone, separated by pauses, rather than a continuous tone. The frequency of the tone itself is set by the 500K potentiometer connected between pins No. 9 and No. 16 of the 4046. The 250K potentiometer in the 4011 portion of the circuit controls the frequency or rate of the bursts.

If the burst rate is made high enough to enter the audible frequency range, the ear will not be able to distinguish between the individual bursts of tone. The tone frequency and the burst frequency will combine to create a single complex continuous tone, with a rather harsh sound. Frequency modulation will occur, producing enharmonic sidebands. The burst rate range can be changed by substituting different capacitor values for the 3.3 µF capacitor shown here. Try a 0.1 µF to 1 µF capacitor for FM effects. Experiment with other resistor and capacitor values throughout the circuit. Different effects can be achieved by adding external filtering circuits, or mixing together the outputs of two tone burst generator circuits.

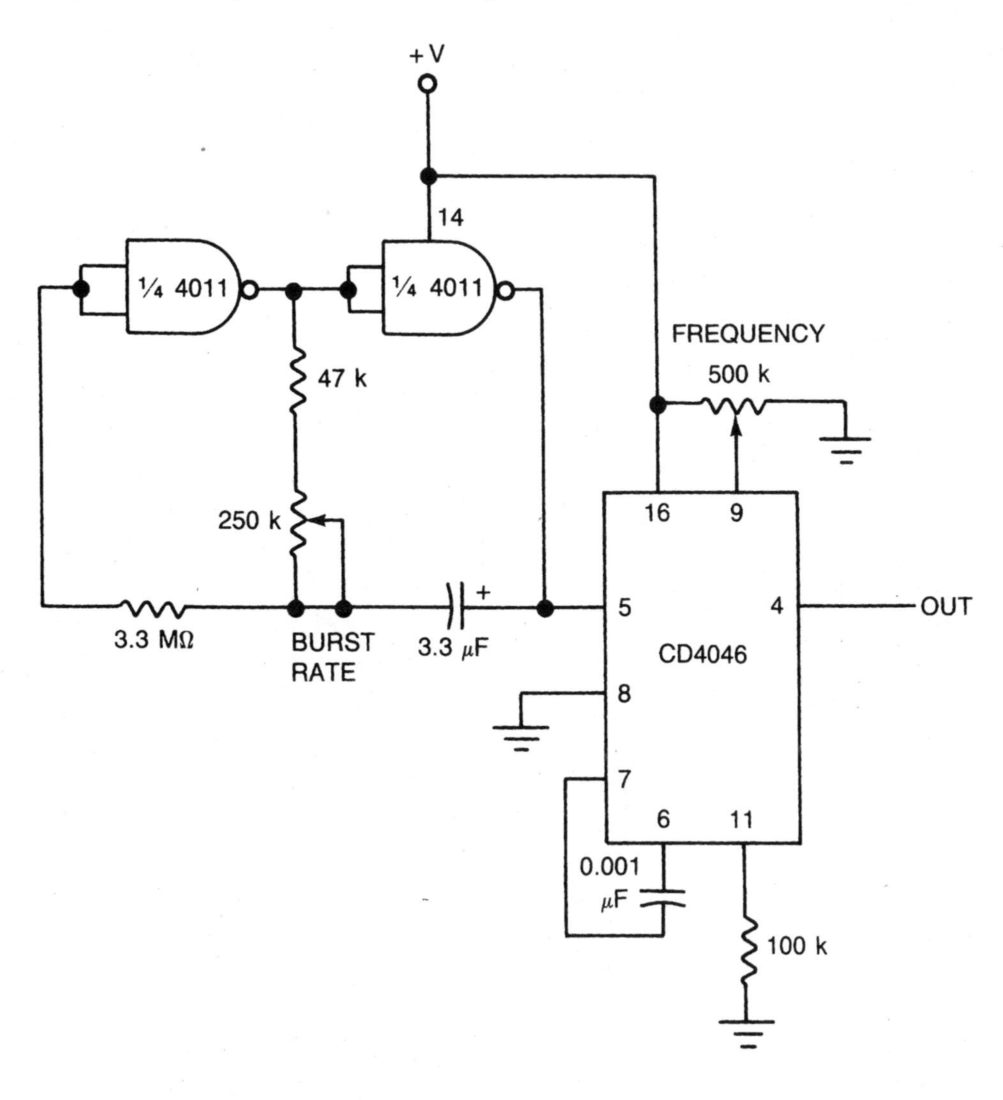

Tunable oscillator

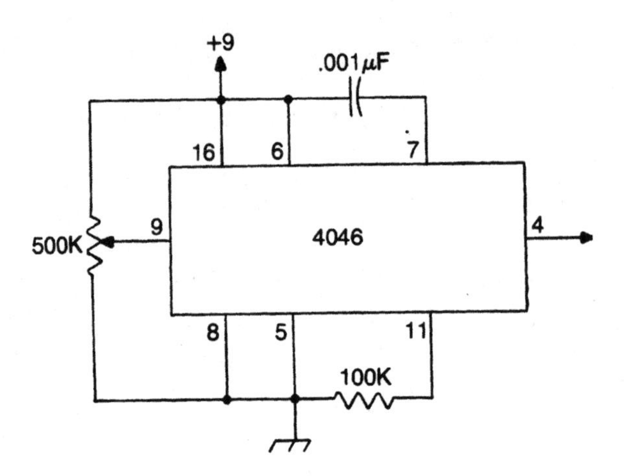

Frequency synthesizer

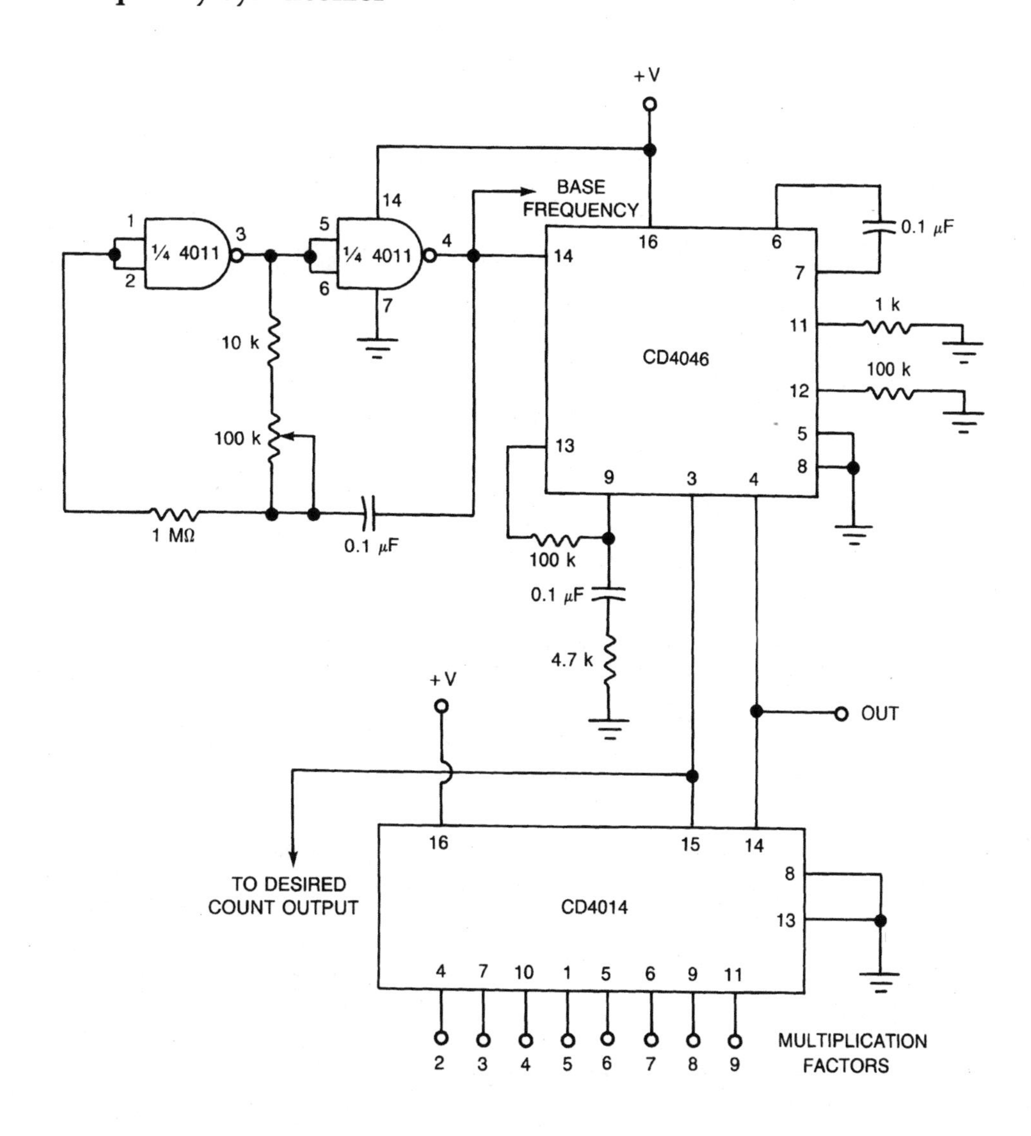

4049 HEX inverting buffer

The 4049 contains six independent inverting buffers. Only the power supply connections are common to all six inverters.

The output of each stage is the opposite state as its input. That is, a LOW input results in a HIGH output, and a HIGH input produces a LOW output.

The 4049 is similar to the 4050, except for the signal inversion performed by this chip.

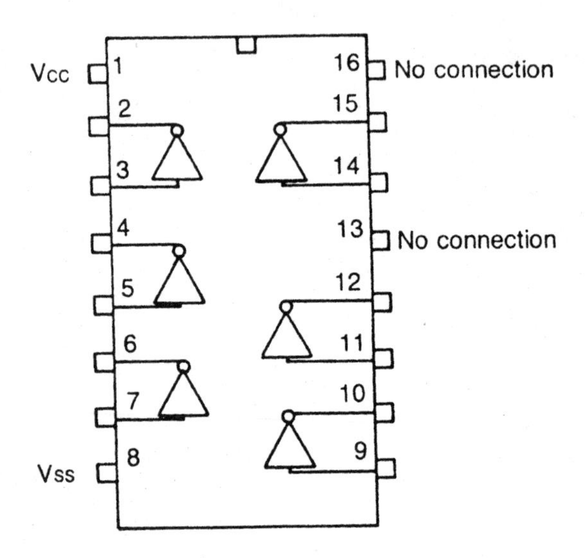

Logic probe

This very simple but effective piece of test equipment can be used to determine the logic state at any given point within a digital CMOS circuit. The ground clip should be connected to the ground of the circuit under test. The probe tip is touched to the specific connection point to be examined. If the LED lights up, there is a logic HIGH signal (or a high-speed pulse) at that connection point in the circuit under test.

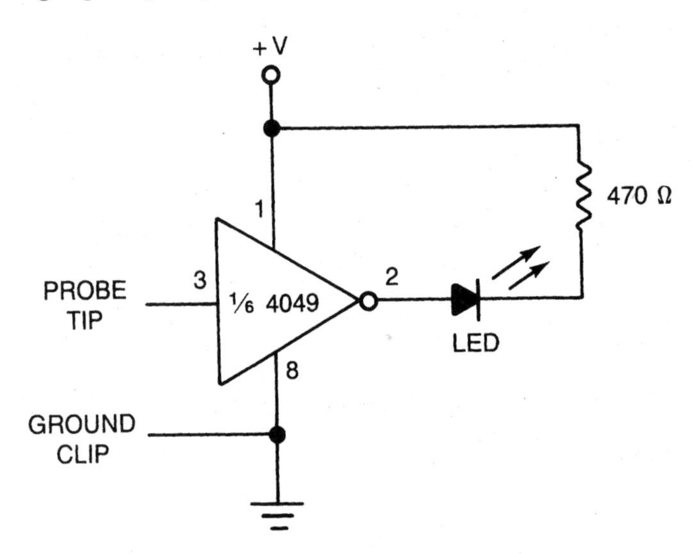

Bounceless switch

Mechanical switches often bounce a little when opened or closed. This is of little or no importance in most linear circuits, but in many digital circuits, such as counters, each bounce can appear to the circuit as a separate switch operation. A debouncing circuit such as this one ensures smooth, reliable switching action, with no data errors due to mechanical bouncing.

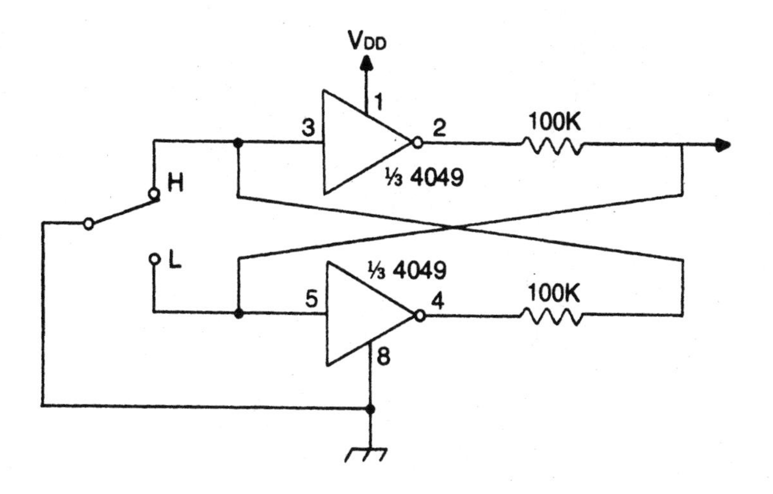

Clock generator

This circuit generates a steady stream of pulses. Using the component values shown here, the clock frequency will be about 370 Hz. Try experimenting with other resistor and capacitor values in this circuit to produce other clock frequencies. Resistor R1 should have a value of about ten times that of resistor R2, or the circuit may not operate reliably.

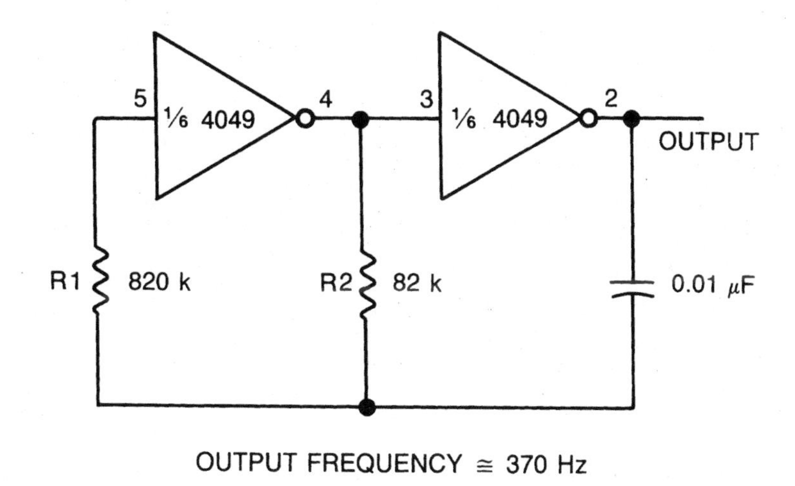

Phase shift oscillator

Try experimenting with alternate resistor and capacitor values in this circuit to obtain other signal frequencies. For proper operation all three resistors should have identical values. The two capacitors should also have equal values.

While this circuit uses digital inverters, it is actually an analog oscillator circuit. The output signal is a somewhat distorted sine wave.

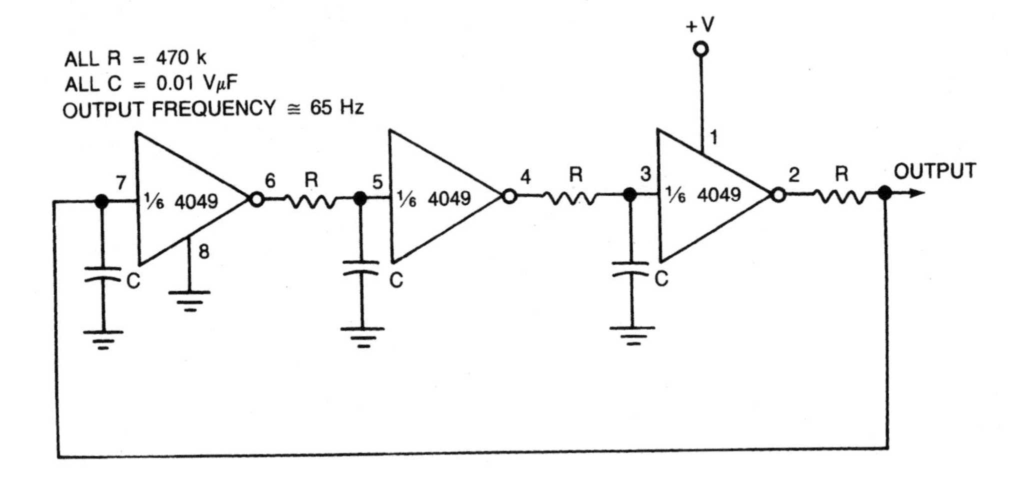

Triangle-wave generator

Usually digital circuits can only generate rectangle waves with varying duty cycles. This circuit uses a little special trickery, including linear feedback, to generate a fair approximation of a linear triangle wave. The output signal frequency is set by the values of the resistor marked "R" and the capacitor marked "C" in the diagram. The frequency formula is

$$F = \frac{1}{(1.4RC)}$$

Changing the value of either the other resistor or the other capacitor in the circuit a small amount will alter the wave shape somewhat. Changing these component values too much, however, may prevent the circuit from functioning at all.

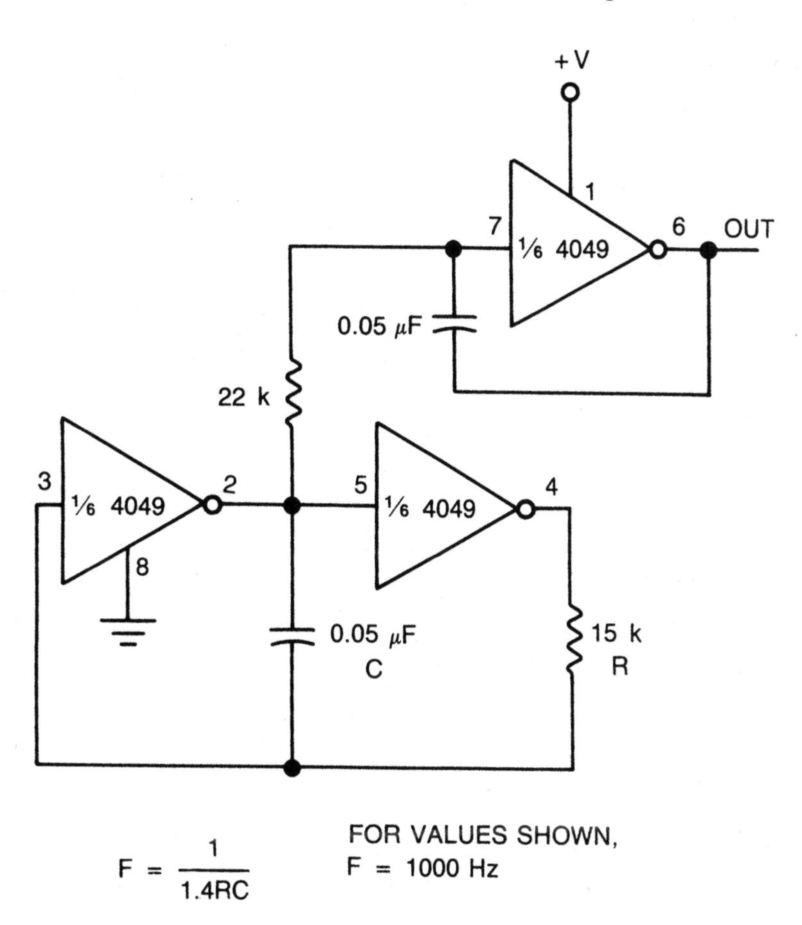

$$F = \frac{1}{1.4RC}$$

2-waveform function generator

This minor modification on the circuit of the last circuit has two outputs. The signal at output A is a triangle wave, as described for the last circuit. A more common rectangle wave of the same frequency is simultaneously available at output B. The signal frequency for both outputs is controlled by the values of resistor R1 and capacitor C1, according to the formula

$$F = \frac{1}{(1.4RC)}$$

Changing the value of either resistor R2 or capacitor C2 a small amount will alter the output A wave shape somewhat. Changing these component values too much, however, may prevent the circuit from functioning at all.

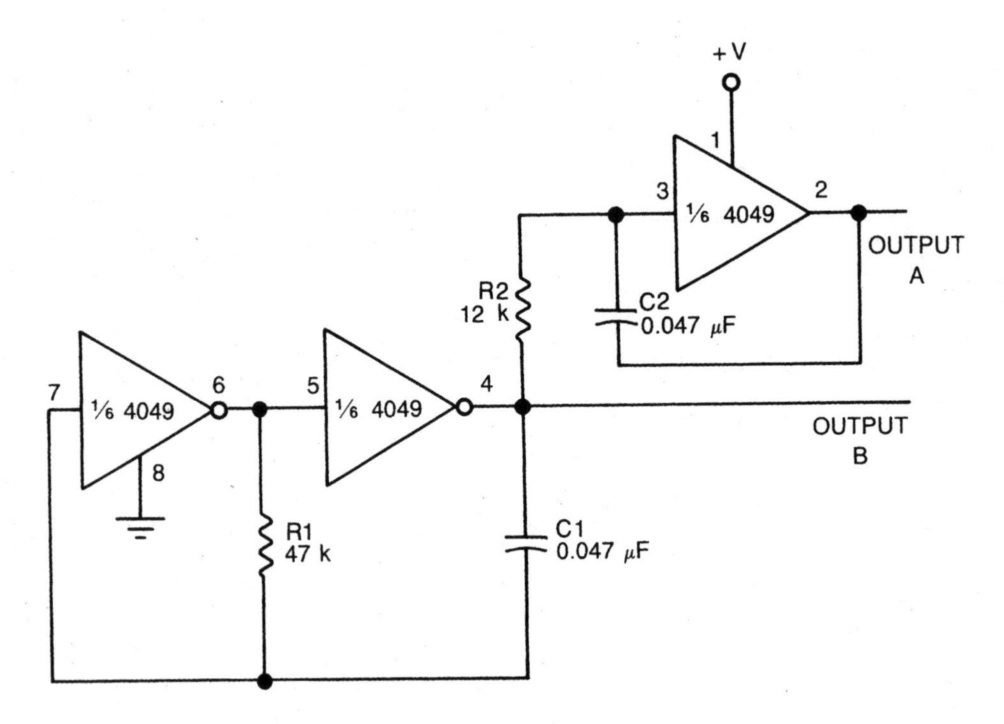

Linear amplifier

While this circuit uses digital inverters, it actually functions as a crude linear amplifier, with a gain of about 10. The input signal can be any analog signal. It is not limited to the usual digital rectangle waves.

This digital linear amplifier circuit really does work, but it is hardly high fidelity. It can be quite useful in many nondemanding applications where the existing digital circuitry has a few otherwise unused inverter stages available.

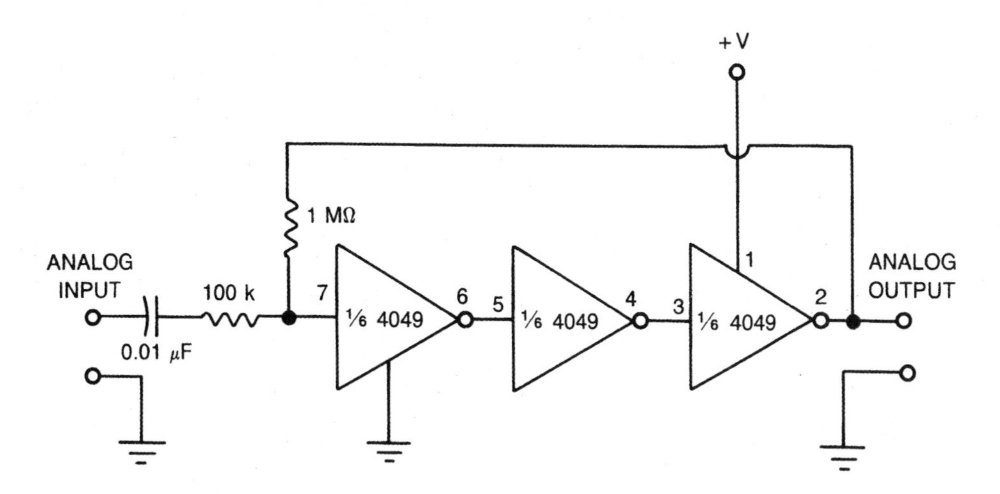

4051 8-channel analog multiplexer

This chip is essentially made up of a set of eight interlinked digitally controlled electronic switches. These switches can carry either digital or analog signals. A 3-bit digital control input (A, B, and C) determines which of eight inputs will feed the single output, or which of eight outputs will carry the single-input signal. The switches can operate in either direction, depending on the rest of the circuitry used.

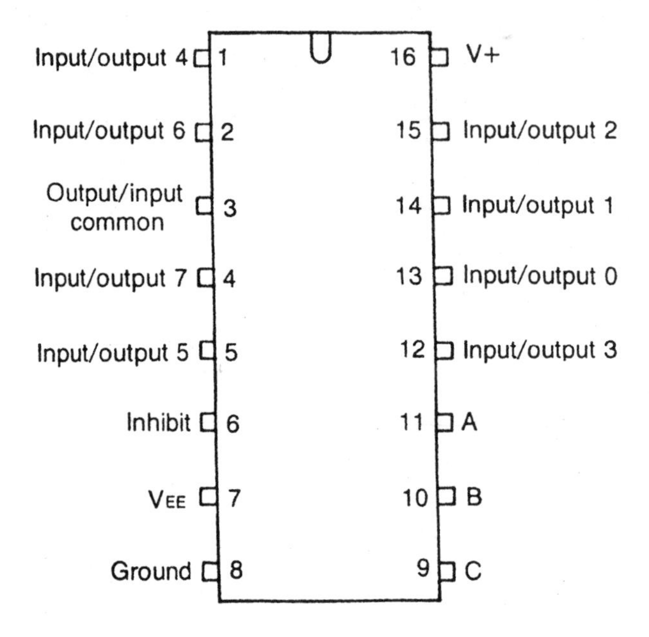

1-of-8 multiplexer

A single input drives any of eight possible output lines, depending on the digital control data on the address lines (A, B, and C).

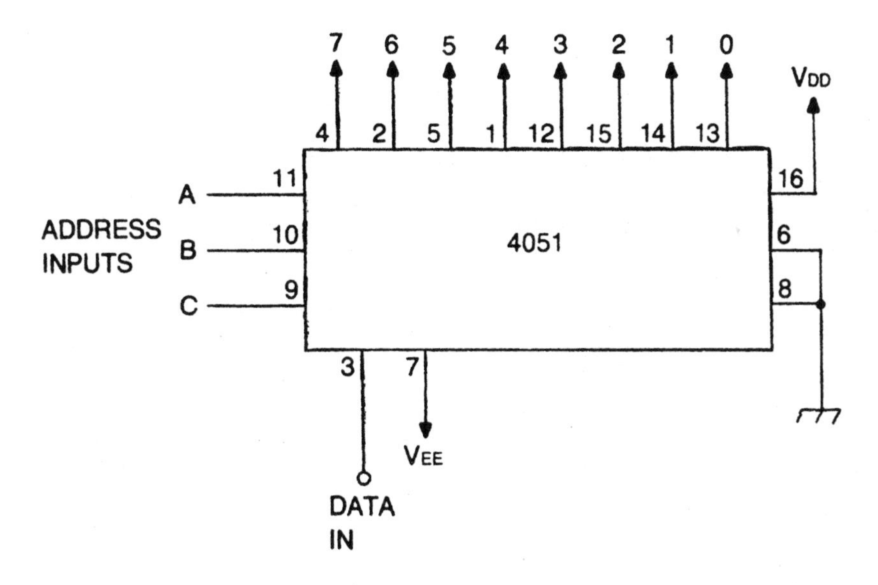

1-of-8 demultiplexer

Any of eight inputs can be directed to the output line, depending on the digital control data on the address lines (A, B, and C).

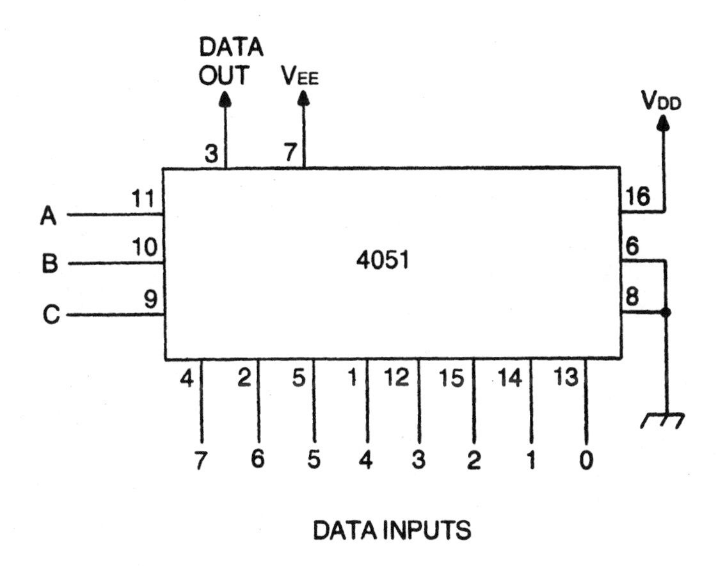

4066 quad bilateral switch

The 4066 is a quad bilateral switch with an extremely high off-resistance and low on-resistance. The switch will pass analog or digital signals in either direction, and is extremely useful in digital switching. In effect, it is the equivalent of an ordinary mechanical switch, except it is operated by logic signals on the appropriate control input pin, rather than physical motion or pressure.

The 4066 is pin-for-pin compatible with the 4016, but it has a much lower on-resistance, and its on-resistance is relatively constant over the input-signal range. Essentially, the 4066 is simply an improved version of the 4016.

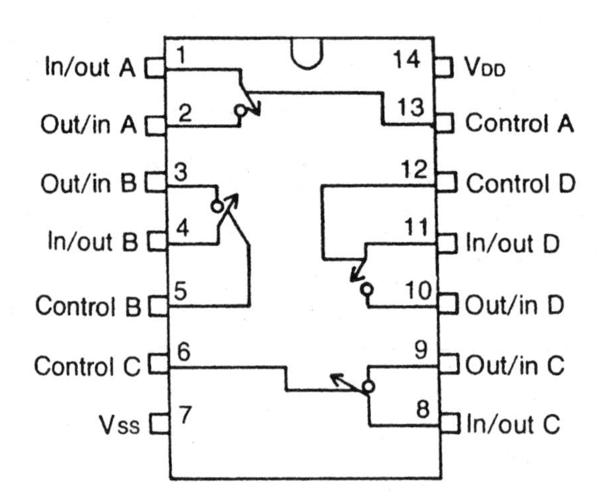

Data selector

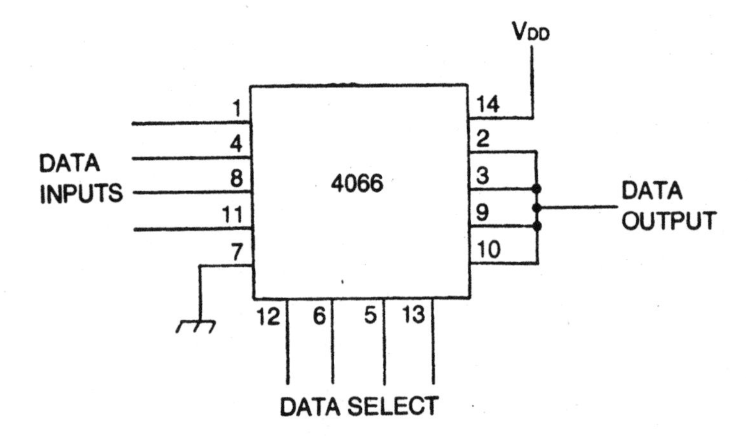

Data bus controller

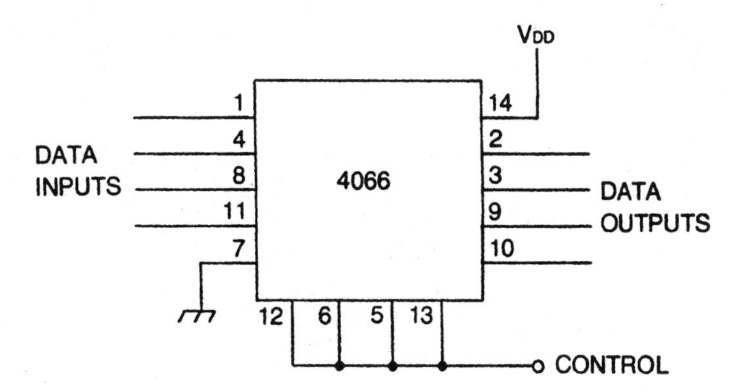

Programmable gain op amp

The gain of an op amp (operational amplifier—see Sec. 4) is determined by the ratio of the input resistance to the feedback resistance. Varying the input resistance while holding the feedback resistor constant will alter the effective gain. In this circuit, four different input resistors can be switched in and out of the signal path by the switches within the 4066. The four digital control inputs (A, B, C, and D) can select any of sixteen possible combinations of the four input resistors, which may be used singly, or in various parallel combinations. If all four 4066 switches are open (control code = 0000), there will be no signal path for the input, effectively disabling the op amp altogether.

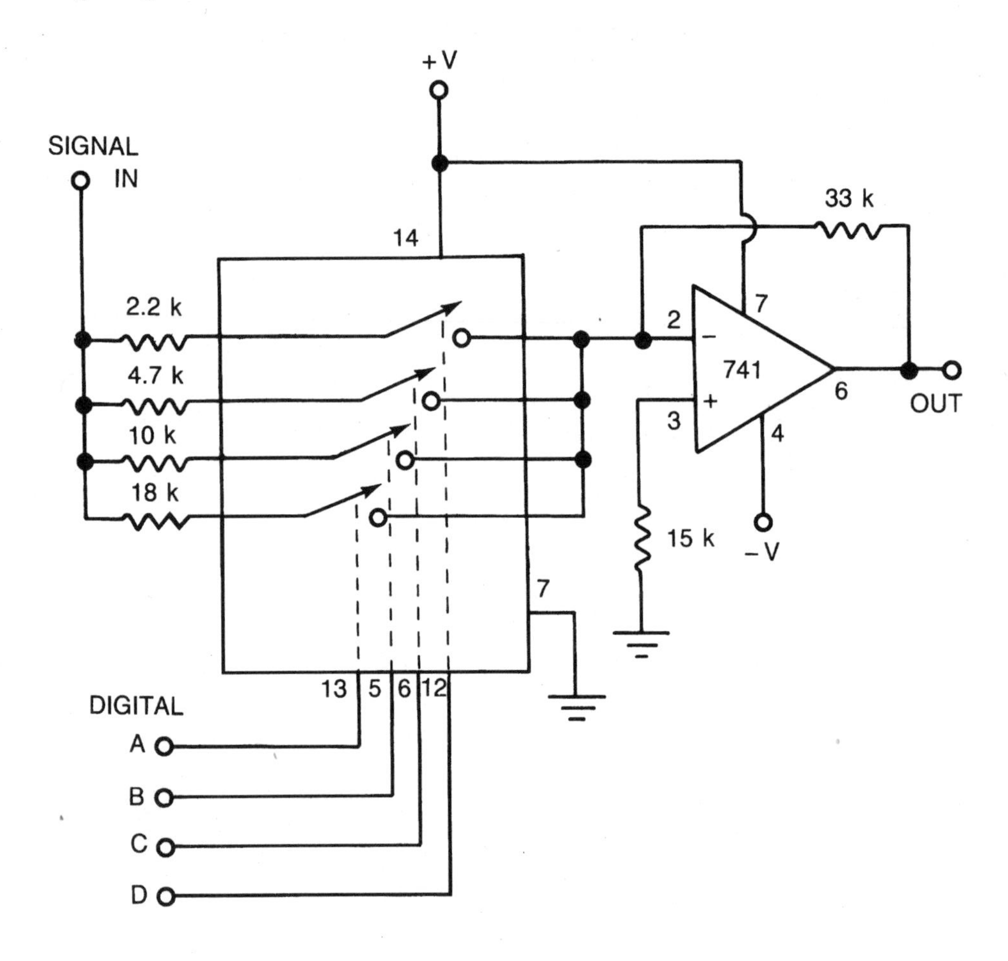

4070 quad 2-input EXCLUSIVE-OR gate

The 4070 contains four EXCLUSIVE-OR (X-OR) gates, each with two inputs. The output of each gate is HIGH if and only if one of the inputs is HIGH, but not both. In other words, as the Truth Table shows, the output is HIGH if the two inputs are at opposite logic states, but LOW if they are the same. An X-OR gate can be used as a 1-bit digital comparator.

4070 Truth Table

Inputs		Output
A	B	
0	0	0
0	1	1
1	0	1
1	1	0

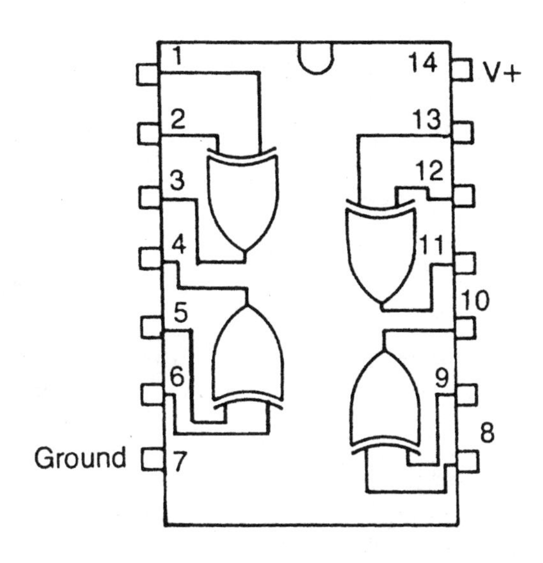

1-bit comparator

The output will be HIGH if A = B, that is, if both inputs are LOW or both inputs are HIGH. If the inputs are at opposite states (and it doesn't matter which is which), the output of the gate will be LOW.

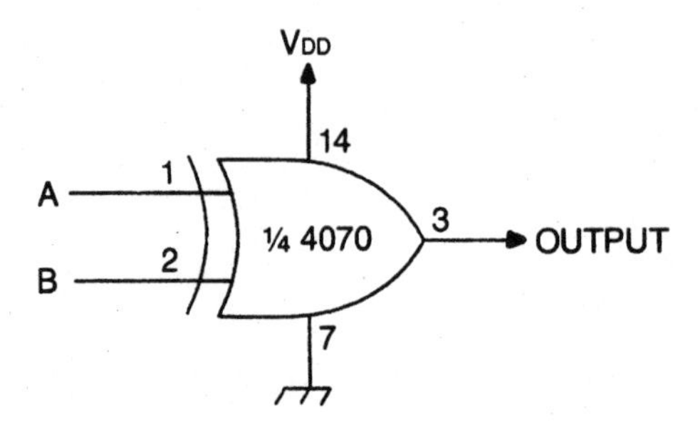

Clock generator

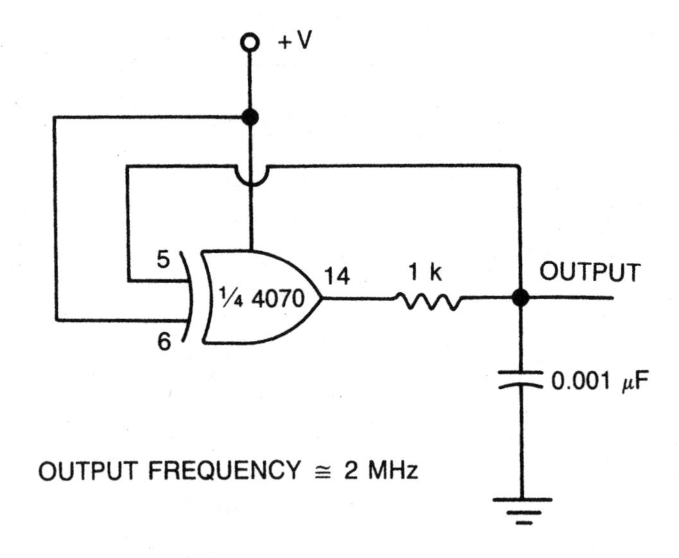

Controlled inverter/buffer

The control input determines how the gate will treat the data input. If the control input is LOW, the device will function as a noninverting buffer. The output will match the data input. If the control input is HIGH, the effect will be that of an inverter. In this case, the output will have the opposite logic state as the data input.

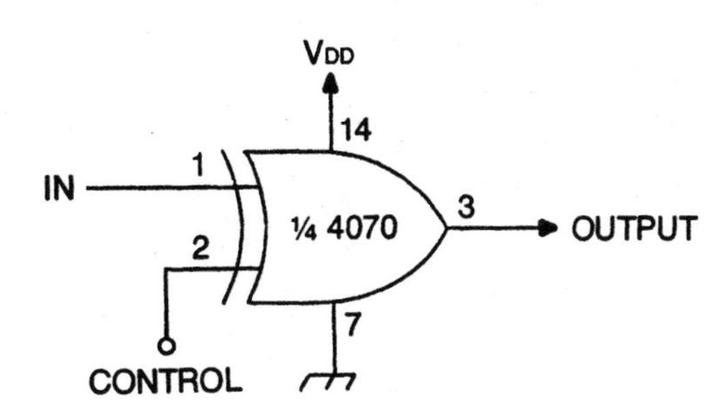

Binary adder

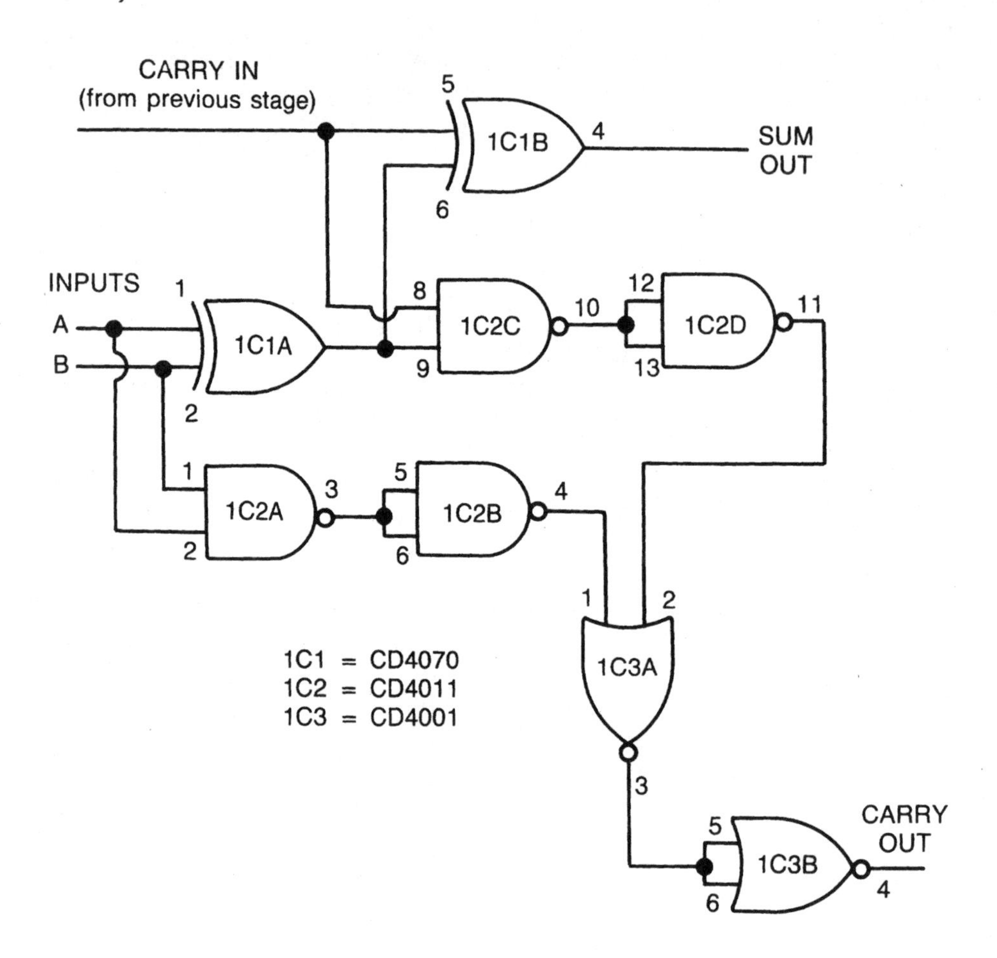

Frequency multiplier

This circuit is designed for use with pulse or rectangle waves. The signal at output A will have twice the frequency as the input signal. The signal at output B is at twice the frequency as the signal at output A, or four times the frequency of the original input signal. These frequency relationships are fixed results of the basic circuit design, and cannot be adjusted. The exact value of the two resistors is not critical, and there would be little point in experimenting with alternate values here.

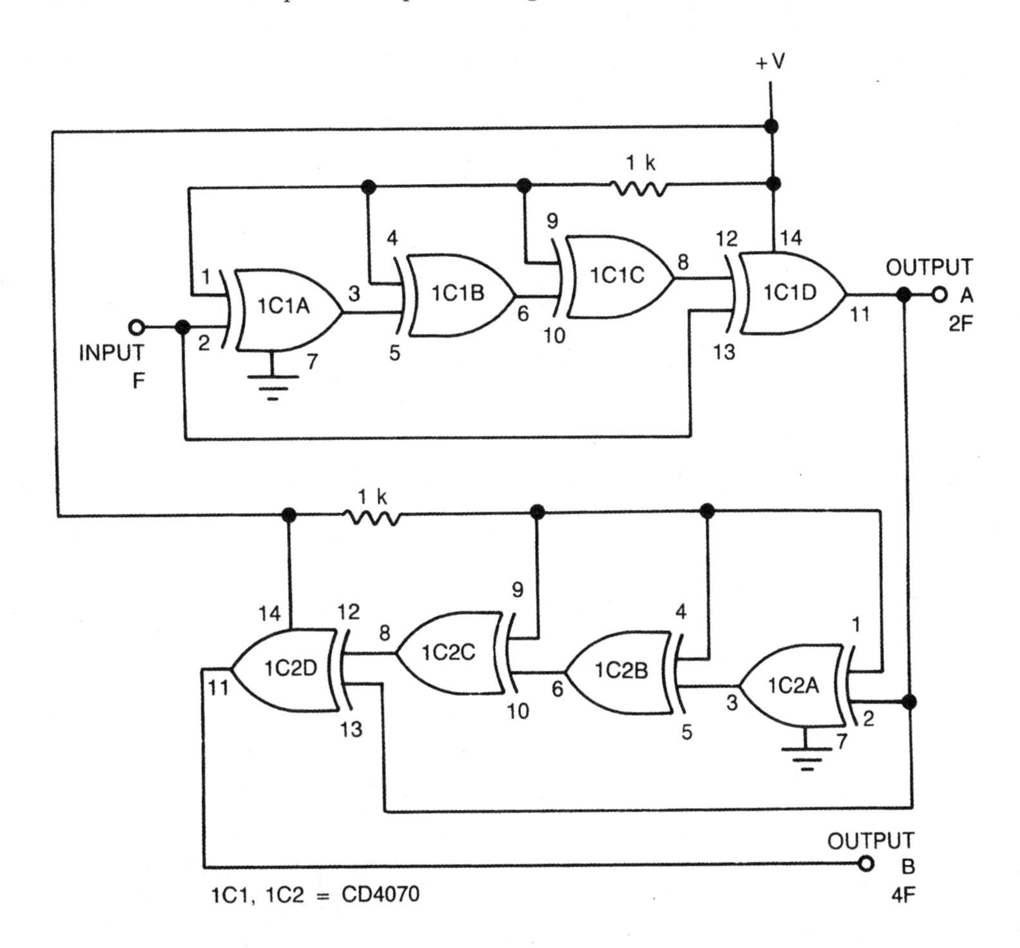

4081 quad 2-input and buffered B-series gate

The 4081 contains four AND gates with two inputs each. The output of each gate is HIGH if and only if both of the inputs are HIGH. The output is LOW if either input (or both) is LOW.

The 4081's gates also have buffered outputs that improve transfer characteristics by providing very high gain.

4071 Truth Table

Inputs		Output
A	B	
0	0	0
0	1	0
1	0	0
1	1	1

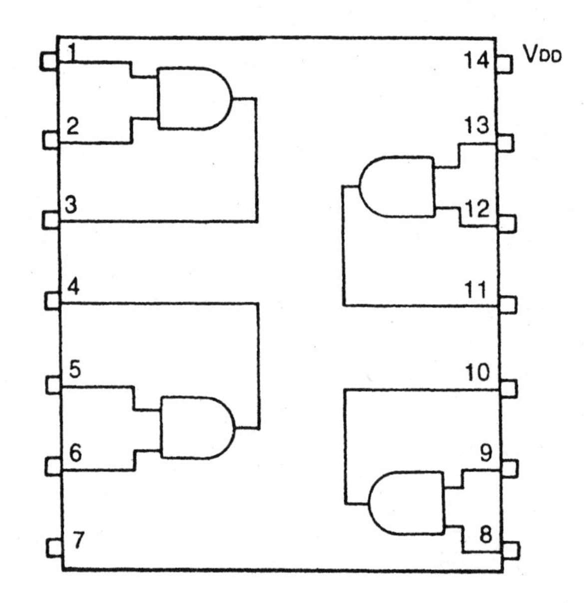

4-input AND gate

All four inputs must be HIGH to give a HIGH output. If any one (or more) of the inputs is LOW, the output of this gating circuit will be LOW.

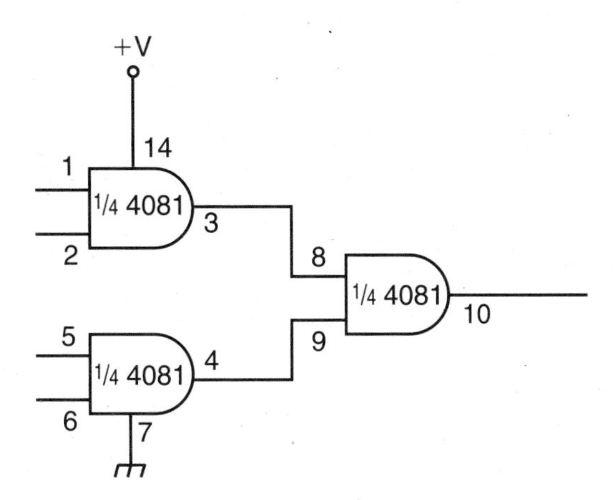

Noninverting buffer

The output logic state equals the input logic state, but the signal is effectively boosted, for greater fan-out.

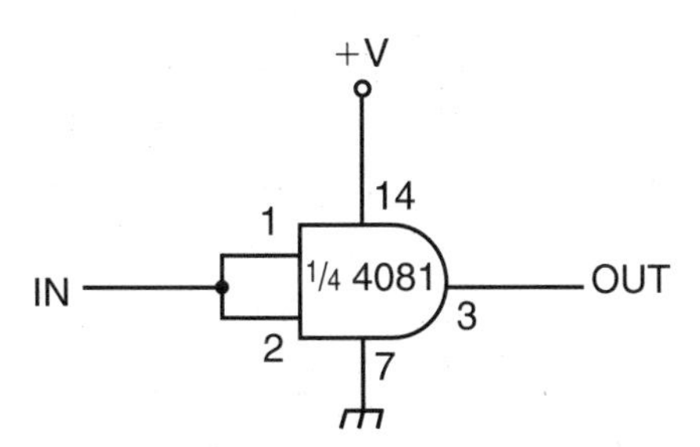

NOR gate

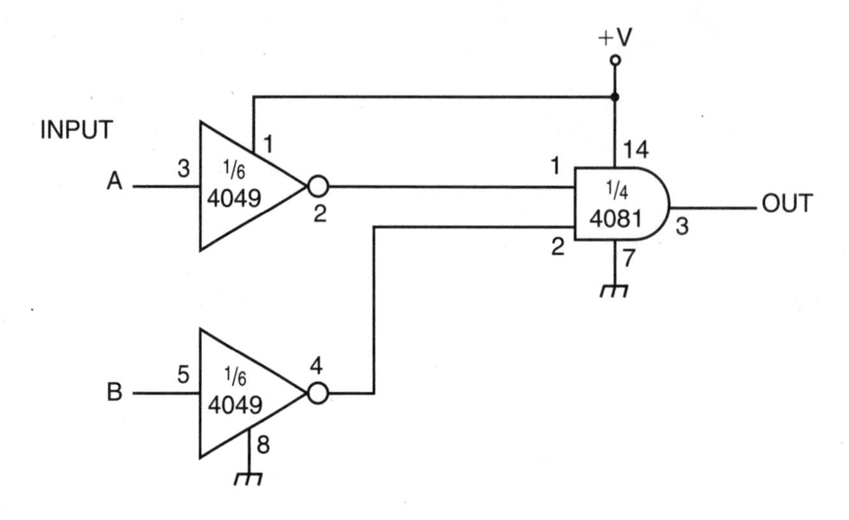

Monostable multivibrator

The timing period is set by the values of capacitor C and resistor R. Increasing either (or both) of these component values will result in a longer timing period.

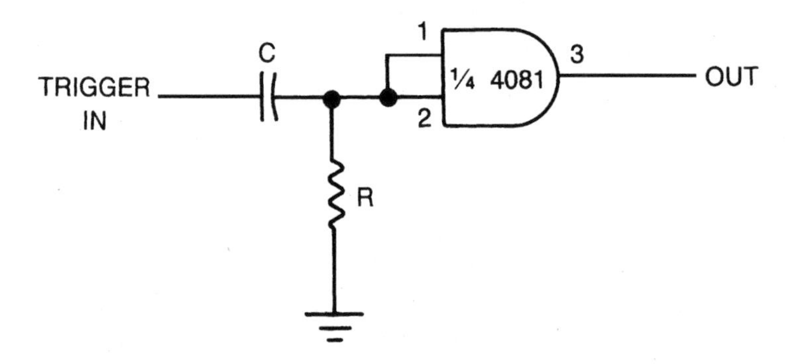

4511 BCD-to-7-segment display latch/decoder/driver

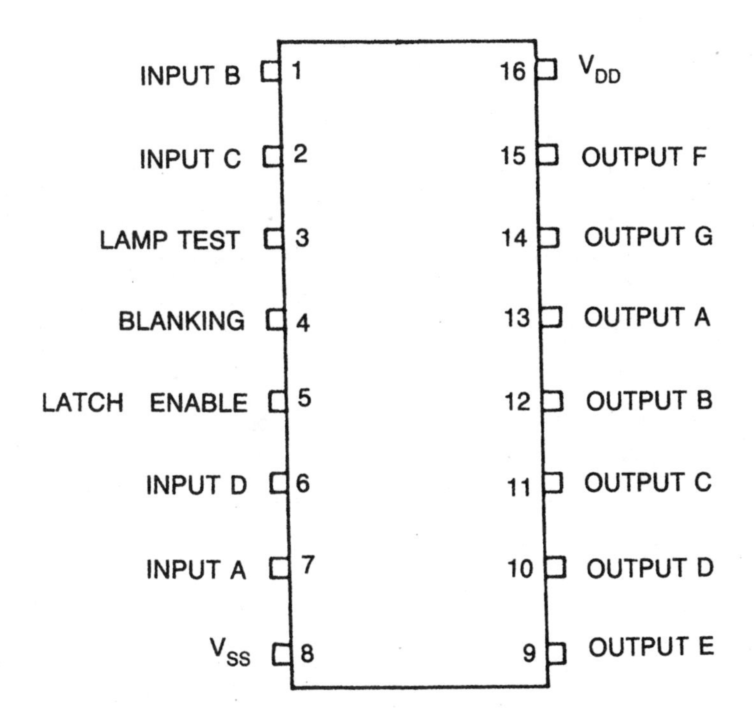

Random number generator

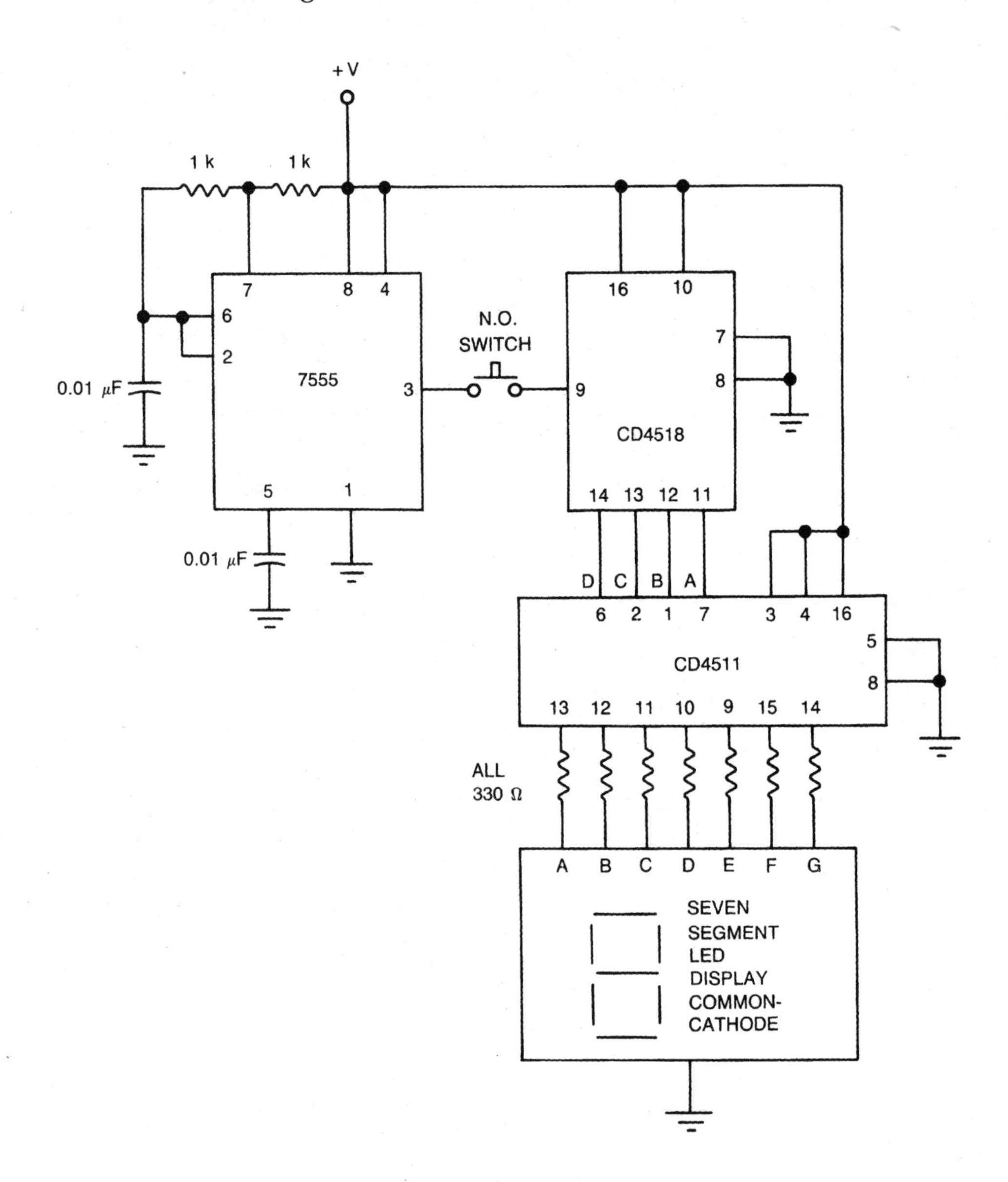

4518 dual BCD counter

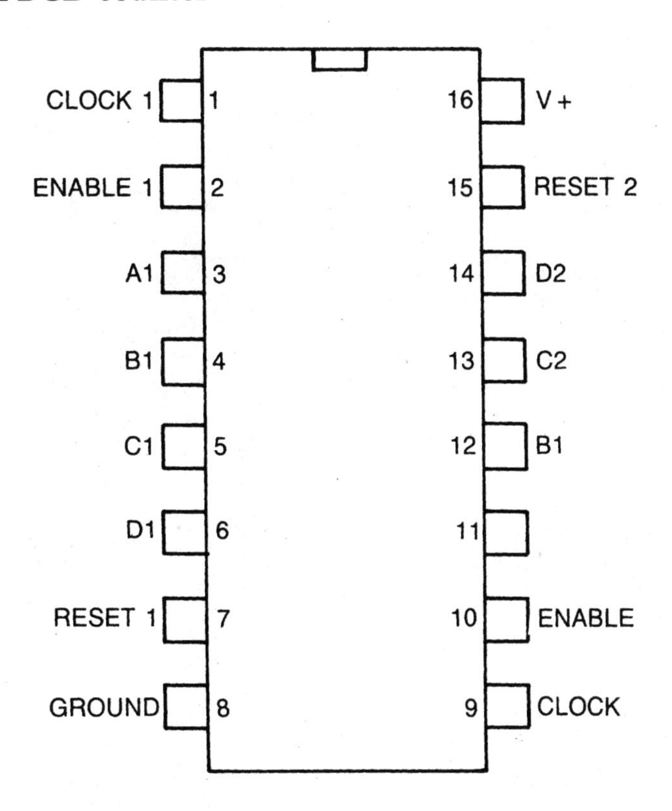

Cascaded BCD counters for multidigit counts

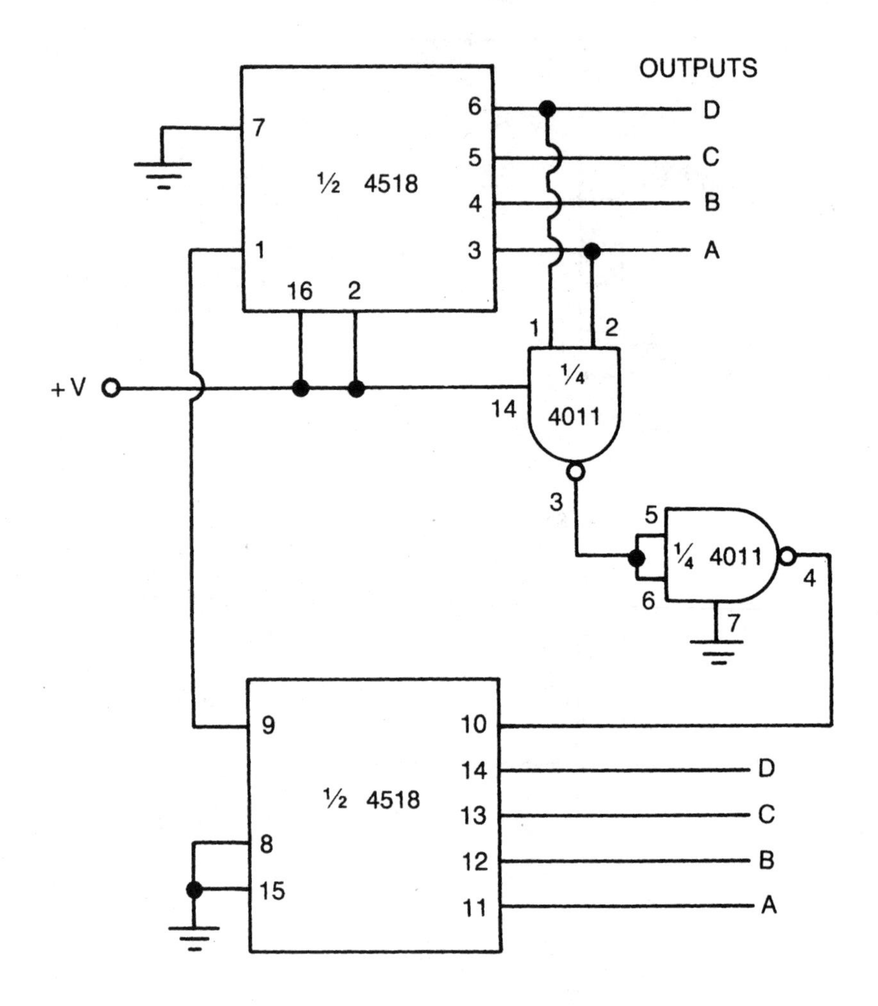

Dual one-shot

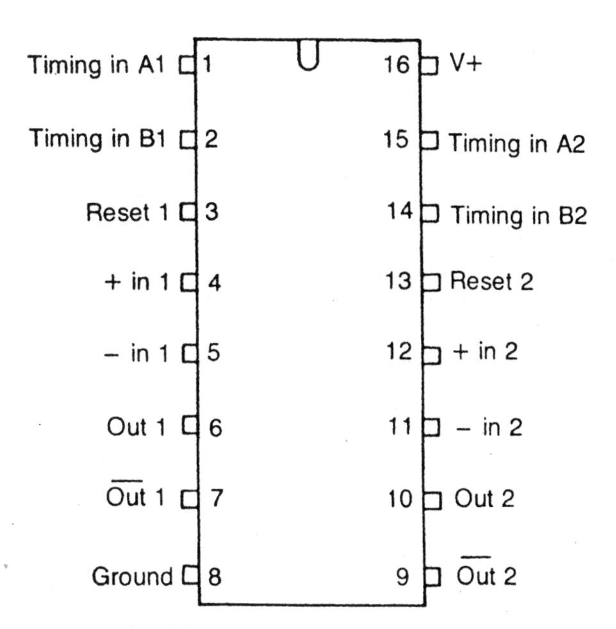

Stepped wave generator

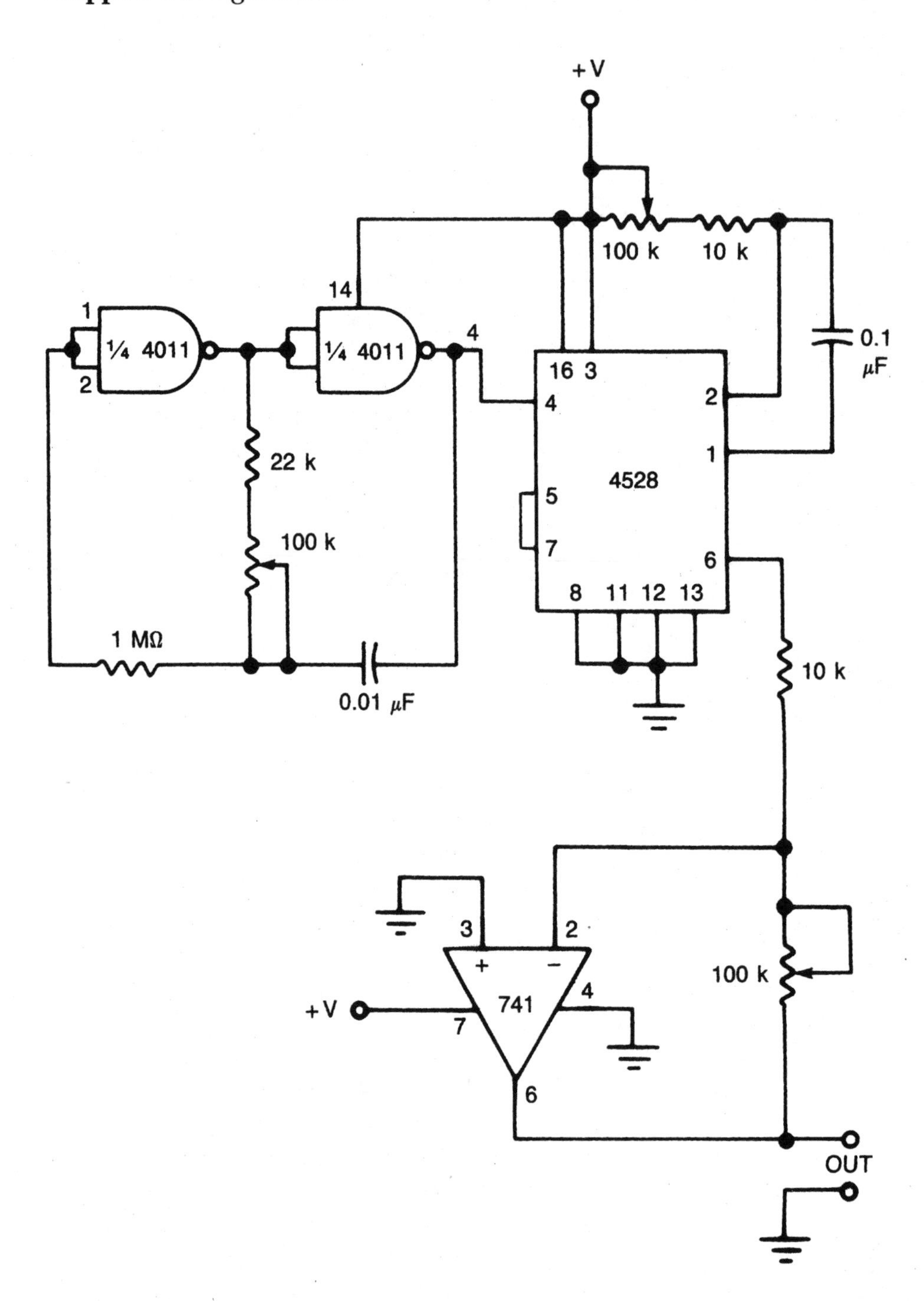

3
SECTION
Special-purpose digital

Some special-purpose digital devices don't fit into either the 74xx (see Sec. 1) or 4xxx (see Sec. 2) numbering schemes. These devices (mostly CMOS), which are presented in this section, are generally more sophisticated than those in the preceding sections.

DS1386/DS1486 Watchdog Time-keeper

Dallas Semiconductor's Watchdog Time-keeper family of devices provides both an accurate real-time clock and interrupt capabilities at specific, user-programmable times. It also provides an upgradable nonvolatile RAM path.

The DS1386-08 module (shown here) has 8K of nonvolatile RAM, the DS1386-32 has 32K, and the DS1486 has 128K. Except for the memory size, these modules have the same features and functions.

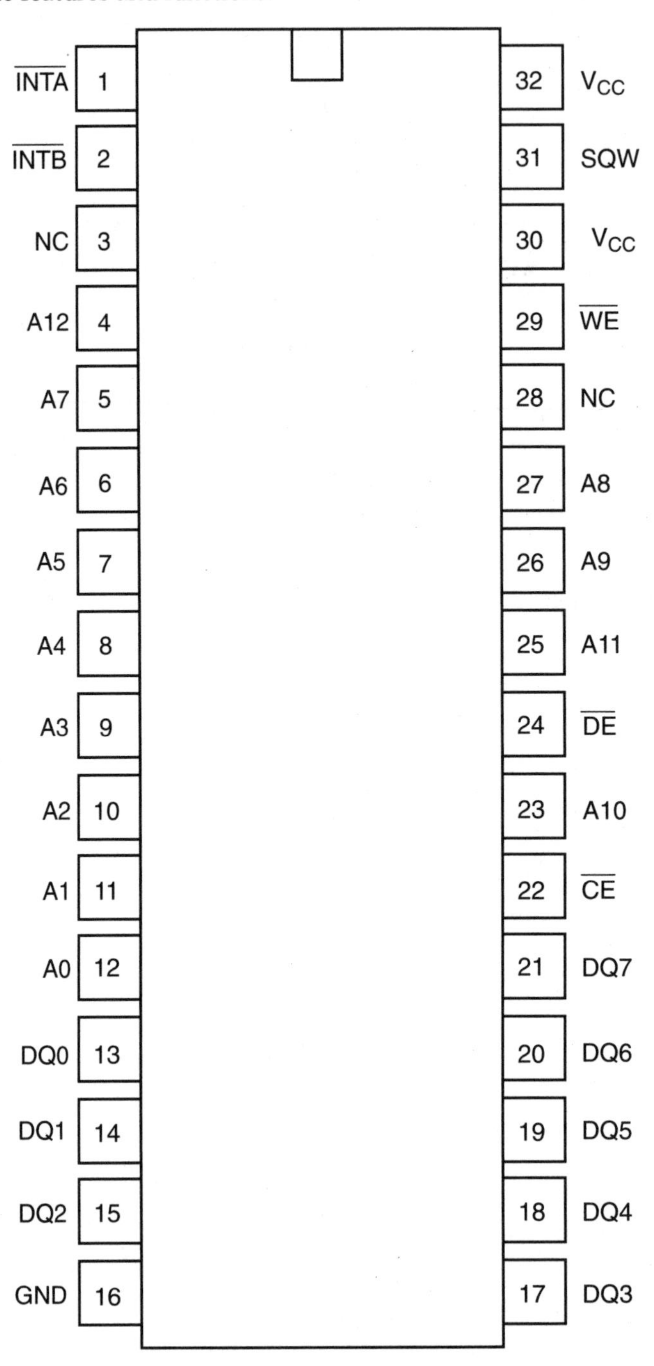

Watchdog Time-keeper interface with 8051 microcontroller

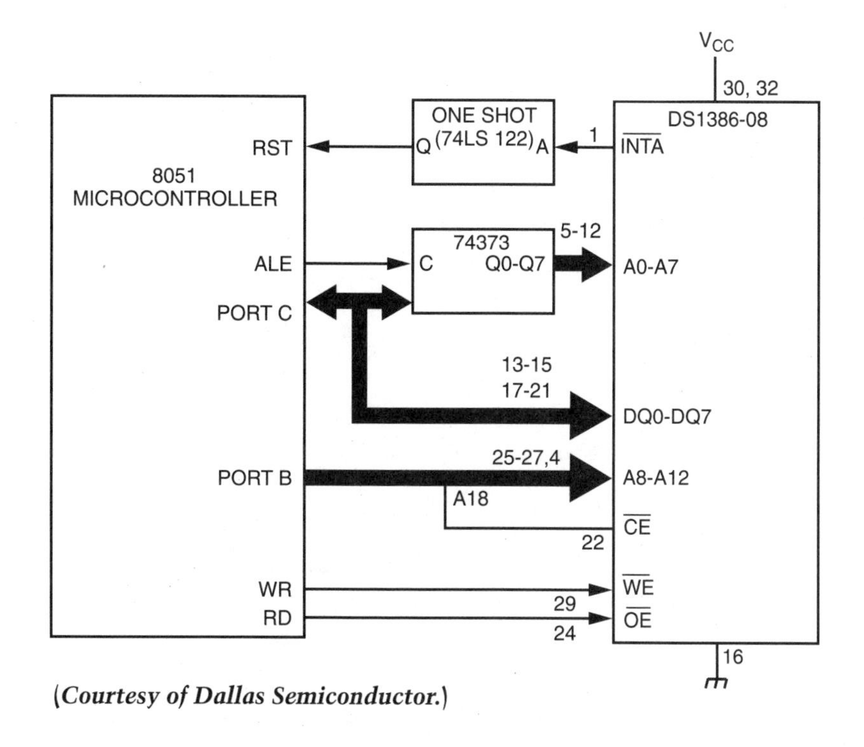

(*Courtesy of Dallas Semiconductor.*)

Watchdog Time-keeper interface with 68HC11 microcontroller

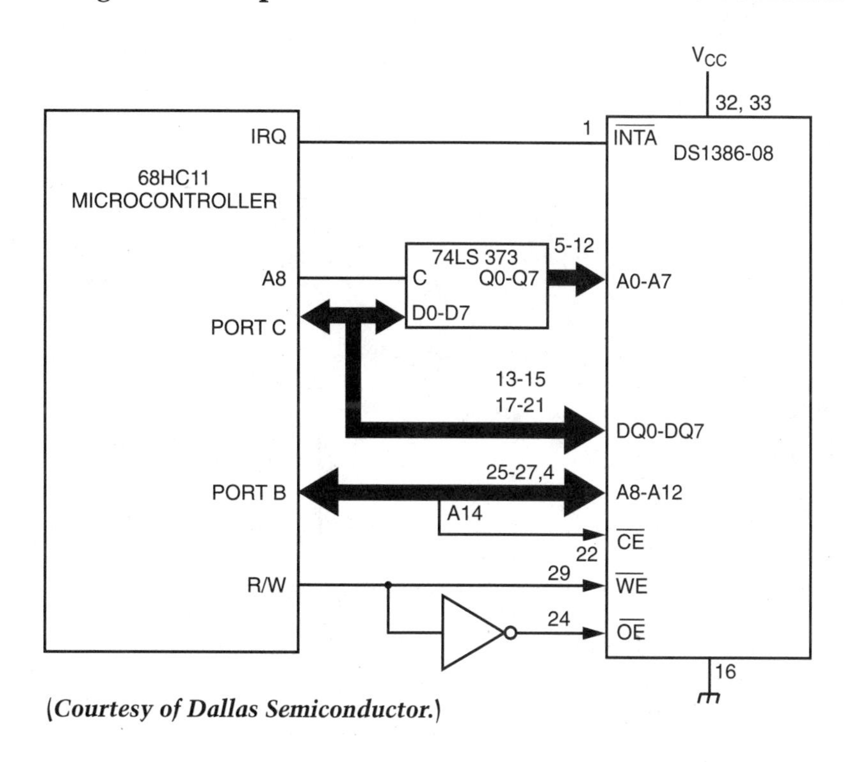

(*Courtesy of Dallas Semiconductor.*)

DS1620 digital thermometer/thermostat

The DS1620 is a digital thermometer/thermostat, giving 9-bit temperature readings as its output. It features three-alarm outputs and user-programmed switching temperatures, making the DS1620 well suited in thermostat as well as simple thermometer applications.

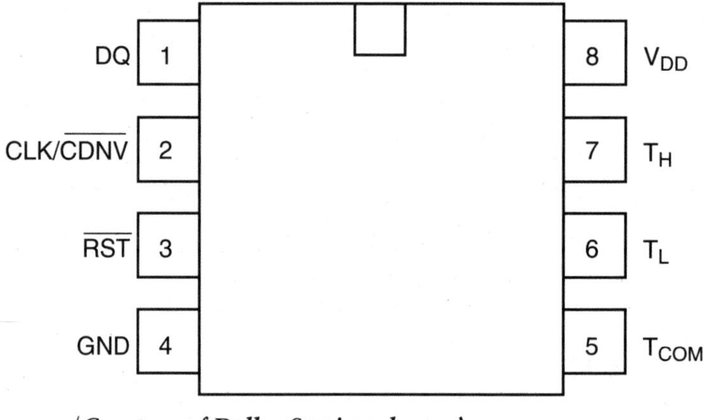

(*Courtesy of Dallas Semiconductor.*)

Fan driver

The fan is turned on when the sensed temperature exceeds the T_{high} trip-point programmed into the DS1620. Any resistive electrical load can be controlled by this circuit.

The drive transistor should be selected to handle the desired load. The 2N7000 suggested in the diagram would be suitable for driving a 12 volt, 100 mA fan.

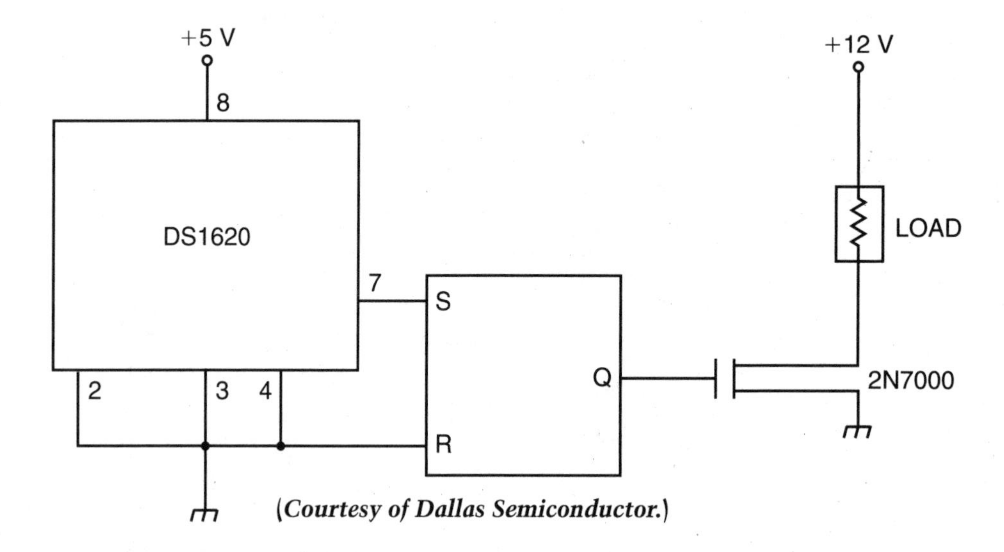

(*Courtesy of Dallas Semiconductor.*)

Fan driver

This alternate fan driver circuit uses the DS1620's T_{com} output, rather than the T_{high} output, which can simultaneously be utilized for some other application. Any resistive electrical load can be controlled by this circuit.

The drive transistor should be selected to handle the desired load. The 2N7000 suggested in the diagram would be suitable for driving a 12 volt, 100 mA fan.

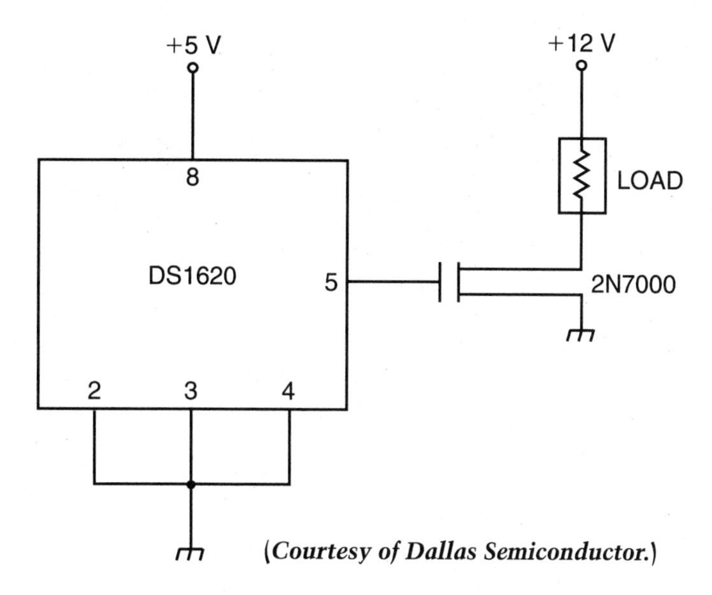

(*Courtesy of Dallas Semiconductor.*)

Fan controller with external latch

This fan driver/controller circuit uses an SR-type flip-flop to latch the fan on or off, to prevent too frequent false switching when the sensed temperature is very close to the switch point. In other words, this is a fan driver circuit with hysteresis. The fan is switched on when the sensed temperature exceeds the preprogrammed T_{high} level. The fan will remain on, even if the sensed temperature drops below the T_{high} value. Only when the sensed temperature drops below the T_{low} level is the fan turned off. It will then remain off until the sensed temperature exceeds T_{high} again.

The drive transistor should be selected to handle the desired load. The 2N7000 suggested in the diagram would be suitable for driving a 12 volt, 100 mA fan. Any resistive electrical load can be controlled by this circuit.

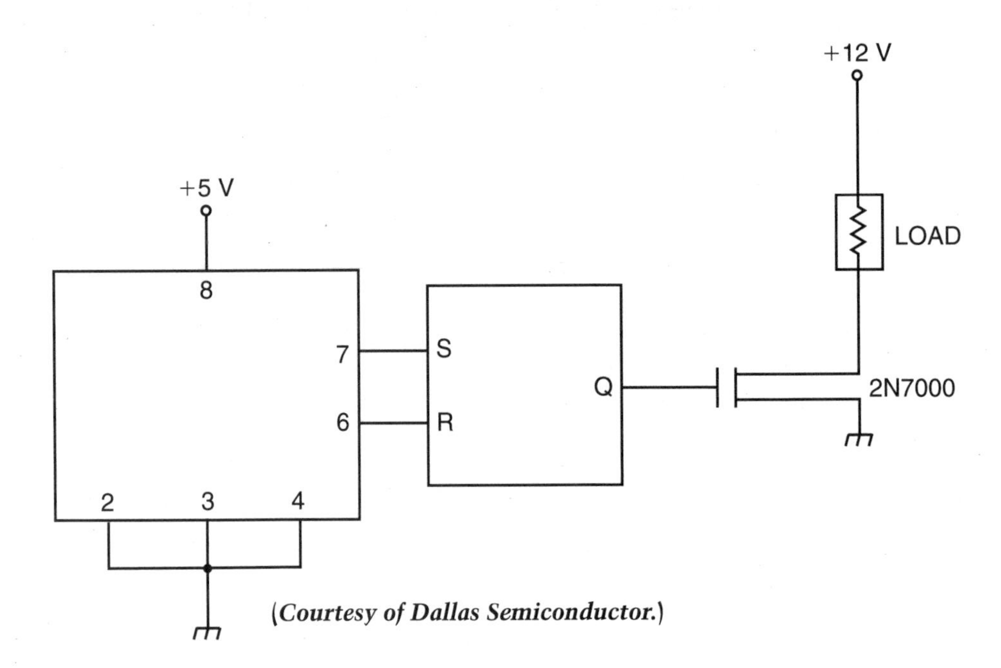

(*Courtesy of Dallas Semiconductor.*)

Heater driver

The heater is turned on when the sensed temperature drops below the T_{low} trip-point programmed into the DS1620. Any resistive electrical load can be controlled by this circuit.

The drive transistor should be selected to handle the desired load.

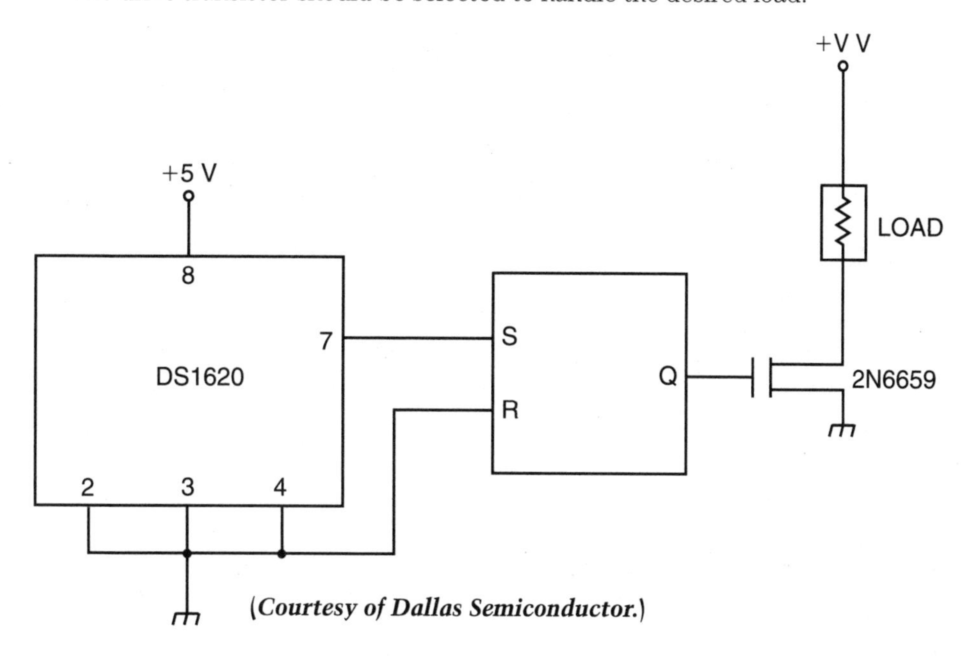

(*Courtesy of Dallas Semiconductor.*)

MC14490 HEX contact bounce eliminator

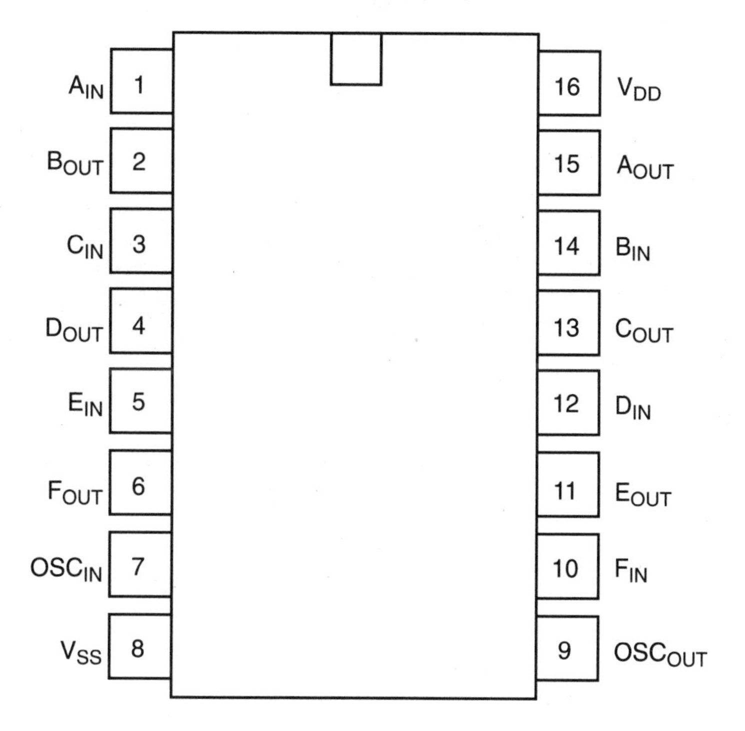

Latched output circuit

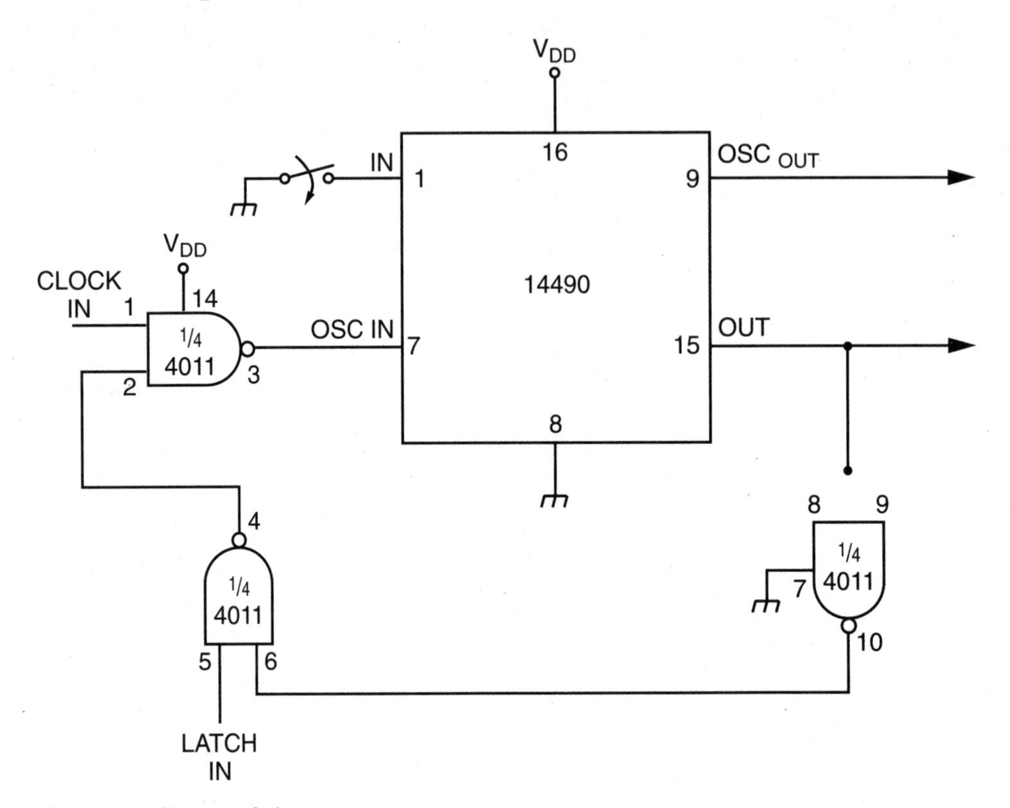

(*Courtesy of Motorola.*)

Fast attack/slow release circuit

This circuit is useful in application where different leading and trailing edge delays are required. The box labelled "-N" is a suitable divider/counter stage, with the value of N determining the difference between the attack and release times to be exhibited by this circuit.

To get the opposite response, i.e., a slow attack, and fast release, simply reverse the connections marked "A" and "B" in the diagram.

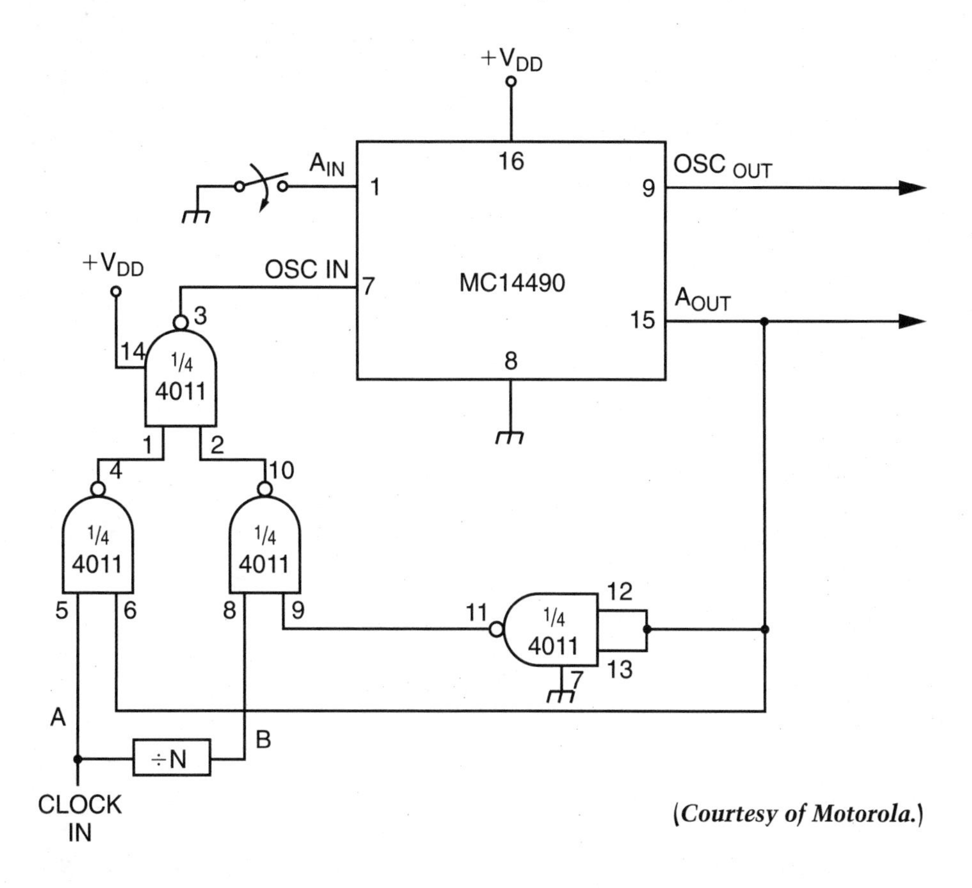

(*Courtesy of Motorola.*)

4
SECTION
Operational amplifiers and comparators

One of the most common elements in modern linear electronic circuits is the operational amplifier, or *op amp*. At one time, this was a fairly esoteric and uncommon type of circuit, used only in highly sophisticated analog computer circuits. It was expensive and difficult to design and build op amps using discrete components. The advent of the integrated circuit changed all that. Today, the op amp is probably the most widely used single type of linear IC around.

Very simply, an operational amplifier is an analog amplifier with two inputs. The signal at one input is inverted, while the signal to the second input is noninverted.

Assuming there is no feedback path from output to input, the theoretical open-loop gain of an operational amplifier is infinite. Since the output voltage can't exceed the supply voltage, a very, very small differential input signal can saturate the output, so in most cases, we can reasonably assume the open loop gain is functionally equivalent to infinity.

In most (but not all) practical op amp applications, a lower gain would be desired, if not essential. This can be accomplished by connecting a resistance element in a feedback path between the output and the inverting input, creating a closed negative feedback loop. The closed-loop gain will be determined by the feedback resistance, and the basic circuit configuration. Using an ordinary fixed resistor, the gain will be linear, and frequency independent. However, other, nonlinear components can be used in the feedback path to create special responses. For example, a capacitor in the feedback loop will result in a gain that varies with signal frequency (useful in filter and oscillator circuits). A transistor in the feedback path will give a logarithmic gain response curve, rather than the straight-line linear response of a simple resistor.

Countless circuits have been designed around operational amplifiers.

To avoid unnecessary clutter in the schematics and possible confusion, the power supply connections are not shown in most of the circuit diagrams in this section. Of course, each op amp must be properly powered to function. The power supply connections must always be assumed, if not explicitly shown.

TL062 dual low-power JFET input operational amplifier

The TL062 contains two JFET input type operational amplifiers designed for low power applications. The supply current drawn by this device is typically just 200 µA per amplifier.

The operational amplifiers in the TL062 offer very good specifications, including high slew rate, gain bandwidth product, input impedance, and output voltage swing, as well as low input bias and input offset currents.

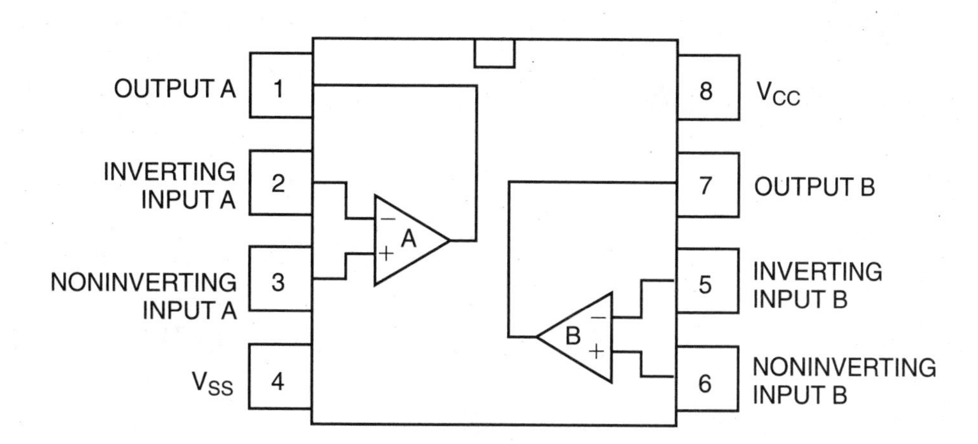

Fast summing amplifier

The output will be the inverted sum of all three inputs. This circuit can easily be adapted to include more or fewer inputs. If all input resistors have equal values, all of the input signals will see the same amount of gain, and will be equally represented in the output. Nonequal input resistors can be used to weigh the inputs, so some inputs will have a greater influence on the output than others. Using the component values shown in the diagram, each input will be subjected to a gain of 2. Changing the value of the feedback resistor will affect all of the inputs proportionately.

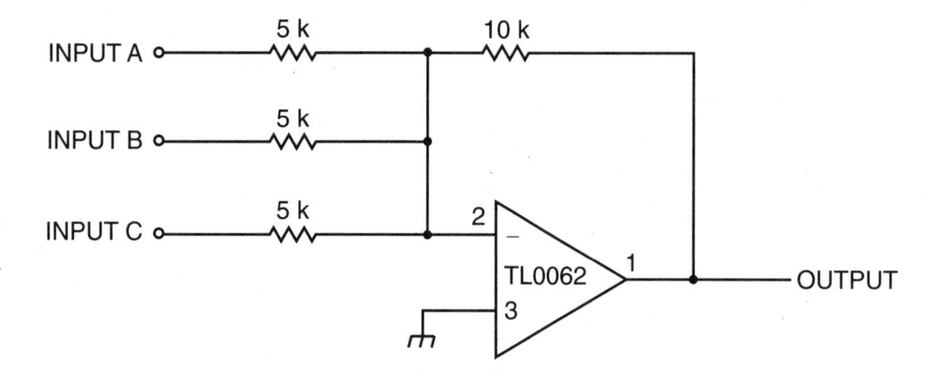

Differential amplifier

The output voltage equals the difference between the two input voltages. Using the resistor values shown in the diagram, the amplifier's gain is unity. To change the gain equally for both inputs, change the value of the feedback resistor. To weight the inputs (so one has greater influence than the other), use unequal input resistor values.

The capacitors in the circuit add stability, particularly at high frequencies. For many applications, these capacitors can be omitted without affecting the circuit's operation.

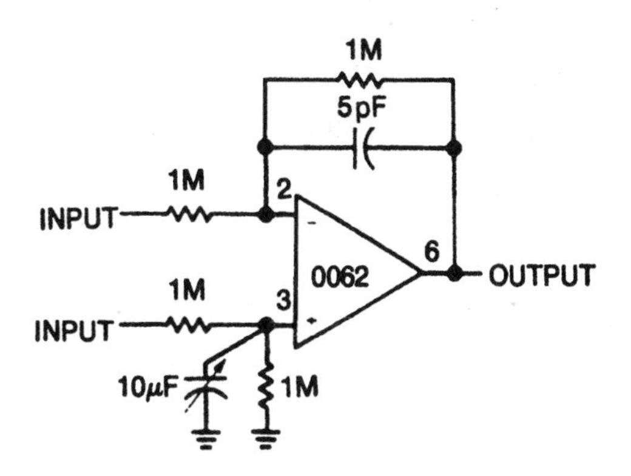

Fast precision voltage comparator

The output of this circuit is TTL compatible.

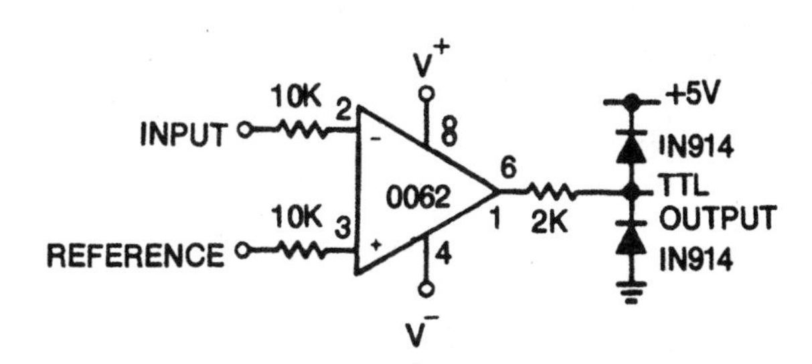

Fast voltage follower

The output voltage equals the input voltage. This circuit is used in buffering, isolation, and impedance matching applications. The small feedback capacitor gives a very fast response to changes in the input signal.

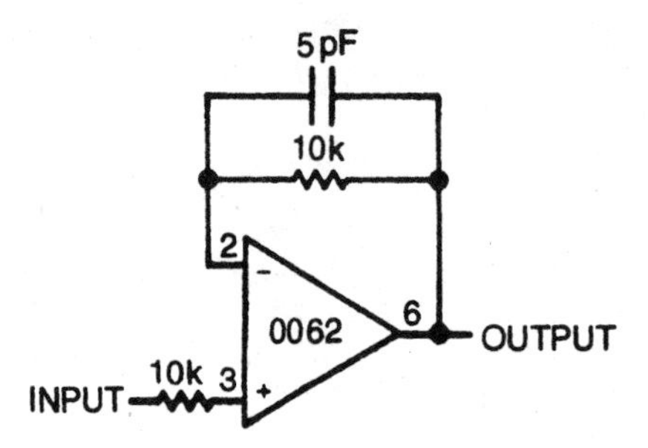

Fast inverting dc amplifier

The output voltage has the opposite polarity as the input voltage. That is, a positive input voltage results in a negative output voltage, and vice versa. The gain is equal to the value of the feedback resistor (RB) divided by the value of the input resistor (RA).

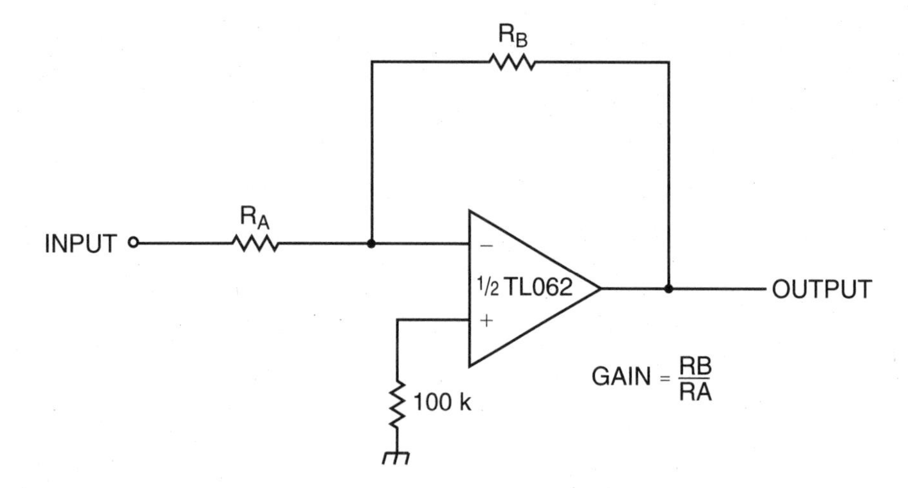

Low-input impedance preamplifier

This circuit is designed to boost very low-level input signals to a more usable amplitude. A low-noise op amp device such as the TL062 is absolutely essential in this type of application. If a lower grade op amp is used, the desired input signal may be partially lost under the internal noise generated by the op amp itself.

Notice that the input signal is not referenced to ground in this circuit. It is fed across both the inverting and noninverting inputs. The output signal, however, is referenced to ground.

The 100K potentiometer in the feedback loop controls the amplifier's gain. Using the component values shown, gains of from a little over unity up to about 100 are possible. The 1K resistor in series with the potentiometer prevents the feedback resistance from being set to zero. If the intended application doesn't require a volume control, a fixed gain level can be set by using an appropriate fixed value resistor in the feedback path.

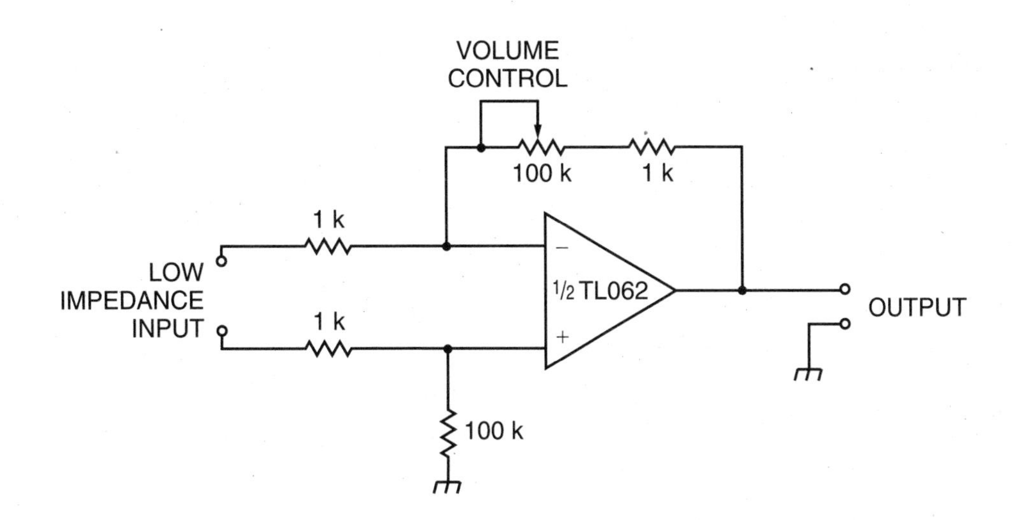

Dynamic microphone preamplifier

This preamplifier circuit is designed for use with a dynamic microphone. It raises the relatively low input signal from the microphone to a more usable amplitude. Further amplifier stages will undoubtedly be used in most applications.

A small speaker can be used as a dynamic microphone in noncritical applications.

The 500K potentiometer in the feedback loop functions as a volume control. If this adjustability is not required in the intended application, a single fixed resistor of an appropriate value can be substituted in the op amp's feedback path.

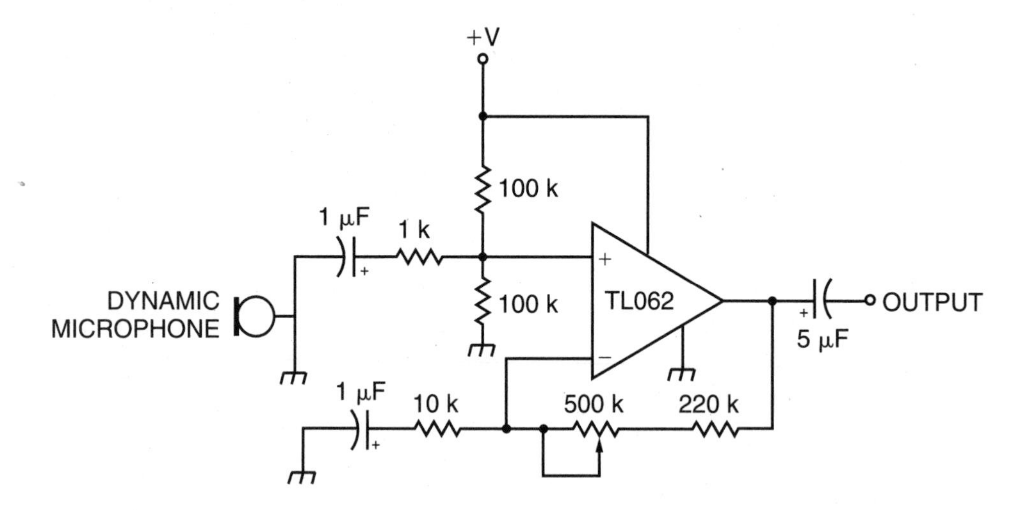

124 quad operational amplifier

This chip contains four independent standard grade operational amplifiers in a single package, convenient for medium to high density circuits. Only the power supply connections are common to all four operational amplifiers on this chip. Each operational amplifier in this IC is roughly similar to the 741, although some of the specifications aren't quite as good.

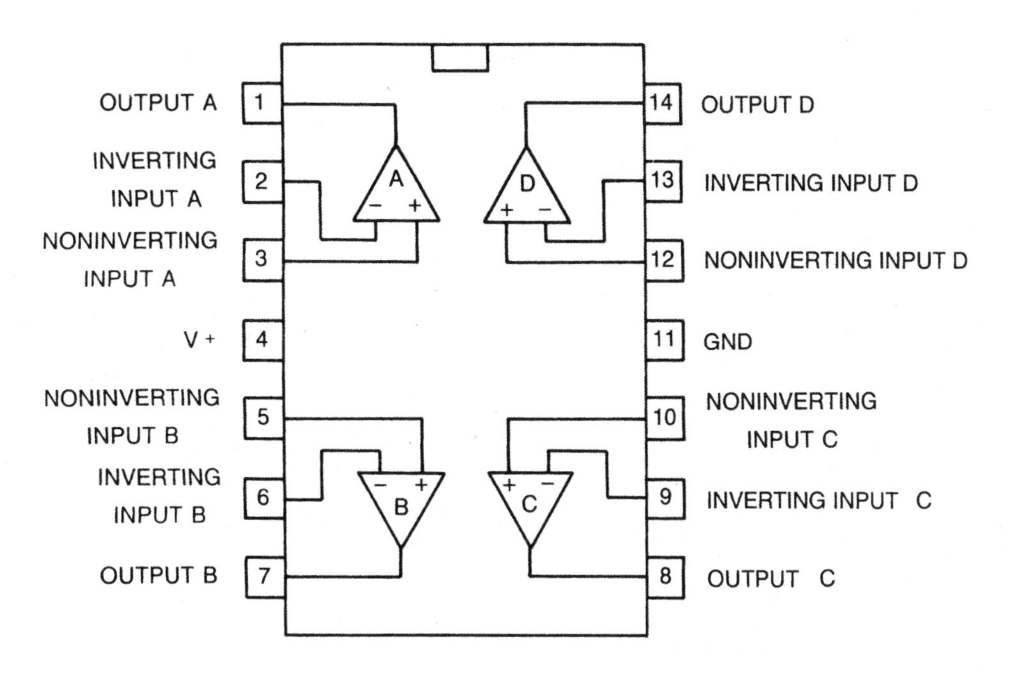

Voltage follower

This simple circuit is also known as a unity-gain buffer amplifier. The output signal will be at the same amplitude and polarity as the input signal.

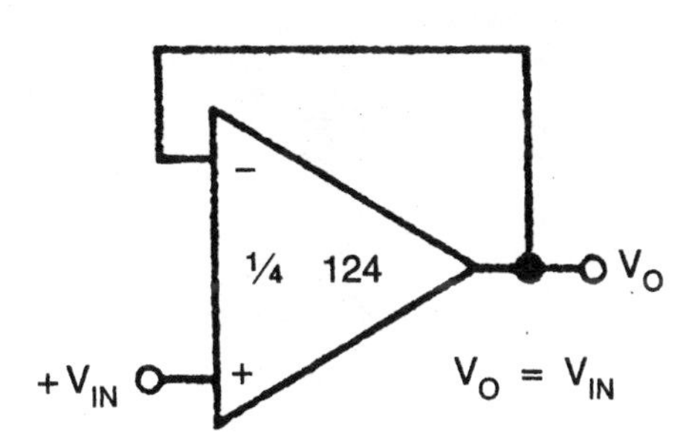

Comparator with hysteresis

The built-in hysteresis in this circuit prevents output chatter when the input voltage is very close to the reference voltage, and there is the possibility of noise, distortion, or other minor fluctuations in the input signal line.

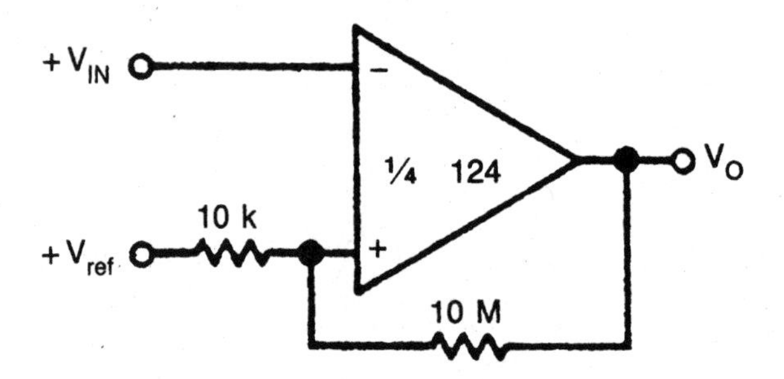

Current monitor

The NPN transistor in this circuit should be selected to suit the intended application. Almost any NPN transistor should work in this circuit.

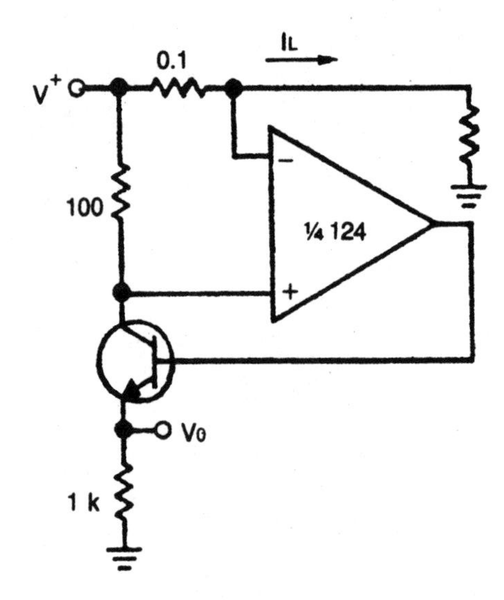

DC summing amplifier

The sum of input voltages V_3 and V_4 is subtracted from the sum of input voltages V_1 and V_2. Using the component values shown here, all inputs will be given the same unity gain. If the intended application calls for weighting of one or more inputs, use appropriate different values for the input resistors.

This circuit can be easily expanded for additional noninverting and/or inverting inputs, simply by adding more input resistors in parallel with those shown here.

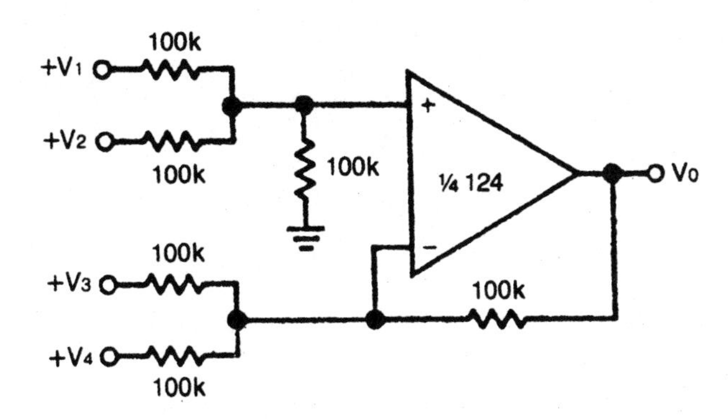

Full-wave rectifier and averaging filter

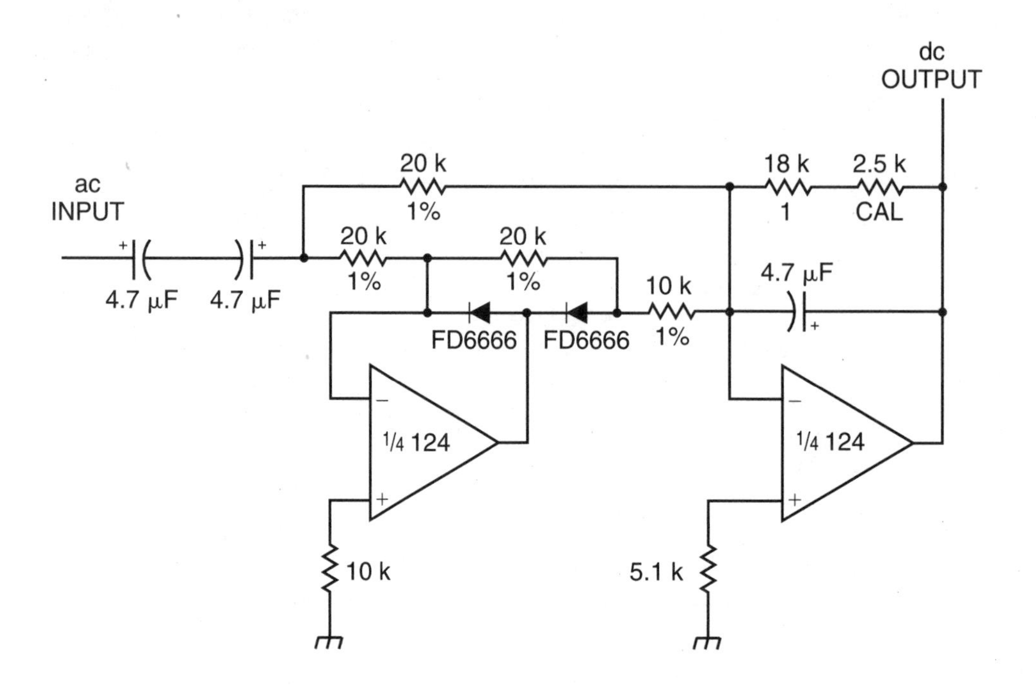

Power amplifier

Additional amplification is provided by the NPN transistor, which is selected to suit the intended application.

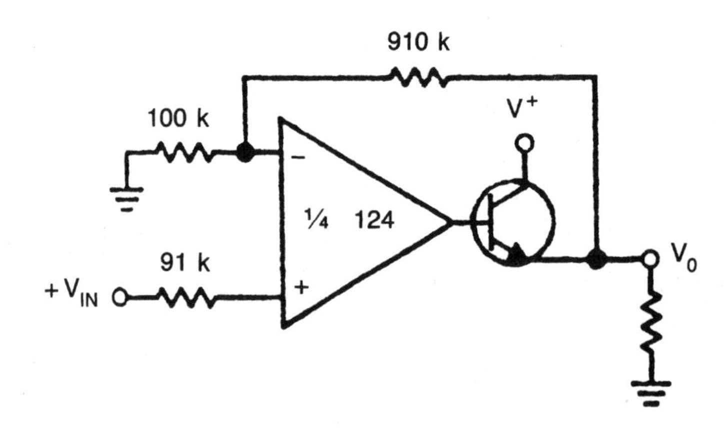

Lamp driver

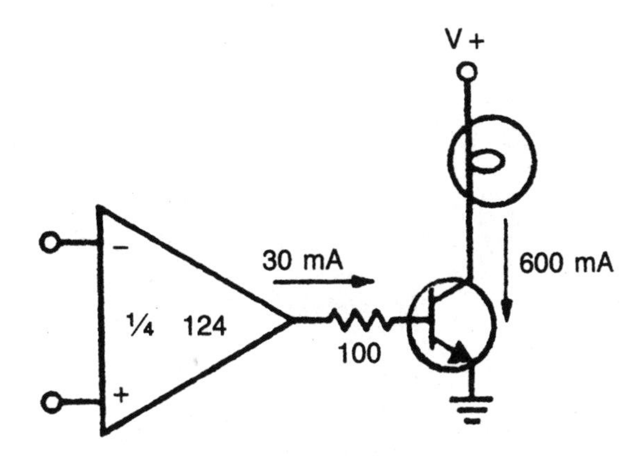

LED driver

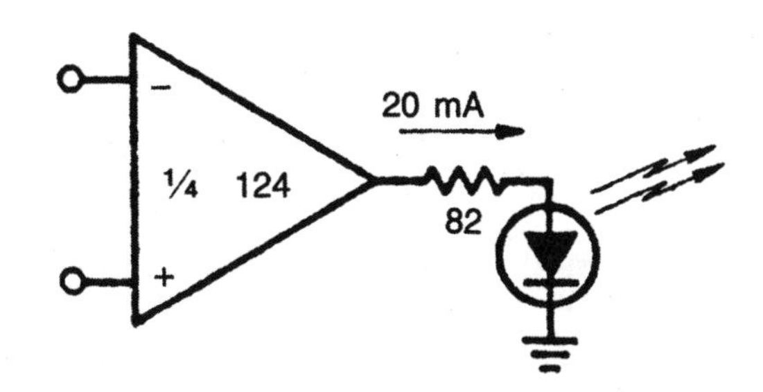

Root extractor

The output voltage of this circuit is equal to the square root of the input voltage.

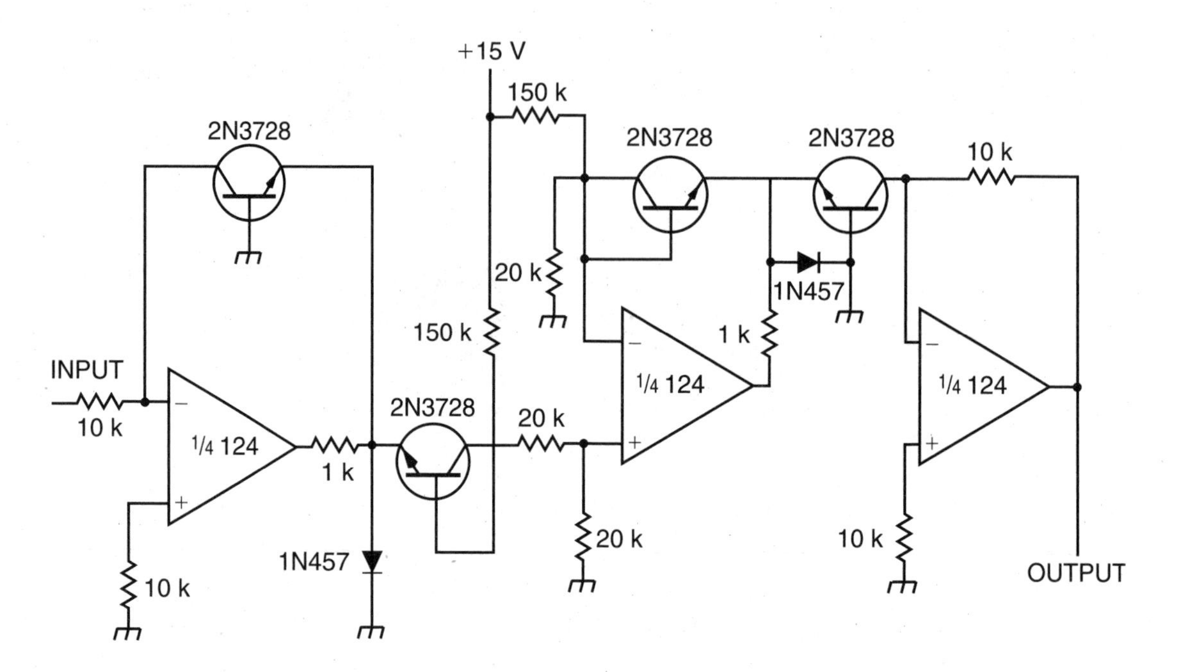

Driving TTL

A pull-down resistor permits an operational amplifier to drive a TTL gate. Of course, the input signal must be some form of rectangle wave or dc voltage at the levels recognized as HIGH and LOW by the digital gates. No surrounding circuitry is shown about the operational amplifier here. The feedback path and other components used with the op amp will depend on the intended application, of course.

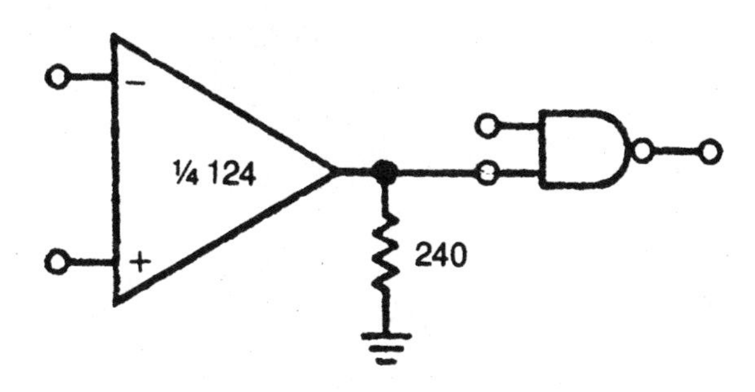

Square-wave oscillator

Changing the value of the feedback resistor, or the capacitor in this circuit, will alter the frequency of the output signal. The output is in the form of a square wave, with a 1:2 duty cycle.

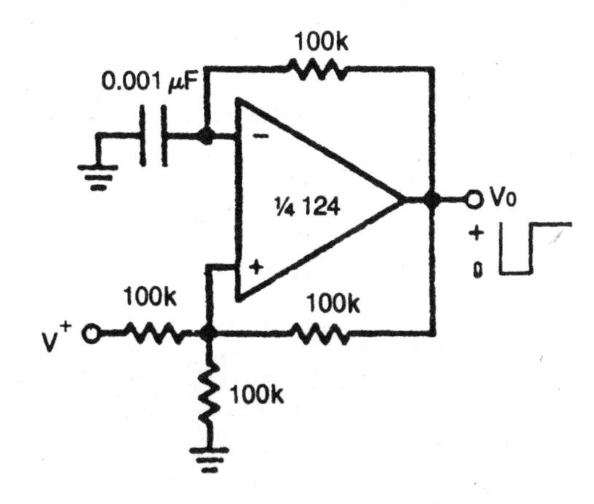

Pulse generator

This simple modification of the square-wave oscillator circuit on page 165 generates a nonsymmetrical rectangle or pulse wave, with a duty cycle other than 1:2. The two diodes set up two different feedback paths for the HIGH and LOW portions of the cycle. Each part of the cycle sees a different feedback resistance, so it has a different timing period.

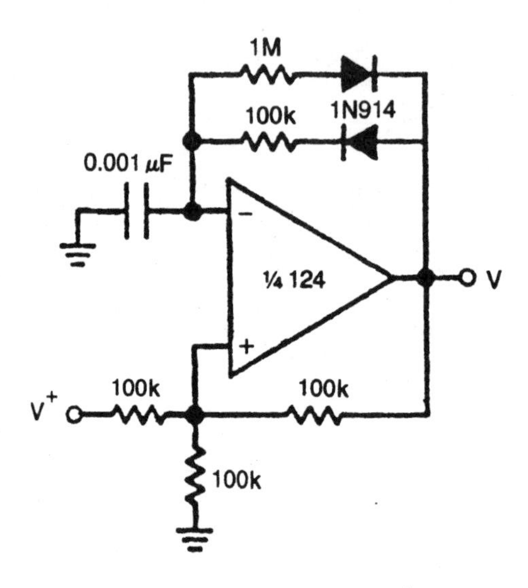

Alternate pulse generator

This pulse generator circuit is a little simpler and cruder than the one shown above, but it may be adequate for some applications, where it is important to keep the component count done, for whatever reason.

There is less control over the duty cycle and signal frequency in the circuit. The diode permits current to flow in only one direction, so one portion of the cycle essentially sees an infinite (actually very high) feedback resistance.

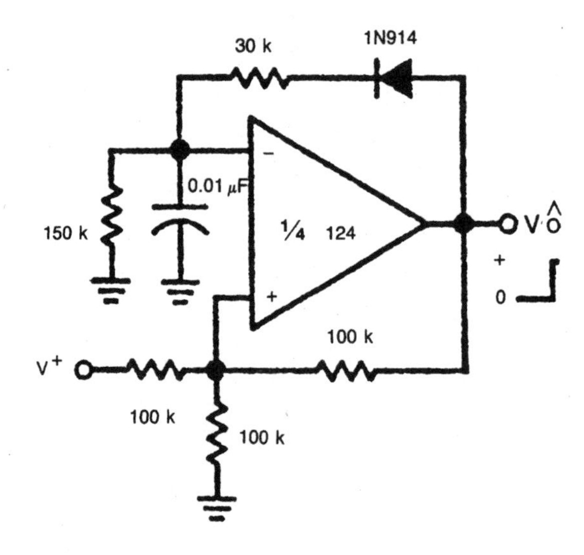

Voltage controlled oscillator

The frequency of the output signal(s) is directly proportional to the level of the control voltage fed to the circuit's input. Usually a dc control voltage will be used, but a very low-frequency ac control signal will produce a vibrato (fluctuating frequency) effect in the output. If the control voltage input has a frequency above about 10 Hz, frequency modulation will occur, and sidebands will appear in the output signal(s).

This VCO circuit has two outputs. One produces a square wave, and the other generates a somewhat modified triangle wave. Both outputs always have identical frequencies and phase.

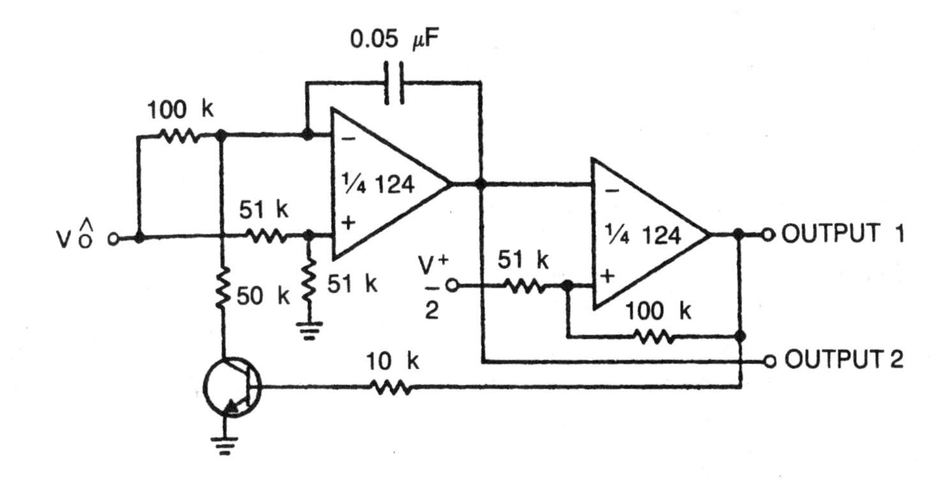

Photovoltaic cell amplifier

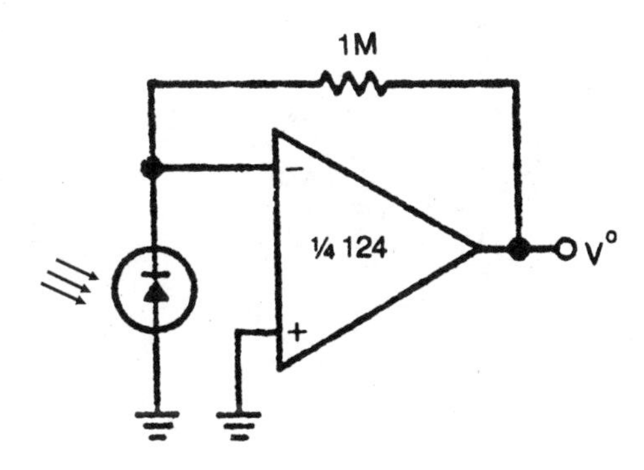

AC-coupled inverting amplifier

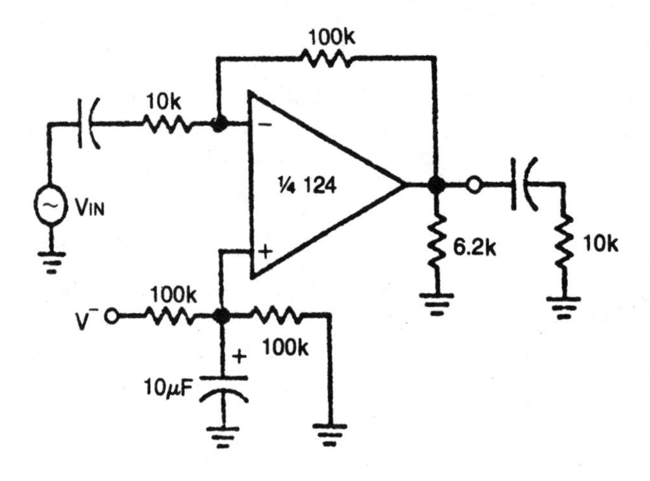

AC-coupled noninverting amplifier

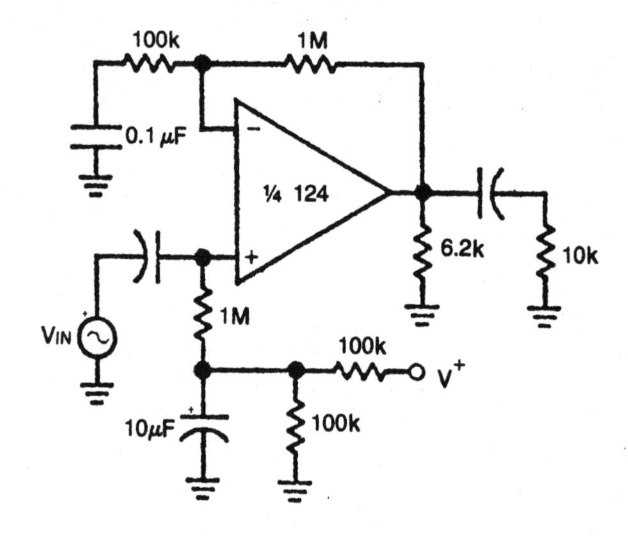

High-input impedance adjustable gain dc instrumentation amplifier

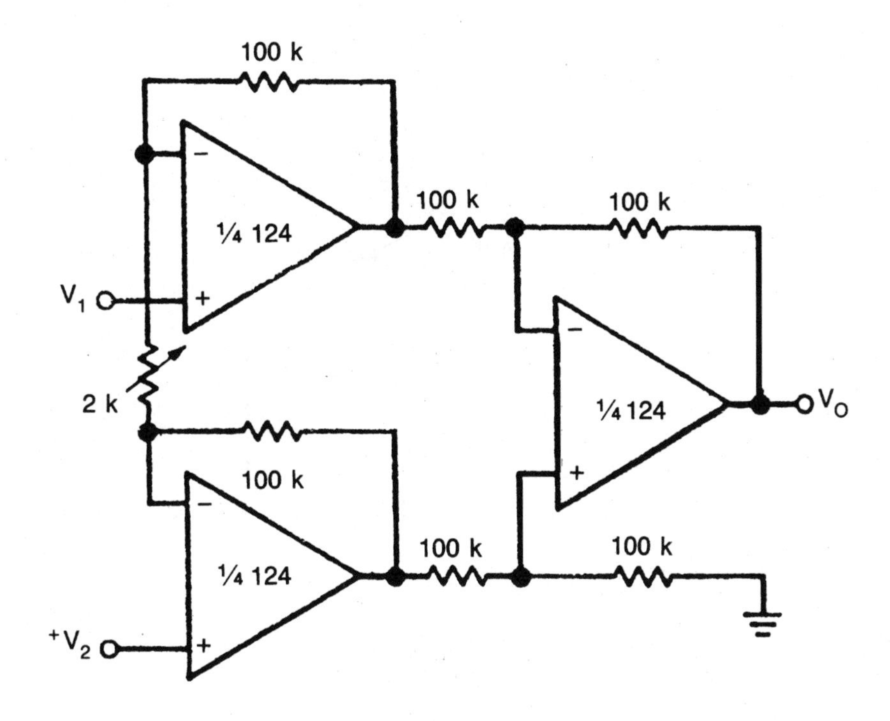

High-input impedance dc differential amplifier

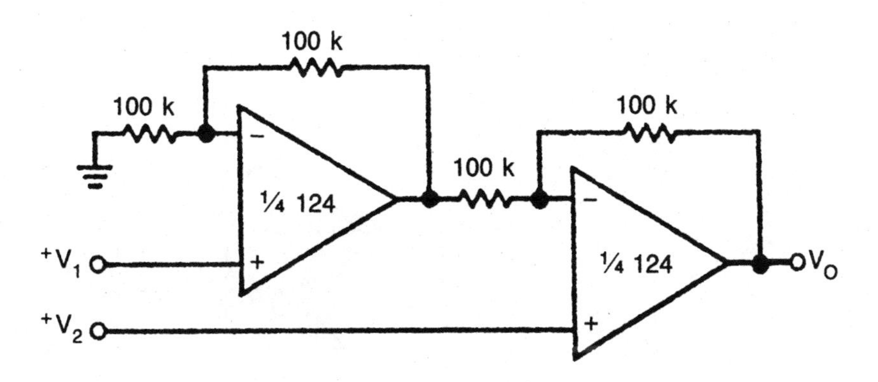

Low drift peak detector

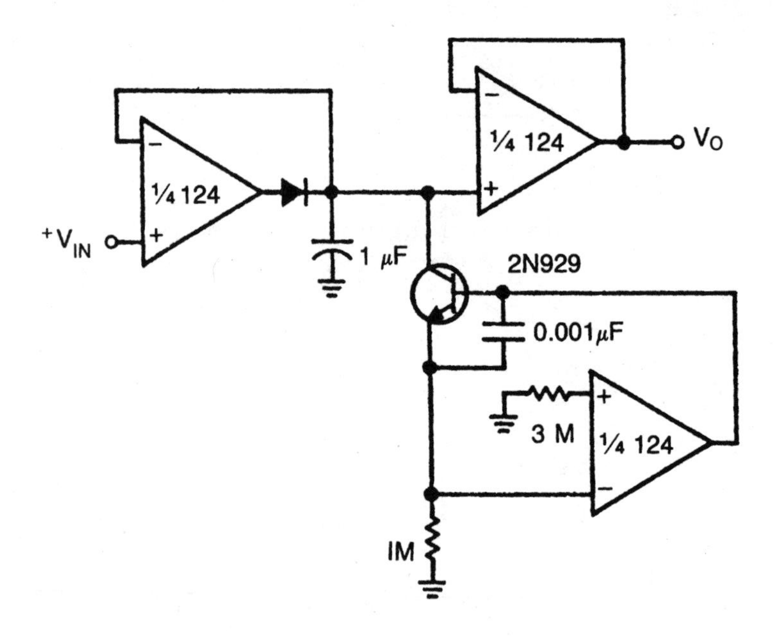

DC-coupled low-pass RC active filter

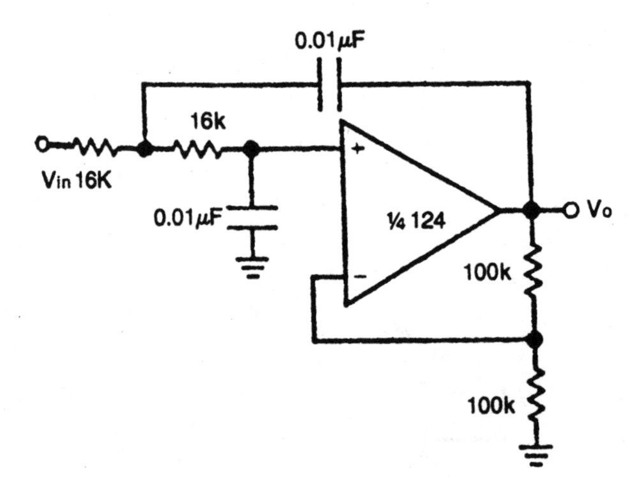

Bandpass active filter

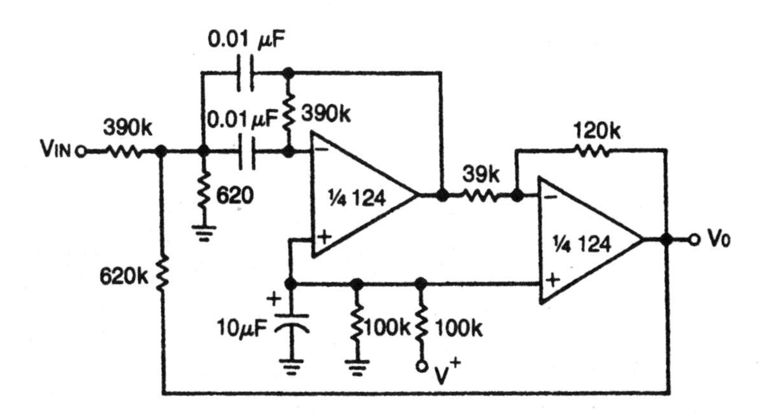

"BI-QUAD" RC active bandpass filter

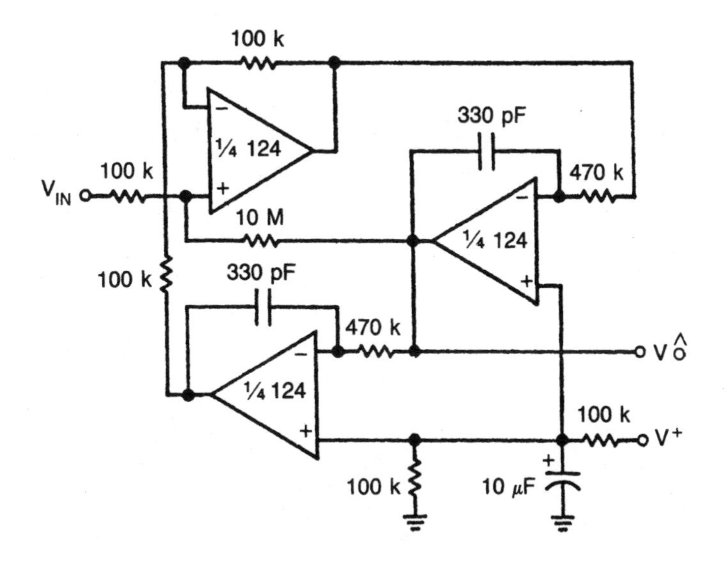

324 quad operational amplifier

The 324 is an improved version of the 124. This chip contains four independent standard grade operational amplifiers in a single package, convenient for medium- to high-density circuits. Only the power supply connections are common to all four operational amplifiers on this chip. Each operational amplifier in this IC is roughly similar to the 741, with similar specifications.

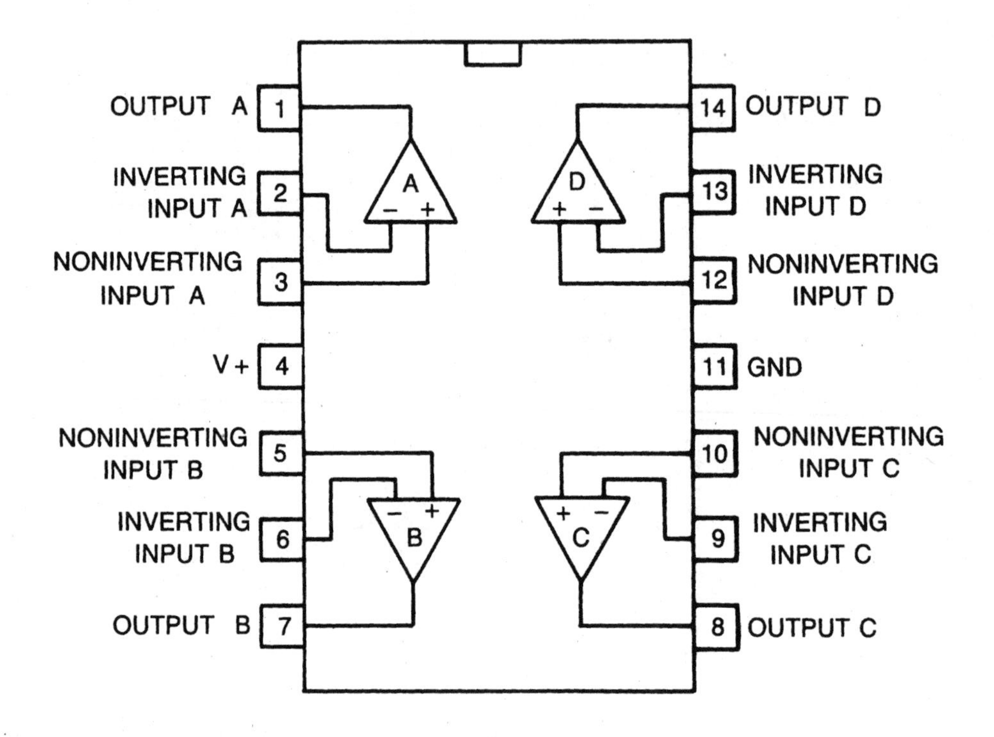

Voltage follower

This simple circuit is also known as a unity-gain buffer amplifier. The output signal will be at the same amplitude and polarity as the input signal.

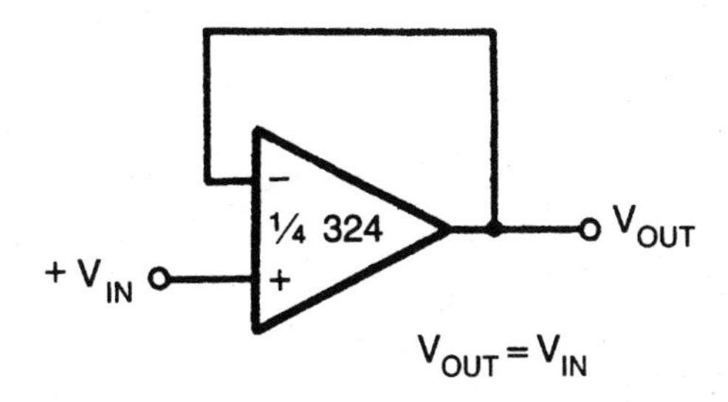

Driving TTL

A pull-down resistor permits an operational amplifier to drive a TTL gate. Of course, the input signal must be some form of rectangle wave or dc voltage at the levels recognized as HIGH and LOW by the digital gates. No surrounding circuitry is shown about the operational amplifier here. The feedback path and other components used with the op amp will depend on the intended application, of course.

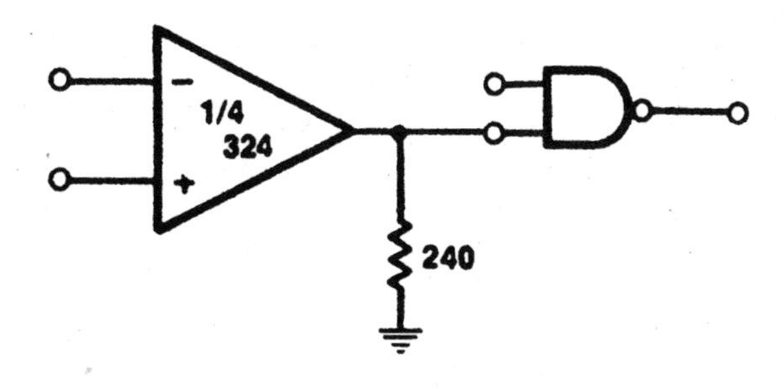

Pulse generator

The HIGH and LOW portions of the output waveform's cycle will be very unequal, because the diode in the feedback loop permits current to flow in only one direction, so one portion of the cycle essentially sees an infinite (actually very high) feedback resistance.

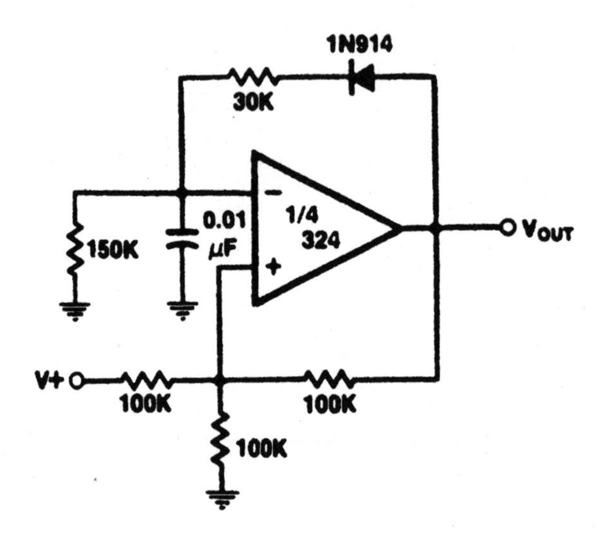

Pulse generator with presettable ON and OFF times

This simple modification of the pulse generator circuit on page 174 produces a non-symmetrical rectangle or pulse wave, with a duty cycle other than 1:2. There is more control over the actual duty cycle in this circuit. Both portions of the cycle can be preprogrammed by selecting appropriate resistor values. The two diodes set up two different feedback paths for the HIGH and LOW portions of the cycle. Each part of the cycle sees a different feedback resistance, so it has a different timing period.

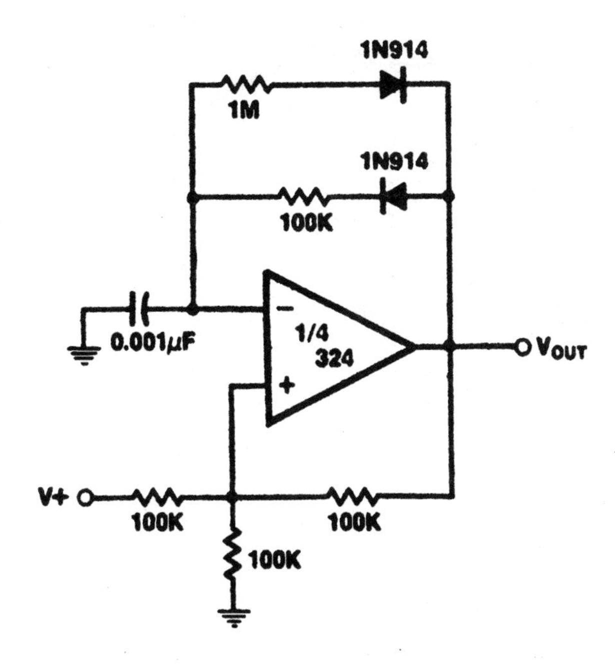

Square-wave oscillator

Since both the LOW and HIGH portions of the cycle pass through the same feedback resistor, they will have equal times, resulting in a true square-wave output with a 1:2 duty cycle.

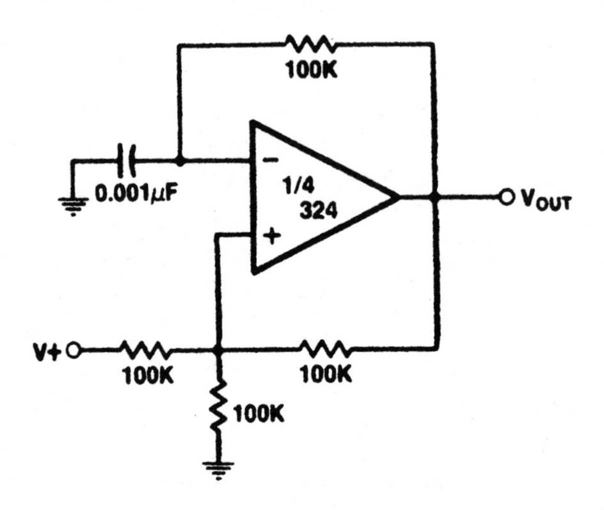

VCO

The output frequency of a voltage-controlled oscillator circuit like this one is directly proportional to the level of the control voltage fed into the circuit. Usually a dc control voltage will be used, but a very low-frequency ac control signal will produce a vibrato (fluctuating frequency) effect in the output. If the control voltage input has a frequency above about 10 Hz, frequency modulation will occur, and sidebands will appear in the output signal(s).

This VCO circuit has two outputs. One produces a square wave, and the other generates a somewhat modified triangle wave. Both outputs always have identical frequencies and phase.

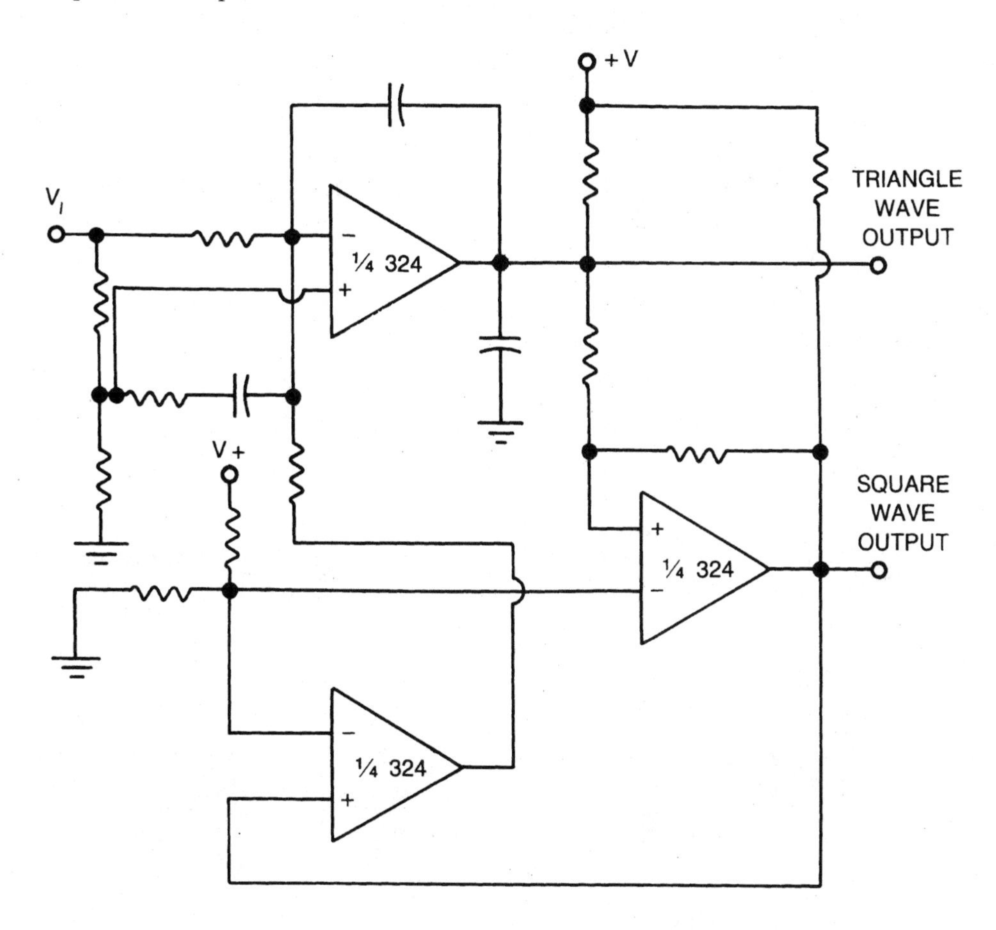

Analog multiplier/divider

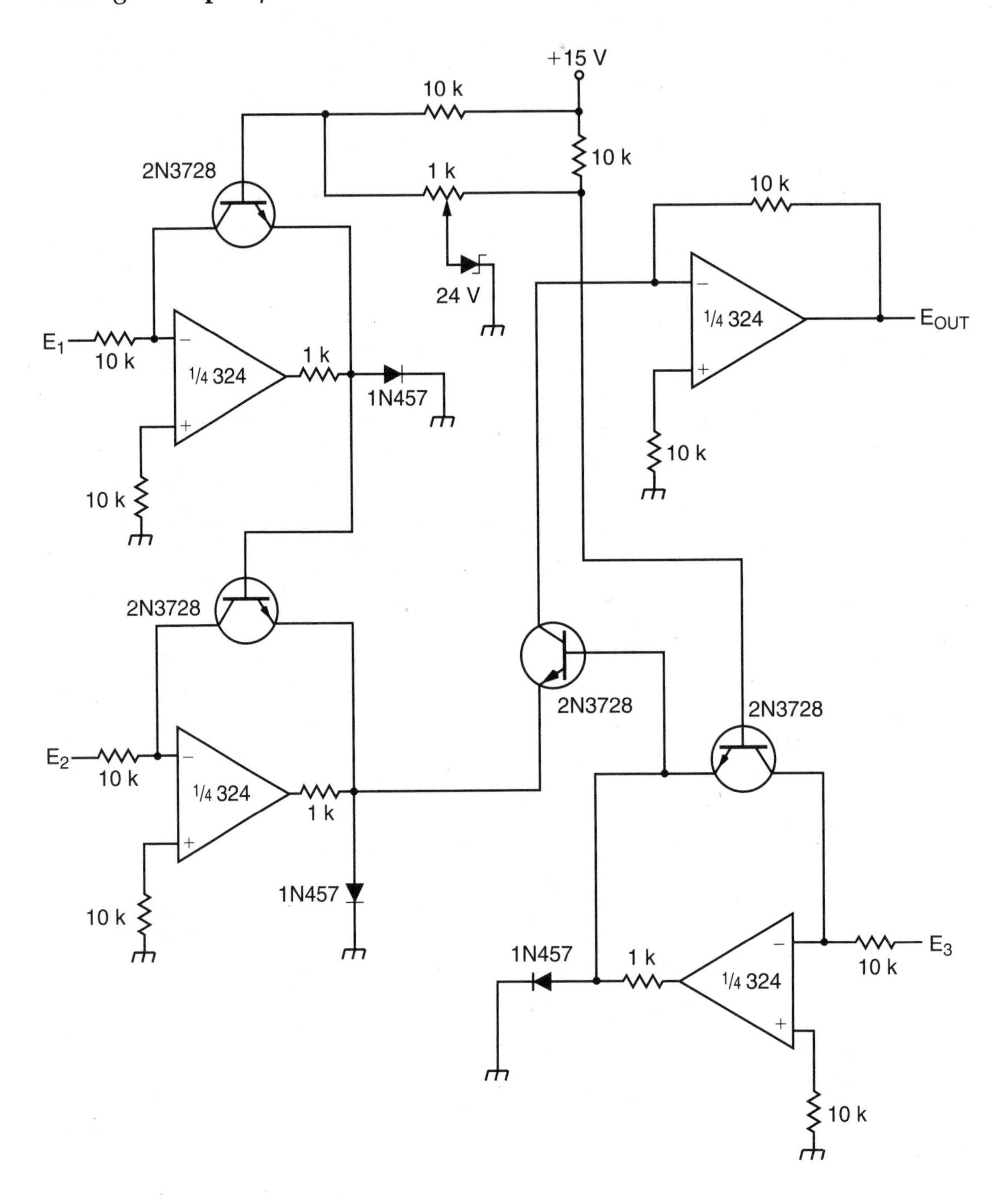

"Fuzz" circuit

This circuit adds a pleasant form of distortion to the output of an electric musical instrument, particularly an electric guitar.

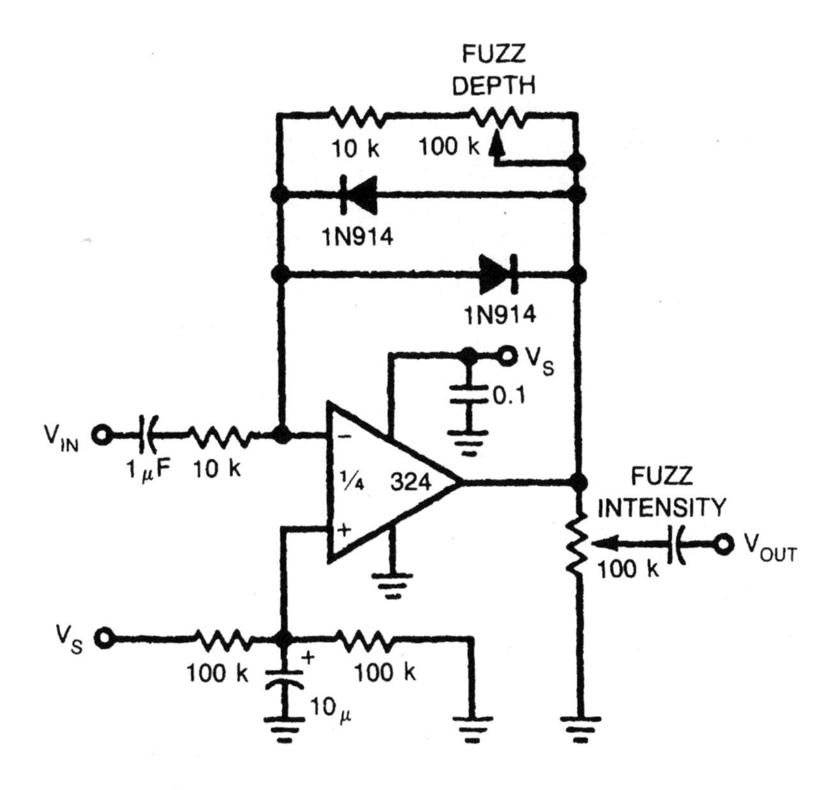

Tremolo circuit

Tremolo is a rapid fluctuation of volume, giving a fluttering effect to the output of an electric musical instrument, particularly an electric guitar. Almost any audio signal can be used as the input to this circuit and treated to the tremolo effects.

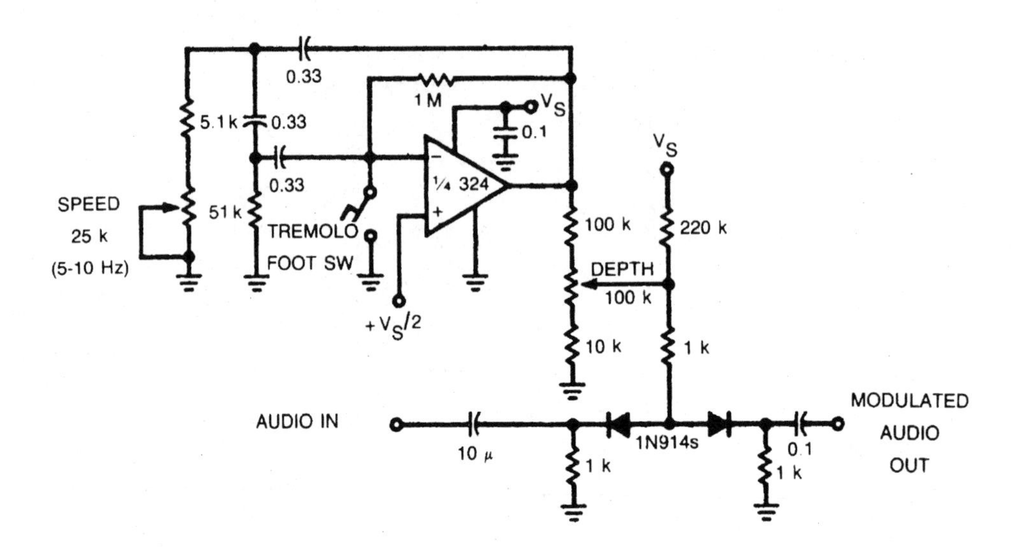

Volume expander circuit

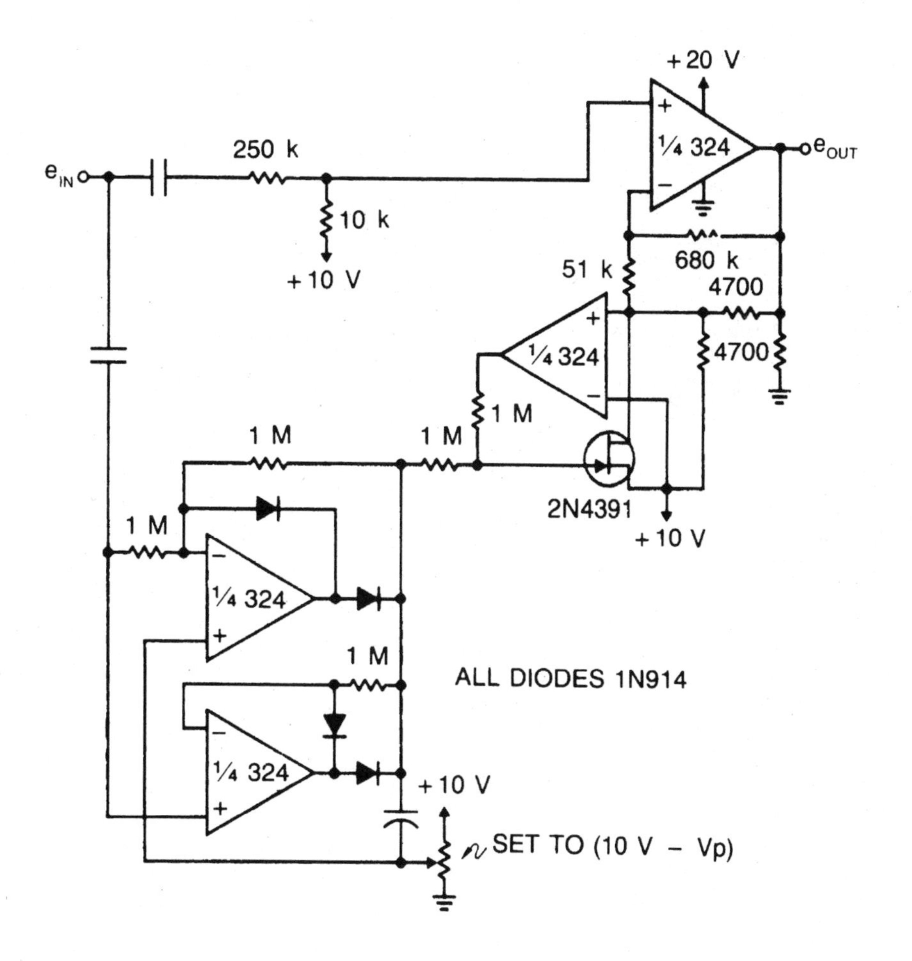

Ascending sawtooth-wave generator

The output waveform of this oscillator circuit begins its cycle at its lowest voltage level, then it smoothly builds up to its maximum voltage level, then it very quickly snaps back down to the minimum voltage level again, and starts the next cycle. Theoretically, this snap-back action is instantaneous.

Because of the appearance of this waveform on an oscilloscope, it is known as a sawtooth wave, or a ramp wave. The ramps climb in an ascending pattern.

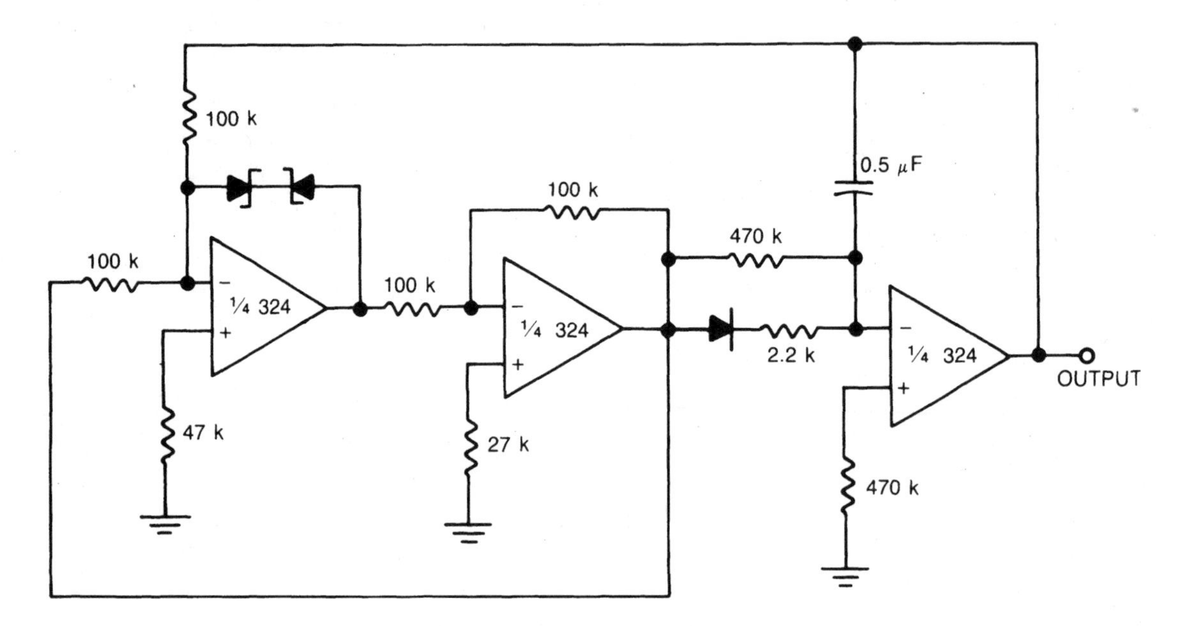

Descending sawtooth-wave generator

This circuit is a variation of the previous one. In this case, the output waveform begins its cycle at its highest voltage level, then it smoothly drops down to its minimum voltage level, then it very quickly snaps back up to the maximum voltage level again, and starts the next cycle. Theoretically, this snap-back action is instantaneous.

Because of the appearance of this waveform on an oscilloscope, it is known as a sawtooth wave, or a ramp wave. The ramps drop in a descending pattern.

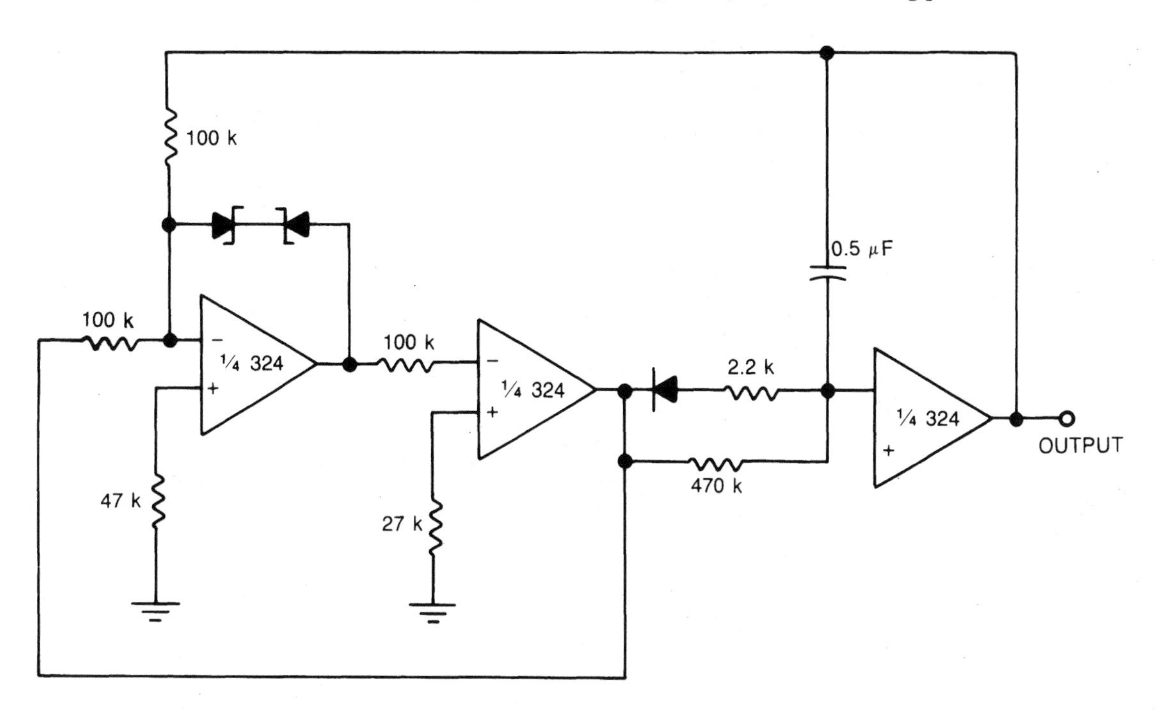

Dead space circuit

This unity-gain amplifier circuit reproduces its input signal at its output, except input signals close to zero are effectively ignored, or treated as if they were true zero volts. This leaves a dead zone in the circuit's response, as illustrated in the graph accompanying the schematic diagram.

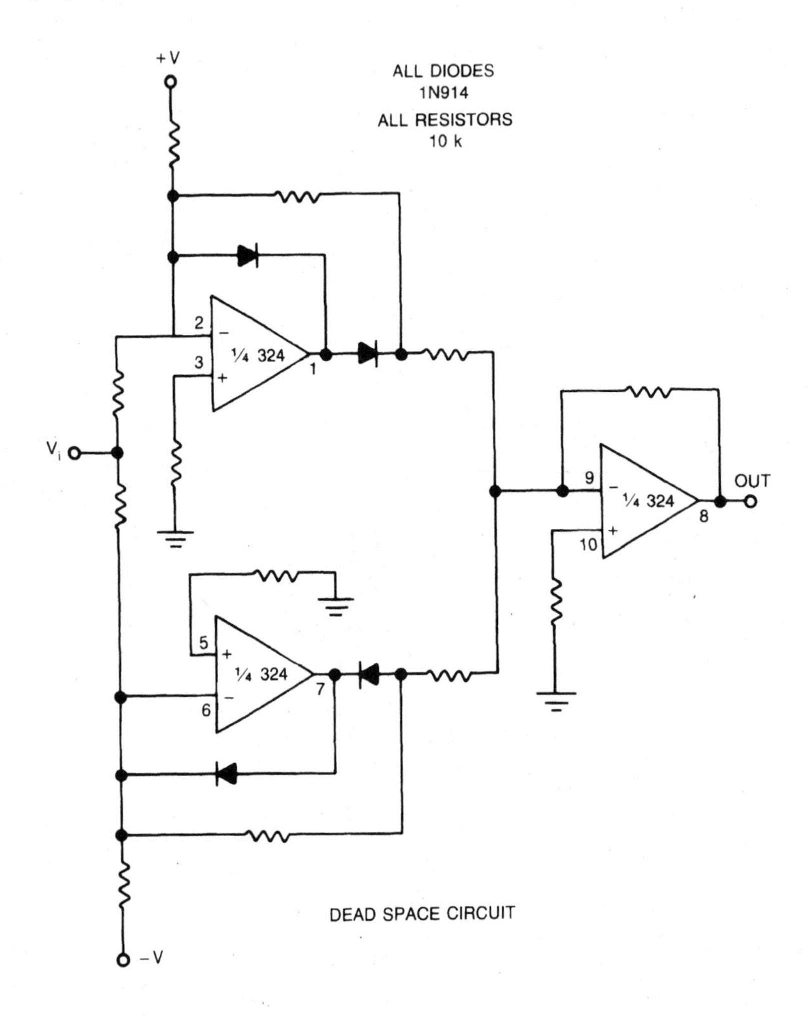

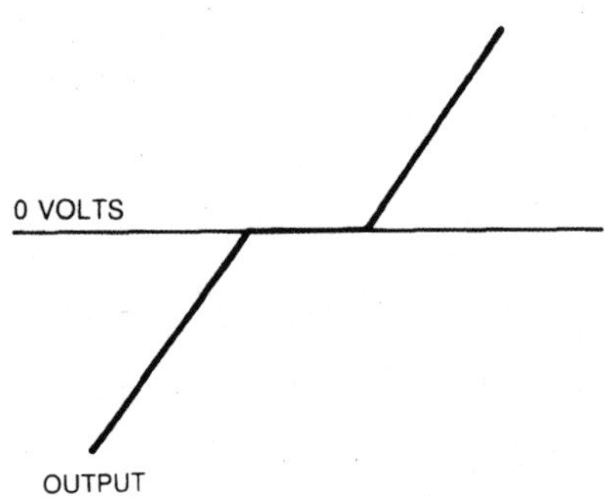

Series limiter

This unity-gain amplifier circuit linearly reproduces the input signal at the output, except when the input voltage goes too far positive or negative. Excessive input voltages are clipped, giving an output response like the one illustrated in the graph accompanying the schematic diagram.

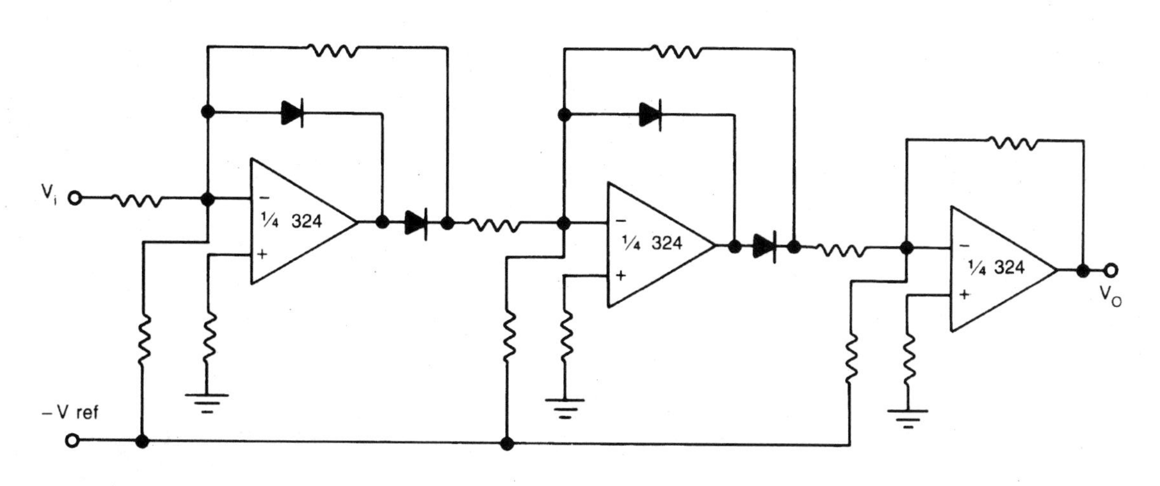

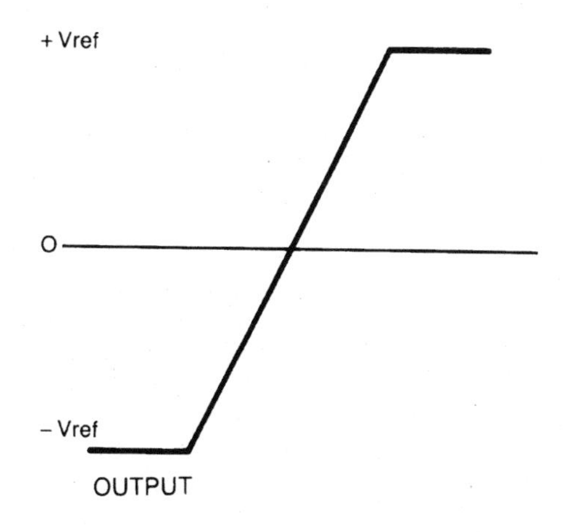

Precision current sink

The output current sinked by this circuit is equal to the input voltage, divided by the value of resistor R_1.

$$I_0 = \frac{V_{in}}{R_1}$$

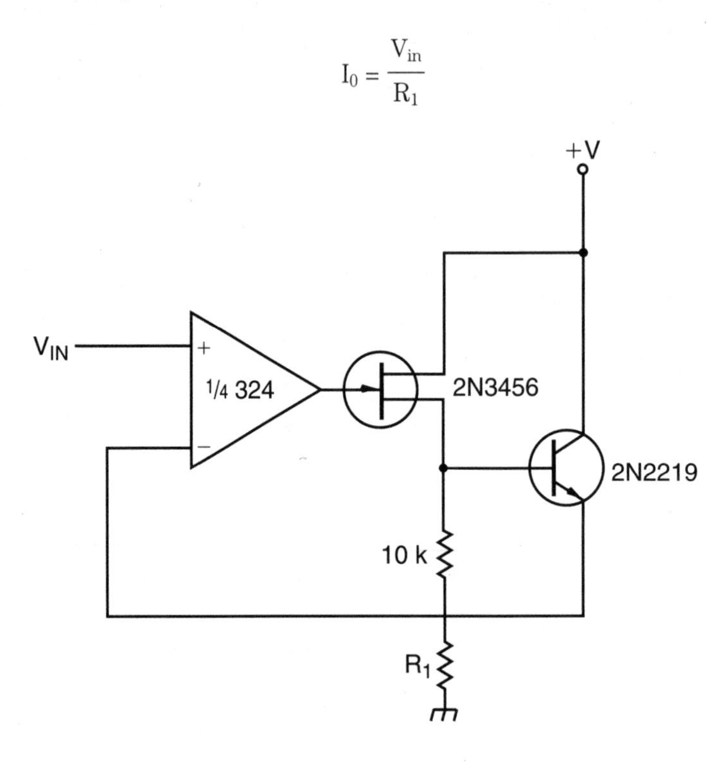

Precision current source

The output current sourced by this circuit is equal to the input voltage, divided by the value of resistor R_1.

$$I_0 = \frac{V_{in}}{R_1}$$

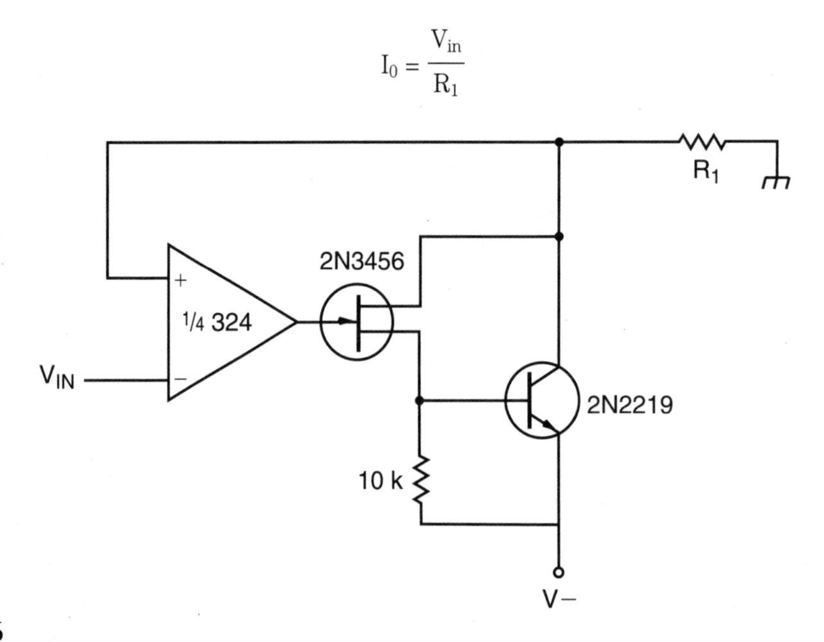

Boosted output amplifier

The pair of output transistors in this amplifier circuit give the output signal a lot of extra drive capability.

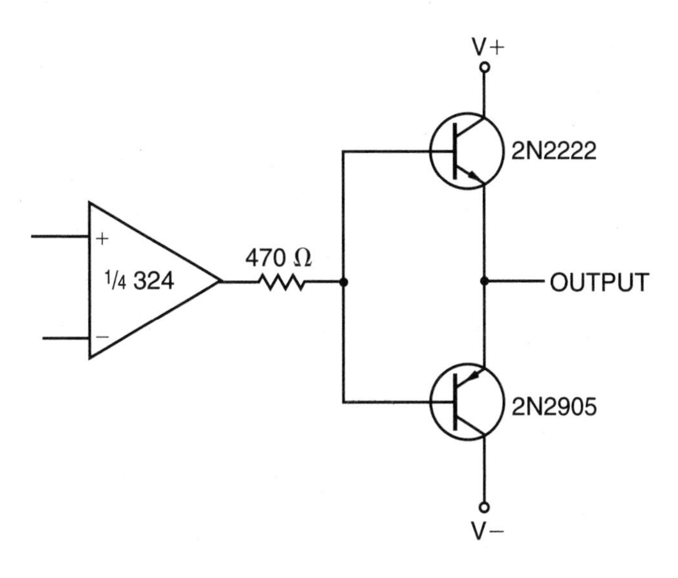

353 wide bandwidth dual JFET input operational amplifier

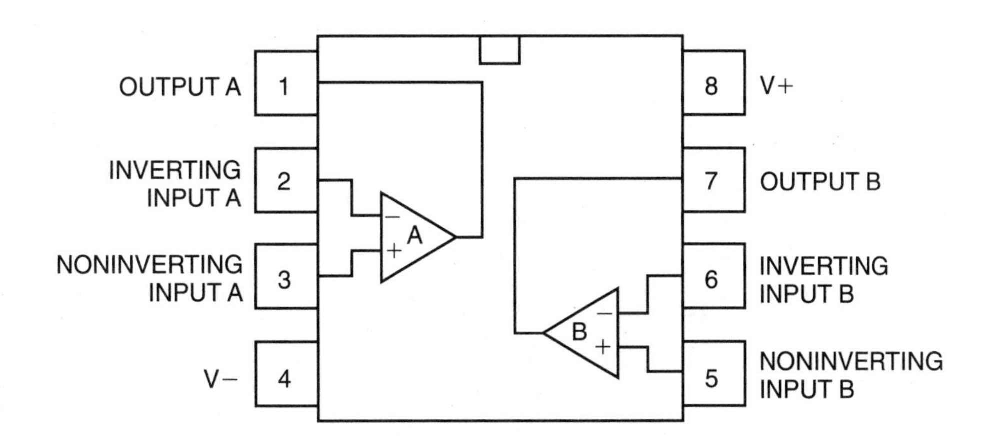

AC-coupled inverting amplifier

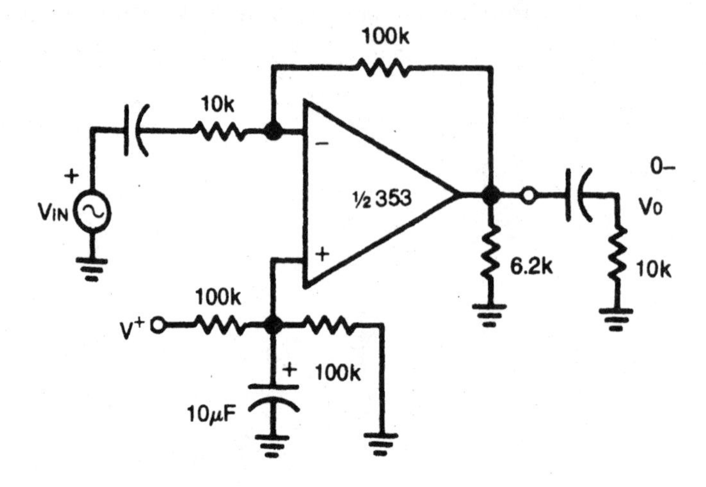

AC-coupled noninverting amplifier

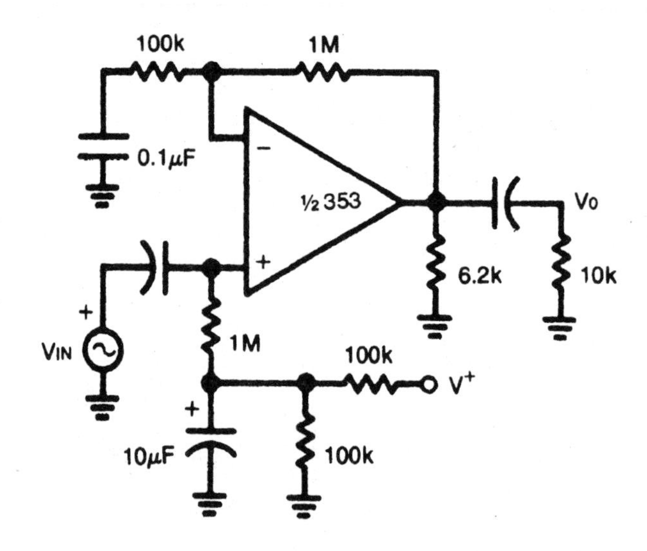

Ohms-to-volts converter

The output voltage of this circuit is proportional to the external resistance placed across its input terminals. The chief application for this circuit would be the ohm-meter section of a VOM or DMM.

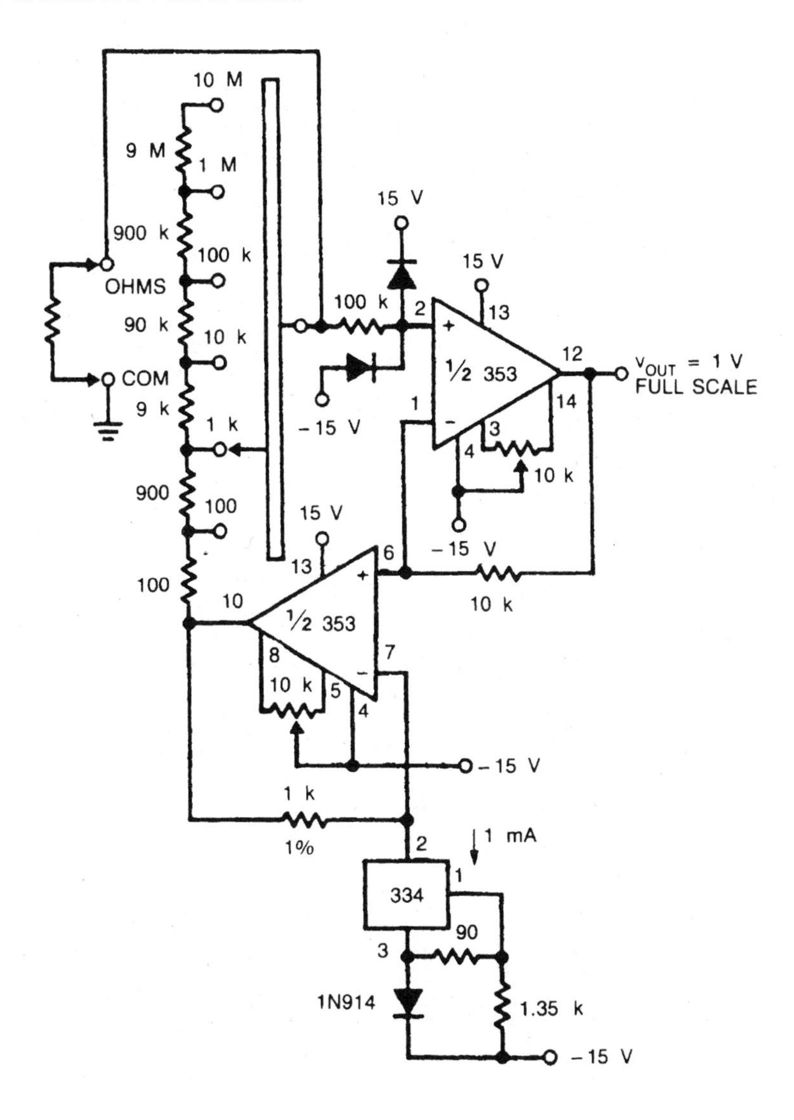

Fourth-order high-pass Butterworth filter

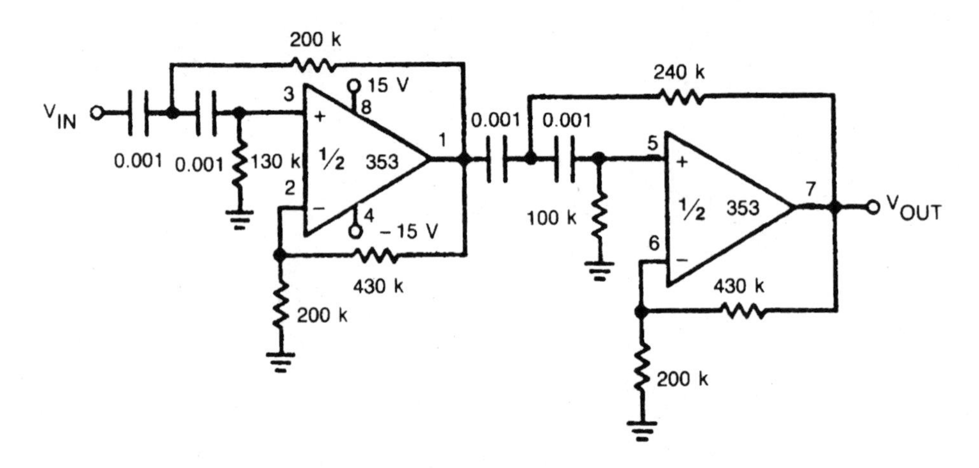

Fourth-order low-pass Butterworth filter

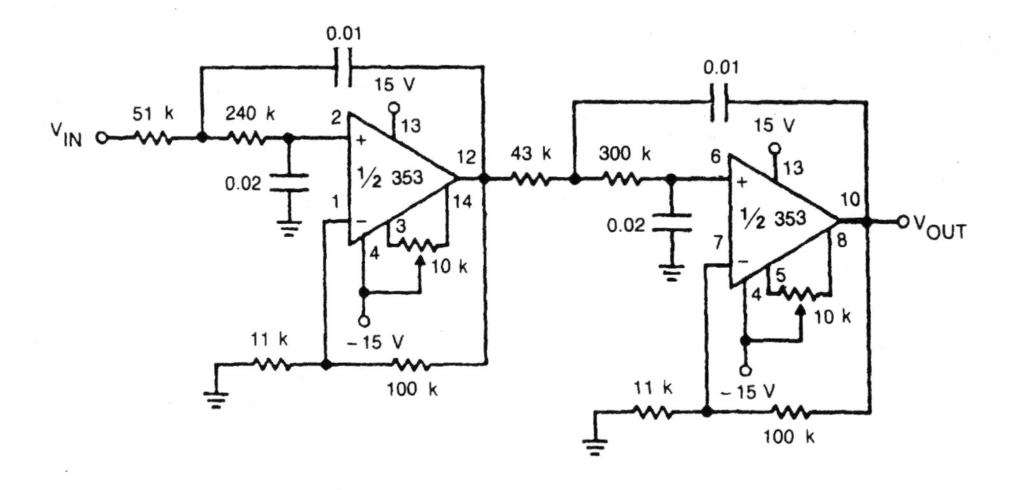

DC-coupled low-pass RC active filter

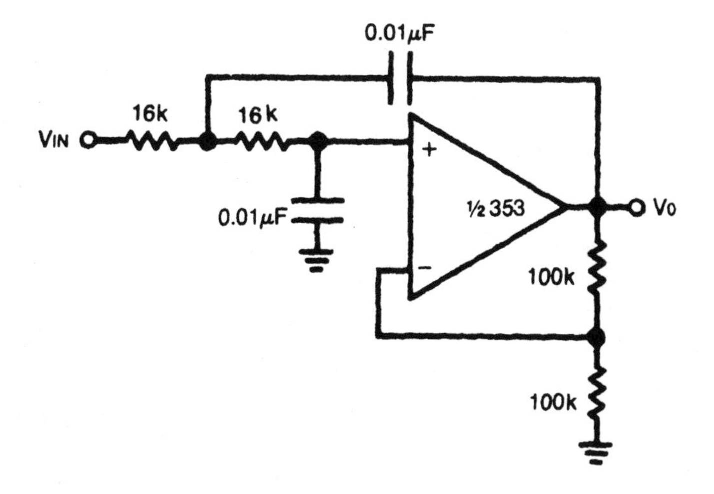

Active bandpass filter

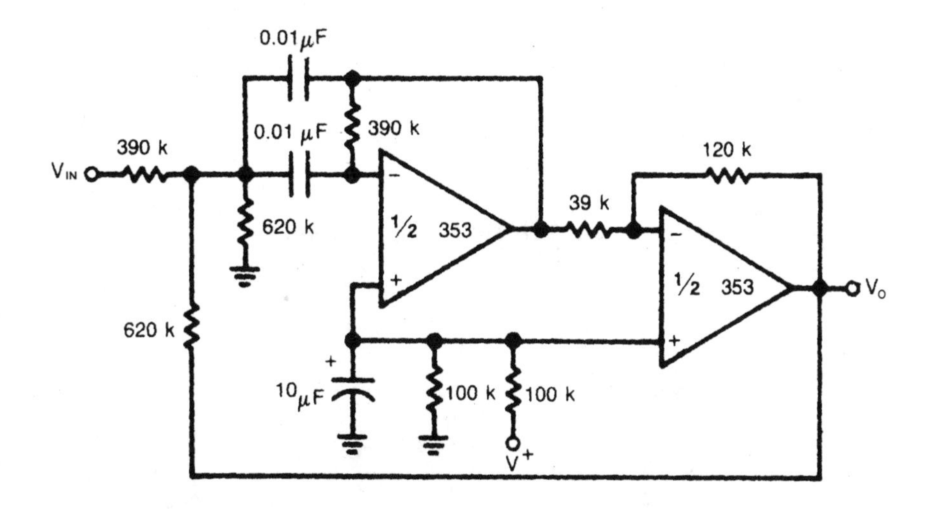

3-band active tone control

This circuit permits the user to customize the tonal response of an audio amplifier to suit individual taste, source quality, and/or environmental conditions. The treble (high frequency), midrange, and bass (low frequency) ranges are individually adjustable.

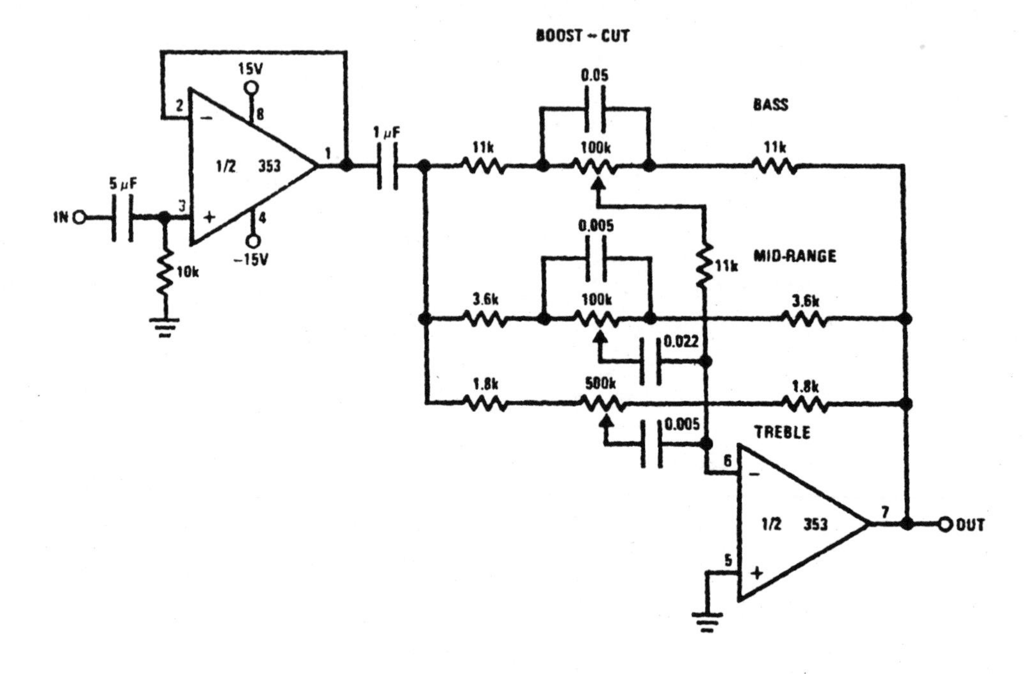

High-input impedance dc differential amplifier

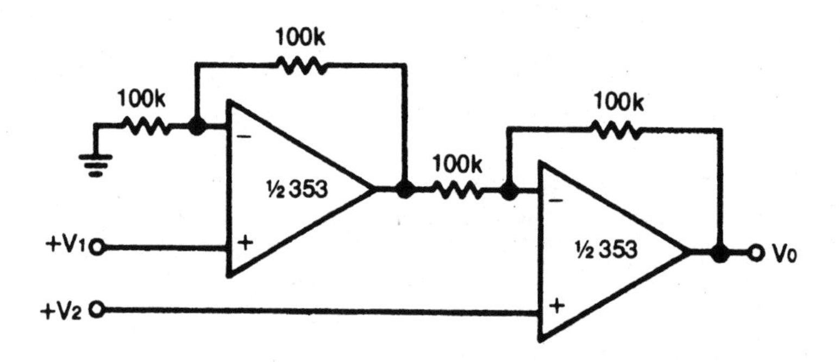

709 monolithic operational amplifier

The 709 is a monolithic operational amplifier intended for general-purpose applications. It can operate over a fairly wide range of voltages. The Class-B output stages give a large output capability with minimal power drain. The design, in addition to providing high gain, minimizes both offset voltage and bias currents.

External components are used to frequency compensate the amplifier.

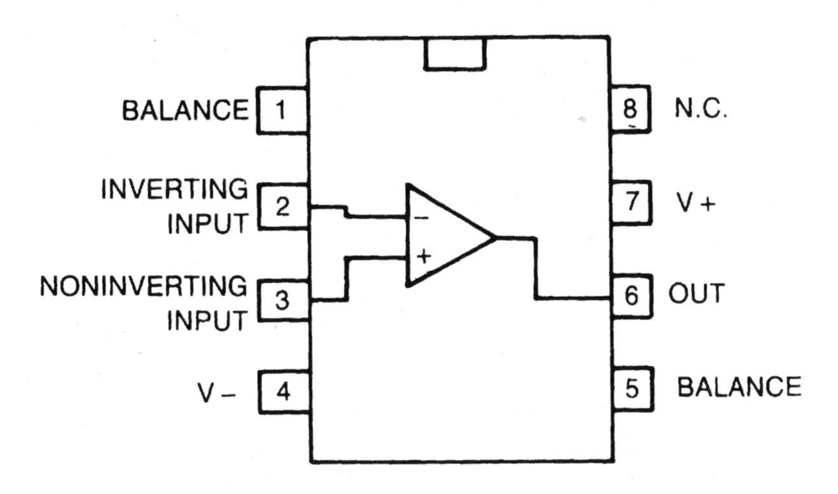

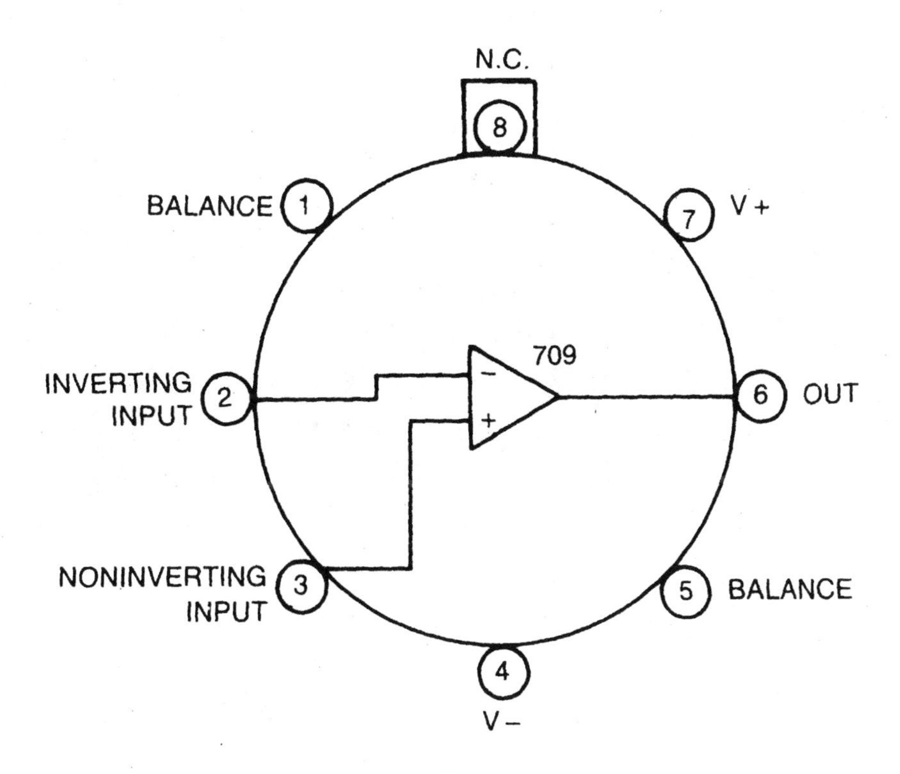

Voltage follower

The output of this unity-gain buffer amplifier matches the input signal. It is used for impedance matching and to minimize loading of the input signal source.

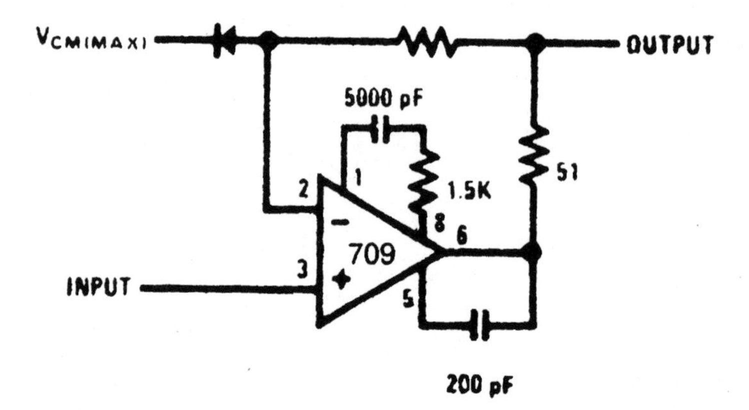

Unity gain inverting amplifier

The output of this unity-gain inverting buffer amplifier matches the input signal, but with the polarity reversed. A positive input voltage becomes a negative output voltage, and vice versa. It is used for impedance matching and to minimize loading of the input signal source, as well as polarity inversion.

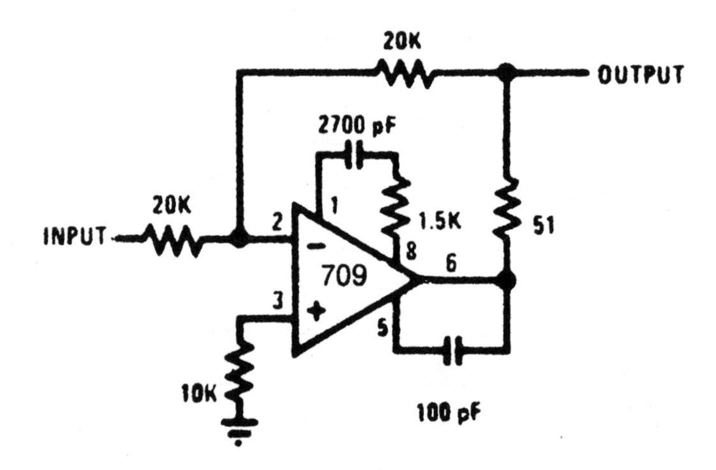

Simple square-wave oscillator

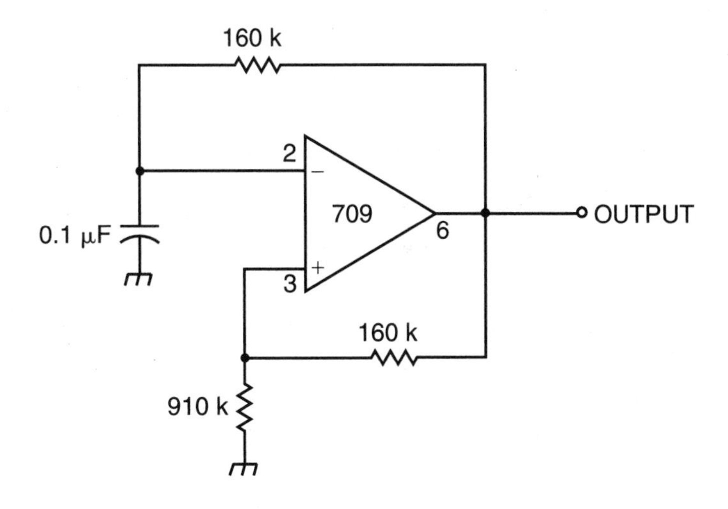

Level shifting differential amplifier

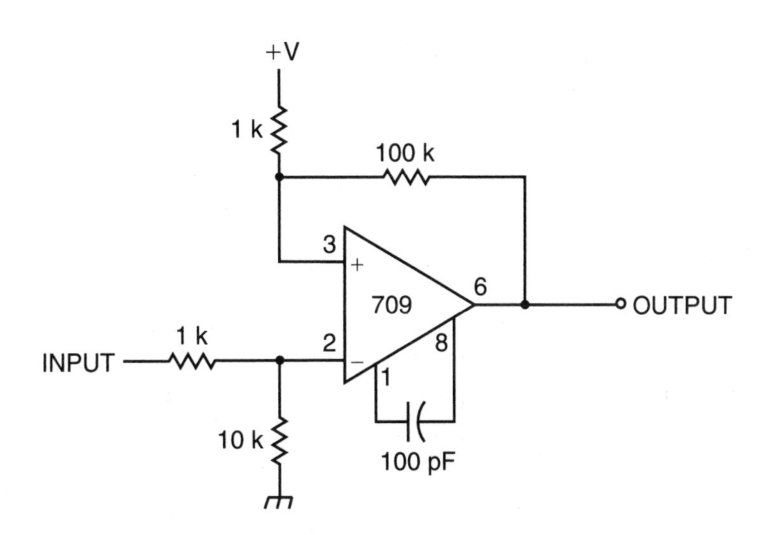

Voltage comparator and lamp driver

The lamp lights up when the input voltage exceeds the reference voltage.

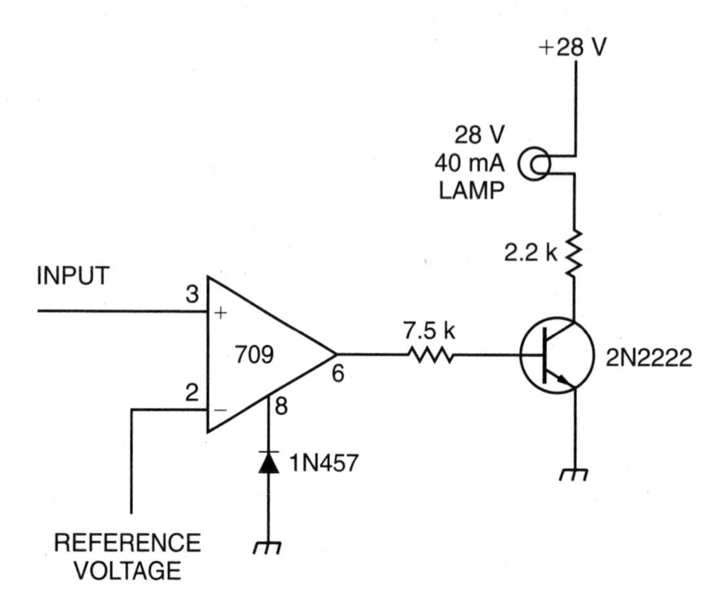

741 internally compensated operational amplifier

The 741 operational amplifier requires no external frequency compensation. It features low power consumption, no latch-up, and wide common-mode and differential voltage ranges. This operational amplifier is one of the most popular devices around. While in many ways it is rather primitive and limited compared to other, more recent operational amplifier devices, it is still effectively the standard other op amps are usually compared to.

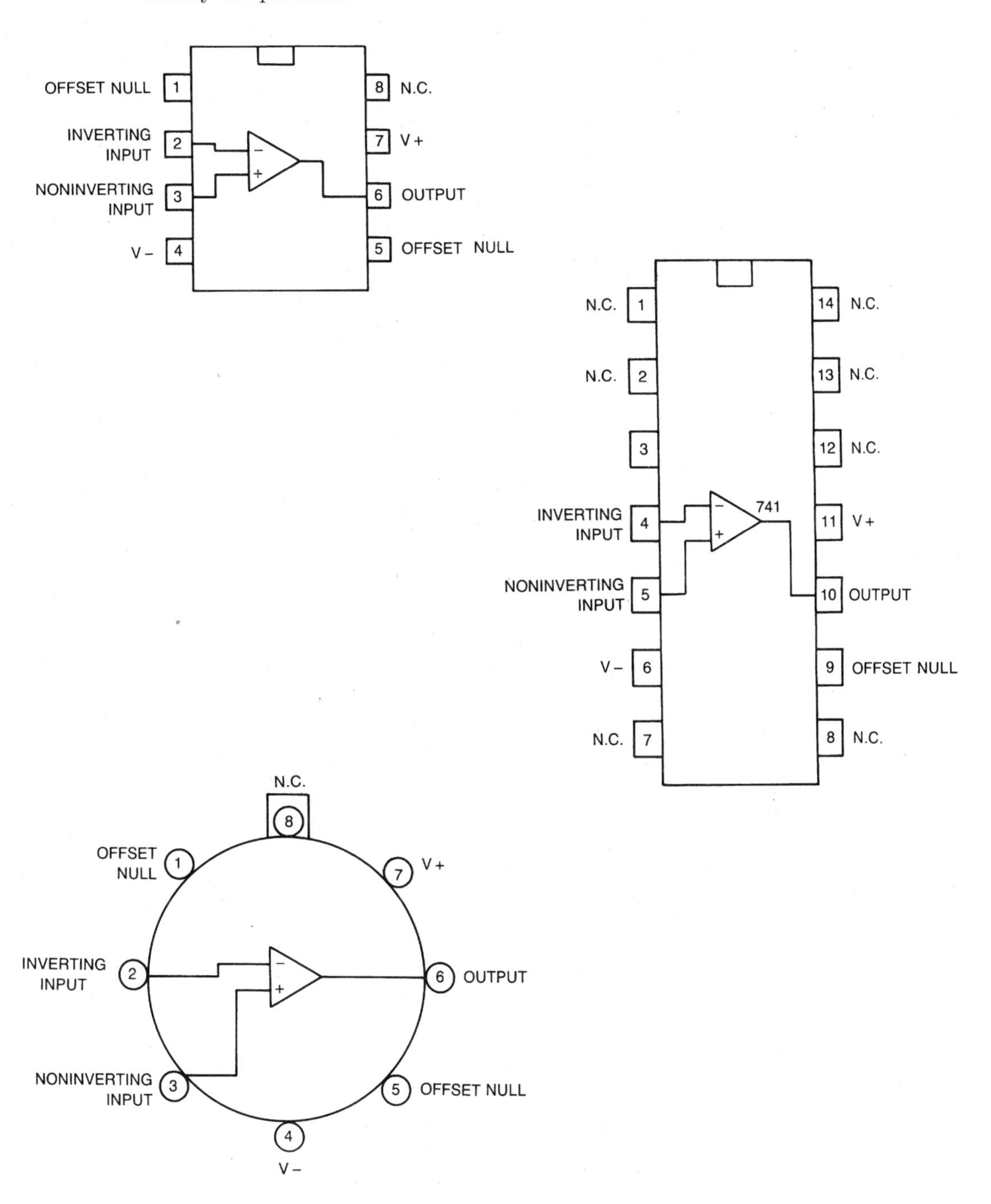

Noninverting amplifier

The output voltage of this amplifier circuit equals the input voltage times the gain, which is determined by the values of the (inverting) input and feedback resistors, using the following formula.

$$Av = \frac{1+R_f}{R_i}$$

Using the component values shown here, the gain is equal to

$$Av = \frac{1+100}{100}$$
$$= 1+1$$
$$= 2$$

The output signal has the same polarity as the input signal.

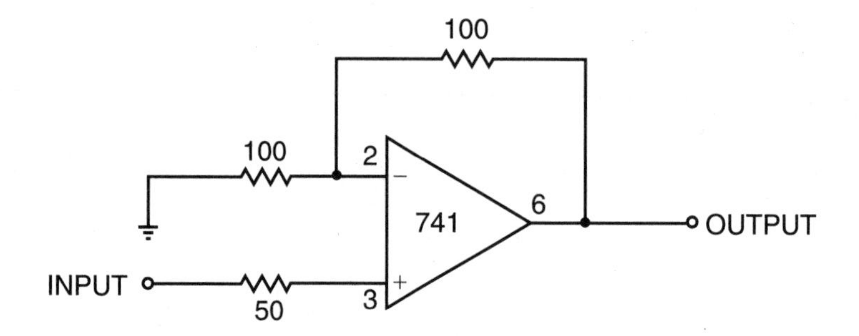

Inverting amplifier

The output voltage of this amplifier circuit equals the input voltage times the gain, with the polarity inverted, or reversed. A negative input voltage results in a positive output voltage, and vice versa. The gain is determined by the values of the input and feedback resistors, using the following formula.

$$Av = \frac{-R_f}{R_i}$$

Using the component values shown here, the gain is equal to

$$Av = \frac{-10000}{1000}$$
$$= -10$$

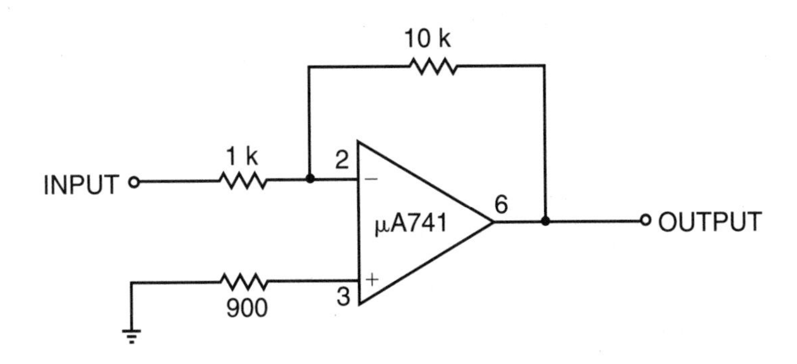

Unity-gain voltage follower

The output of this unity-gain buffer amplifier matches the input signal. It is used for impedance matching and to minimize loading of the input signal source.

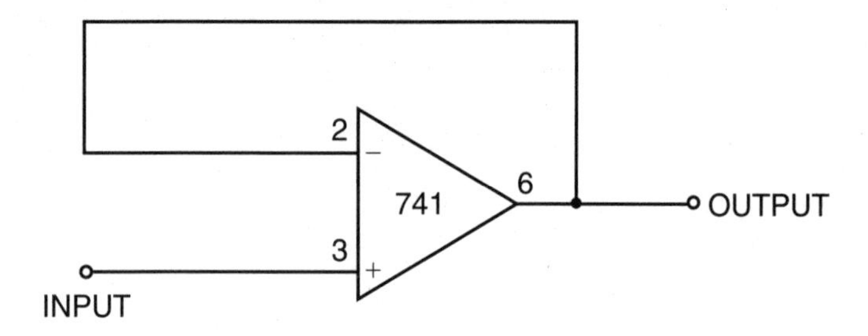

Simple integrator

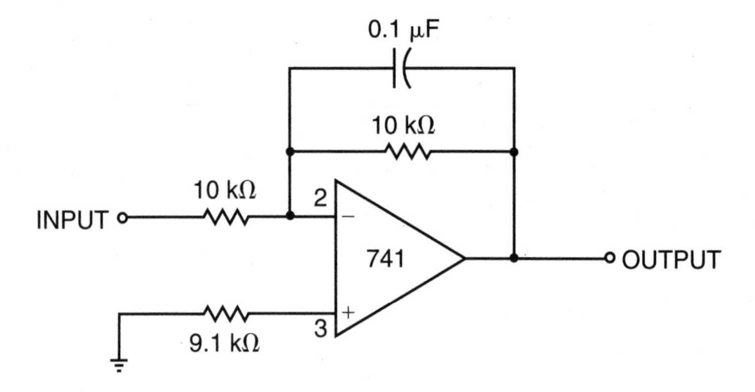

Simple differentiator

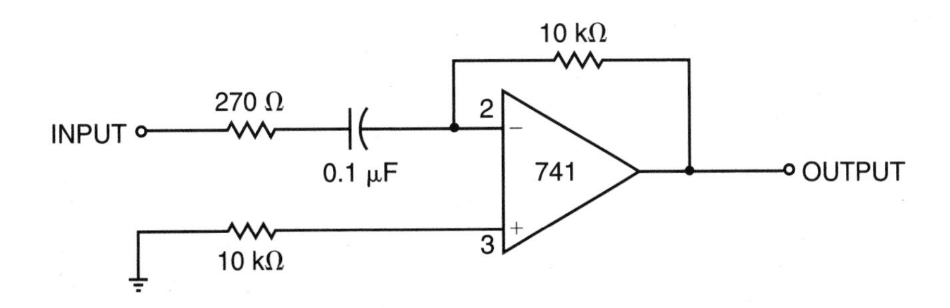

Low-drift, low-noise amplifier

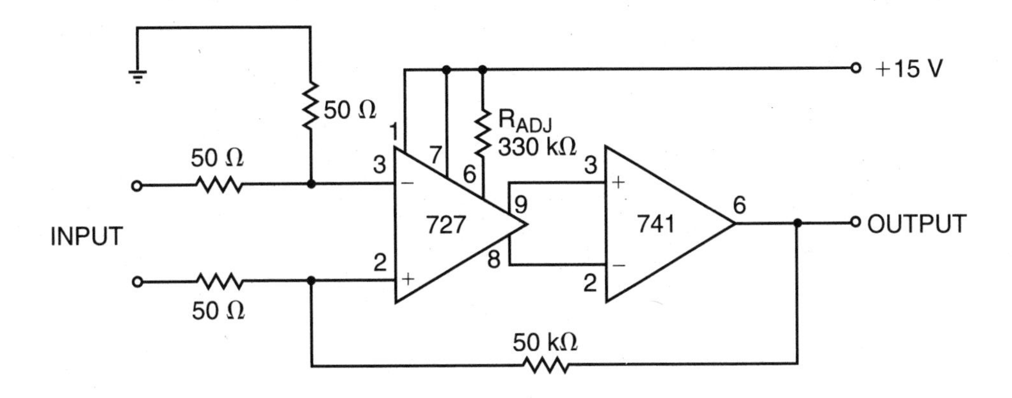

Square-wave generator

This oscillator circuit puts out a true square wave with a duty cycle of 1:2. The output signal frequency is determined by the inverting feedback resistor (R) and capacitor (C), approximately according to the following formula.

$$F = \frac{5}{RC}$$

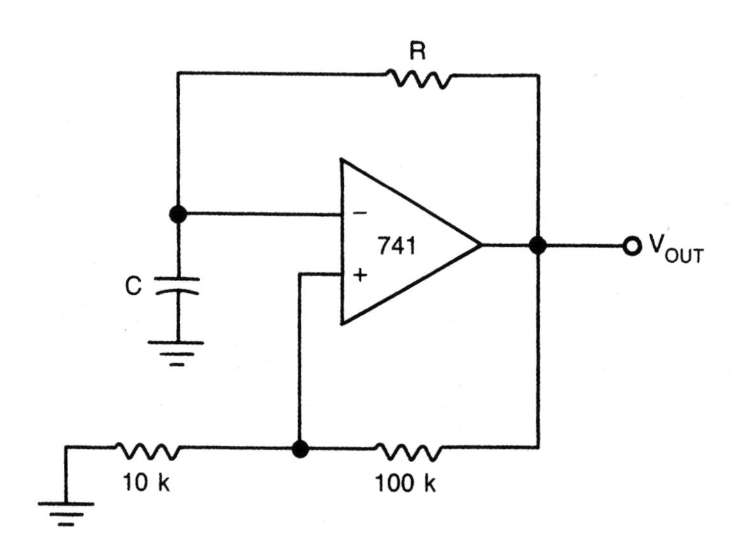

Rectangle-wave generator

This variation on the simple square-wave generator circuit on page 202 puts out a rectangle wave with a duty cycle other than 1:2, providing resistors R_a and R_b have different values. There would be little point to giving these two resistors the same value. The result would be the same as in the previous circuit, with a few extra components that wouldn't accomplish anything in this case.

A diode permits current to flow in only one direction and blocks it in the opposite direction, so during the HIGH portion of the cycle, the feedback path is through resistor R_a, and resistor R_b is effectively switched out of the circuit. During the LOW portion of the cycle we have the exact opposite situation. This time the feedback path is through resistor R_b, and resistor R_a is effectively switched out of the circuit.

The same timing capacitor (C) is used for both portions of the cycle. The timing of the HIGH portion of the cycle is equal to

$$Th = R_aC$$

Similarly, the formula for determining the LOW portion of the cycle is

$$T_1 = R_bC$$

The frequency of the entire cycle works out to

$$F = \frac{1}{(R_a+R_b)C}$$

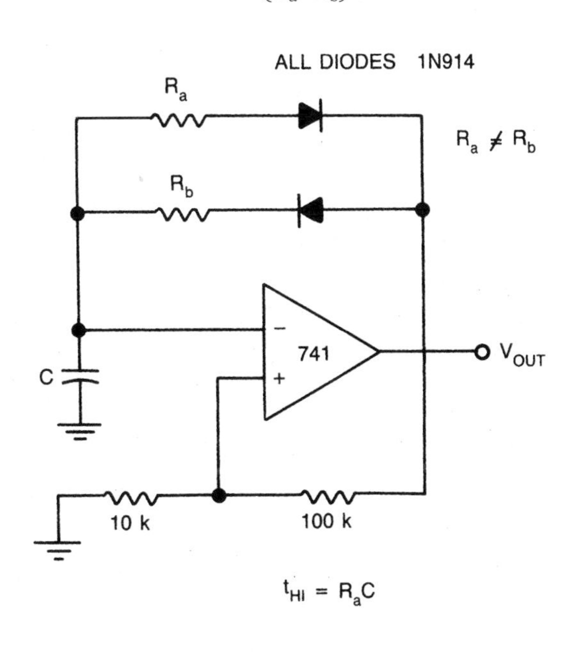

Twin-T sine-wave oscillator

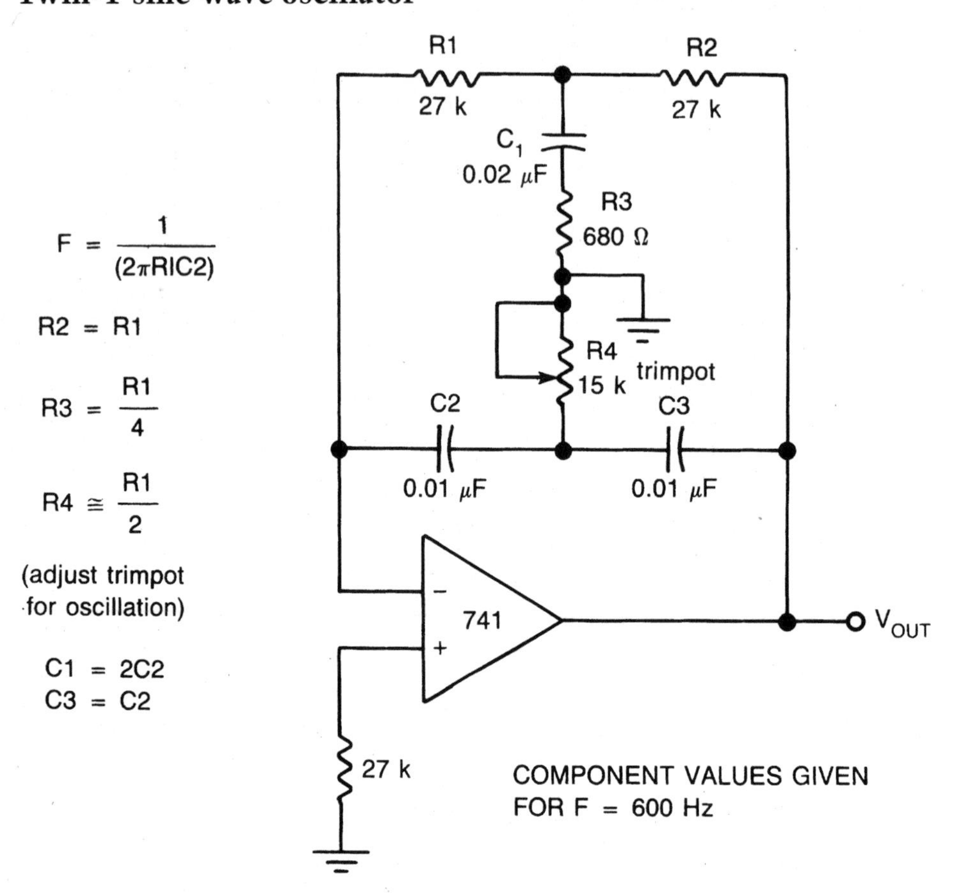

Low-pass active filter

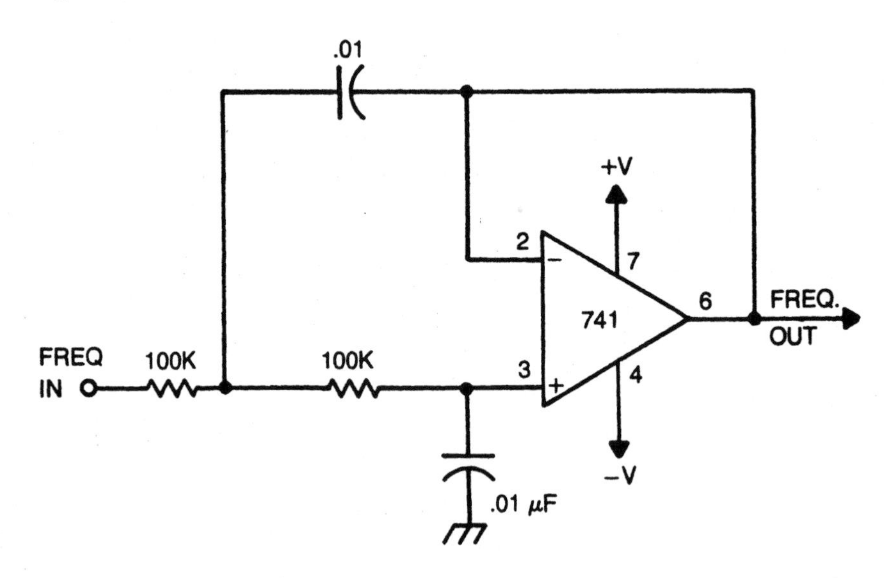

High-pass active filter

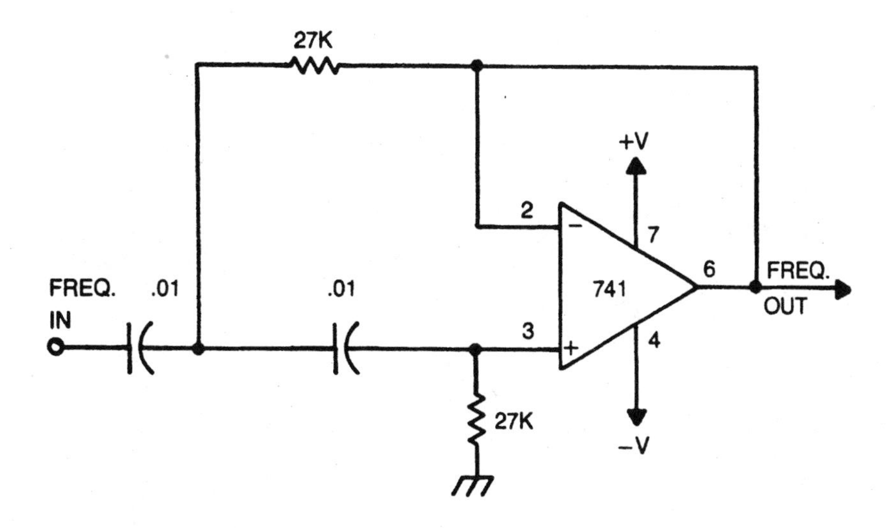

Notch filter

All input frequencies in the input signal are passed on to the output of this filter circuit, except for those within the specific band determined by various component values in the circuit.

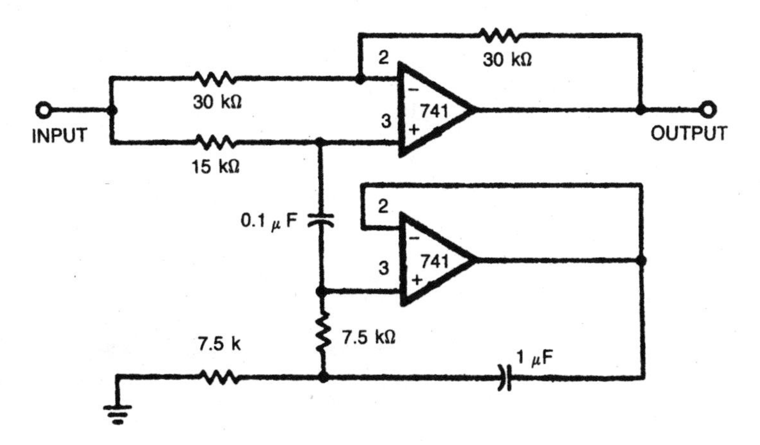

High slew-rate power amplifier

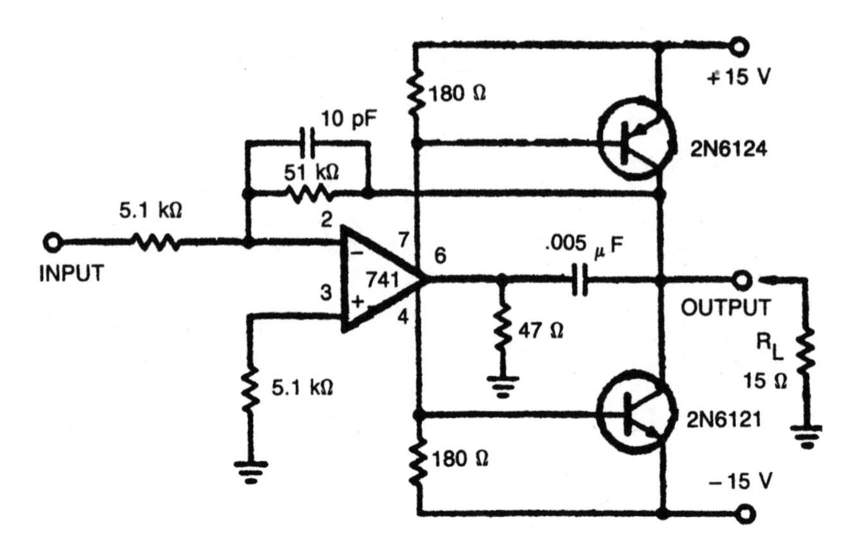

Constant current source

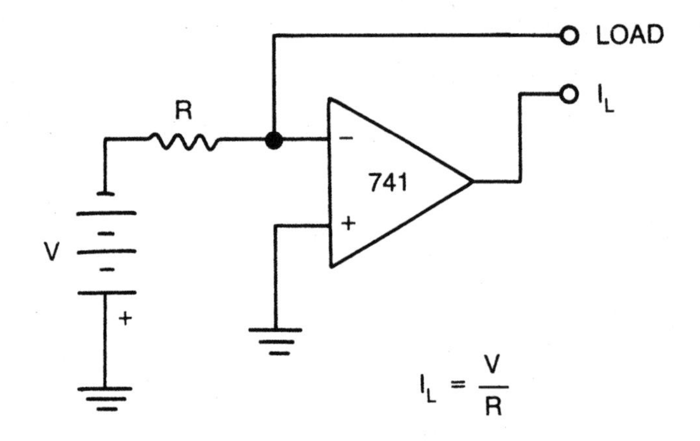

$$I_L = \frac{V}{R}$$

Inverting voltage-to-current converter

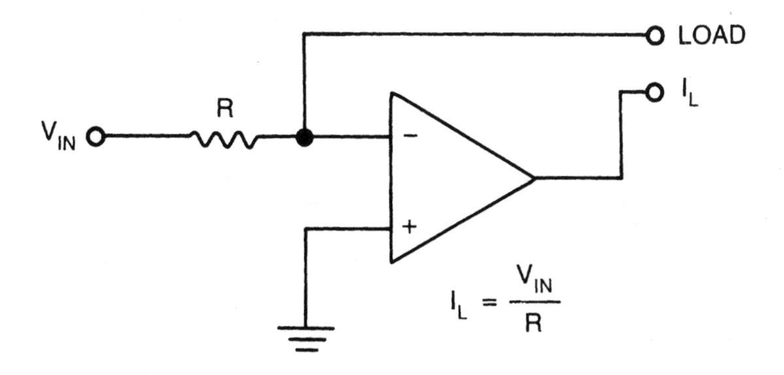

Noninverting voltage-to-current converter

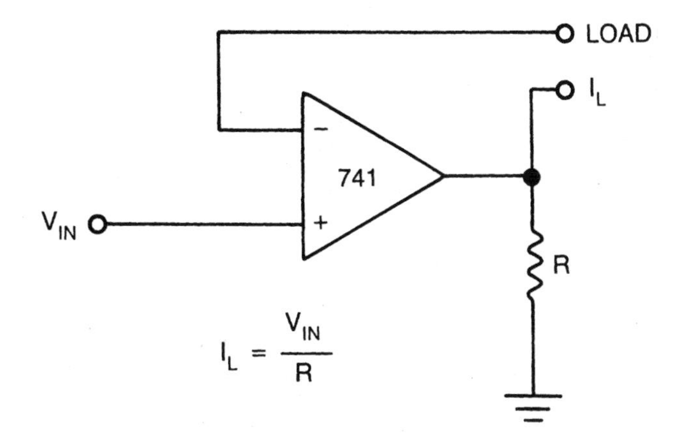

Inverting amplifier with programmable gain

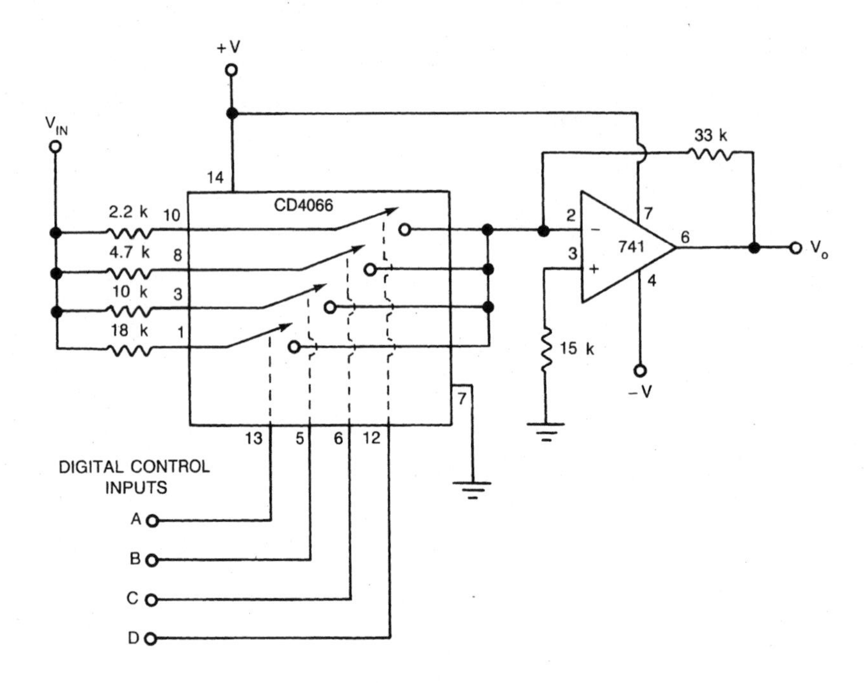

Logarithmic amplifier

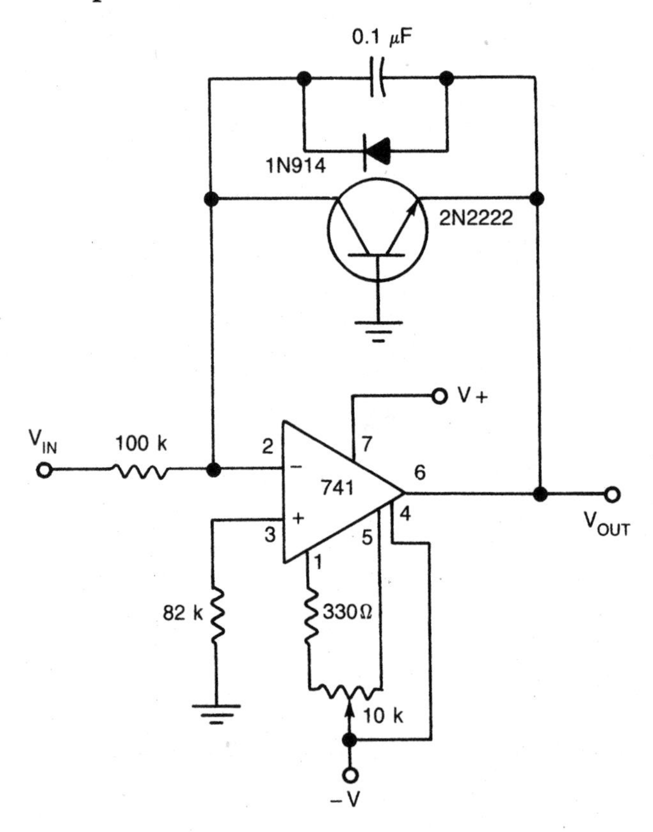

Antilogarithmic amplifier

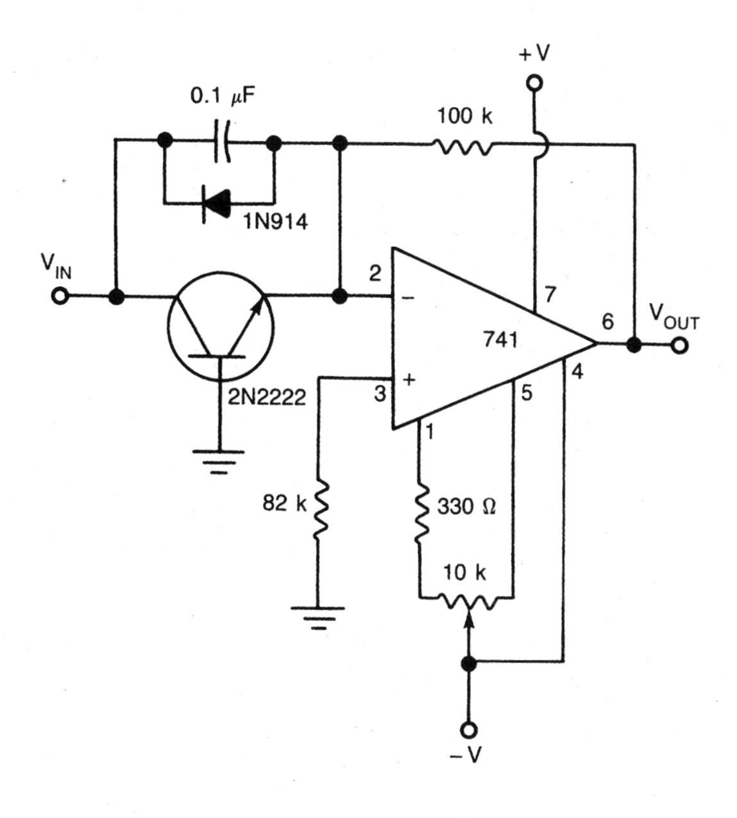

Light-sensitive tone generator

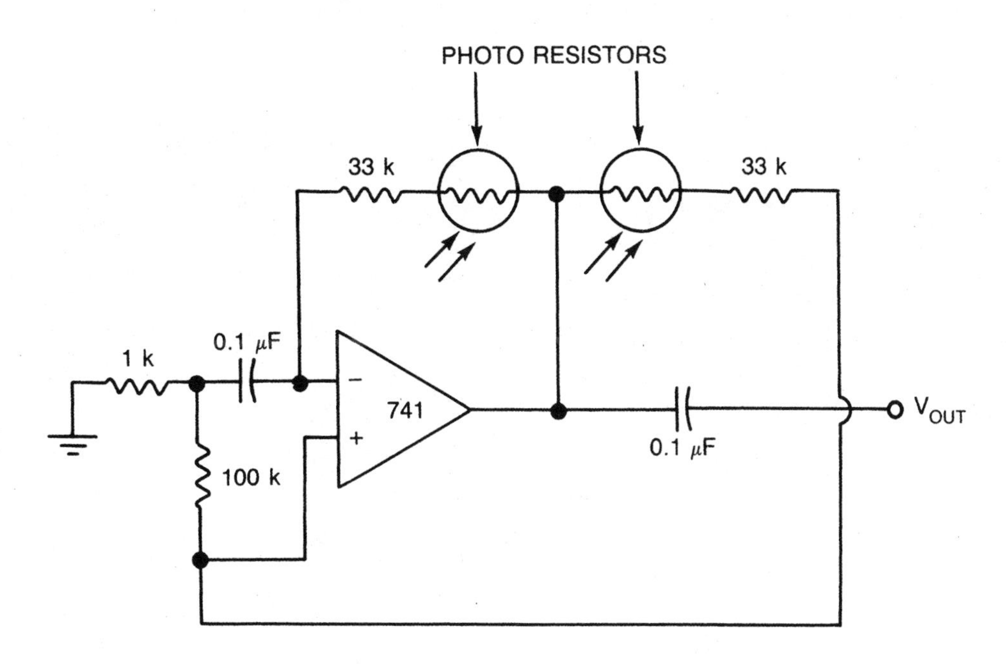

Guitar "fuzz" box

This circuit adds some pleasant distortion to the sound of an electric guitar, or other electronically amplified musical instrument.

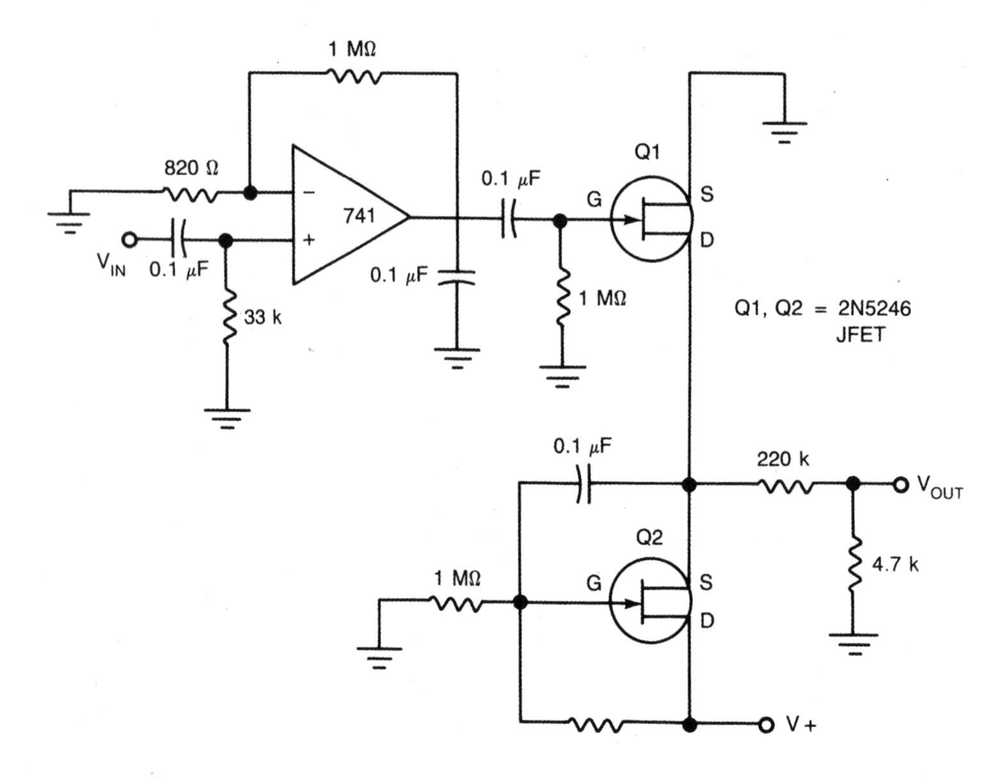

Op amp logic-level shifter

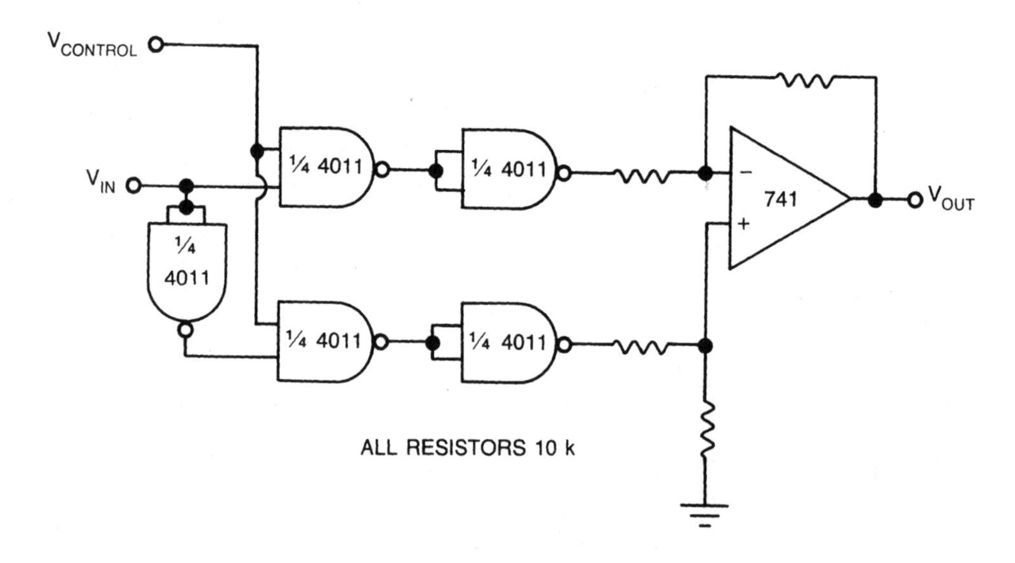

Simple 4-bit D/A converter

A digital input with up to four bits (0000 to 1111) is transformed into a proportional analog voltage by this circuit.

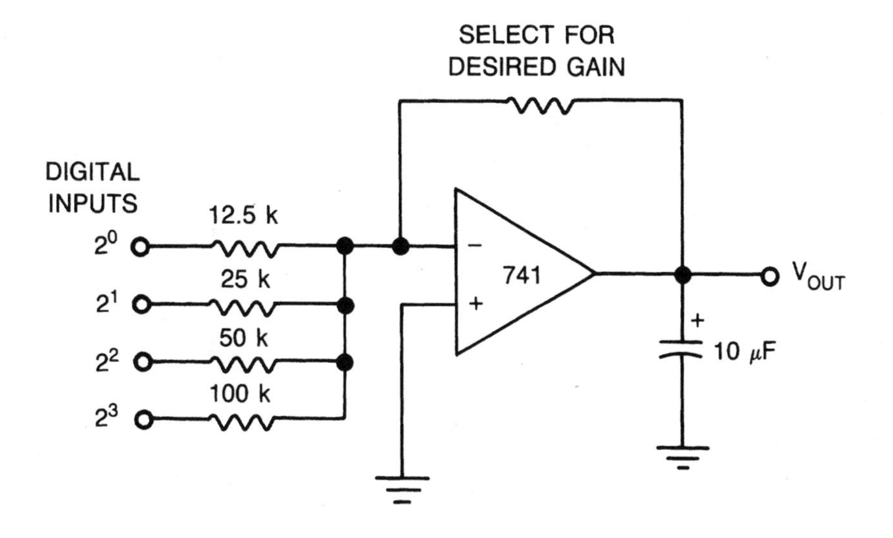

Peak detector

This circuit remembers the highest input voltage it has seen until it is reset by briefly closing the push-button switch to discharge the capacitor.

The peak level cannot be held indefinitely. The charge will leak off the holding capacitor over time, especially if the circuit's output is loaded. A buffer amplifier stage would certainly be advisable in most practical applications to help minimize this leakage.

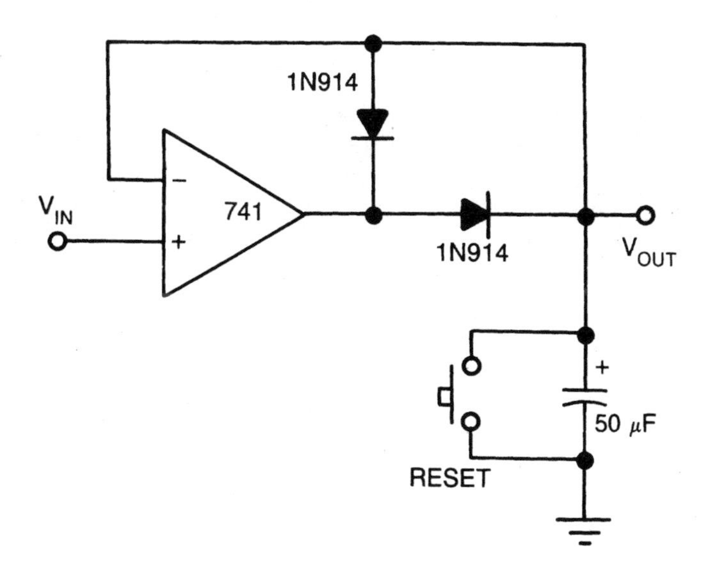

Threshold detector

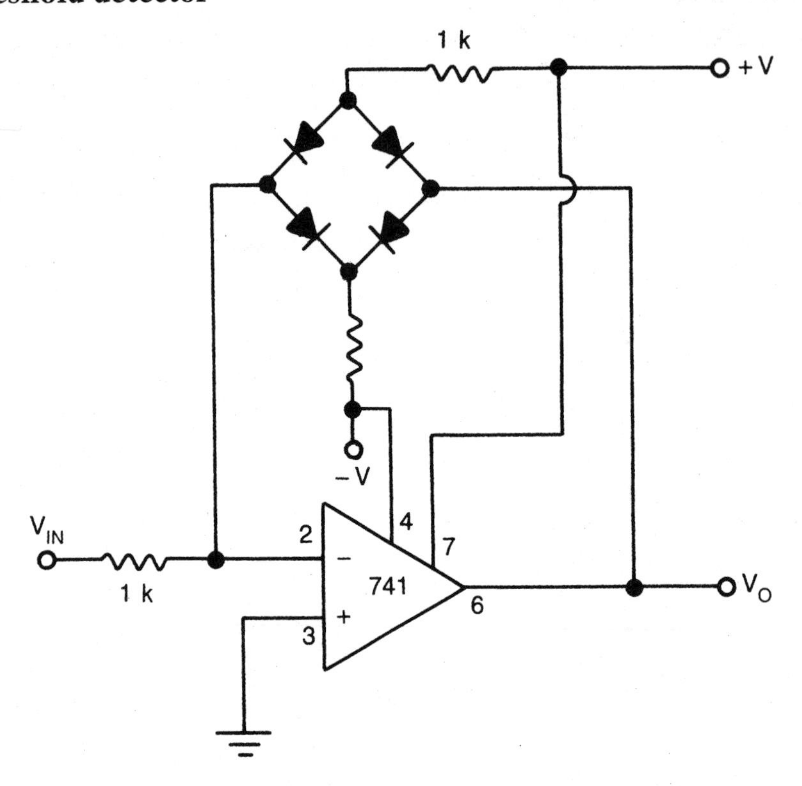

Nonsymmetrical threshold detector

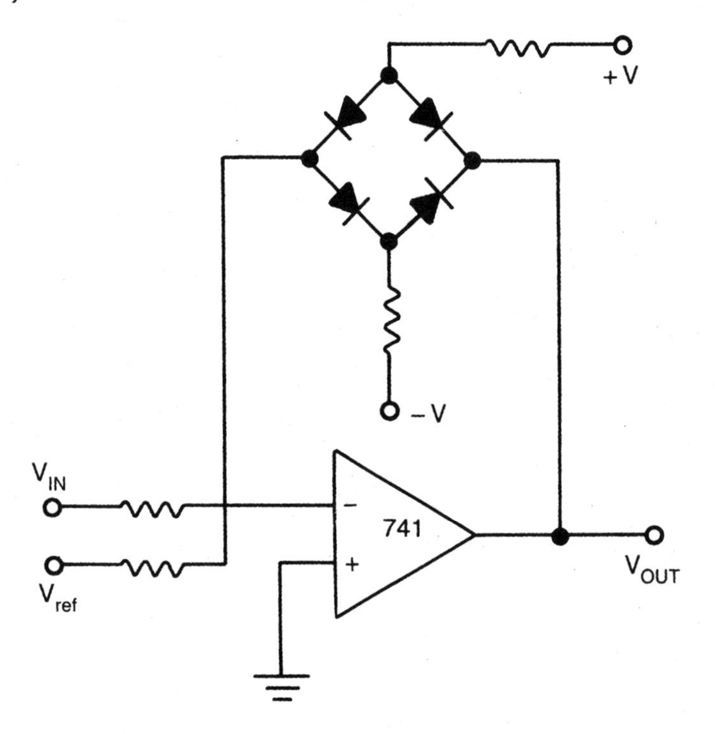

Comparator and lamp driver

The lamp lights up when the input voltage exceeds the reference voltage.

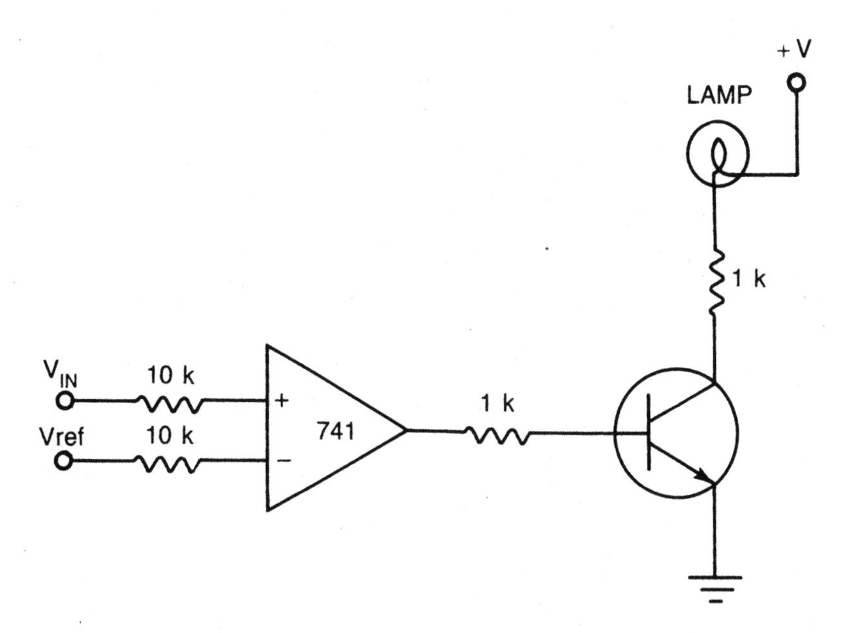

TRANSISTOR SHOULD BE SELECTED TO SUPPLY SUFFICIENT CURRENT TO LOAD LAMP

Current-to-voltage converter

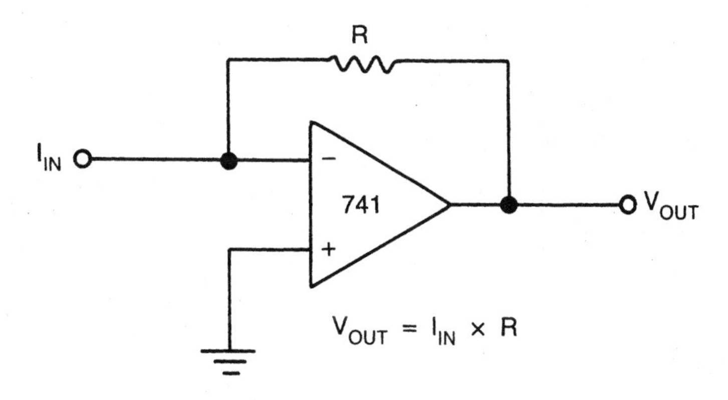

$V_{OUT} = I_{IN} \times R$

Voltage regulator

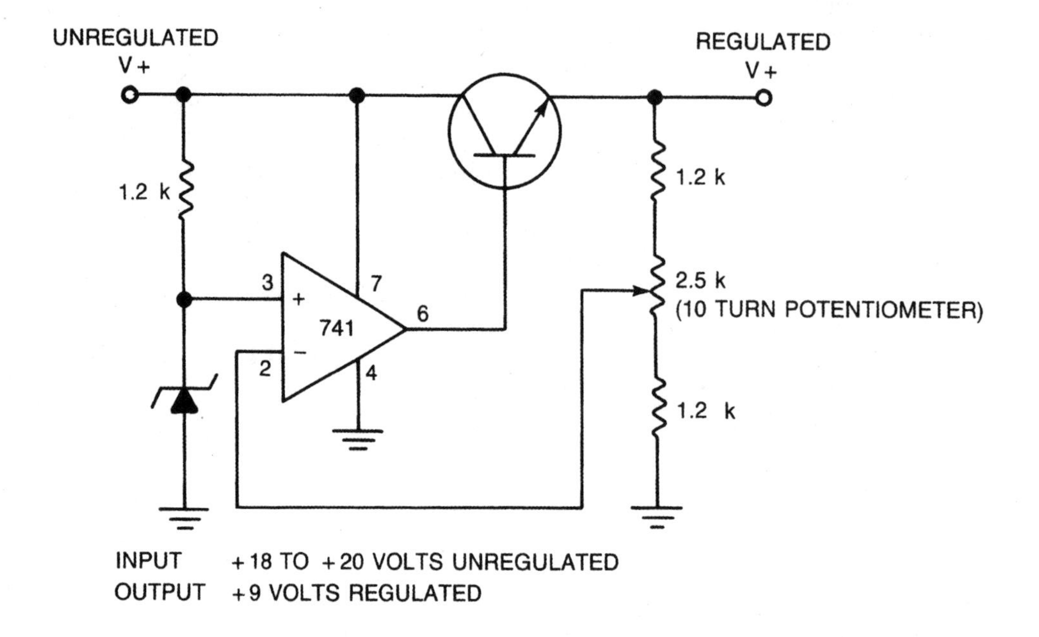

Fast summing amplifier

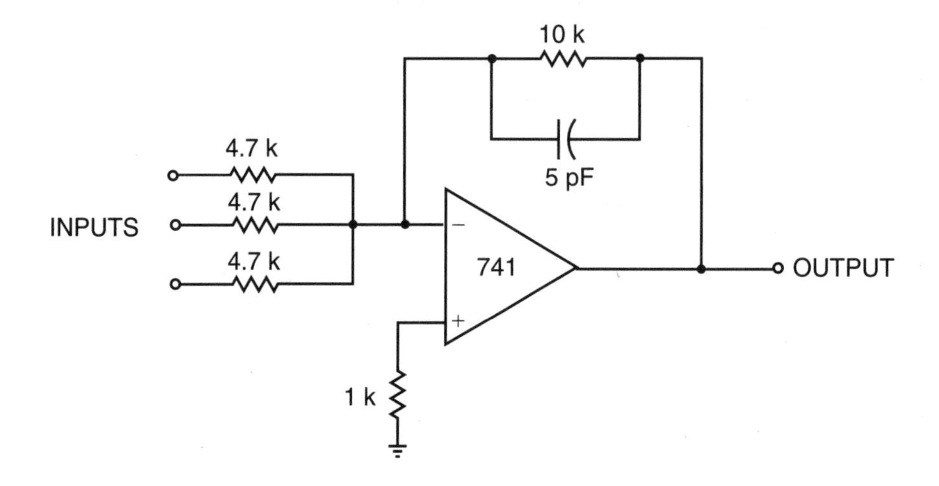

Fast summing amplifier with low-input current

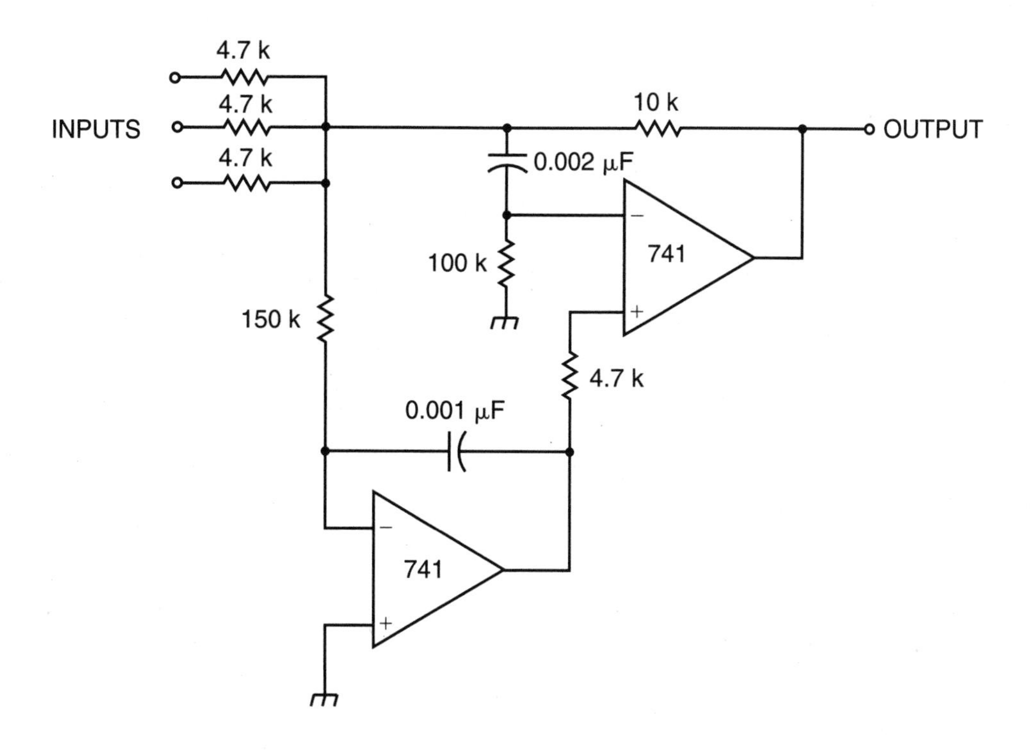

Sine-wave squarer

By using the full open-loop gain of the op amp (theoretically infinite, actually very high), an input signal that deviates even slightly from zero will force the operational amplifier into positive or negative voltage saturation. This means, if the input is a sine wave, the output will be a fairly good approximation of a square wave.

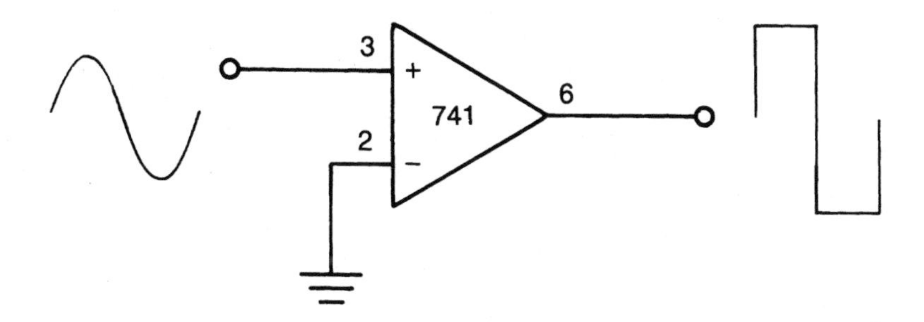

Current monitor

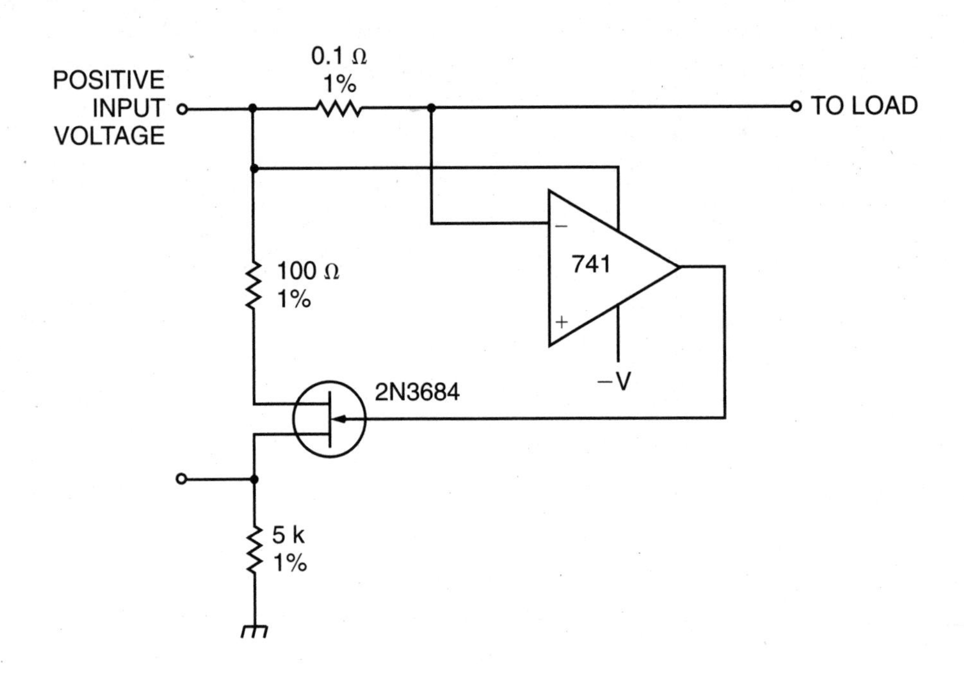

Schmitt trigger

A Schmitt trigger circuit is useful for cleaning up noisy signals, especially rectangle, square, and pulse waves. This circuit features variable hysteresis, which can be adjusted to suit the needs of the specific application.

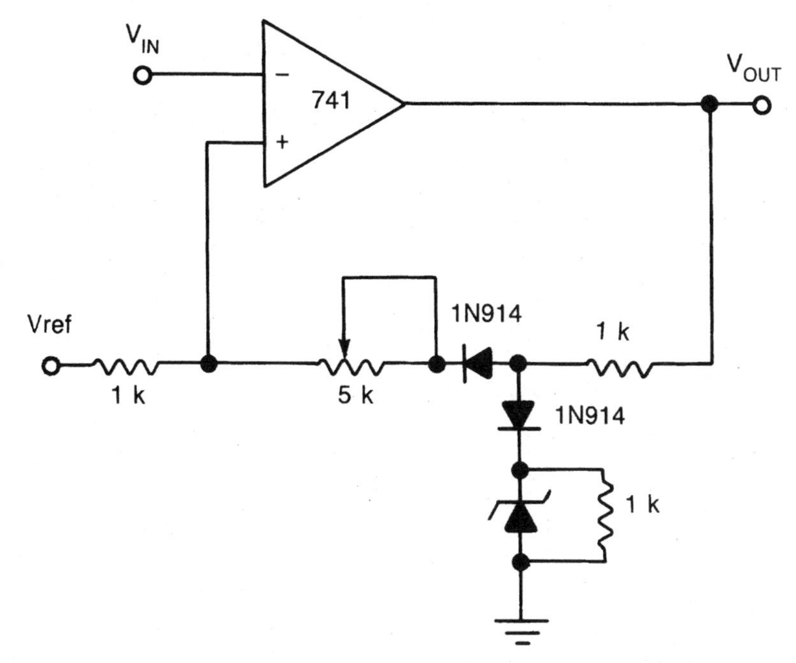

Monostable multivibrator

This type of circuit is also known as a timer. The output is normally LOW. When a suitable trigger pulse is detected at the input, the output will go HIGH for a predetermined period of time equal to the product of resistor R and capacitor C. The timing formula is

$$T = R \times C$$

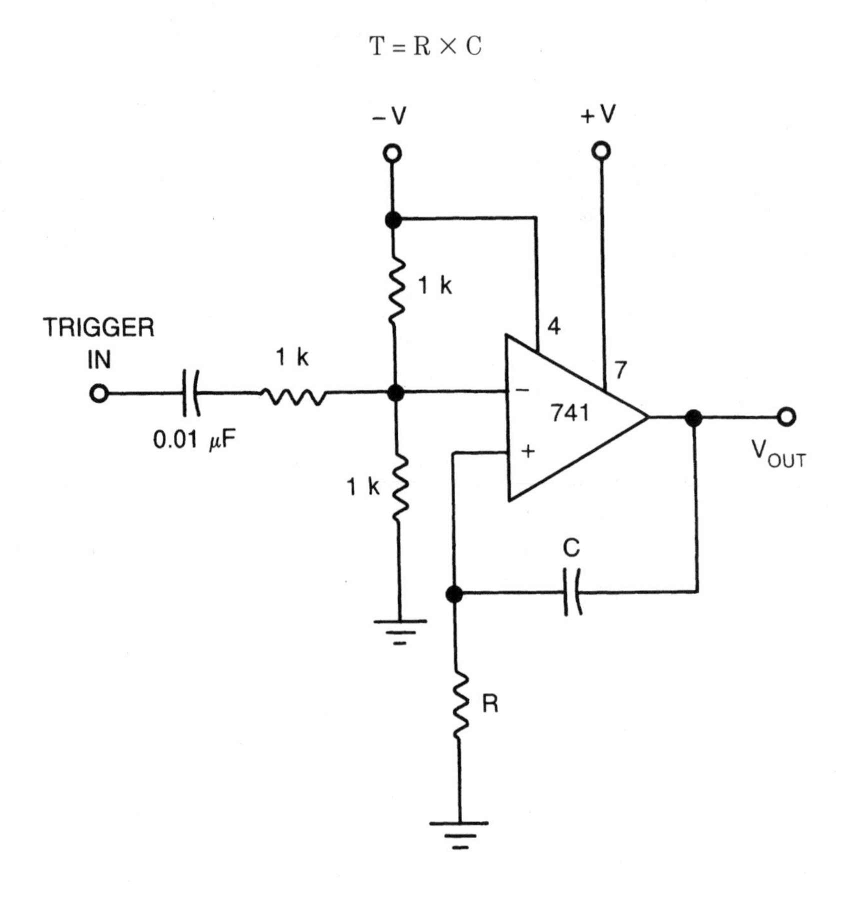

Bistable multivibrator

This type of circuit is also called a flip-flop. Each time the circuit is triggered, the output reverses its level from LOW to HIGH, or from HIGH to LOW. In essence, the circuit remembers and holds the last state it was set to.

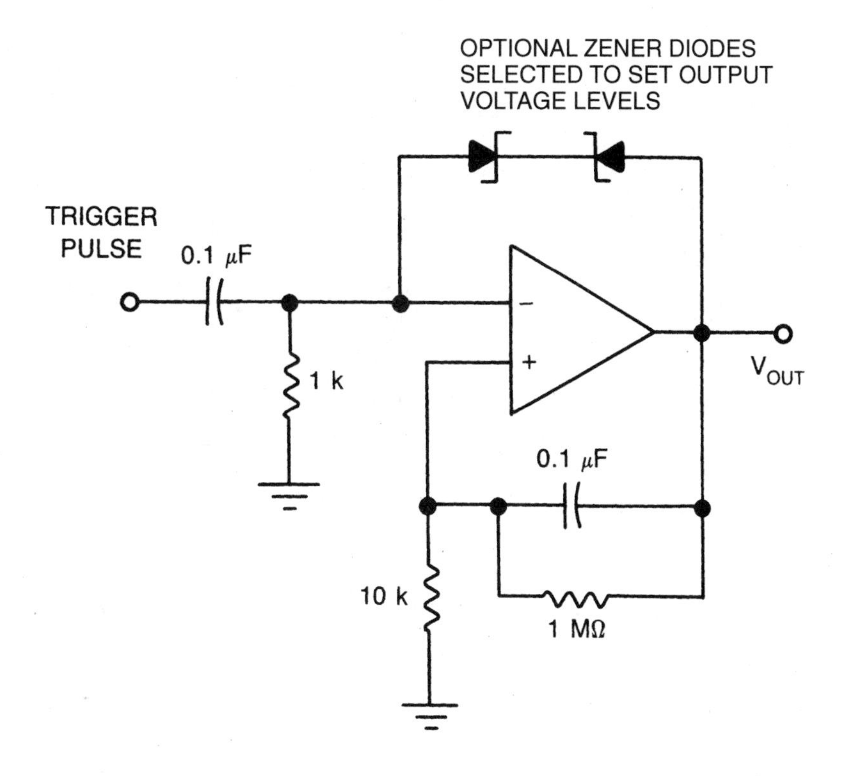

747 dual operational amplifier

The 747 contains two independent (except for the V-connection point) 741-type operational amplifiers. Each op amp requires no external frequency compensation. It features low power consumption, no latch-up, and wide common-mode and differential voltage ranges. This operational amplifier is one of the most popular devices around. While in many ways, it is rather primitive and limited compared to other, more recent operational amplifier devices, it is still effectively the standard other op amps are usually compared to. A 747 chip can be used in place of two 741s in any circuit, simply by correcting the pin numbering.

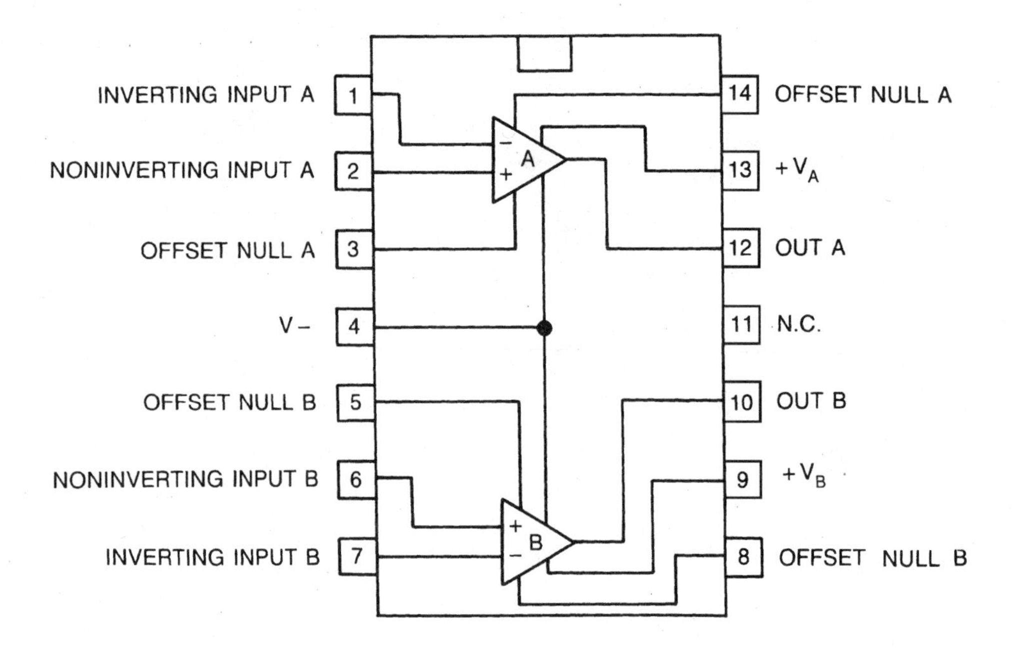

Inverting amplifier

The gain of this amplifier circuit is determined by the values of resistors R_1 and R_2, using this simple formula

$$Av = \frac{-R_2}{R_1}$$

Using the component values shown here, the gain will be −100.

The output polarity is inverted from the input voltage. That is, a negative input voltage will produce a positive output voltage, and vice versa.

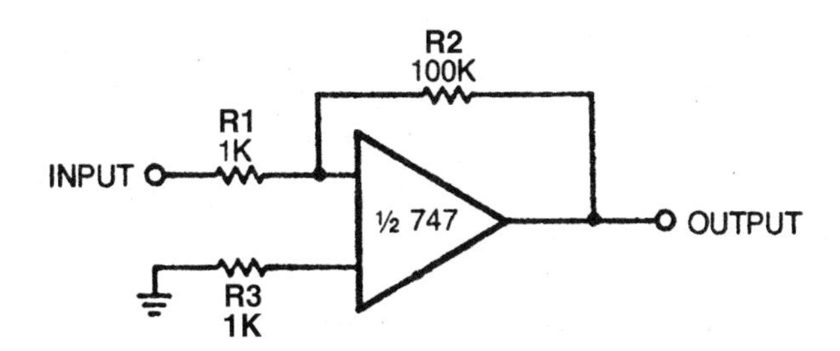

Noninverting amplifier

The gain of this amplifier circuit is determined by the values of resistors R_1 and R_2, using the following formula

$$Av = 1 + \frac{R_2}{R_1}$$

Using the component values shown here, the gain will be 1001.

There is no polarity inversion in this circuit. The output polarity is the same as the input polarity. That is, a positive input voltage will produce a positive output voltage, and a negative input voltage will result in a negative output voltage.

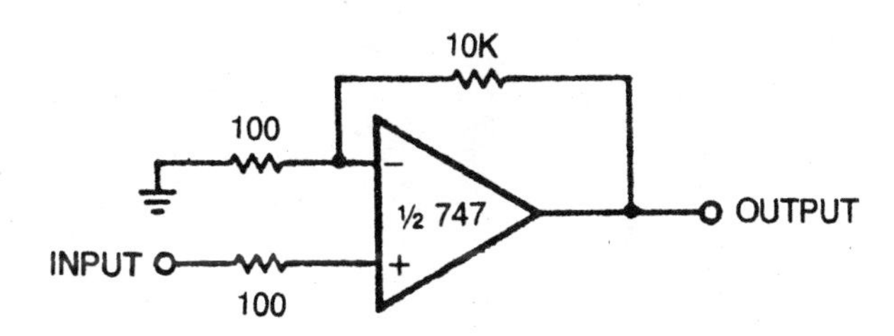

Unity-gain voltage follower

The output voltage of this simple buffer amplifier matches the input voltage with no change in polarity or amplitude. It is used for impedance matching applications, preventing excessive loading of earlier stages, and/or to isolate different sections of a complex circuit.

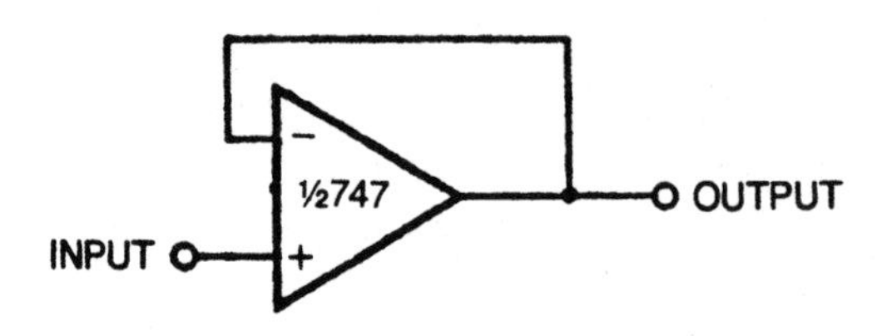

Noninverting summing amplifier

The various input voltages are added together, and their sum is multiplied by the amplifier's gain to produce the output voltage. Using the component voltages shown here, the gain is 10, and all inputs are equally weighted. That is, each input voltage is subjected to the same amount of gain, so each input affects the output equally. To weigh one or more of the inputs, change the value(s) of the appropriate input resistor(s). The overall amplifier gain (common to all inputs) can be changed by substituting a different value for the 100K feedback resistor shown here.

More inputs can be added simply by including the appropriate number of input resistors in parallel. However, loading and stability problems may occur if too many inputs are used. Up to five should work fine.

The first amplifier stage is wired as an inverting amplifier. The output of this stage will have the opposite polarity as the true sum of the input voltages. The second op amp stage reinverts the signal back to its original polarity, and serves as a buffer. Additional gain could be provided in this stage, if appropriate to the intended application.

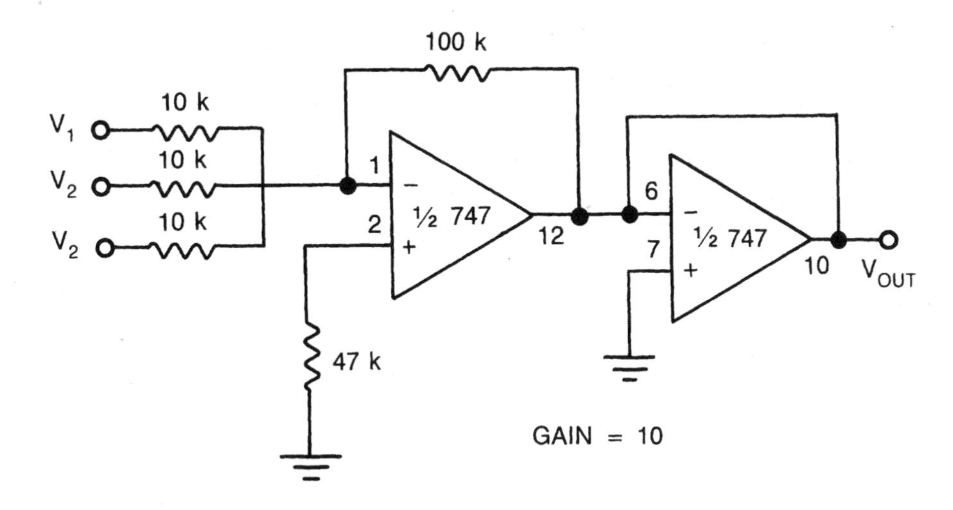

Quadrature oscillator

The cosine-wave output is 90 degrees out of phase with the sine-wave output. Both output signals necessarily have the same frequency.

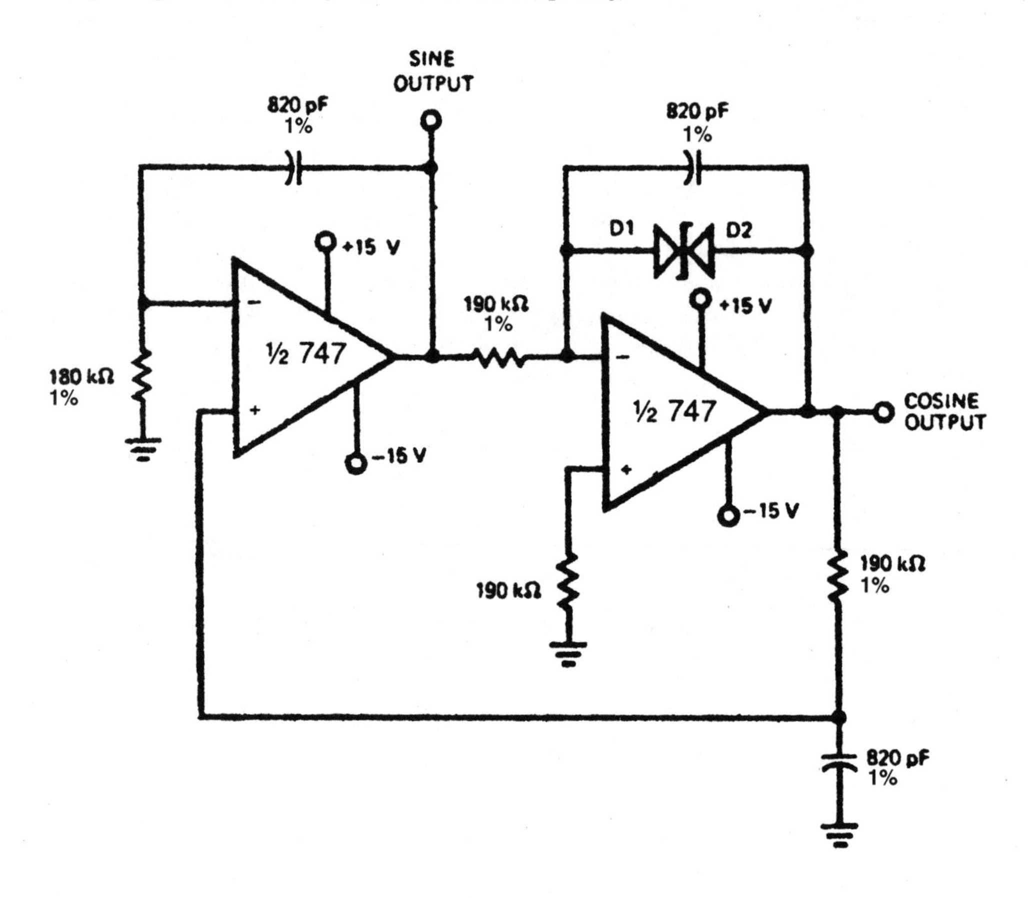

Alternate quadrature oscillator

The cosine-wave output is 90 degrees out of phase with the sine-wave output. Both output signals necessarily have the same frequency.

The signal frequency in this circuit is equal to

$$F = \frac{1}{2\pi RC}$$

(2π is approximately equal to 6.28)

Both R resistors and both C capacitors must have equal values or the circuit may not function reliably, if at all.

The value of resistor R_1 should be slightly less than R.

The exact value of this component is not too critical. Generally, you can use the next standard resistance size down. For example, if R is 10K, then R_1 could be a 9.1K resistor.

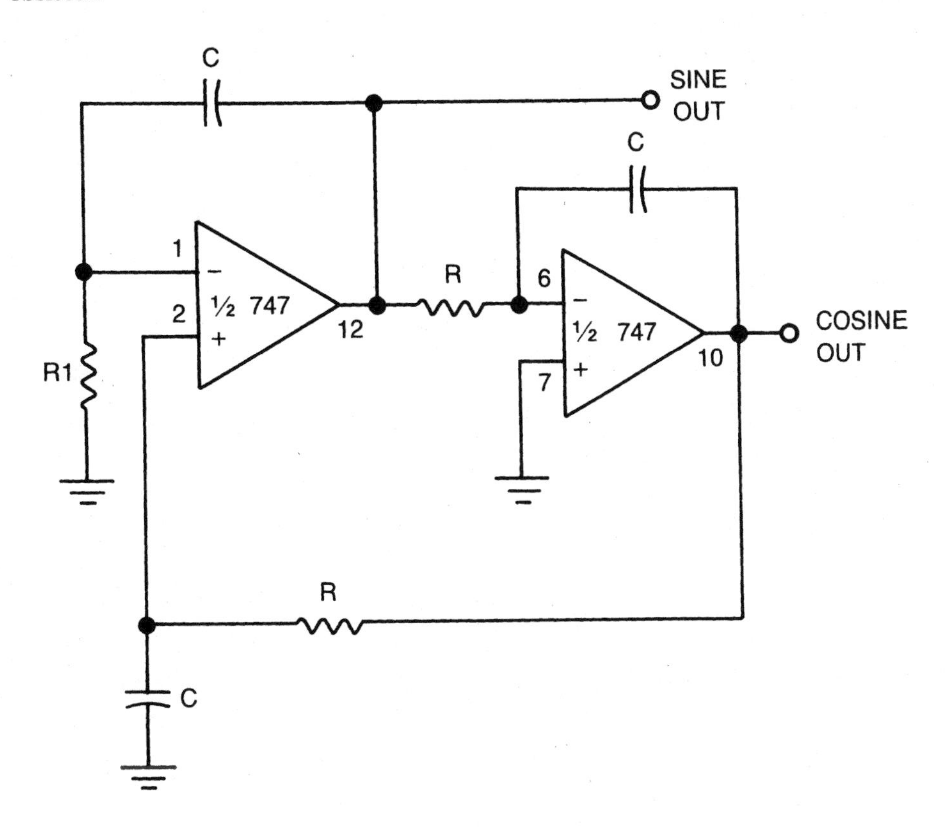

Function generator

The signal generator circuit simultaneously generates two waveforms—a square wave and a triangle wave. Both output signals will always have the same frequency and phase. Essentially, the first op amp acts as a square-wave oscillator, and the second op amp is a low-pass filter to reduce the strength of the odd harmonics, which are contained in both waveforms.

The frequency formula for this circuit is

$$F = \left[\frac{1}{(4R_3C_1)}\right] \times \left(\frac{R_1}{R_2}\right)$$

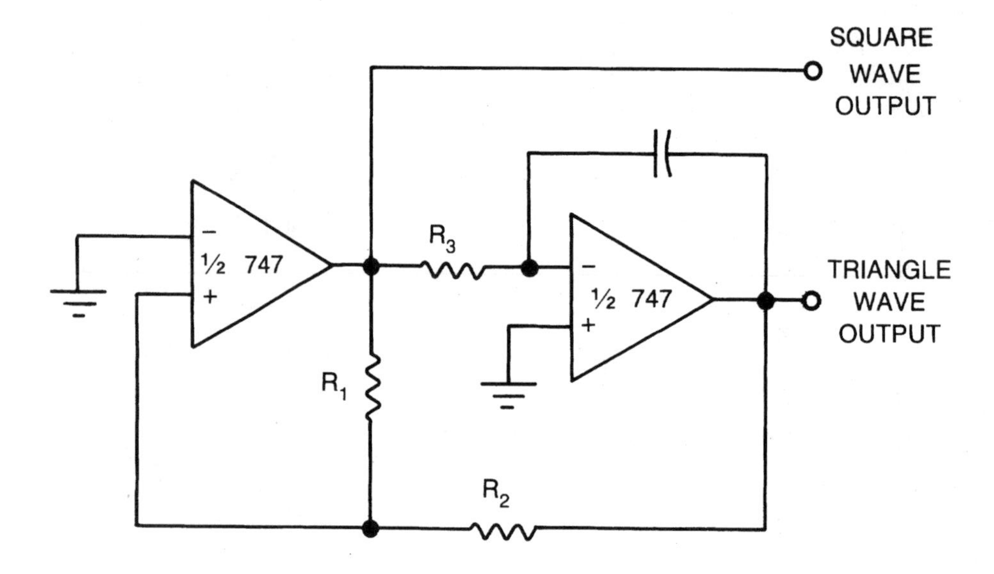

Amplitude modulator

If a low-frequency signal is used as the program input, a tremolo (volume fluctuation) effect will be audible in the square-wave (carrier) signal when it appears at the circuit's output. If an audible frequency program input signal is used, the two inputs will combine in true AM, and complex sidebands will be generated.

This circuit can be used in simple radio transmitters, or in audio sound effects and musical applications.

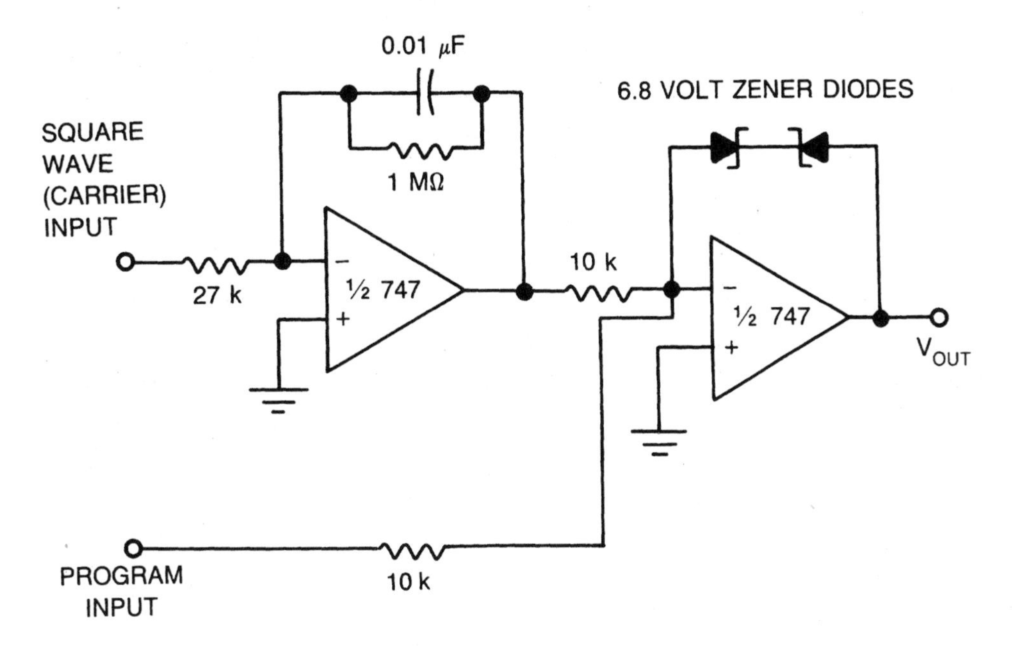

Dual polarity power supply

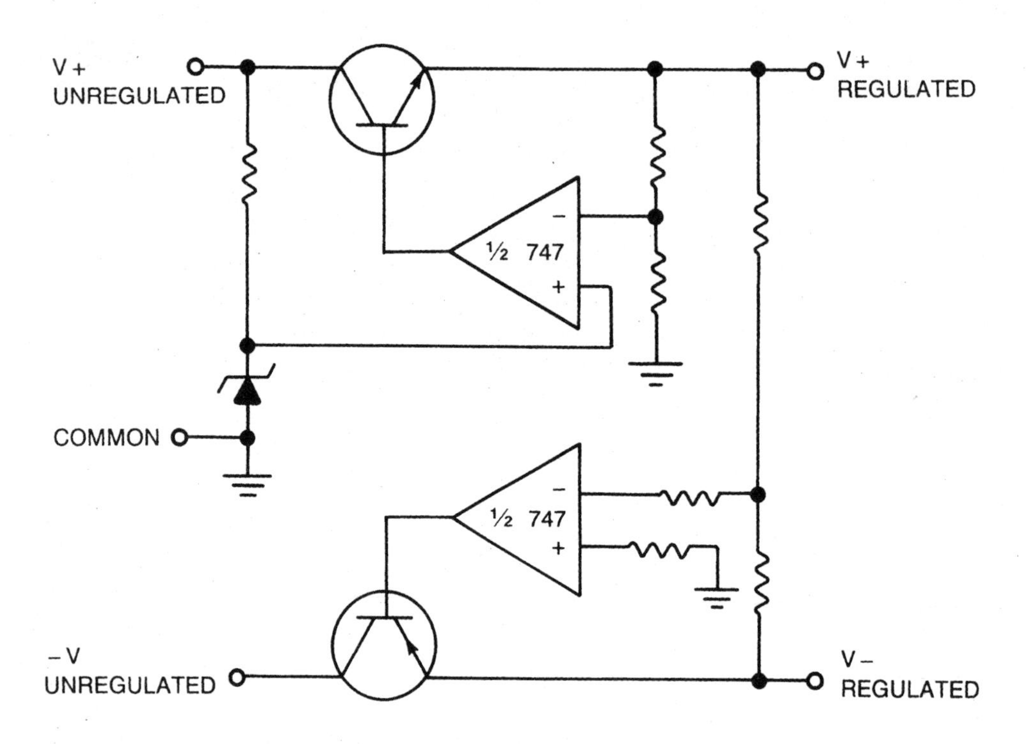

Sample and hold circuit

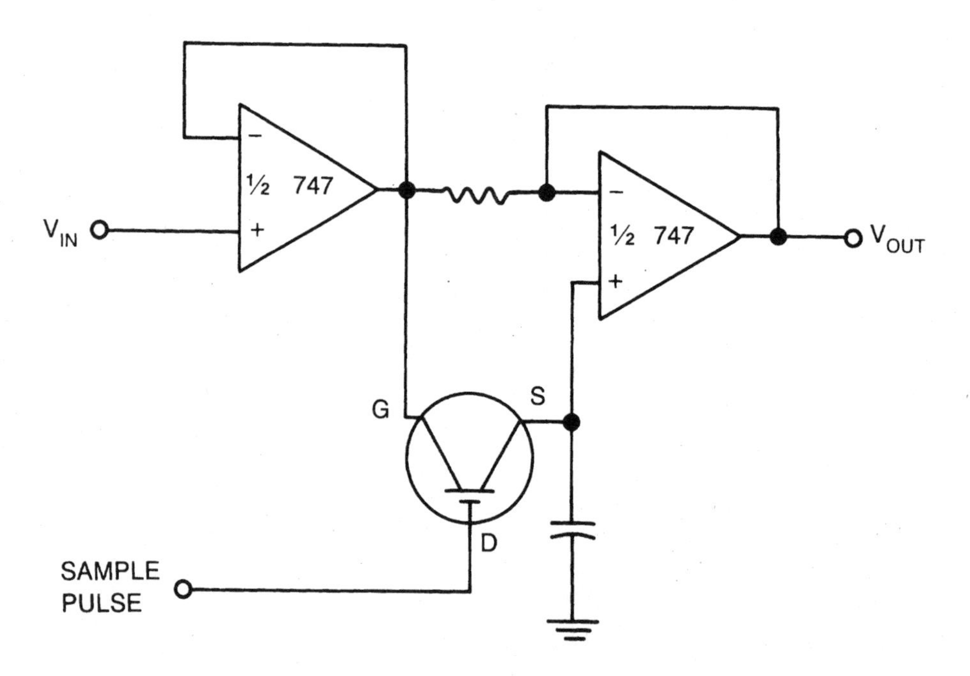

Analog multiplier

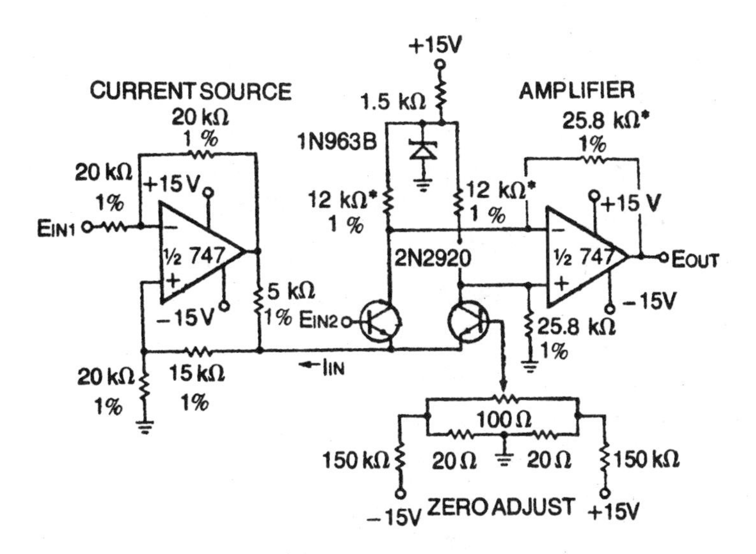

Circuit for tracking positive and negative voltage references

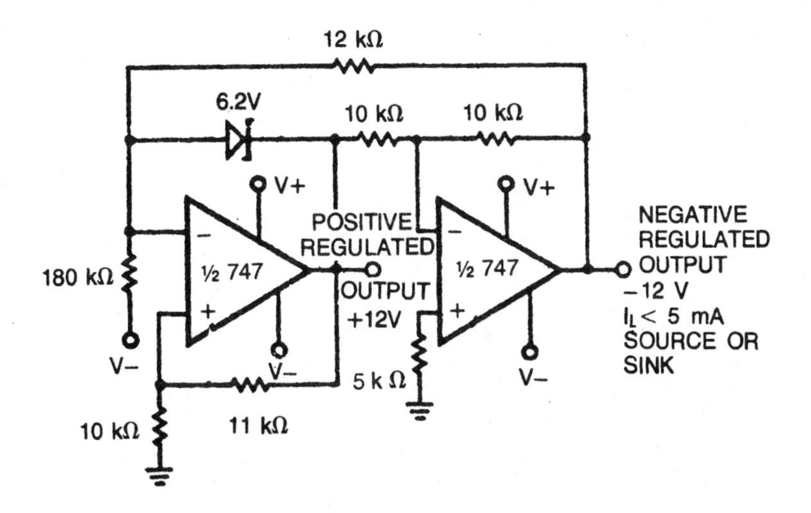

Notch filter

All of the frequency components in the input signal will appear in the output signal, except those within the specified passband. In effect, this circuit bites a notch out of the nominal flat frequency response.

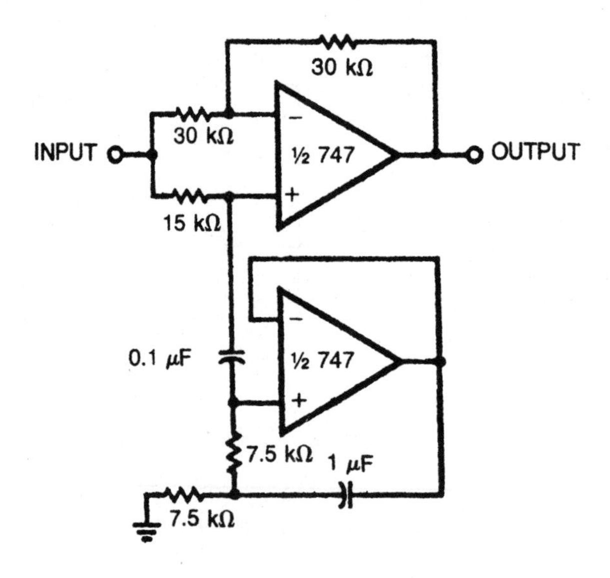

Active bass and treble tone control with buffer

This circuit allows customization of the tonal response of a simple sound system, without excessive loading of the signal source.

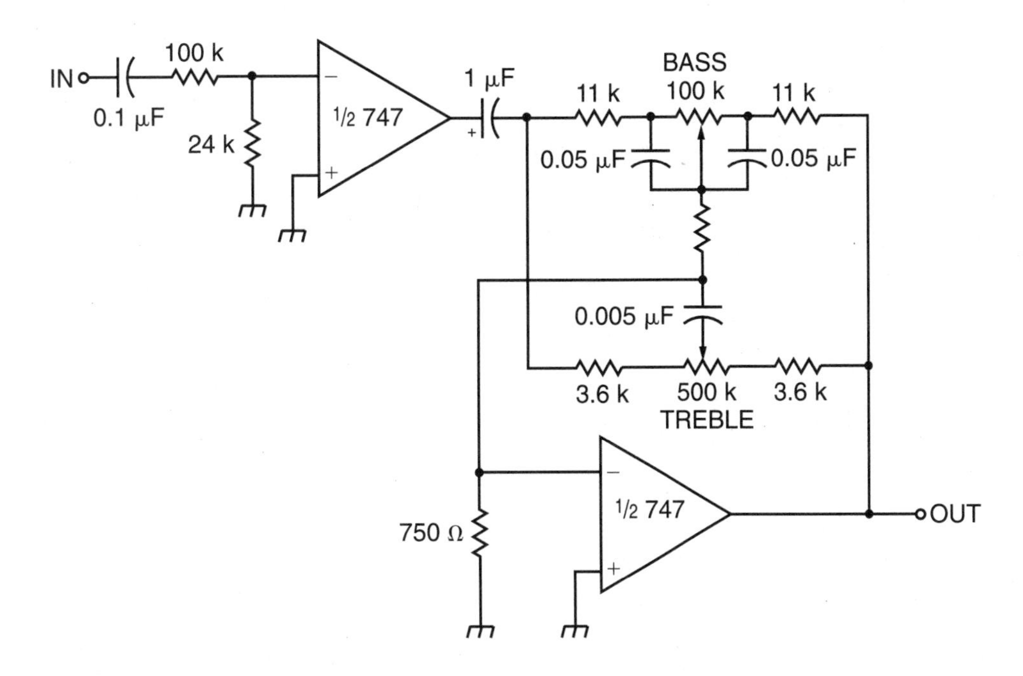

Window comparator

The output voltage from this circuit is normally LOW (about at 0), goes HIGH (slightly below the supply voltage) if and only if the input voltage is within a specific range or window. If the input voltage is above the upper end of the window, or less than the lower end of the window, the output will be LOW.

The window limits are set by the values of the unspecified resistors, R_1 through R_4. R_1 and R_2 set the uppermost end of the window range, while R_3 and R_4 set the lower limit. Each resistor pair is a simple voltage divider, with the resistor values selected to set the desired reference voltage. Obviously, for the circuit to function properly, the R_1/R_2 voltage must be greater than the R_3/R_4 voltage.

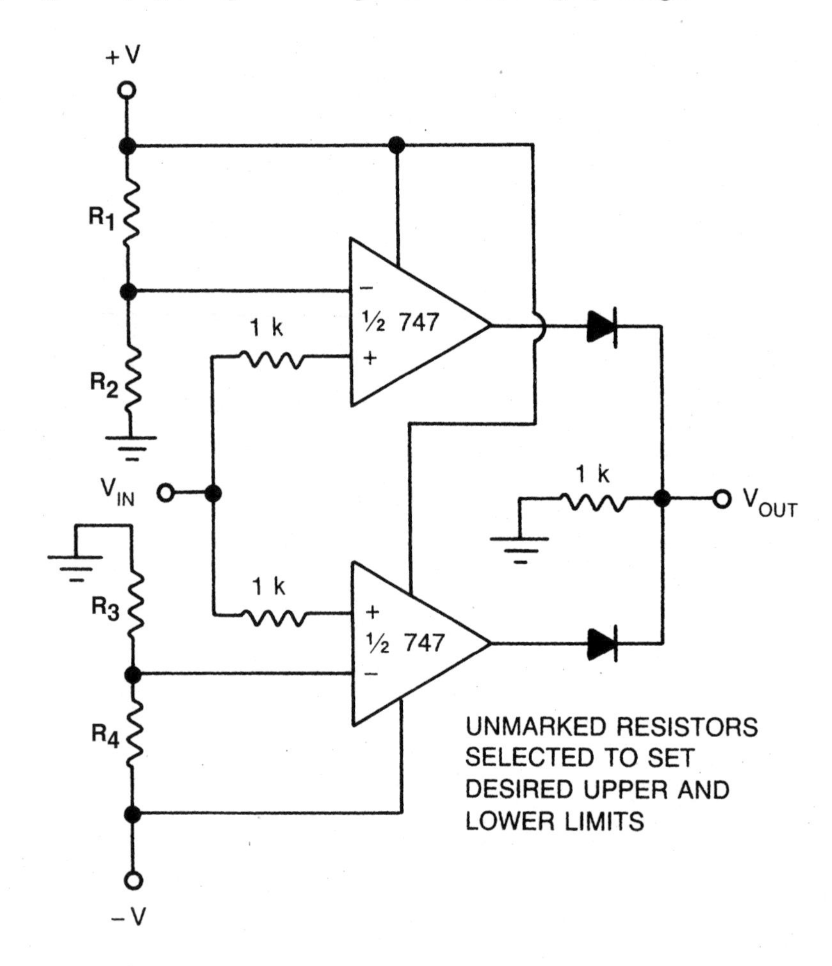

Variable Q notch filter

All of the frequency components in the input signal will appear in the output signal, except those within the specified passband. In effect, this circuit bites a notch out of the nominal flat frequency response.

The filter's Q can be adjusted via the 50K potentiometer. This determines the width of the notch, or how wide a frequency range it covers.

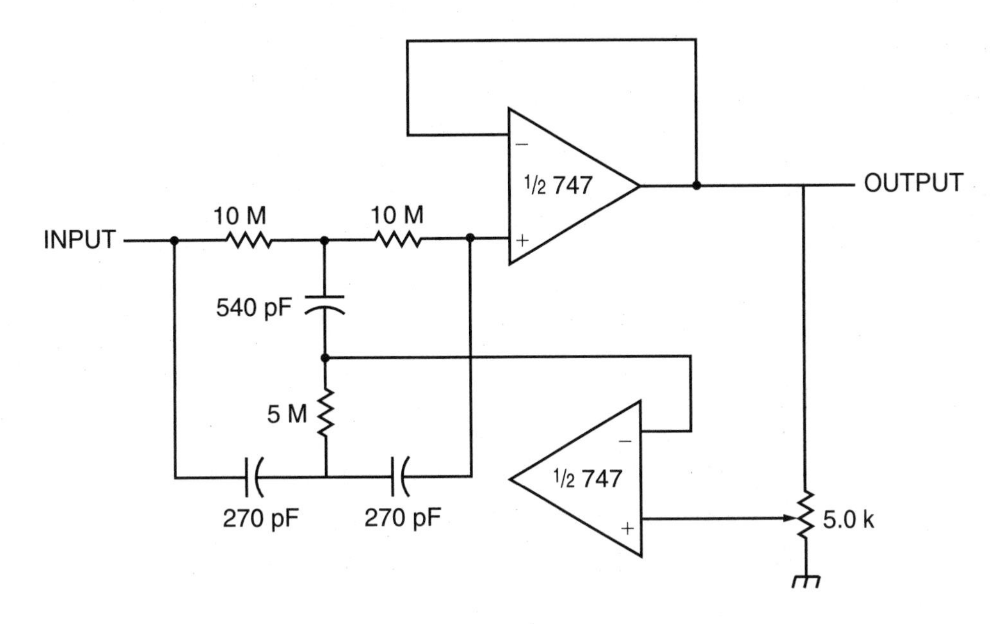

Magnitude detector

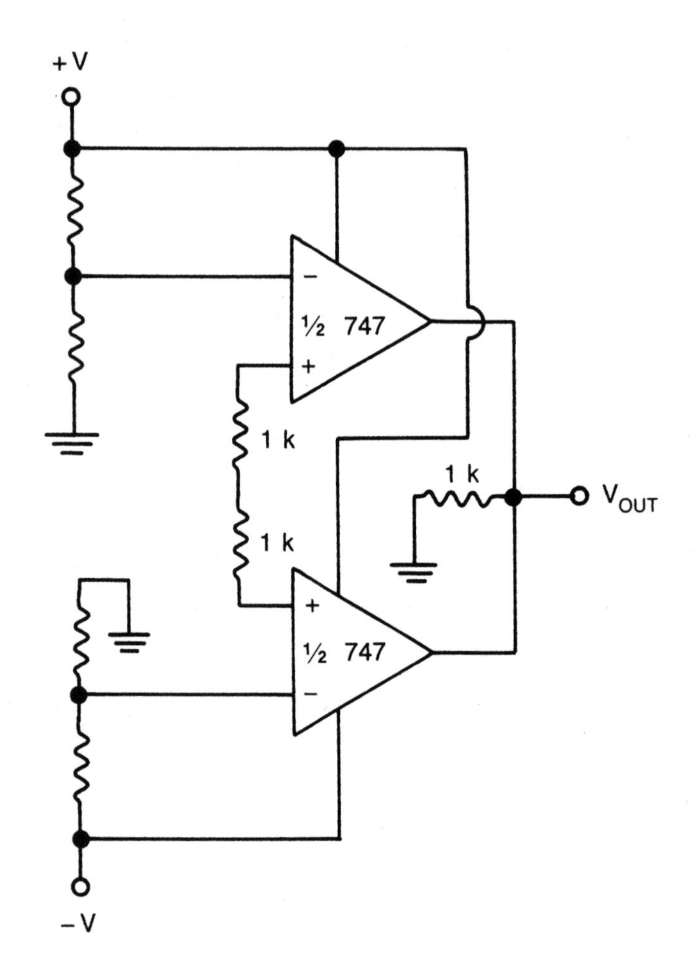

Bilateral current source

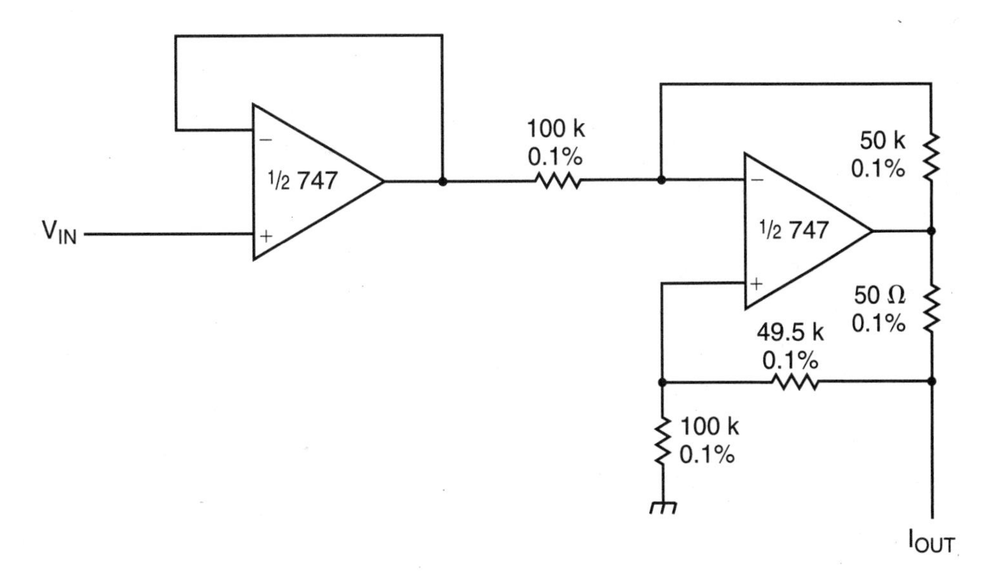

High-pass active filter

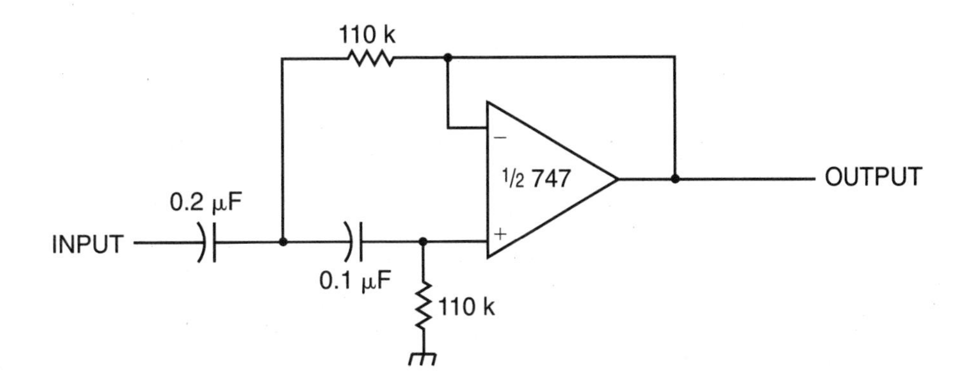

Absolute value extractor

This circuit ignores the polarity of the input voltage to extract the absolute value (distance from zero). The output voltage will always be positive, regardless of whether the input voltage is positive or negative.

Notice the relationships between the resistor values outlined in the circuit diagram. If these relationships are not followed, the circuit will not work properly, if at all. If 10K is selected as the base resistor value (R_1), then the following values should be used:

R_1, R_2, R_5, R_6	10K
R_3, R_4	4.75K (5K nominal)
R_7	2.375K (2.5K nominal)

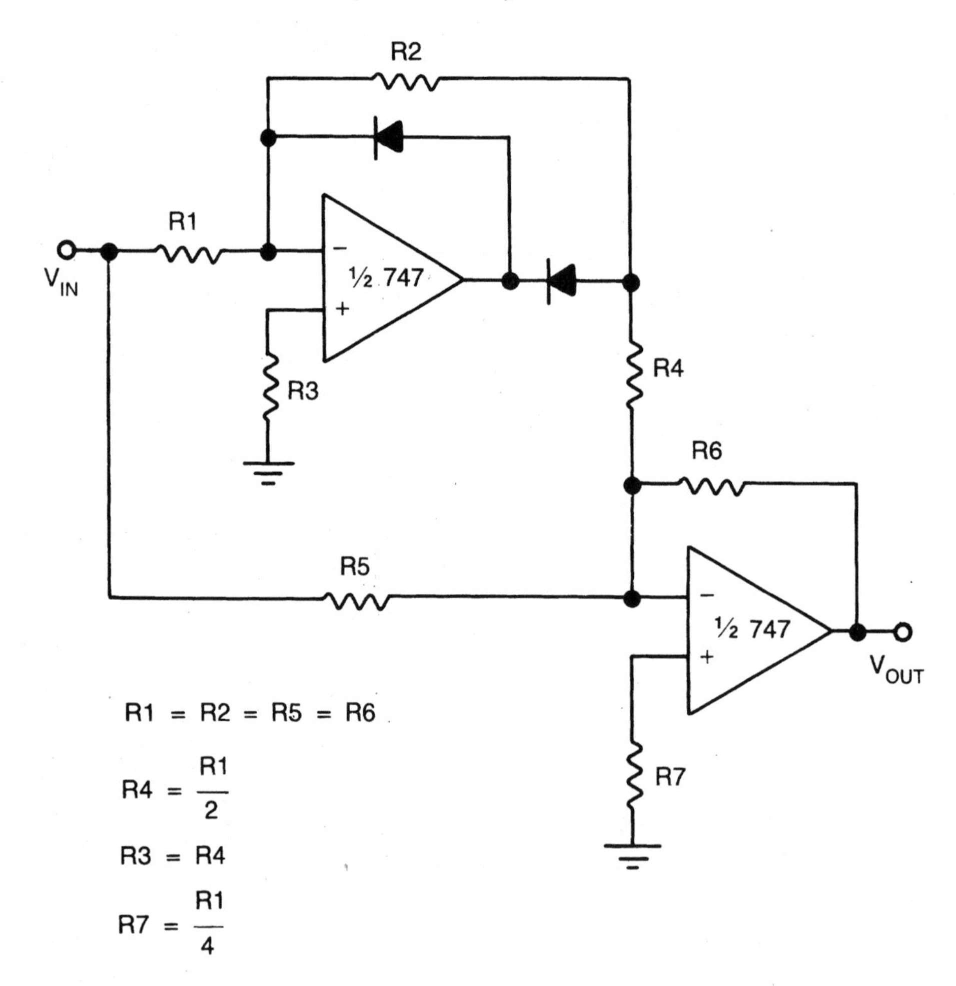

Null indicator

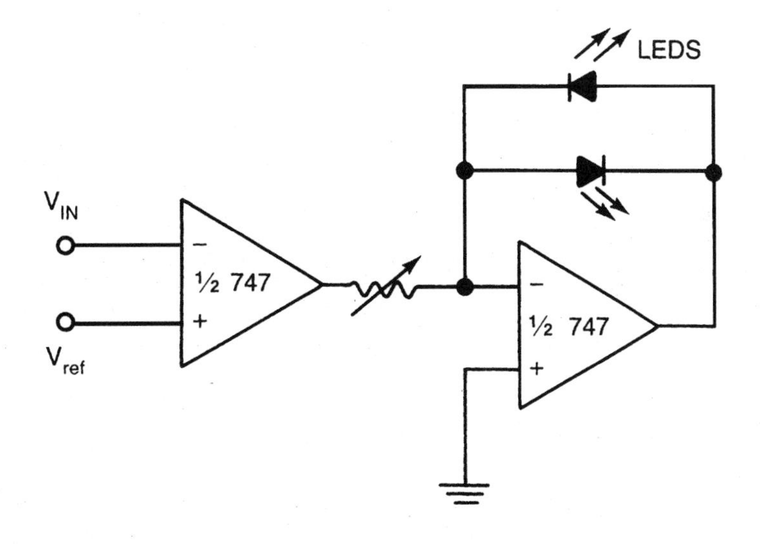

Positive peak detector with buffered output

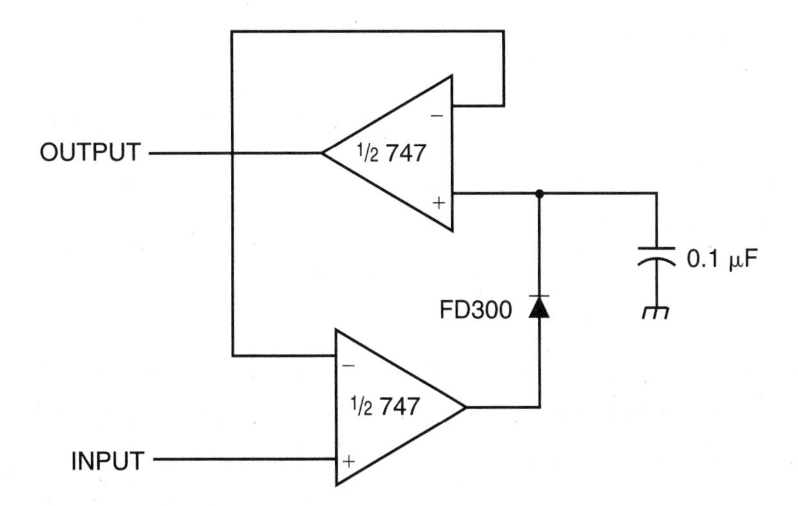

Inverting peak detector

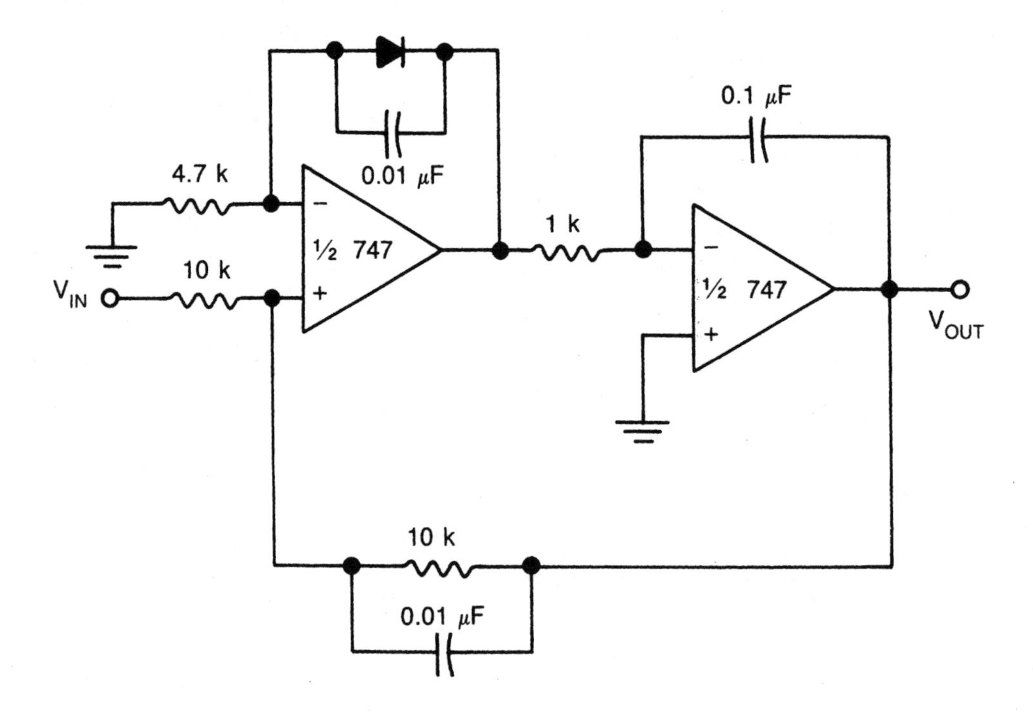

Noninverting peak detector

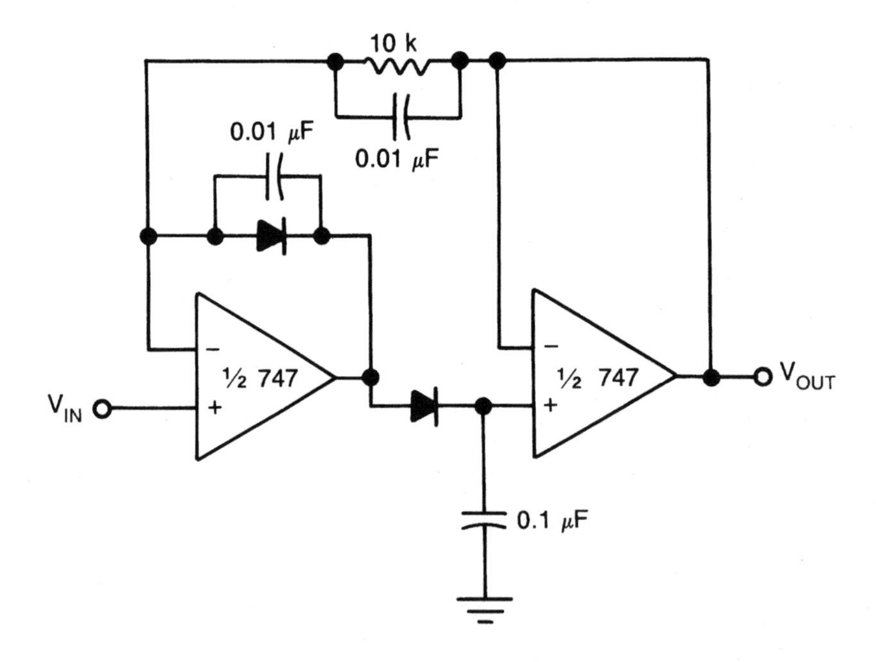

Decibel meter

This circuit reads out the input voltage according to the logarithmic decibel scale. Almost any NPN transistor can be used as the logarithmic feedback device, providing that it can handle the intended signal levels, of course.

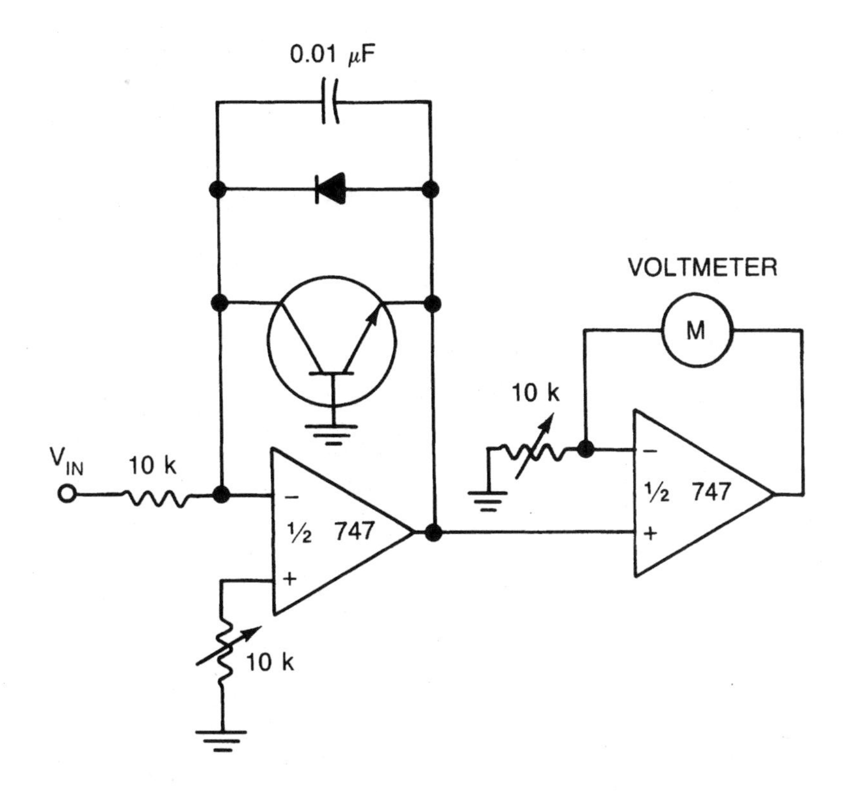

748 high-performance operational amplifier

The 748 is an improved version of the popular 741. It operates similarly, but with somewhat better specifications. It can be frequency compensated with a single 30 pF capacitor.

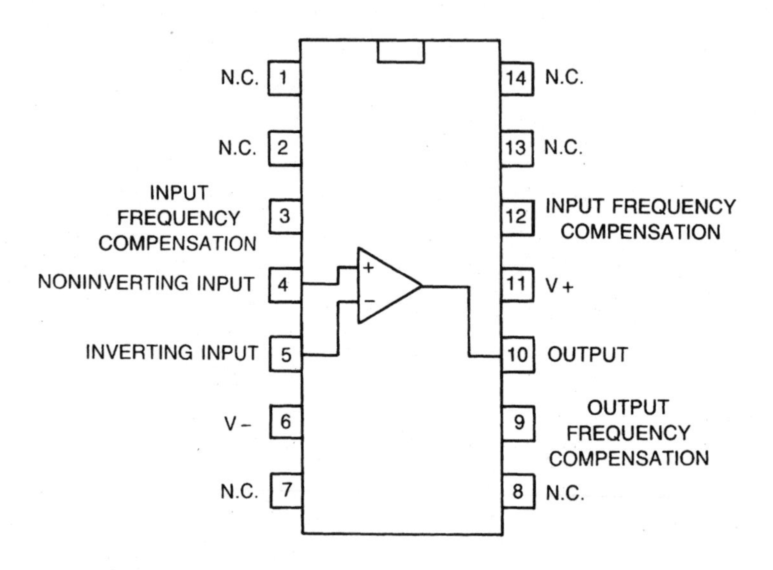

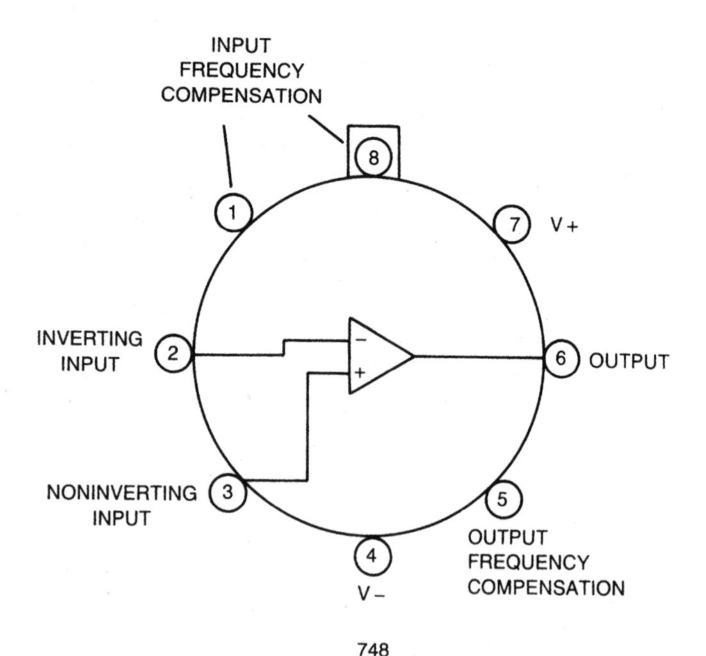

Inverting amplifier with balancing circuit

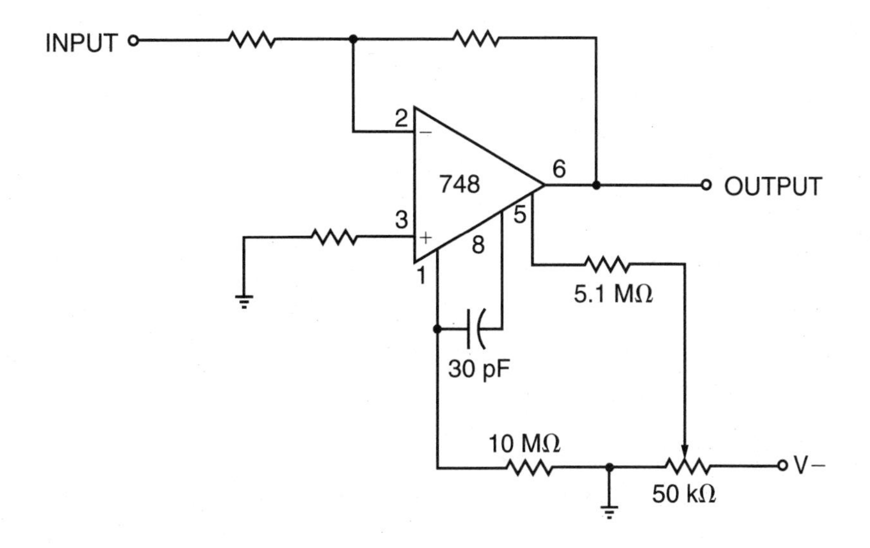

Low-drift sample and hold

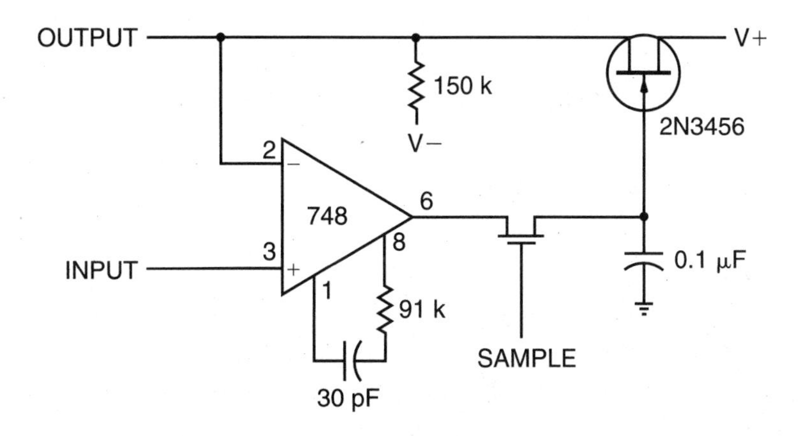

AGC amplifier

AGC is short for Automatic Gain Control.

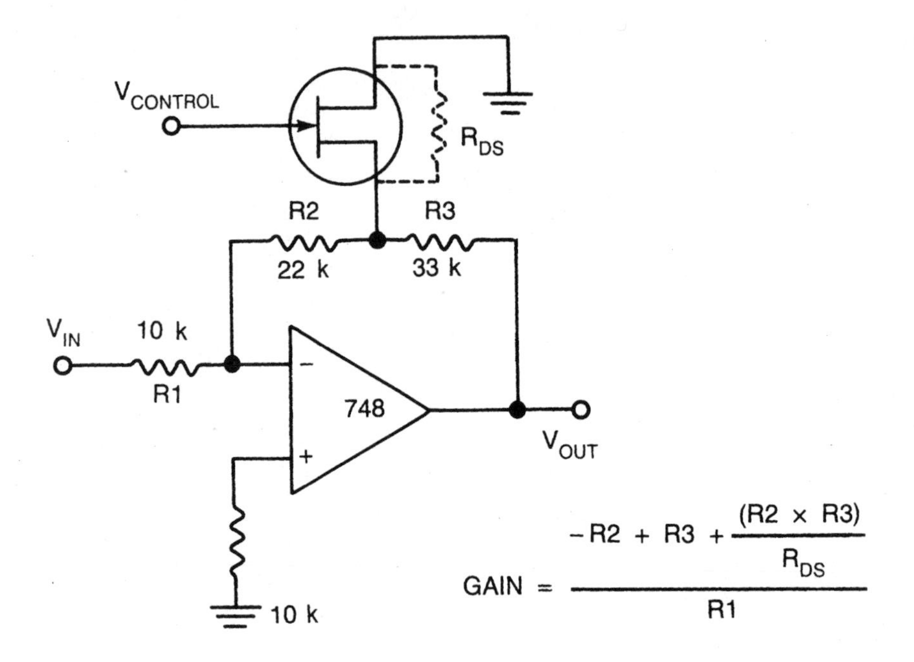

$$\text{GAIN} = \frac{-R2 + R3 + \dfrac{(R2 \times R3)}{R_{DS}}}{R1}$$

Voltage comparator for driving RTL logic or high-current driver

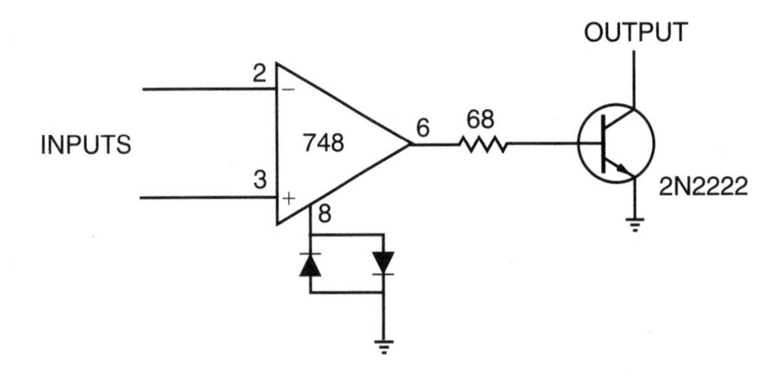

Voltage comparator for driving DTL or TTL ICS

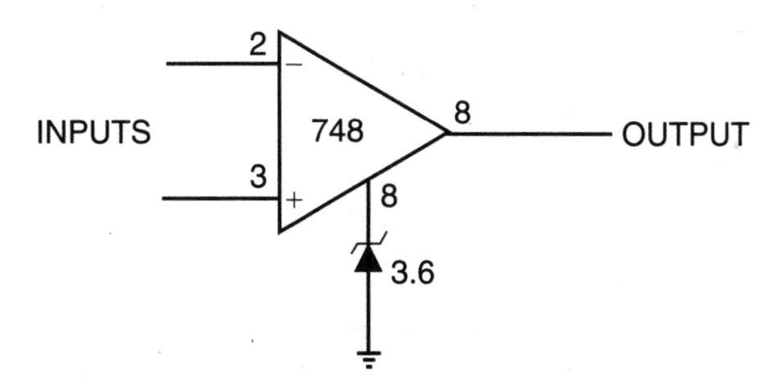

Audio preamplifier with tone controls

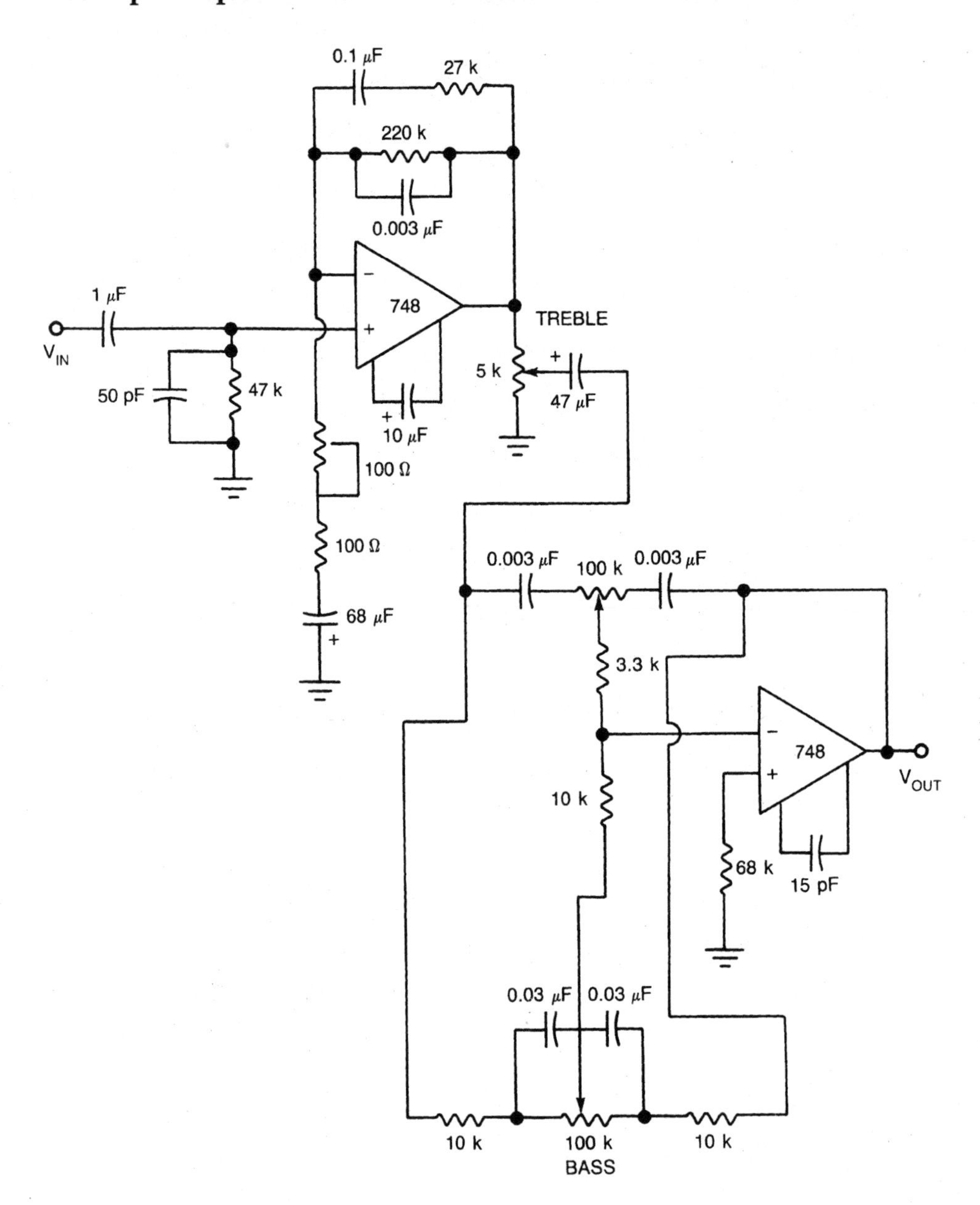

Feed-forward compensation

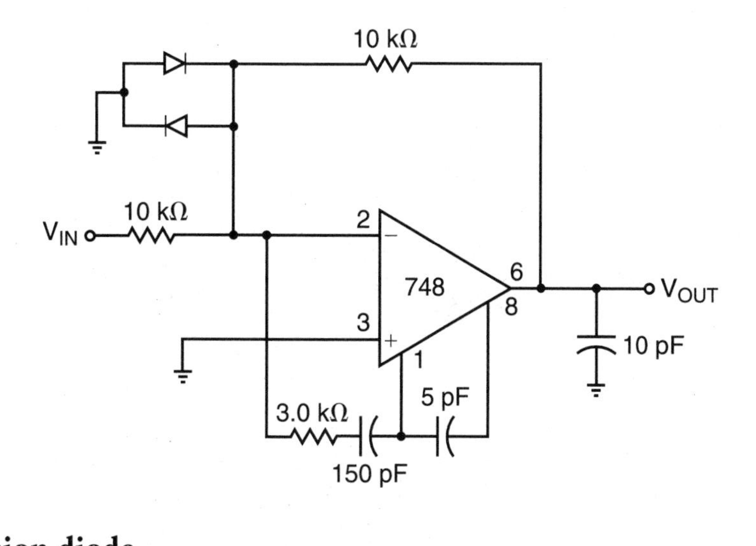

Precision diode

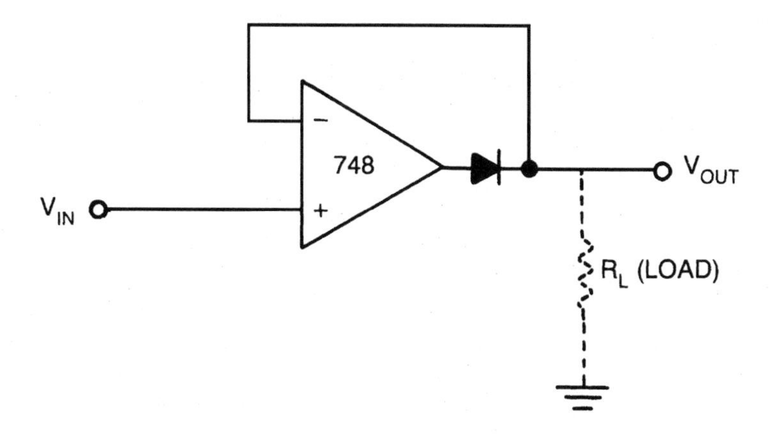

Pulse width modulator

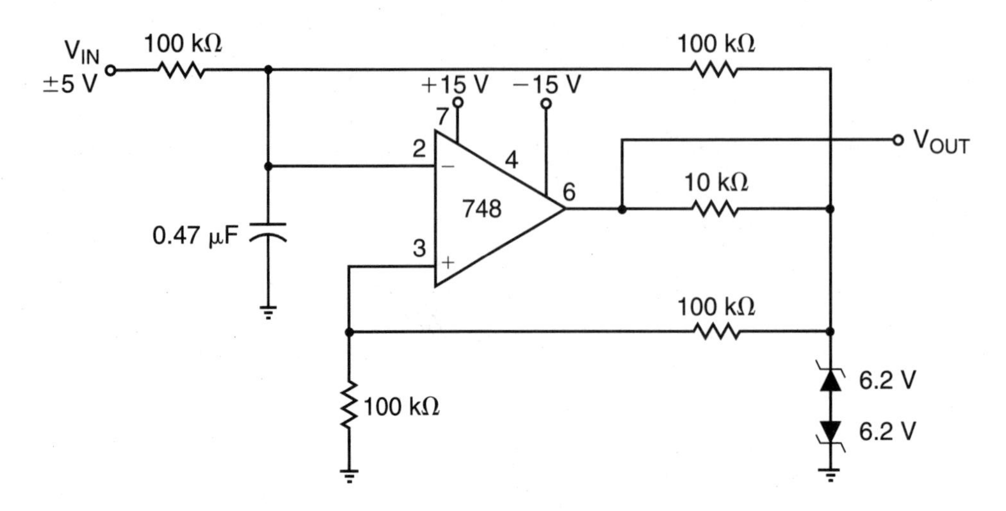

Zero-crossing detector

The output of this circuit is active when the input voltage passes from positive to negative, or from negative to positive. The zener diodes in the feedback loop should be selected with a voltage appropriate to the intended application. If, for example, 2.1 V zener diodes are used, the output voltage will follow the input voltage if it is between −2.1 V and +2.1 V. If the input voltage is greater than +2.1 V, or less than −2.1 V, the output will be saturated.

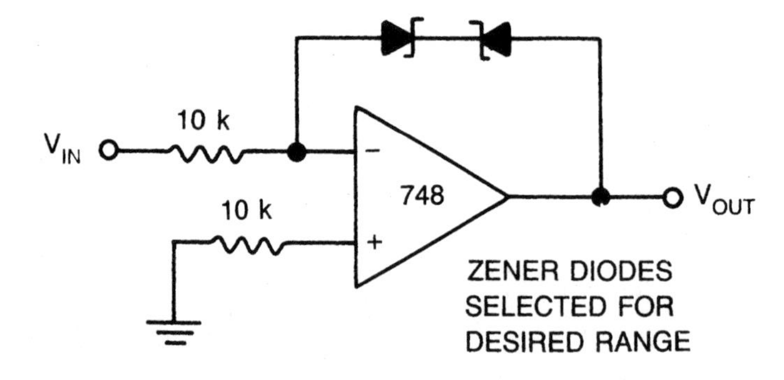

R-2R D/A converter

The analog output voltage will be proportionate to the 4-bit digital value presented at the inputs. The input resistors must keep the R-2R relationship. Two 10K resistors in series can be used for each 20K resistor shown in the schematic.

The 50K potentiometer adjusts the amplifier's gain, and therefore, the analog voltage range of the output. If such control is not called for in the intended application, this potentiometer can be replaced with a fixed resistor of the appropriate value to give the desired gain.

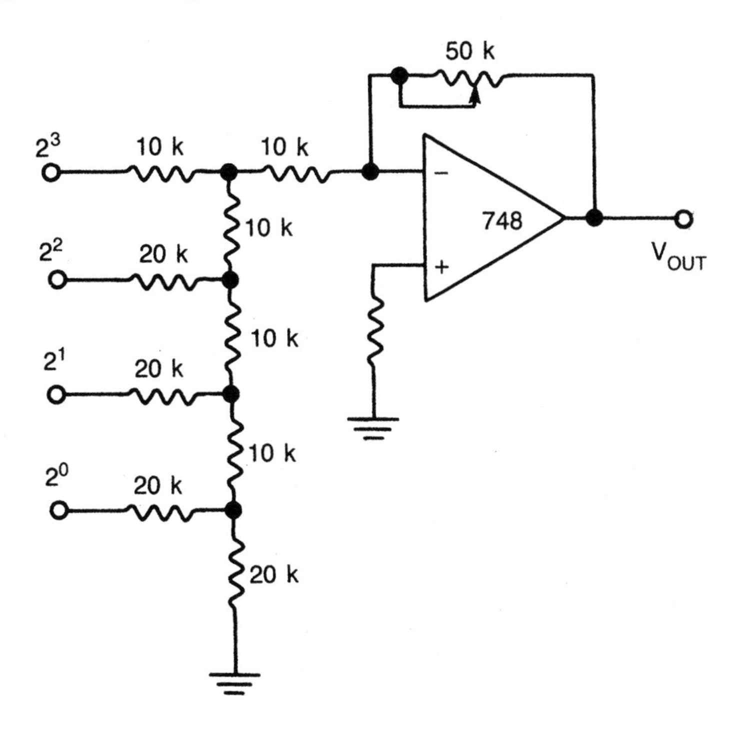

Sawtooth-wave generator

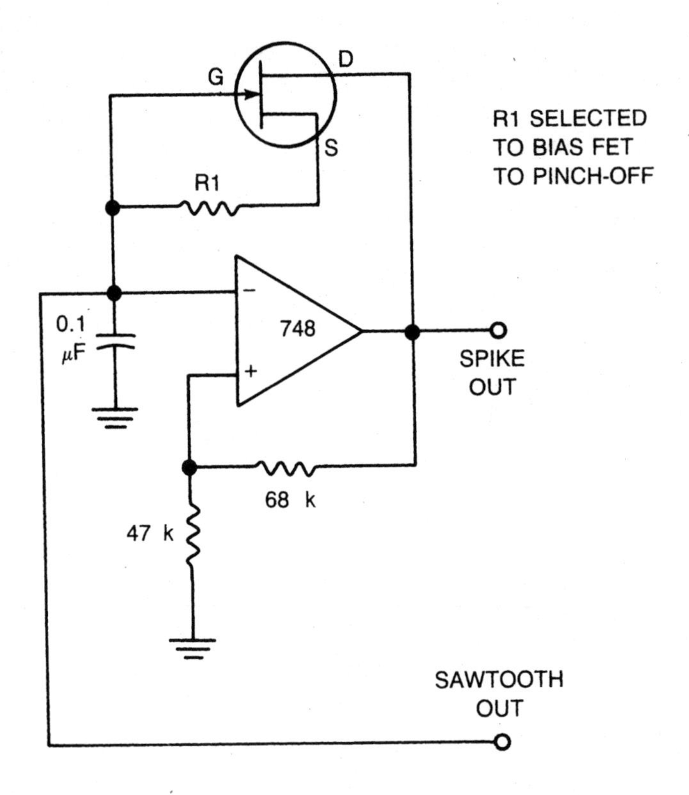

Low-pass filter

Any frequency component in the input signal below the cut-off frequency determined by the resistor and capacitor values in the circuit passes through to the output. Input frequency components above the cut-off frequency are increasingly attenuated at the output.

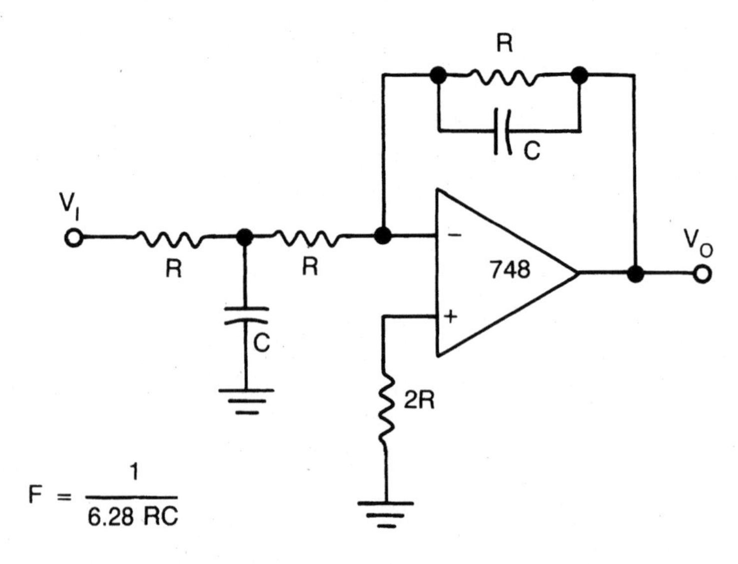

Record scratch filter

This circuit helps reduce the annoying pops and ticks heard when a scratched record is played. Of course, it won't help much if the record is severely scratched. But for mild scratches, this filter can result in a vast improvement in the sound quality.

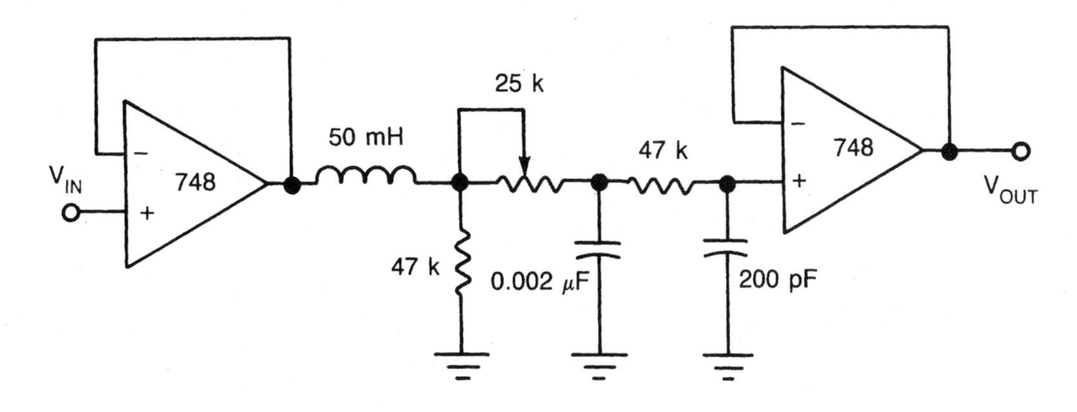

1458 Dual-compensated operational amplifier

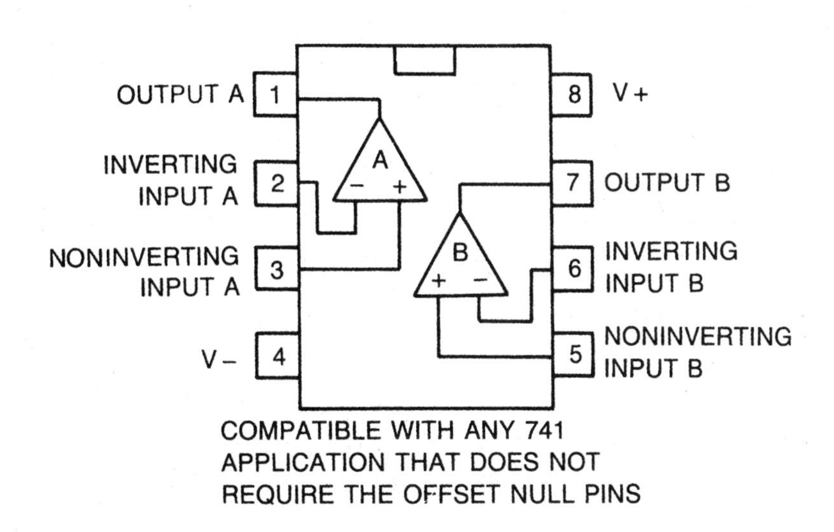

Peak detector

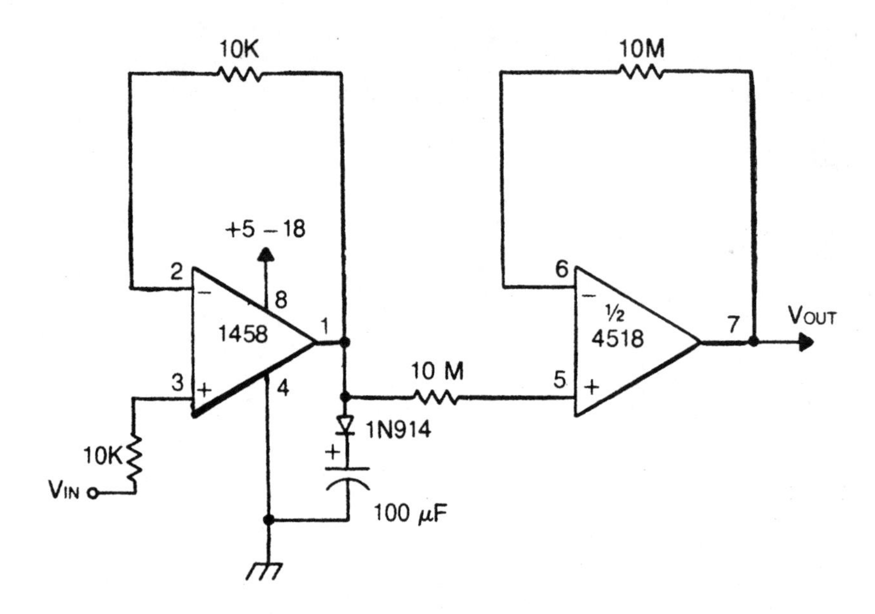

Pulse generator

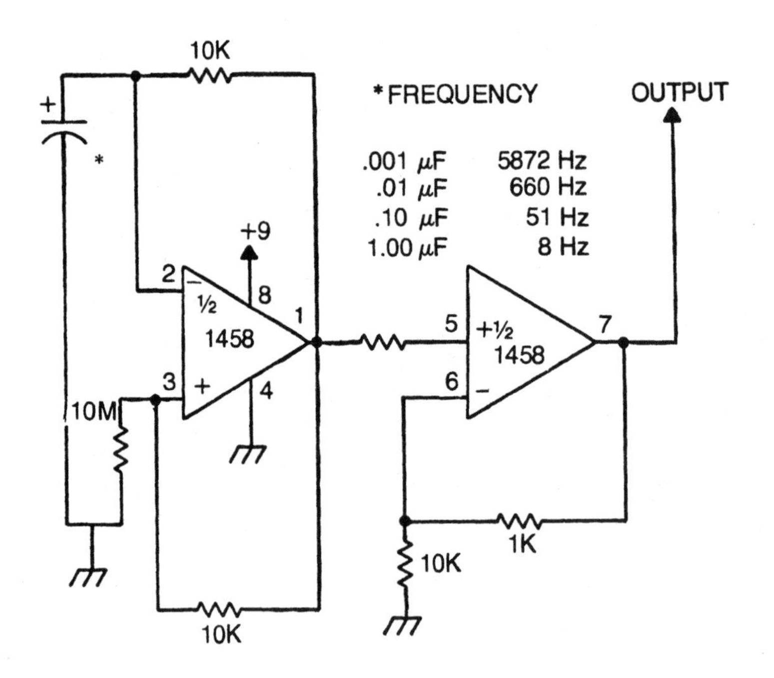

3301 quad single-supply operational amplifier

3-output function generator

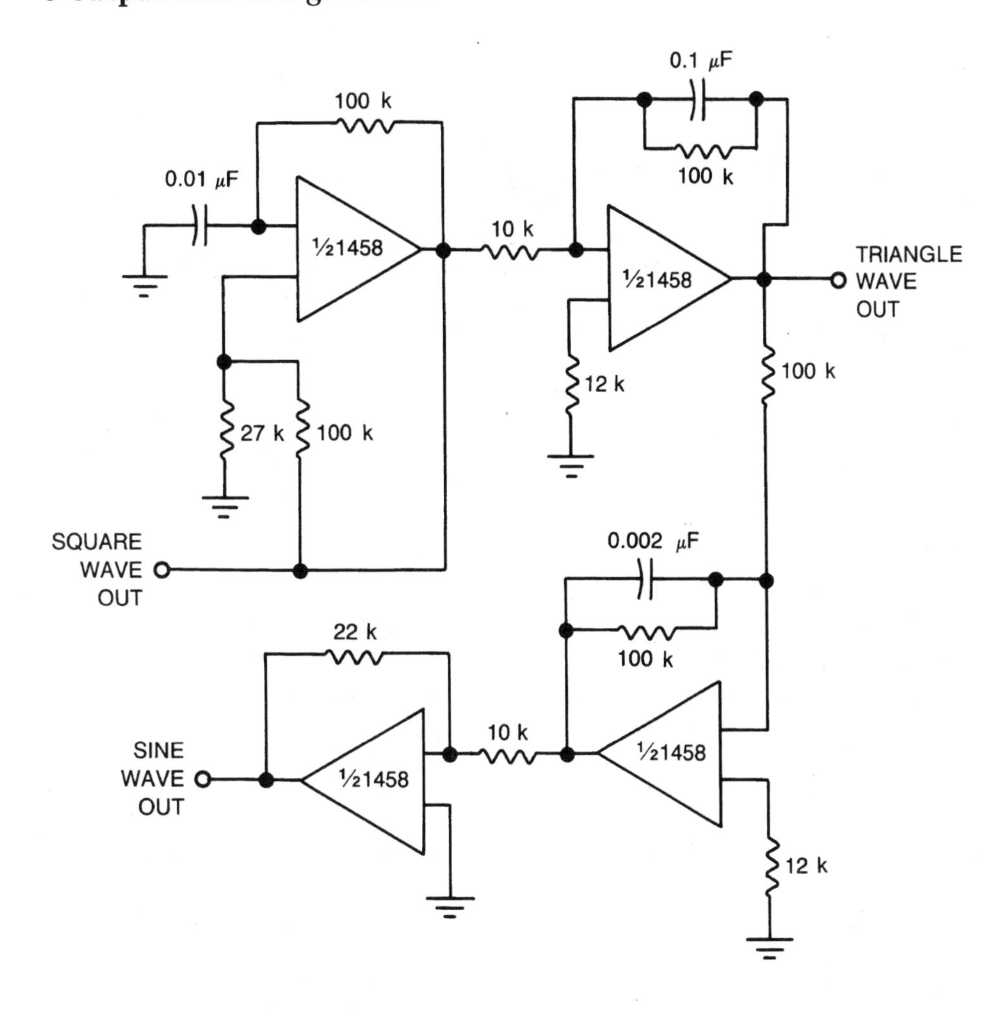

Component values shown for output frequency of about 100 Hz

Noninverting amplifier

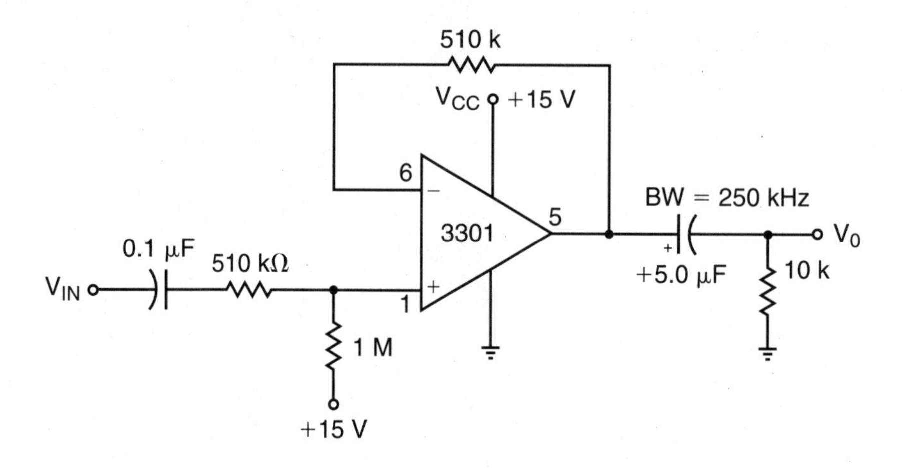

Inverting amplifier

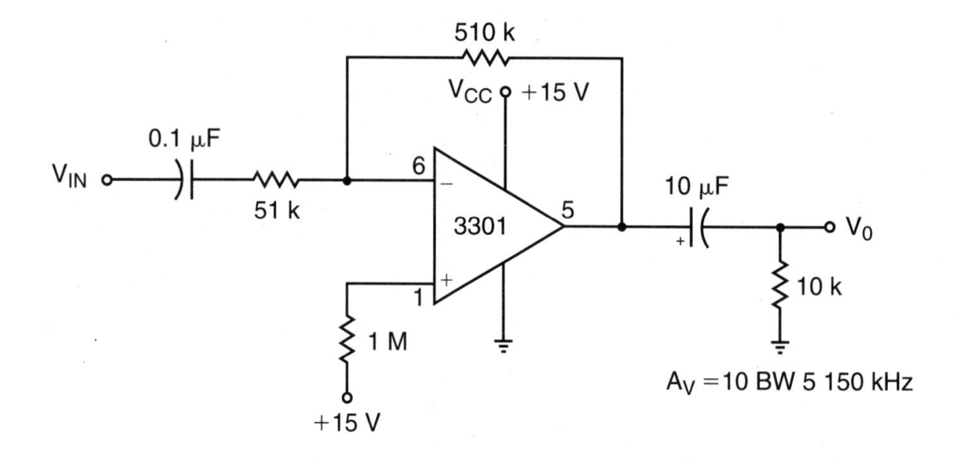

Astable multivibrator

This type of circuit is also known as a rectangle-wave generator.

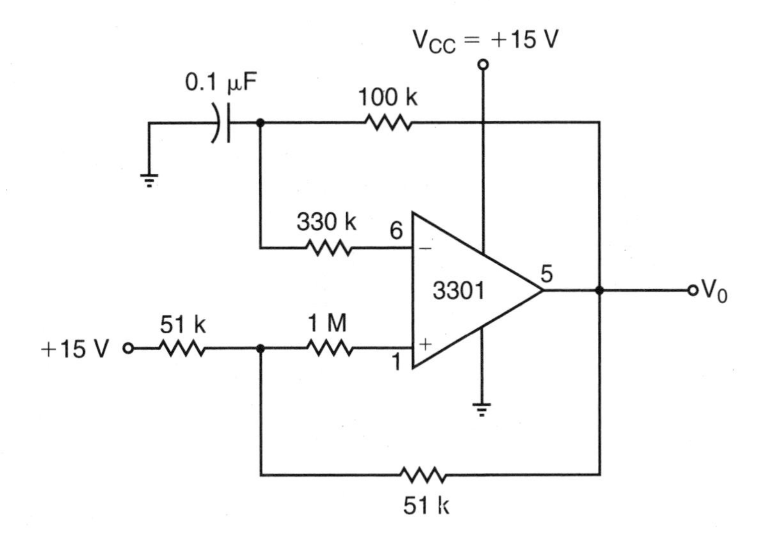

Positive-edge differentiator

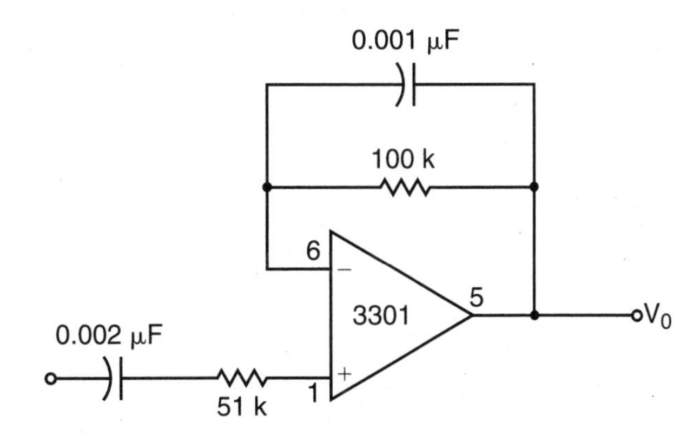

Negative-edge differentiator

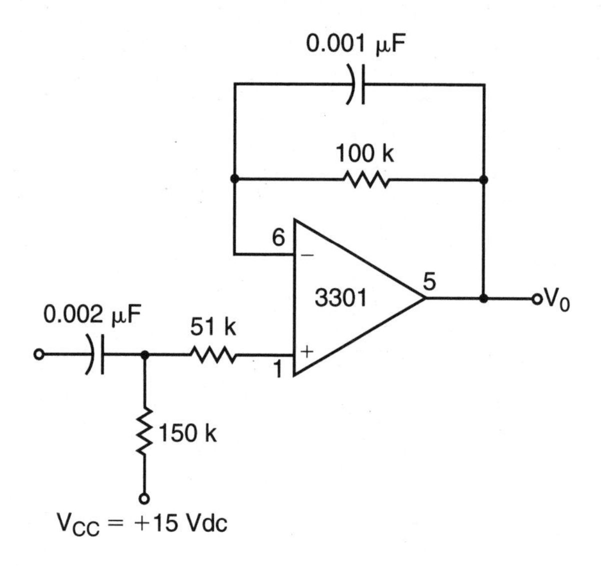

Logic NOR gate

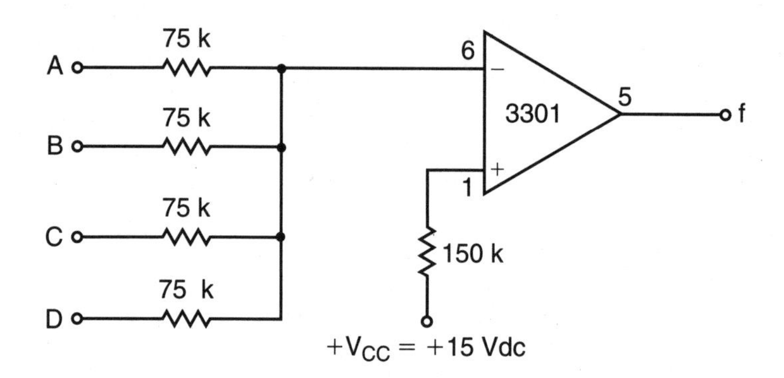

Logic OR gate

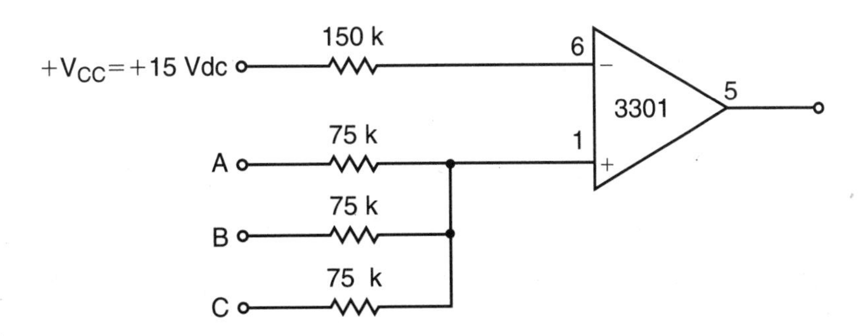

Logic NAND gate (with large fan-in)

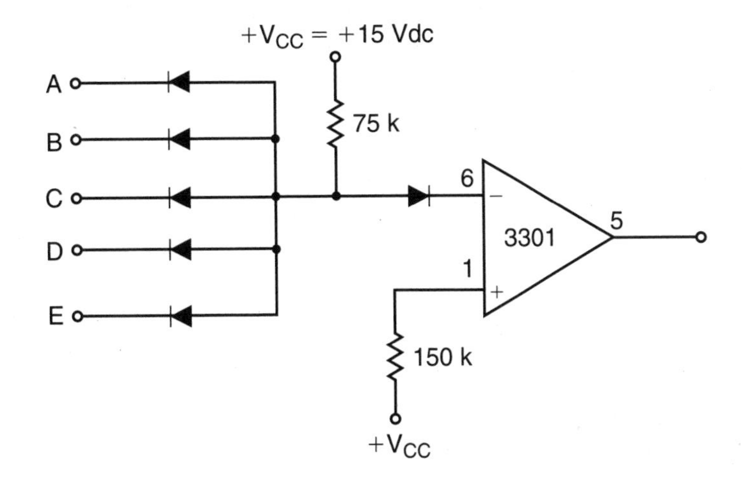

R-S flip-flop

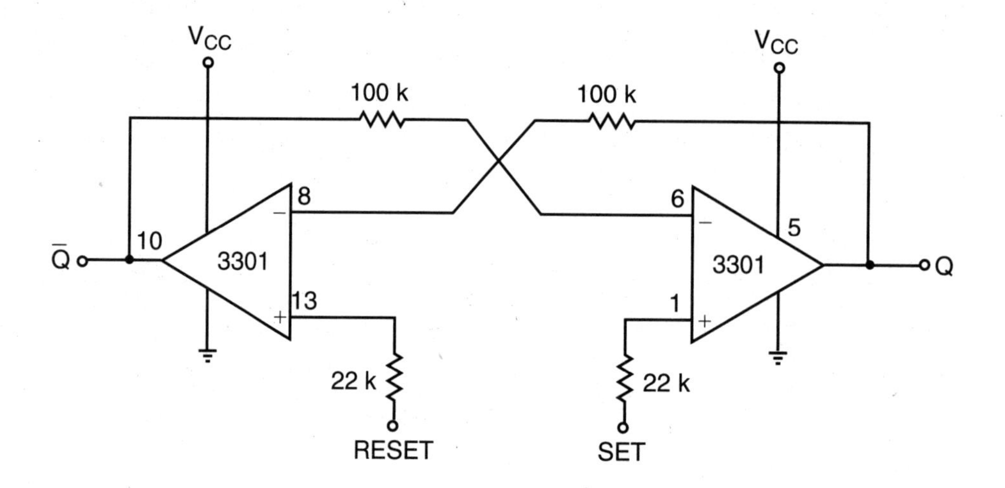

LM3900 quad Norton amplifier

The 3900 is a specialized type of operational amplifier known as the Norton amplifier. This is indicated by the arrowhead between the inputs of its schematic symbol. A Norton amplifier uses a current mirror to achieve the noninverting input function.

Four independent (except for the power supply connections) Norton amplifiers are included on the 3900. These devices were designed to operate from a single-ended power supply, and to provide a large output voltage swing.

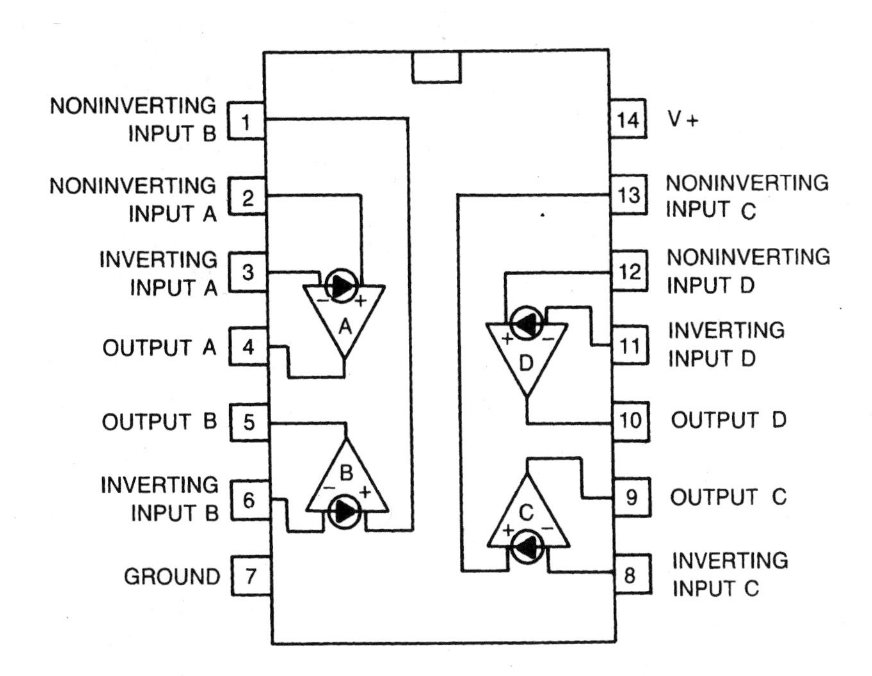

Inverting amplifier

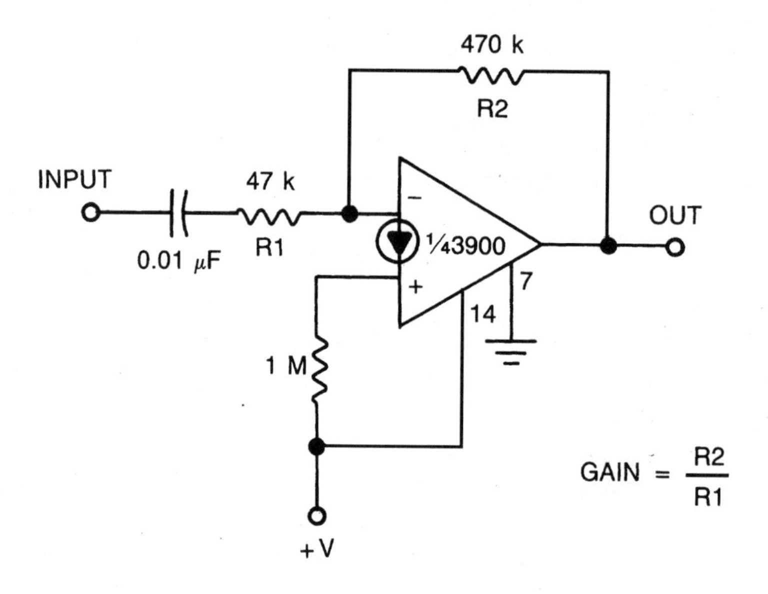

Flip-flop

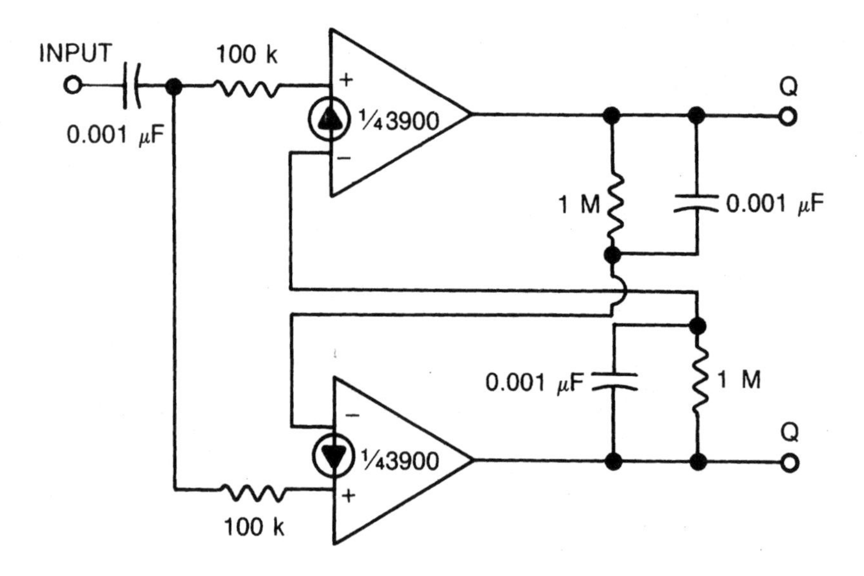

Square-wave generator

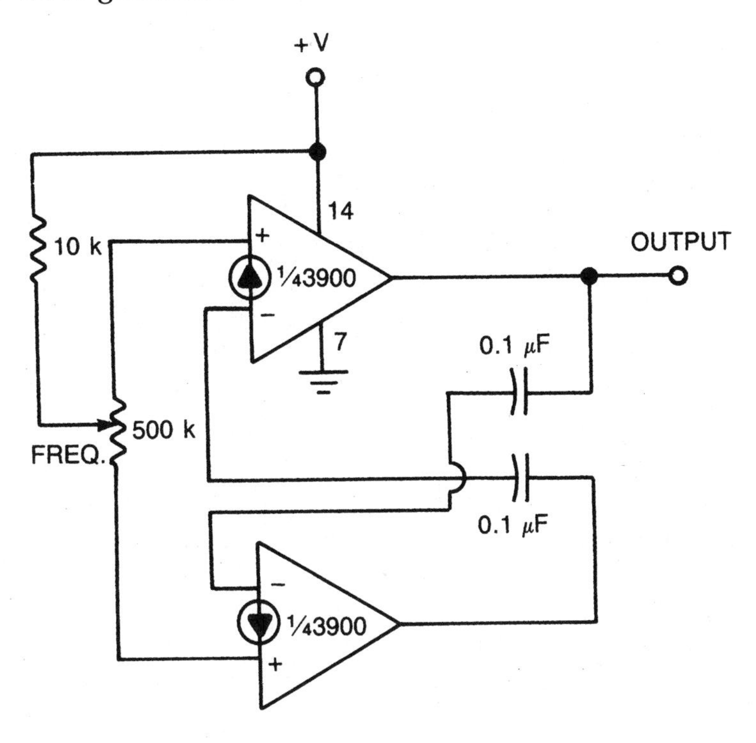

Function generator

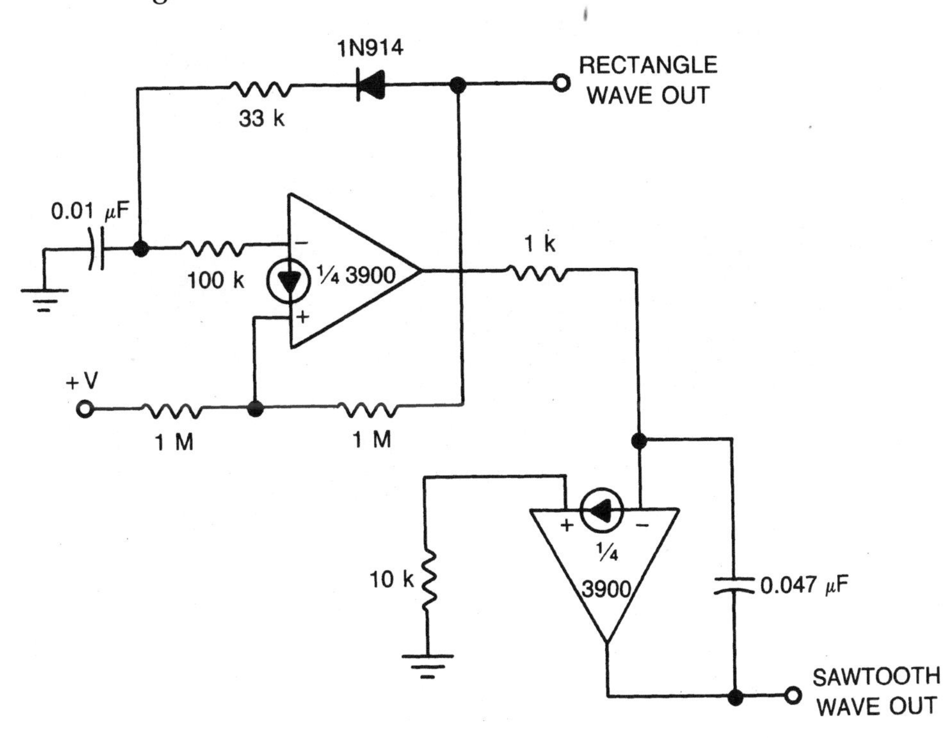

Dual LED flasher

When one LED flashes on, the other flashes off, and vice versa. The LEDs will alternate as long as power is applied to the circuit.

The flash rate can be altered by changing the value of the two feedback capacitors. If these two capacitors have equal values, the flashes will be symmetrical, each of an equal length. If unequal capacitor values are used, one LED will stay on longer during each cycle than its partner.

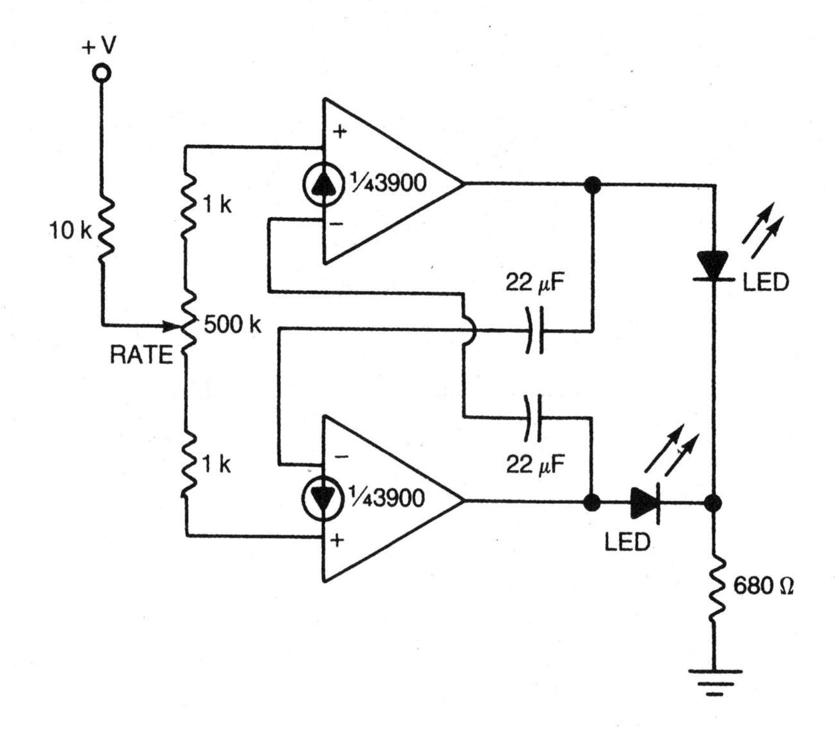

Siren with programmable frequency and rate adjustment

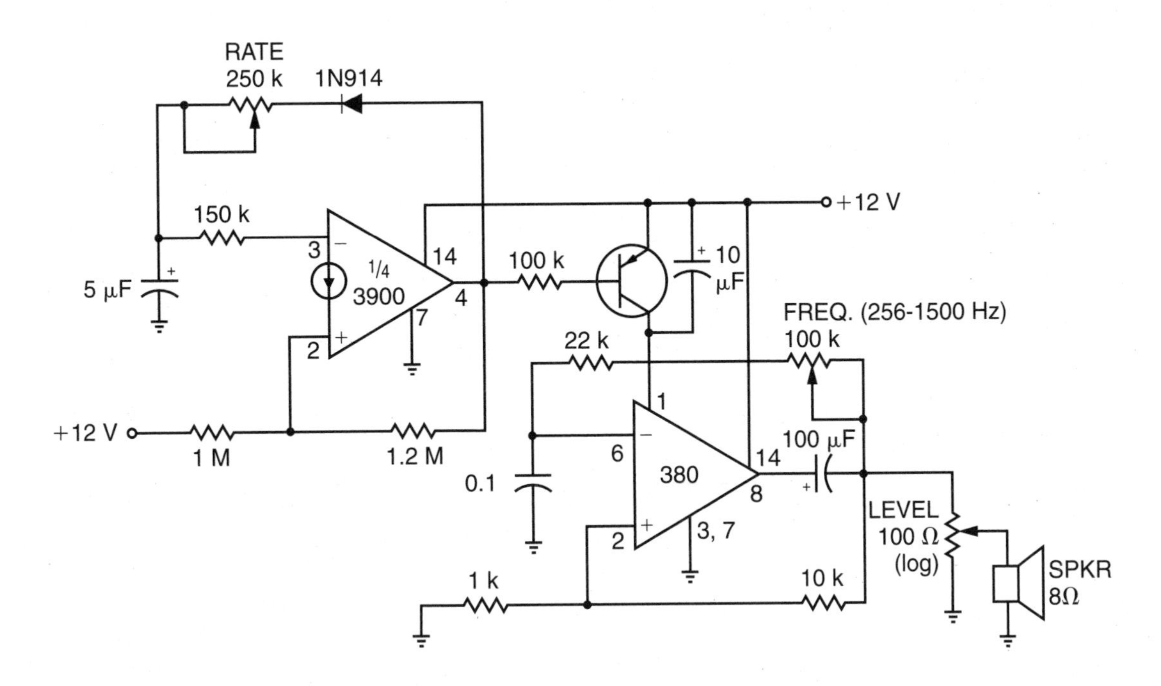

5

SECTION

Audio amplifiers

Perhaps the most commonly used type of active electronic circuit is the amplifier. Usually, an amplifier accepts an input signal, and produces a higher amplitude replica of the input signal at the output. The amount of increase is called the *gain*. To be absolutely literal, the term *amplifier* implies positive gain—that is, the output signal is larger (higher amplitude) than the input signal. In practical electronics, this is not always the case. Some amplifiers actually have negative gain, so that the output signal is lower in amplitude than the input signal. This type of reverse amplification is also known as *attenuation*. Some amplifier circuits have unity gain, which means the output signal has the same amplitude as the input signal. This might sound like a useless circuit, but it can be very useful in interfacing equipment or circuits. A unity-gain amplifier limits potential loading problems, and/or can provide impedance matching. A unity-gain amplifier is also sometimes referred to as a *buffer*.

In this section, we will look at ICs designed primarily for the amplification of signals within the audio-frequency (af) range (nominally about 20 Hz to 20 kHz). In Sec. 6, we will move on to radio-frequency (rf) amplifiers, and related devices.

Incidentally, the operational amplifiers discussed in Sec. 4 are special variations on the basic amplifier. Op amps are sometimes used in audio or rf amplifier applications.

LM380N audio power amplifier

The 380 is a power audio amplifier chip designed for consumer applications. In order to hold system cost to a minimum, gain is internally fixed at 34 dB. A unique input stage allows the inputs to be ground referenced. The output is automatically self-centering to one-half the supply voltage.

The output of the 380 is short-circuit proof, with internal thermal limiting. A copper lead frame is used with the center three pins on either side, compromising a built-in heat sink. This makes the device easy to use in a standard pc layout.

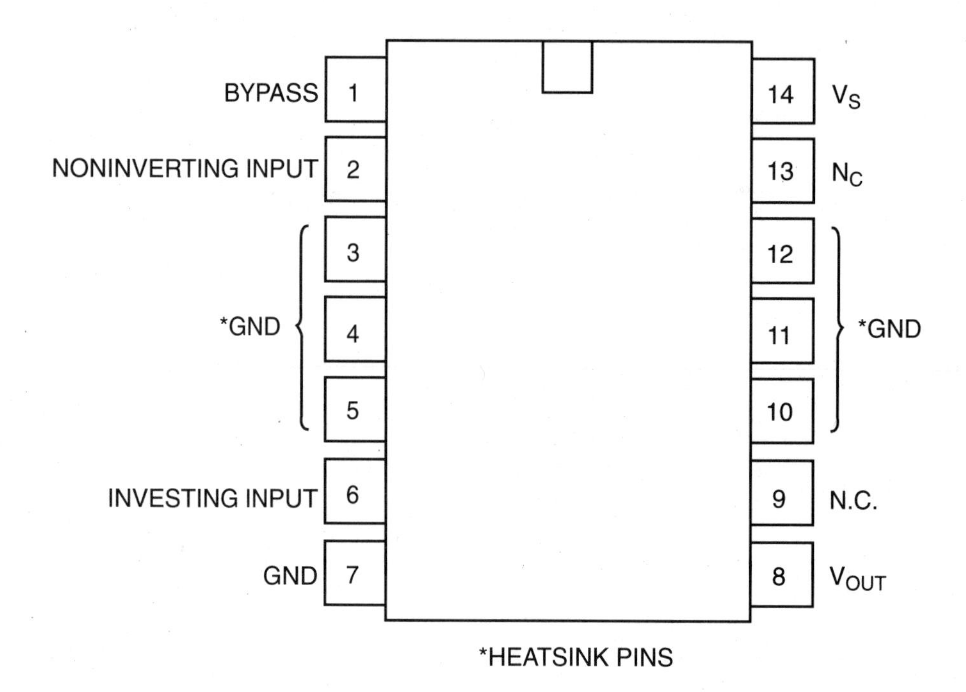

LM380N-8 audio power amplifier

The 380N-8 is identical to the 380N shown in the previous circuit, except this device is enclosed in an 8-pin DIP instead of a 14-pin DIP housing. The built-in heat-sinking pins are eliminated.

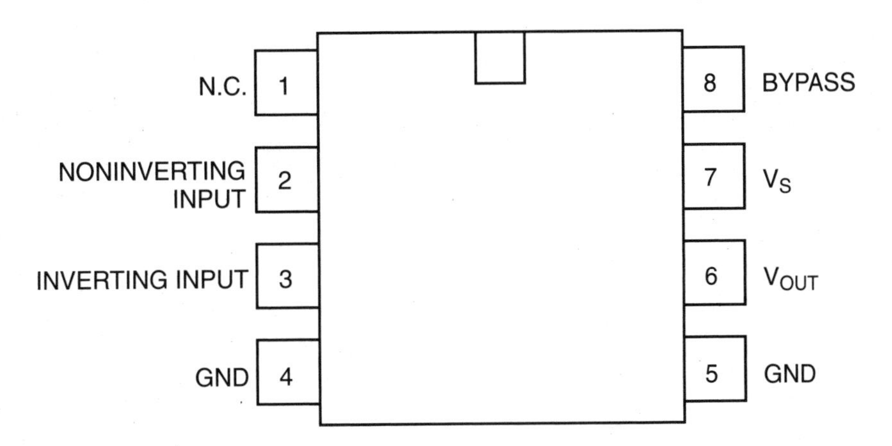

Basic audio amplifier circuit

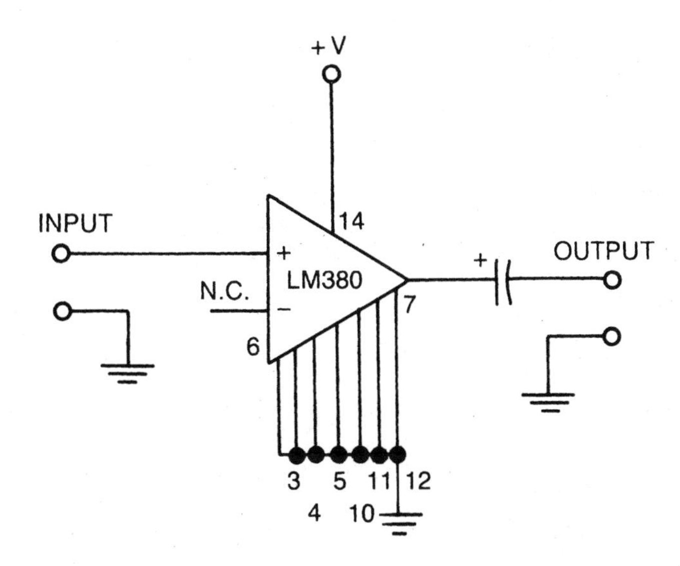

Audio amplifier with dual power supply

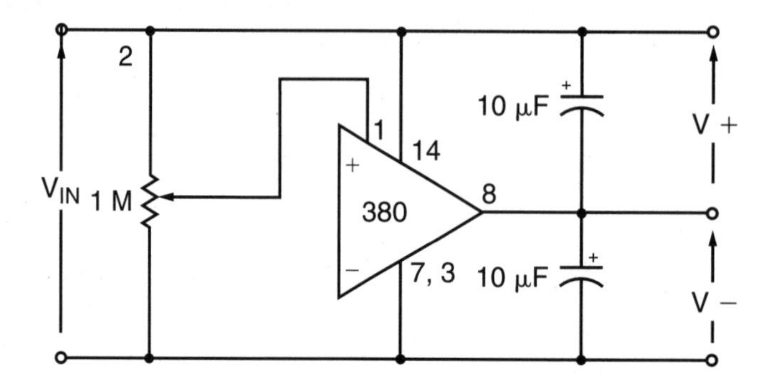

Phono amplifier with tone control

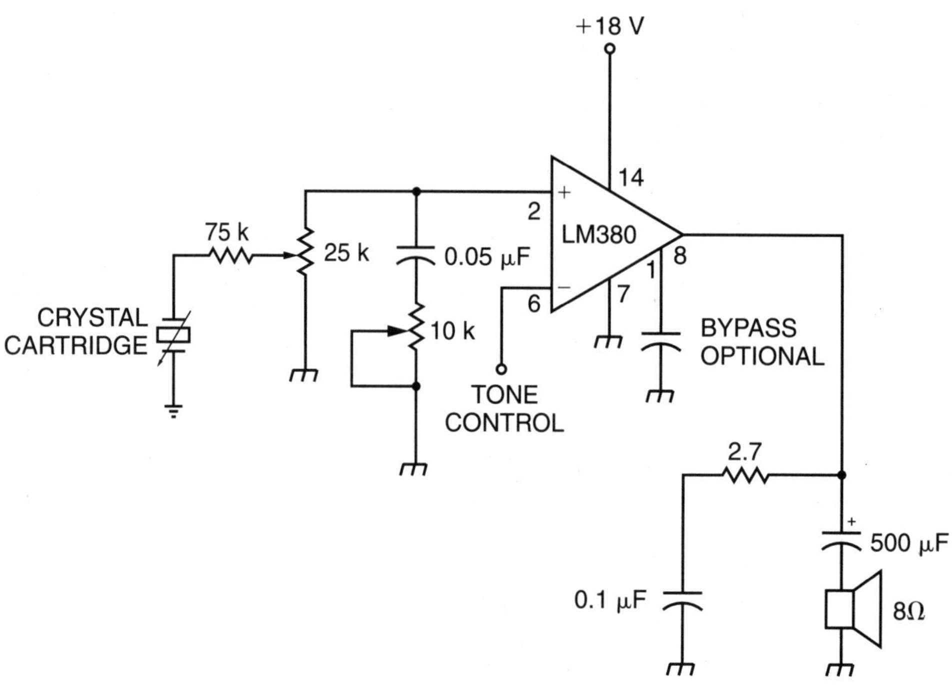

(*Courtesy of National Semiconductor.*)

RIAA phono amplifier

This phonograph amplifier has the standard RIAA frequency response compensation built in.

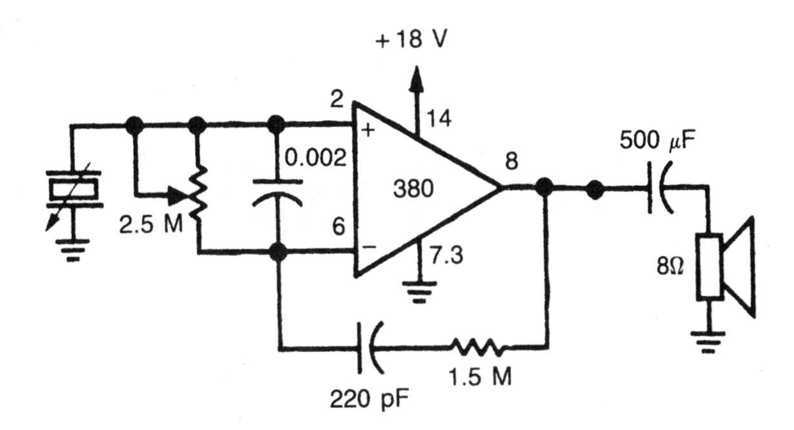

Boosted gain of 22 using positive feedback

The 380's internal gain is fixed at 34 dB, but higher gains can be achieved with external components, as illustrated here.

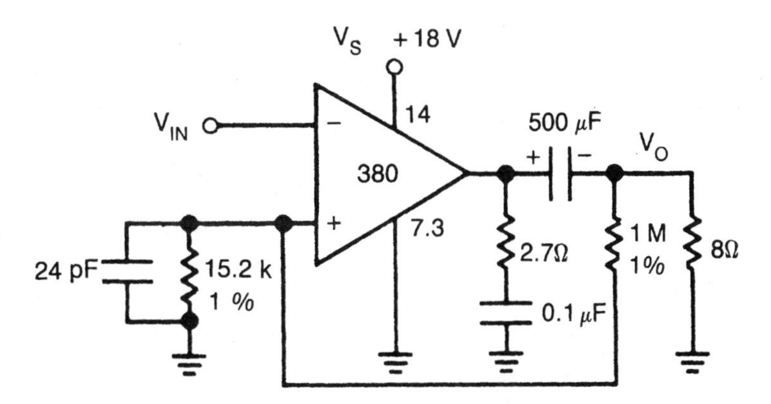

Audio amplifier with high-input impedance

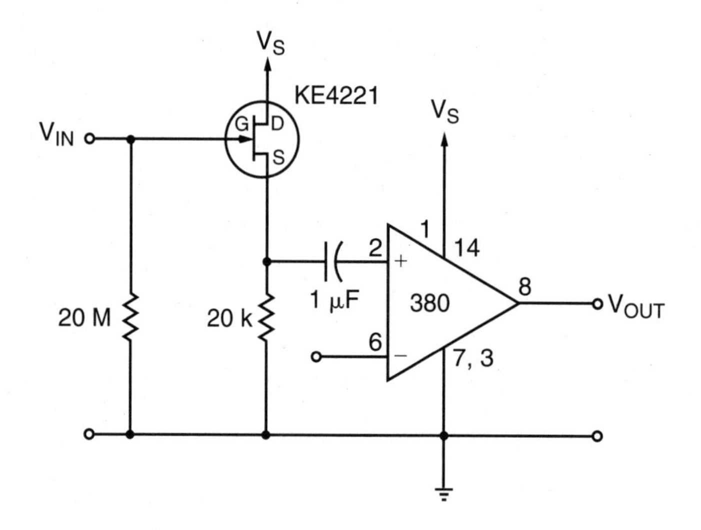

Intercom pair

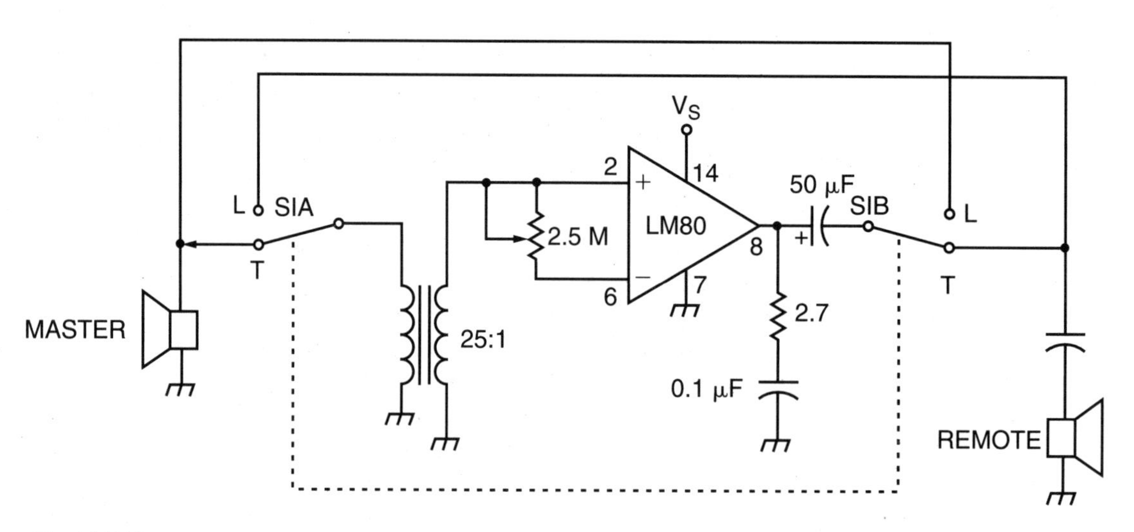

T = TALK
L = LISTEN

(*Courtesy of National Semiconductor.*)

Audio amplifier with voltage divider input

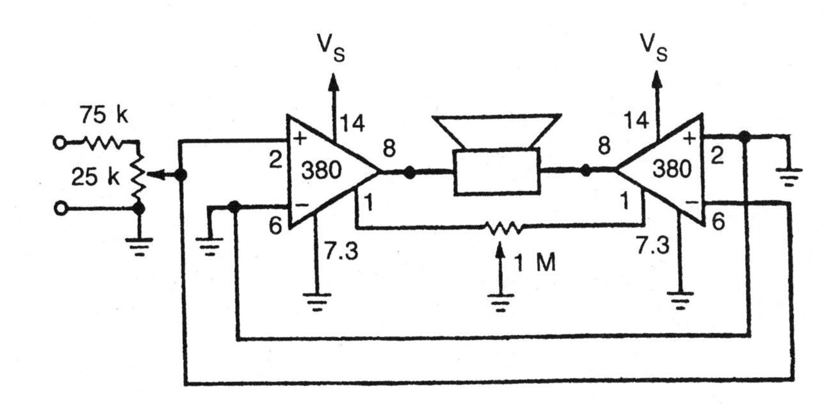

Quiescent balance control

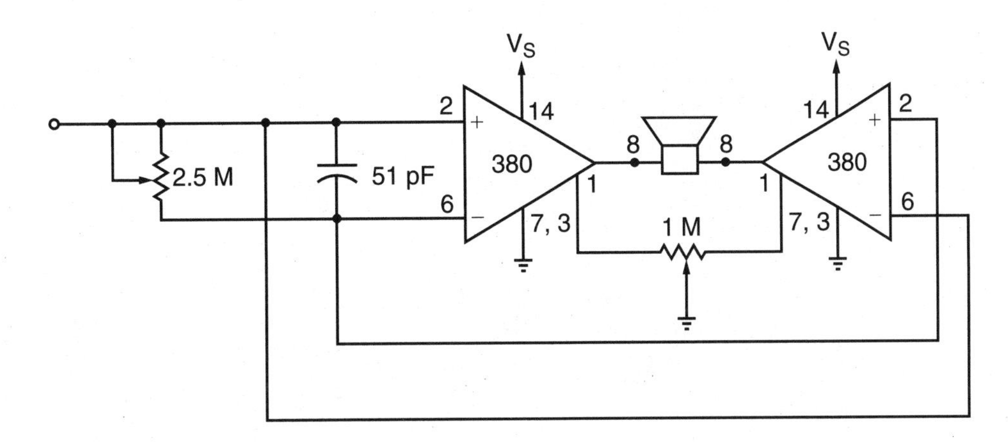

Phase shift oscillator

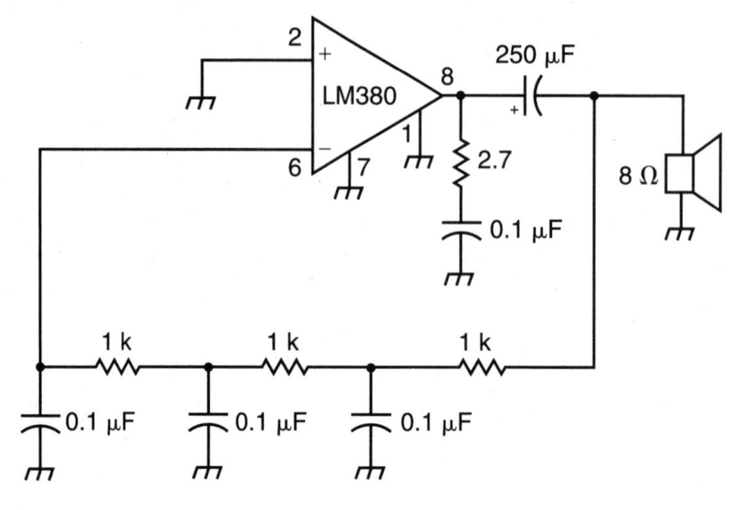

(*Courtesy of National Semiconductor.*)

Power voltage-to-current converter

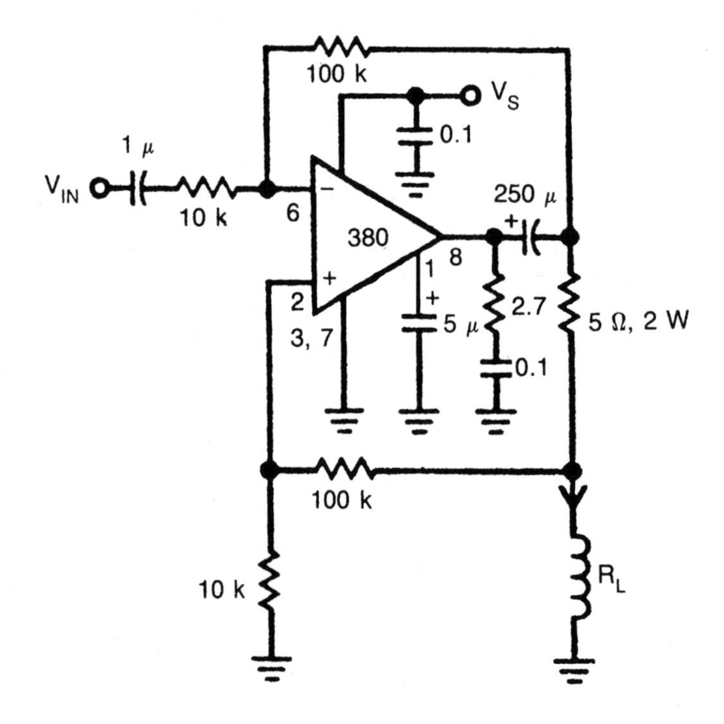

LM384 5-watt audio power amplifier

The 384 is an inexpensive, medium-power audio amplifier intended for use in general consumer applications. The gain is internally fixed at 34 dB. The voltage gain is fixed at 50.

This chip can be operated from a fairly wide range of supply voltages, from 12 V to 26 V (28 V, absolute maximum). It has a high peak current capability, over 1 ampere, but low quiescent power drain. The internal circuitry is short-circuit proof, and thermally limited. The amplifier has high input impedance, and low distortion.

A unique input stage permits the inputs to be referenced to ground. The output signal automatically centers itself to one-half the supply voltage.

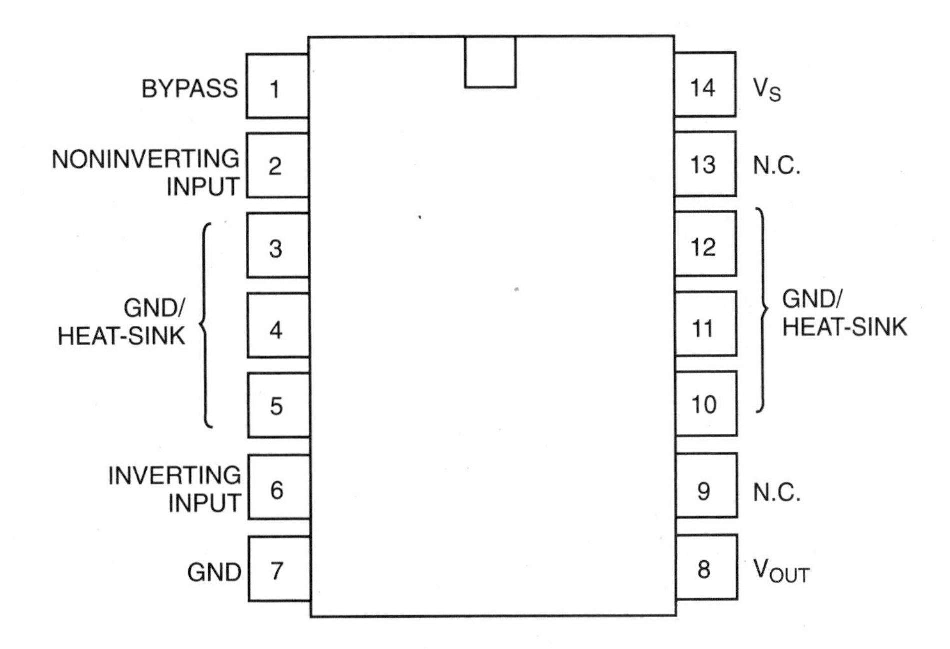

5-watt audio amplifier

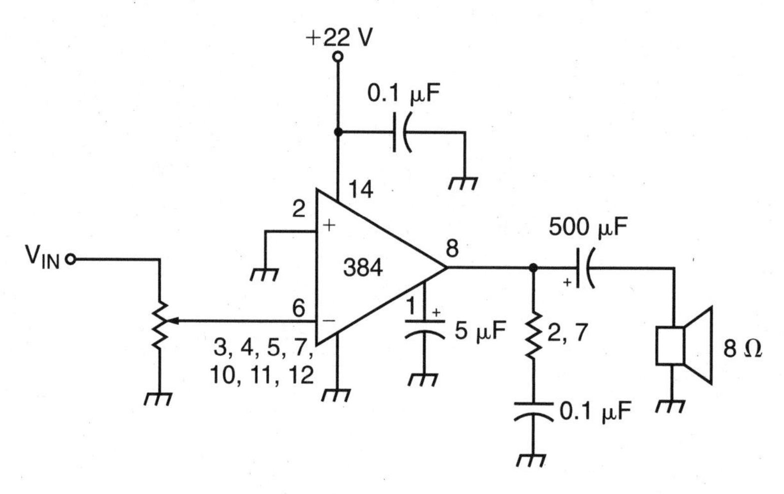

(*Courtesy of National Semiconductor.*)

Bridge amplifier

Bridging two amplifier stages as shown here permits greater output power at minimal cost.

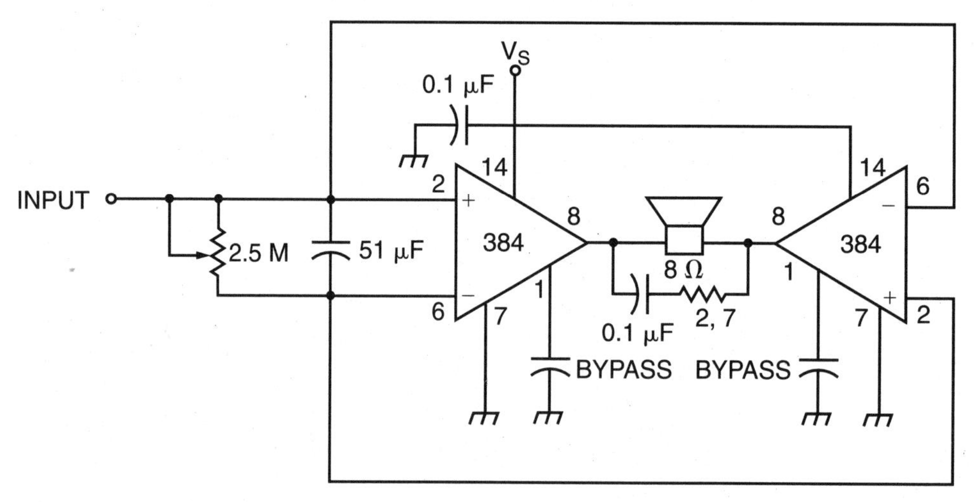

(*Courtesy of National Semiconductor.*)

Intercom

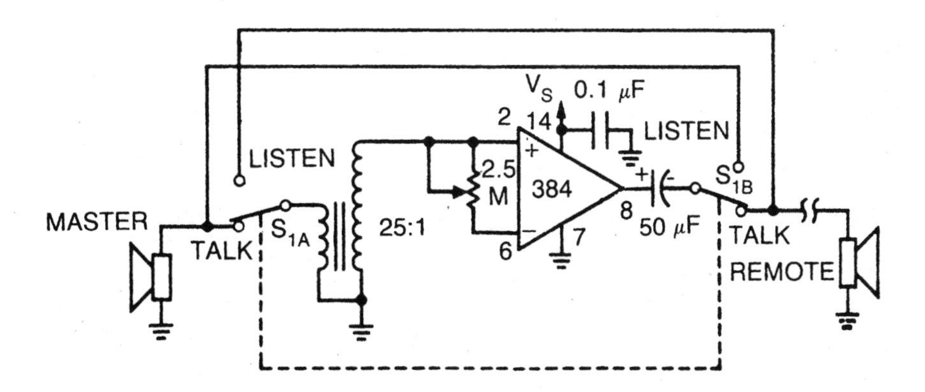

Phase shift oscillator

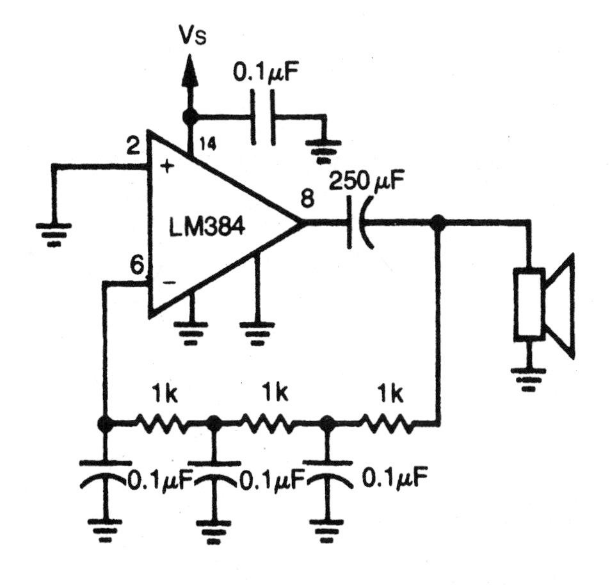

LM386N low-voltage audio power amplifier

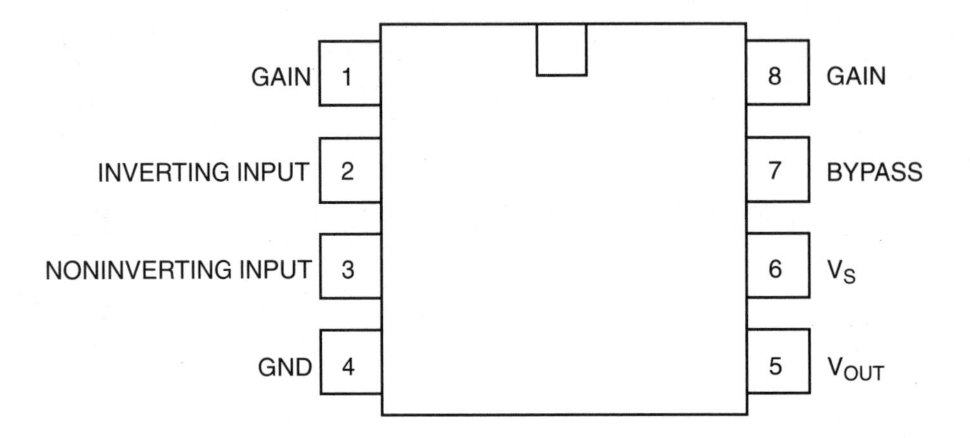

Amplifier with gain of 20 (26 dB)

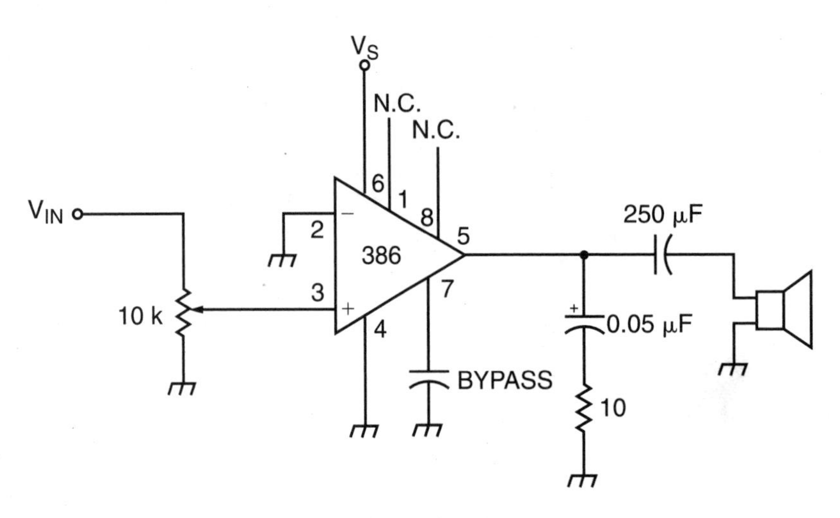

(*Courtesy of National Semiconductor.*)

Amplifier with gain of 50 (34 dB)

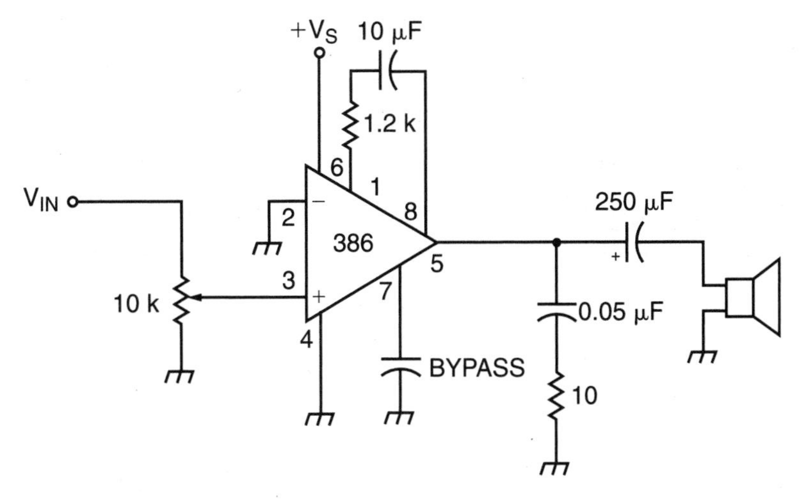

(*Courtesy of National Semiconductor.*)

Amplifier with gain of 200 (46 dB)

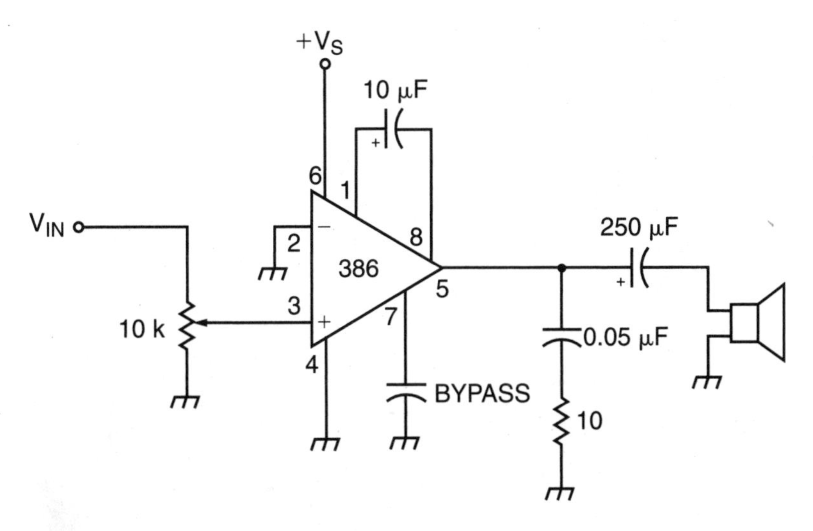

(*Courtesy of National Semiconductor.*)

Amplifier with bass boost

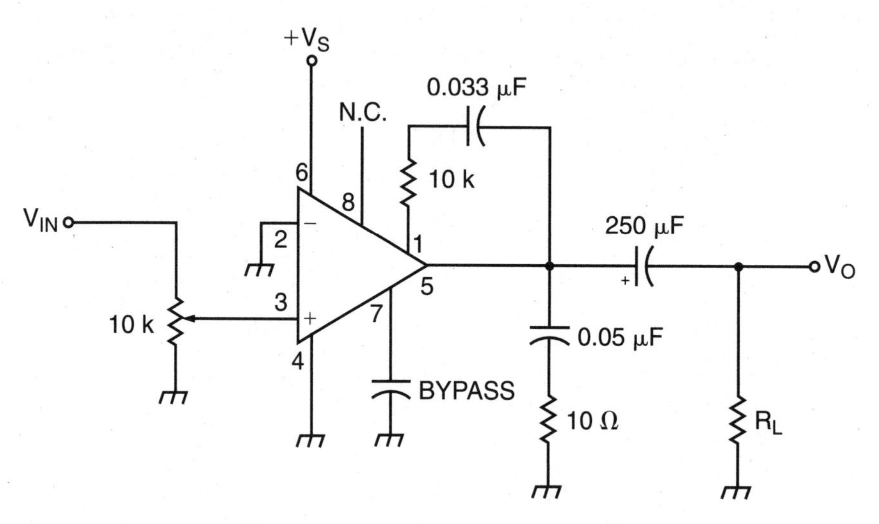

(*Courtesy of National Semiconductor.*)

High-volume warning siren

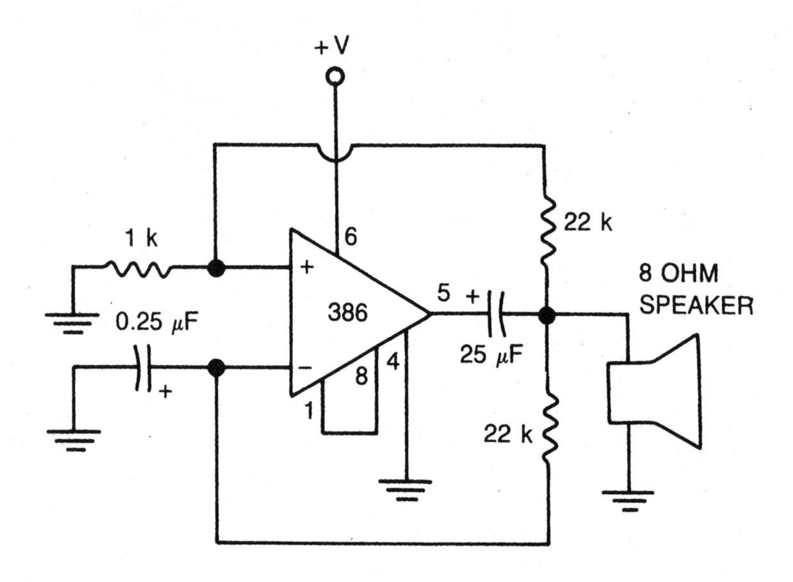

Square-wave oscillator
LM387 low-noise dual preamplifier

The 387 is designed to amplify low-level signals in two channels (stereo), in applications where very low noise is required. The total input noise for this device is rated for a typical value of 1.0 µV.

To achieve this excellent noise performance, each of the two amplifiers on this chip is completely independent, with an internal power supply decoupler-regulator. As a result, the supply rejection is 110 dB, and the channel separation is 60 dB.

These preamplifiers offer high gain of 104 dB, and a large output voltage swing up to 2 V under the supply voltage, peak-to-peak. A single-ended supply voltage of 9 V to 30 V can be used to power this chip, so, for example, if a 15-V supply voltage is used, the output voltage can swing up to 13 V peak-to-peak.

The amplifiers in the 387 are internally compensated for gains greater than 10.

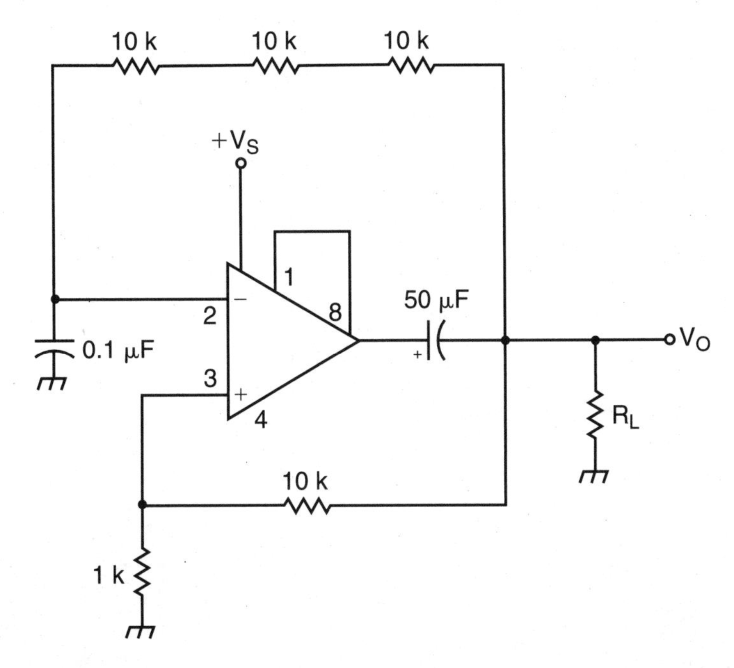

(*Courtesy of National Semiconductor.*)

Phono preamp

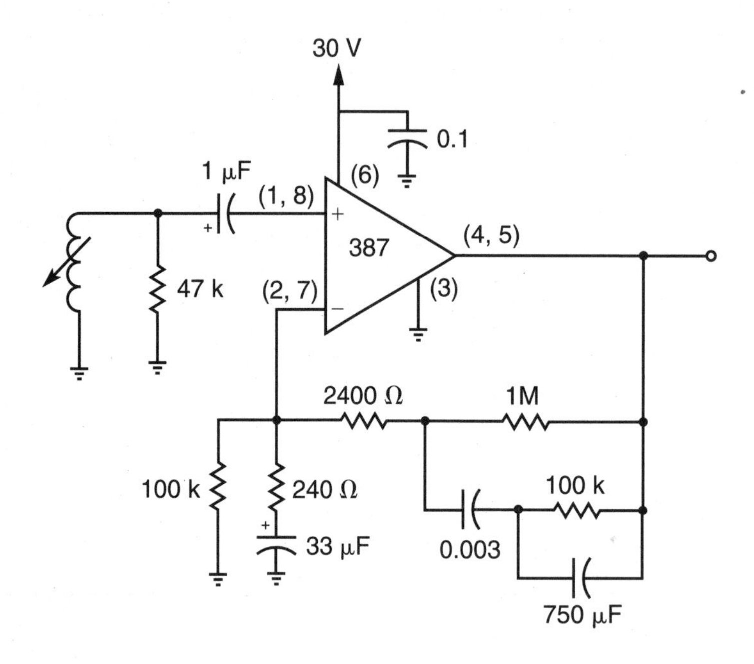

Speech filter

This bandpass filter is designed to pass frequencies in the 300 Hz to 3 kHz (3000 Hz) range. Most of the frequency components that make speech intelligible fall within this range, and most common forms of interference and noise fall above or below this band. This helps emphasize the spoken content of a noisy signal which might otherwise be difficult to understand clearly.

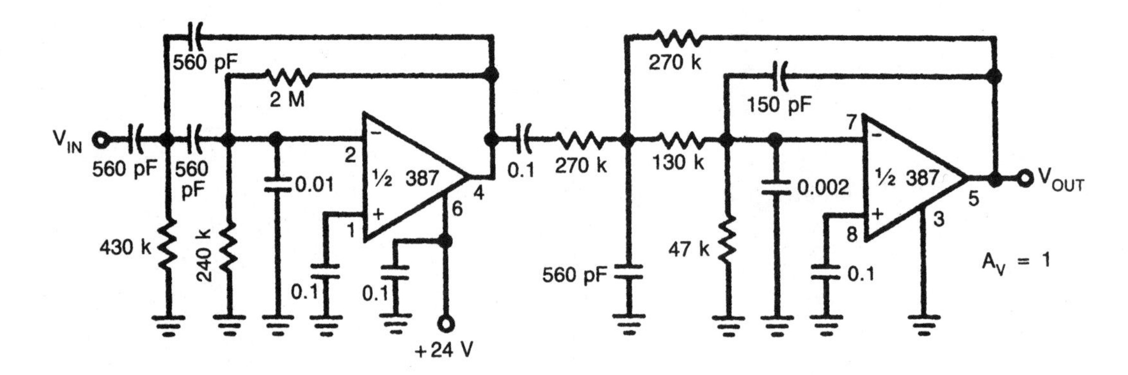

Inverse RIAA response generator

Most vinyl records use a form of frequency compensation with specifications defined by the RIAA. This circuit is designed to reflatten the frequency response upon playback.

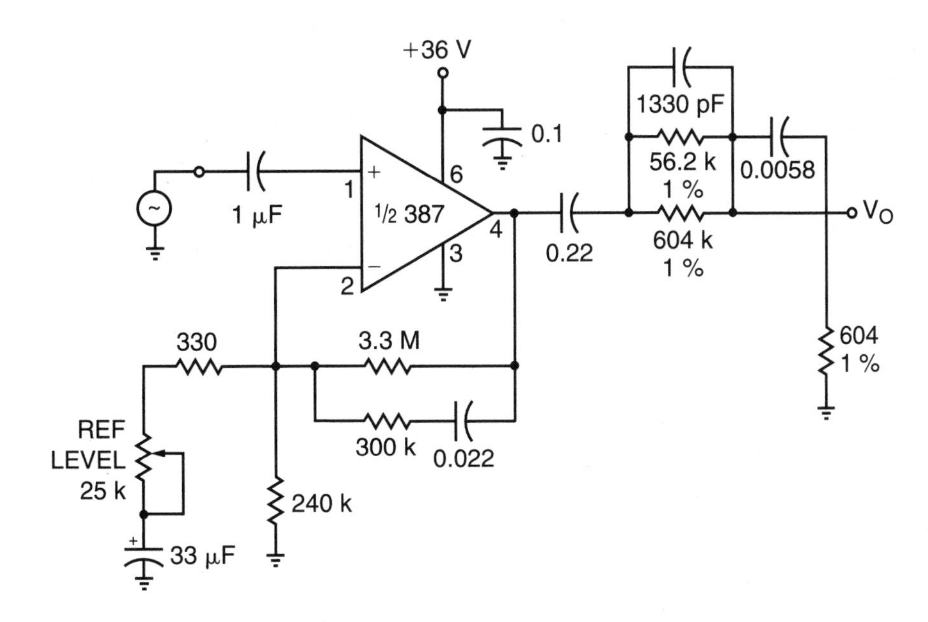

Acoustic pick-up preamp

This preamplifier circuit is well-suited for boosting the signal from a pick-up microphone to amplify the sound of an acoustic musical instrument.

Treble and bass tone controls are provided, as well as three gain ranges selected by a DP3T rotary switch.

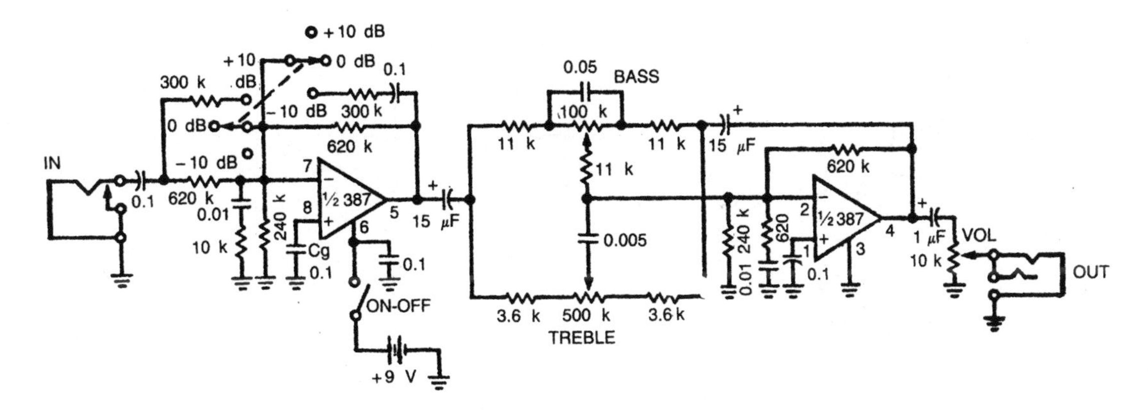

Scratch filter

This filter is designed to help reduce the annoying pops and crackles usually heard when an old vinyl record is played. It will not help if there are severe scratches, of course, but in most cases of mild to moderate wear, this filter will make a significant improvement in the sound. However, there will be some dulling of the high treble range, because minor scratch noise usually occurs at such frequencies, and that is what is filtered out.

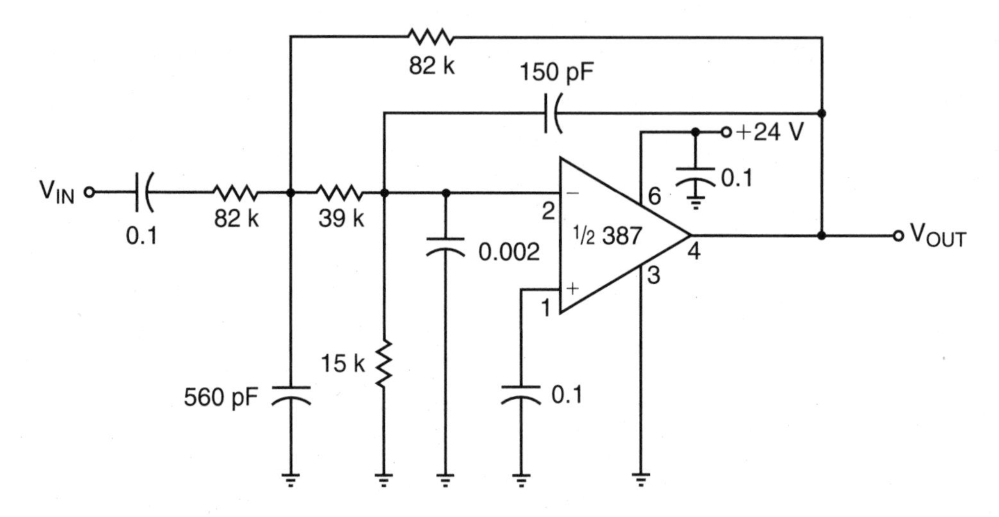

Rumble filter

When playing vinyl records, small inaccuracies in the turntable speed or minor warpage of the record itself can result in a form of very low-frequency noise and distortion known as *rumble.* This circuit is designed to significantly reduce the annoying rumble, while permitting most of the musical content to pass through to the speakers.

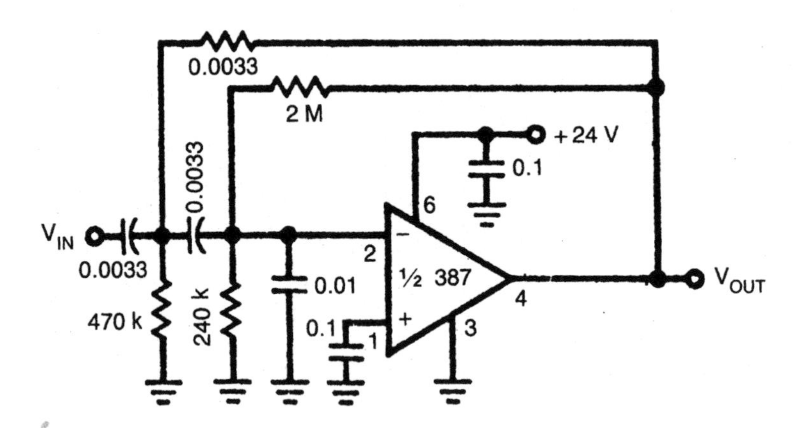

2-channel panning circuit

This circuit permits an audio signal to be cross-faded from one channel to another. This is called *panning*. It can be put to dramatic use in stereo.

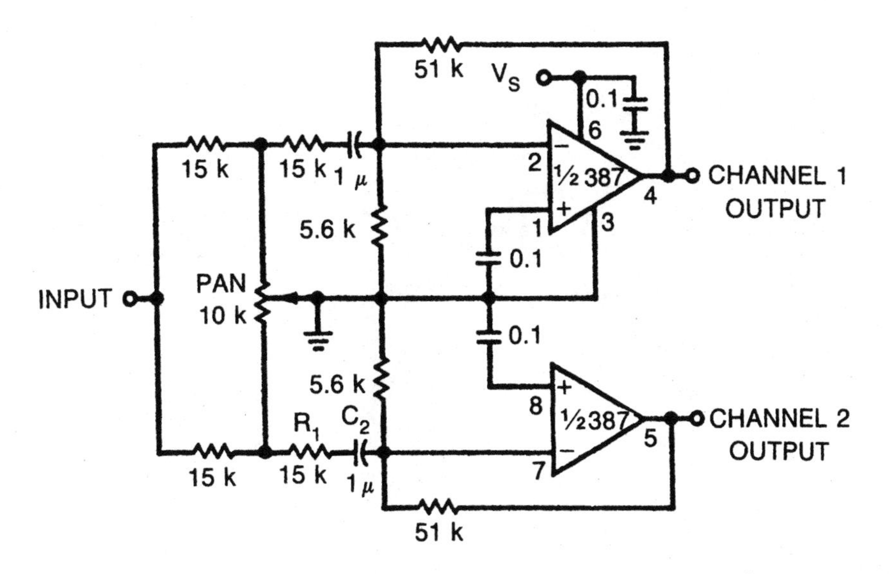

20 kHz bandpass active filter
387A low-noise preamplifier

The 387A is similar to the 387, but with improved specifications.

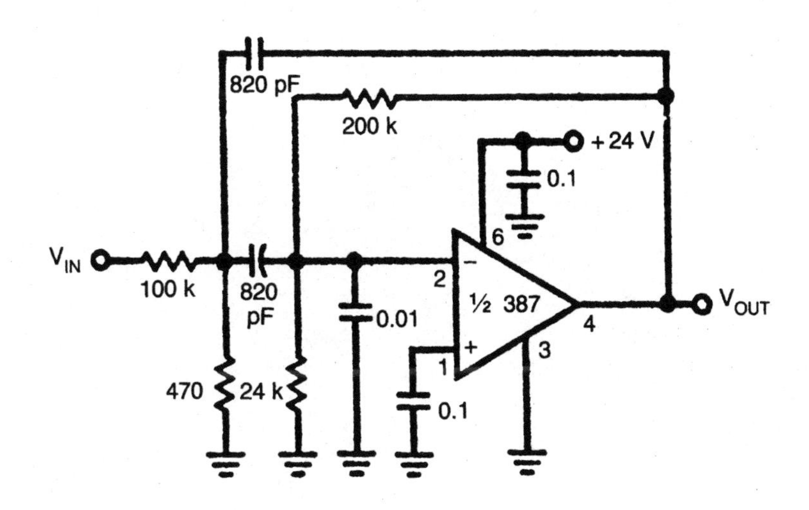

Balanced input microphone preamp

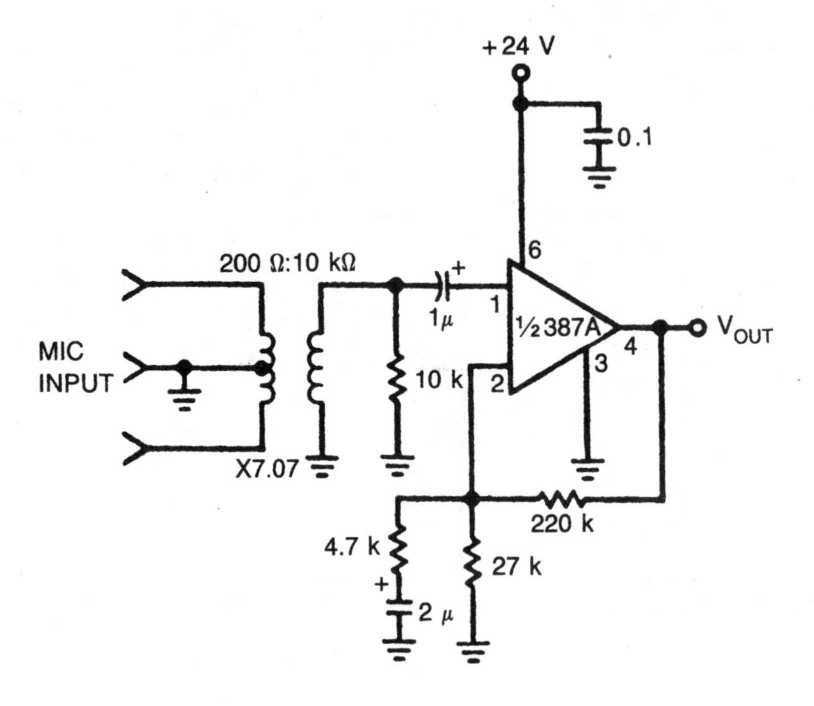

Low-noise transformerless balanced input microphone preamplifier

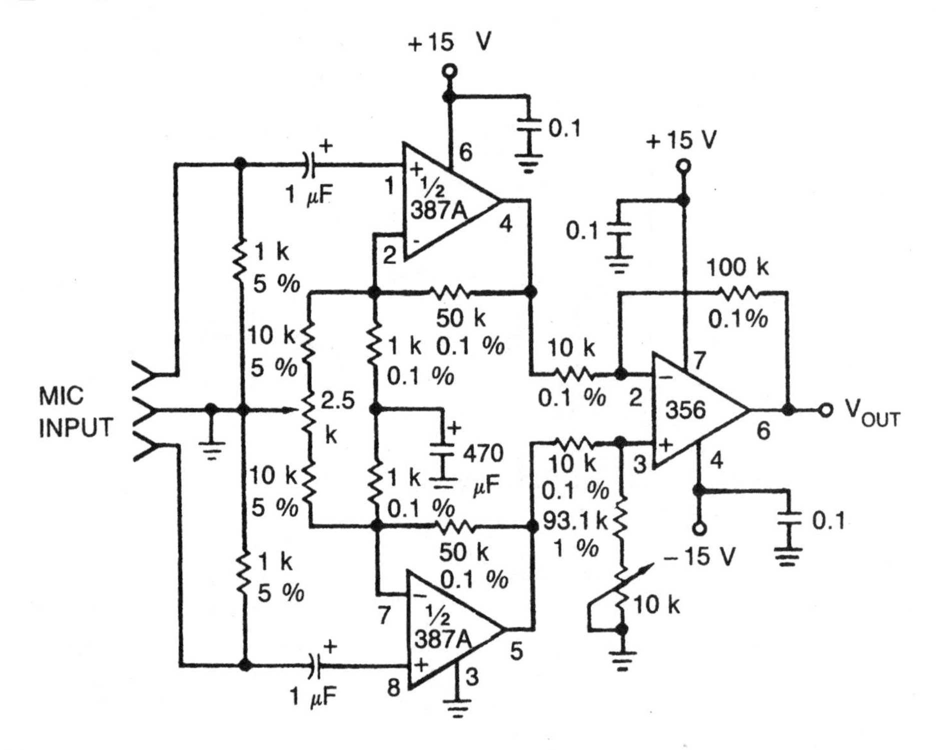

LM389 low-voltage audio power amplifier with NPN transistor array

The 389 contains essentially the same amplifier as the 386, with the addition of an array of three NPN transistors on the same substrate. These transistors have high gain and excellent matching characteristics. They can operate from 1 µA to 25 mA at frequencies ranging from dc (0 Hz) to 100 MHz.

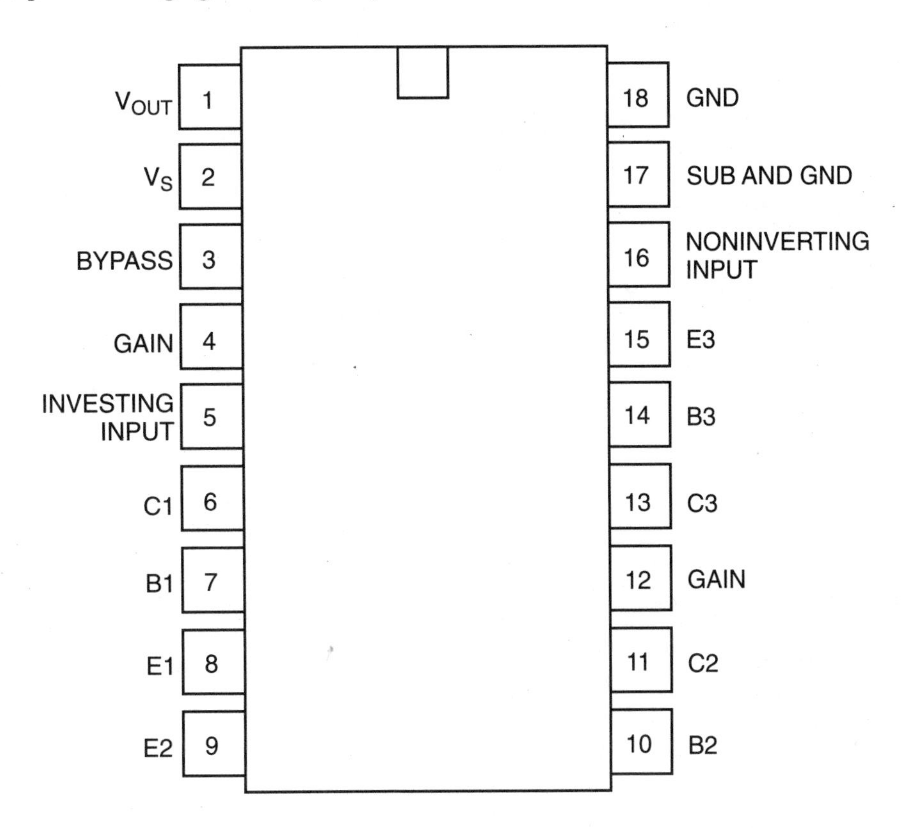

Tape recorder

Because of the relatively large size of this circuit, it is shown in two parts. The points marked "A," "B," or "C" in one-half of the diagram should be connected to the similarly labeled points in the other half of the diagram.

All of the switches in this circuit are shown in the record position. The other switch positions are for playback. The various switch sections should be ganged together for simultaneous operation.

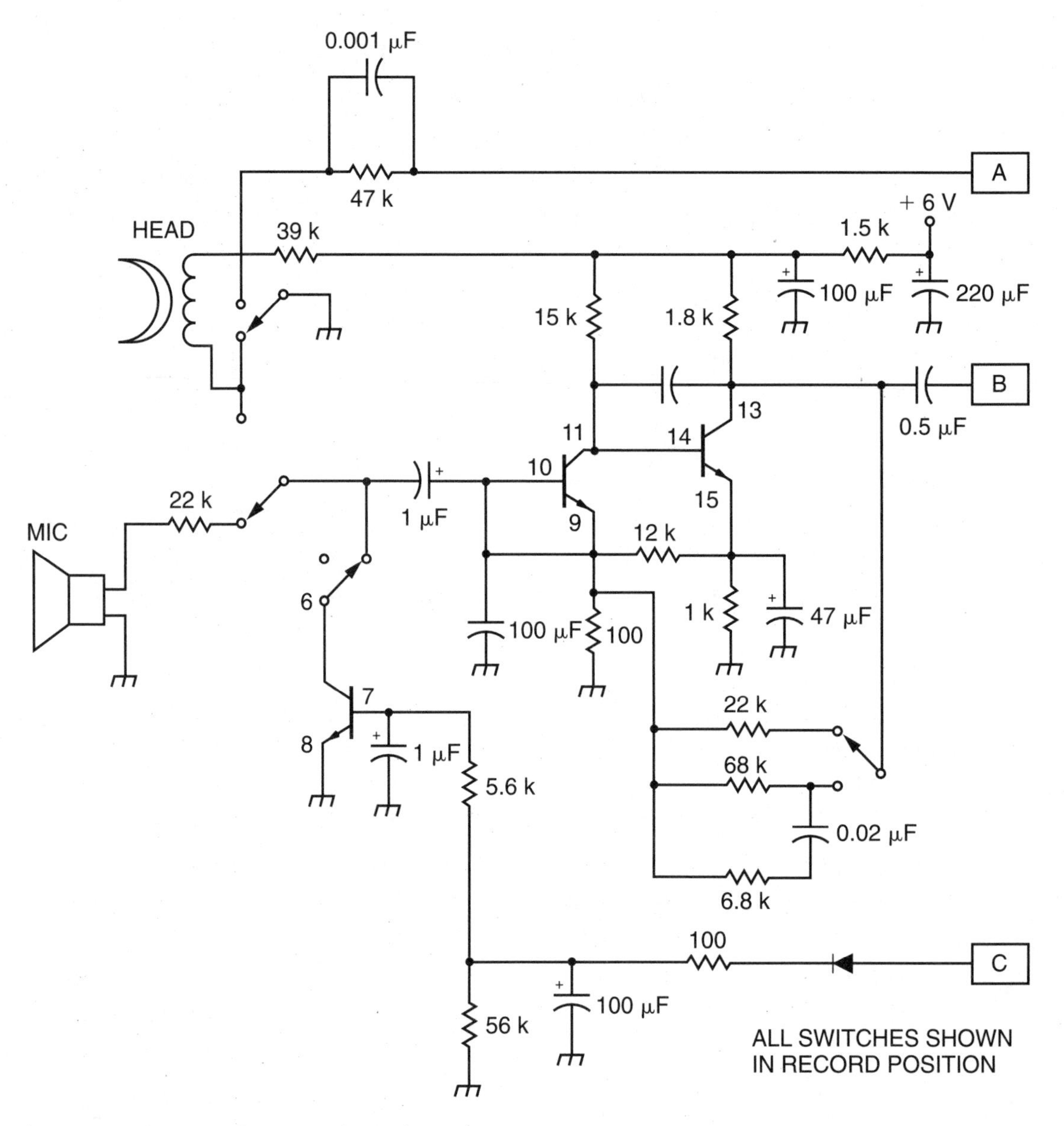

(***Courtesy of National Semiconductor.***)

Tape recorder continued

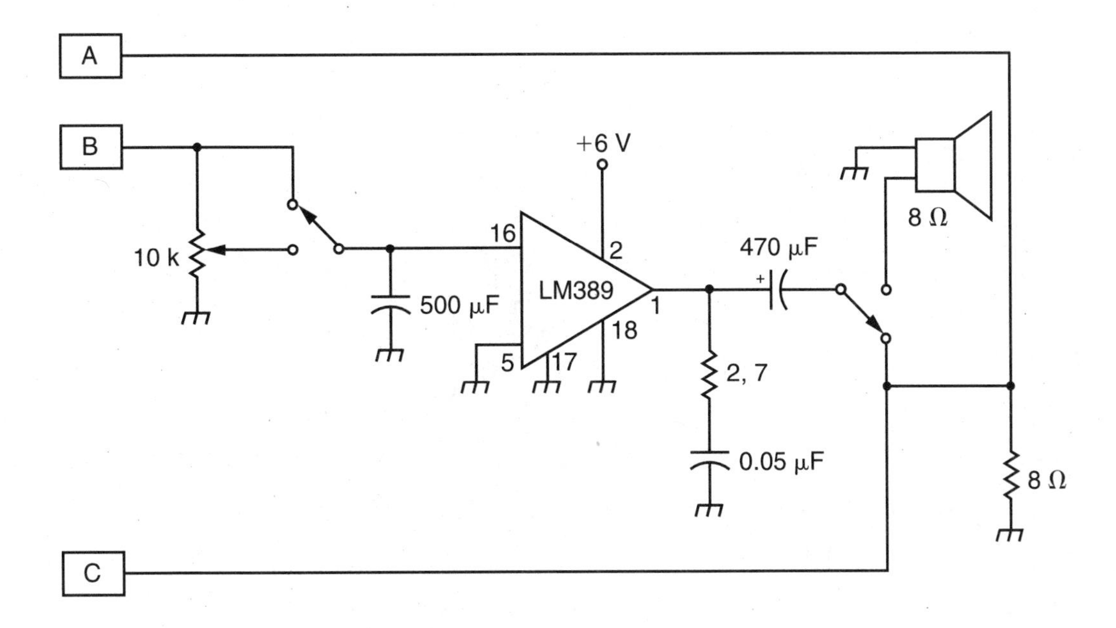

Ceramic cartridge phono amplifier with tone controls

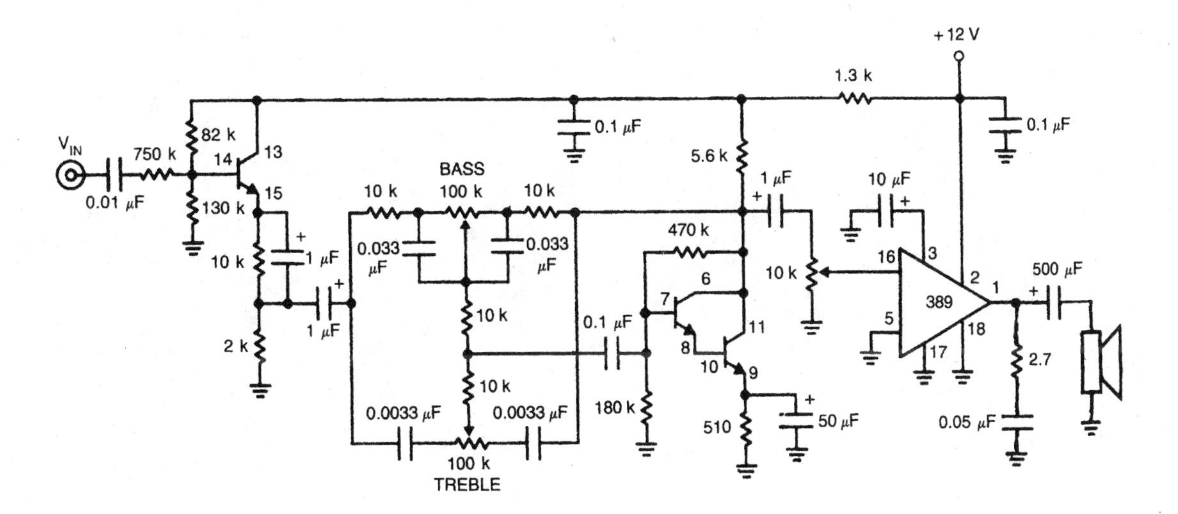

Noise generator

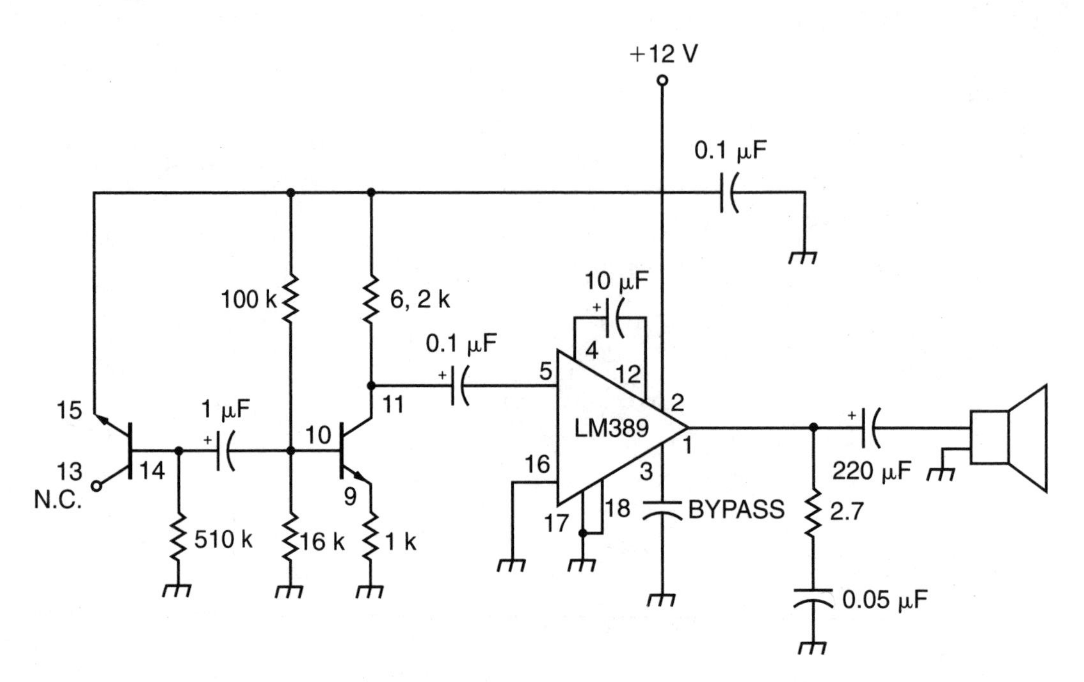

(*Courtesy of National Semiconductor.*)

Siren

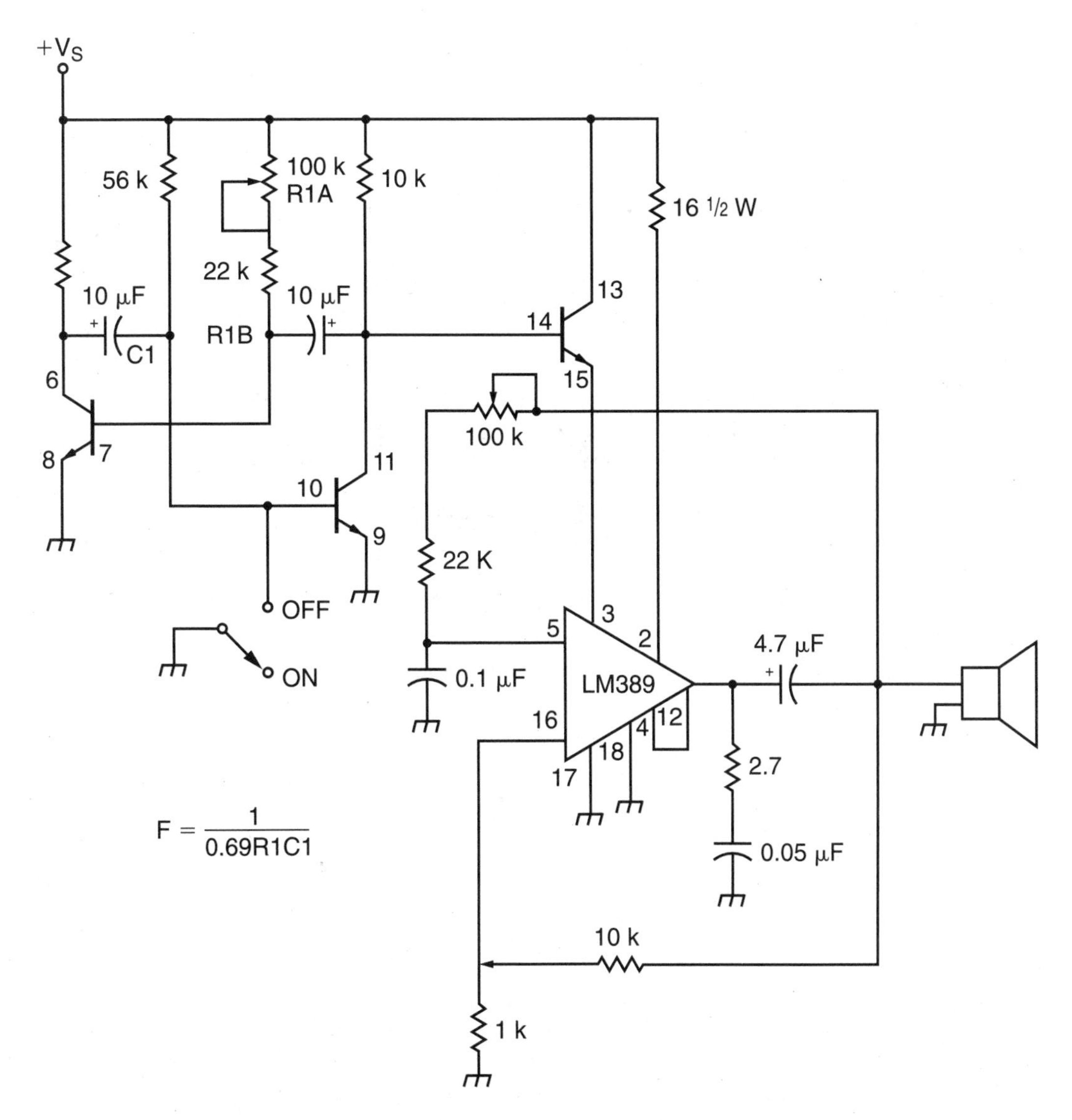

$$F = \frac{1}{0.69R1C1}$$

(Courtesy of National Semiconductor.)

FM scanner noise squelch circuit

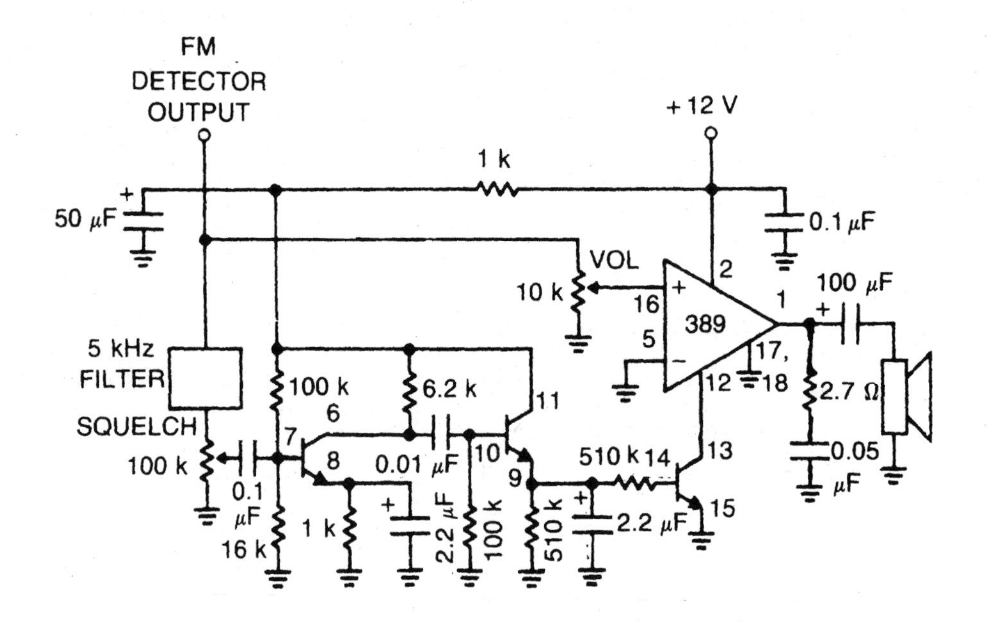

Voltage-controlled amplifier

The gain of a VCA is adjusted by an external voltage applied to a special control input.

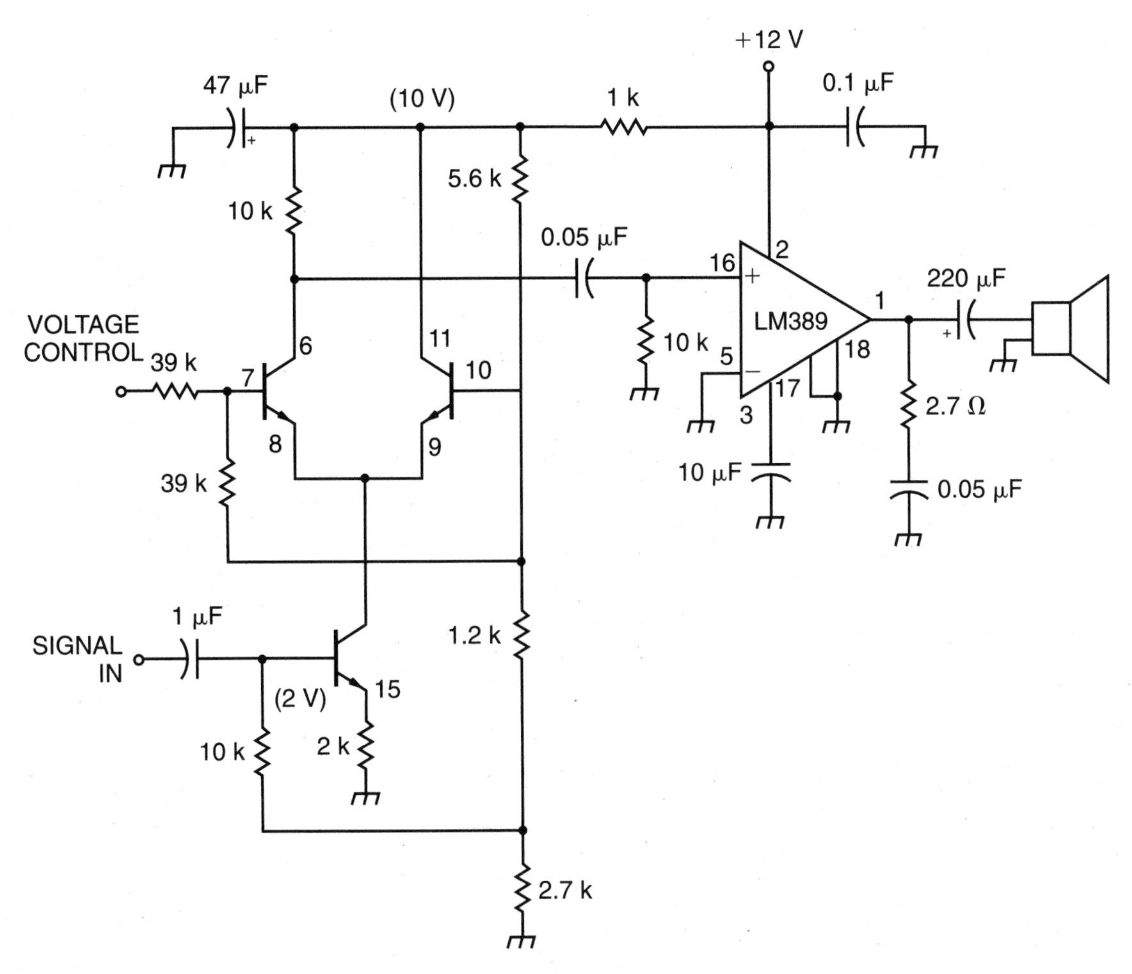

(*Courtesy of National Semiconductor.*)

Tremolo circuit

Tremolo is a musical effect created by rapid but audibly distinguishable fluctuations in the volume of a note.

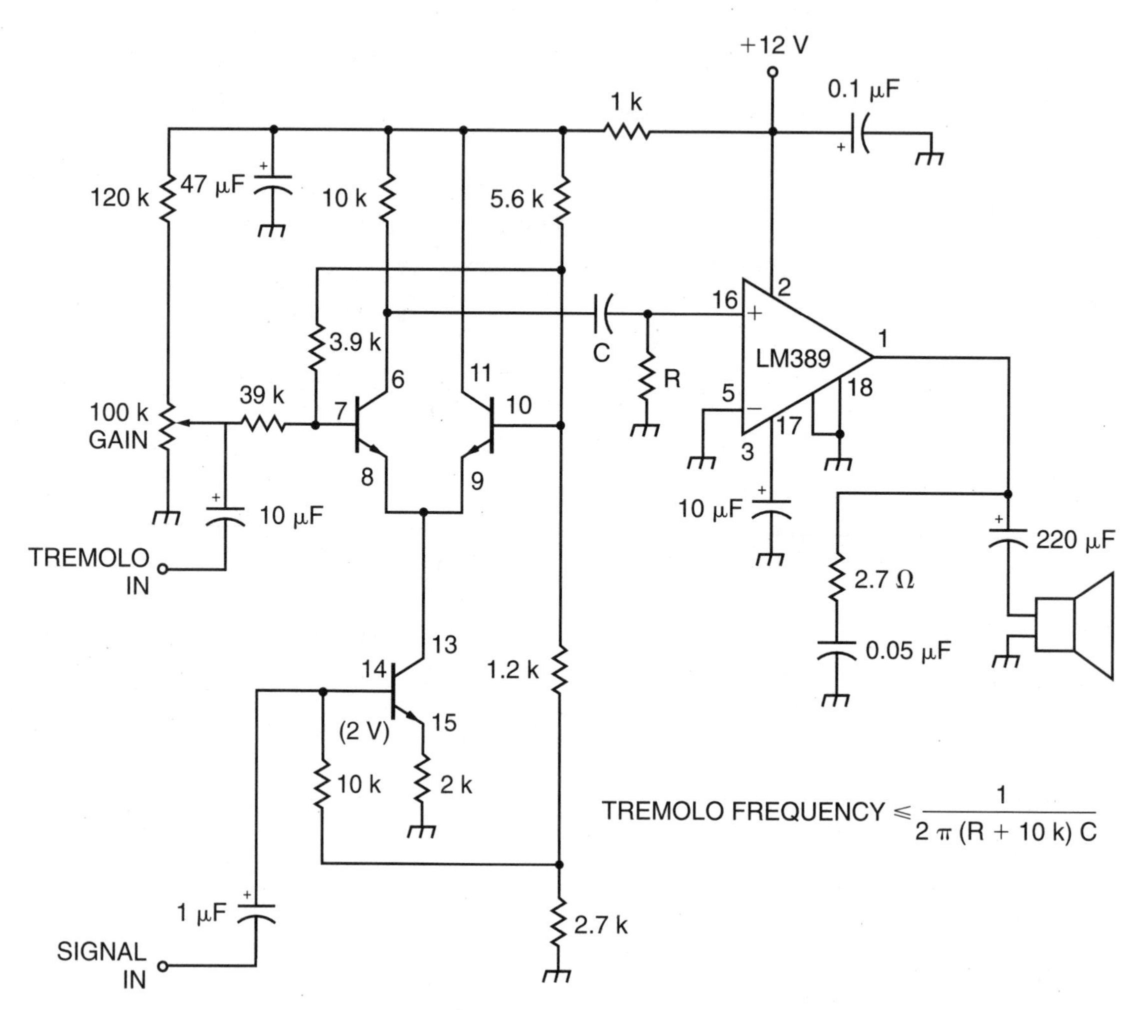

(*Courtesy of National Semiconductor.*)

LM390 1-watt battery operated audio power amplifier

The LM390 is a moderate-power audio amplifier designed specifically for battery powered operation. It is designed to operate off one of the following standard battery voltages: 6 V, 7.5 V, or 9 V.

The amplifier's gain is internally set at 20. Accepting this default gain permits operation with a minimum of external parts. However, if the application requires, the gain can be increased to any value up to 200 by adding an external resistor and capacitor between pins No. 2 and No. 6.

This 1-watt audio amplifier offers excellent supply rejection and low distortion.

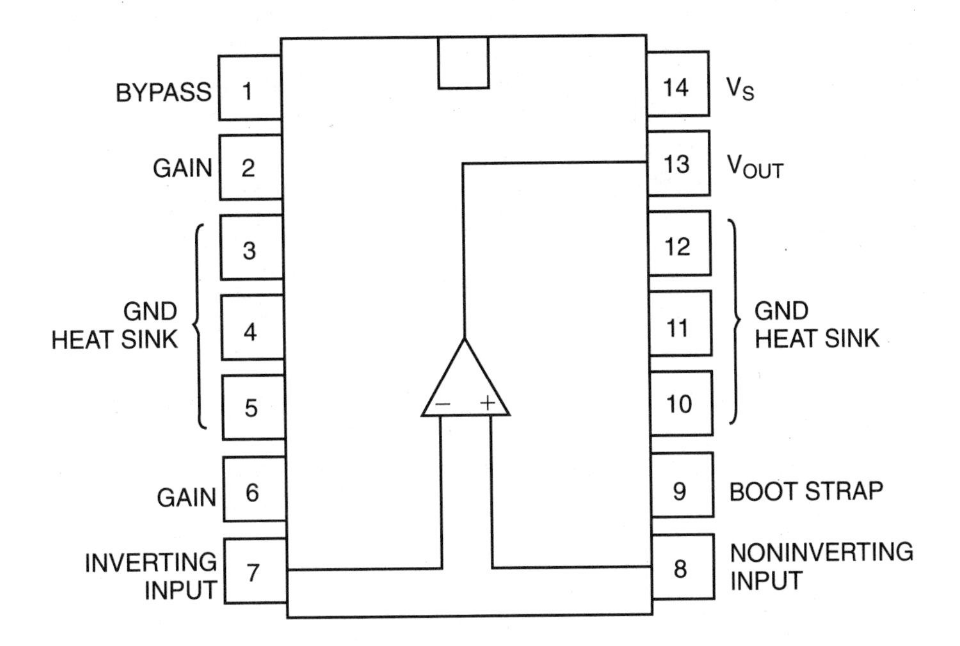

Amplifier with load returned to ground

The gain of this amplifier circuit is 20. The input and output signals are referenced to ground. The 10K potentiometer across the input serves as a volume control. For applications that do not require adjustable signal levels, this potentiometer can be replaced by a pair of fixed resistors arranged as a simple voltage divider network.

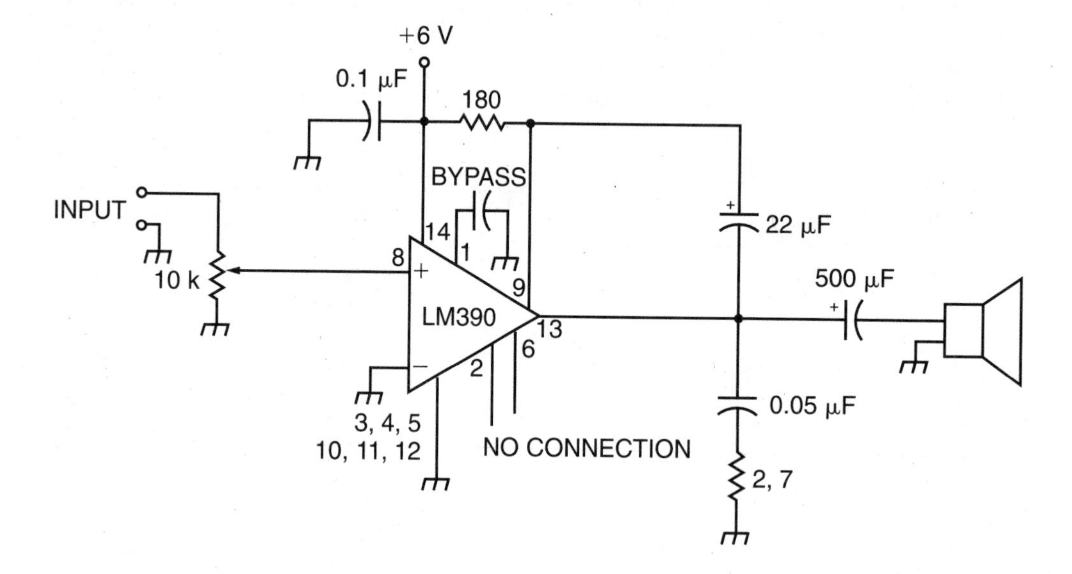

Amplifier with load returned to supply

The gain of this amplifier circuit is 20. The input signal is referenced to ground. The 10K potentiometer across the input serves as a volume control. For applications that do not require adjustable signal levels, this potentiometer can be replaced by a pair of fixed resistors arranged as a simple voltage divider network.

The output signal is not referenced to ground in this circuit, but to the positive supply voltage.

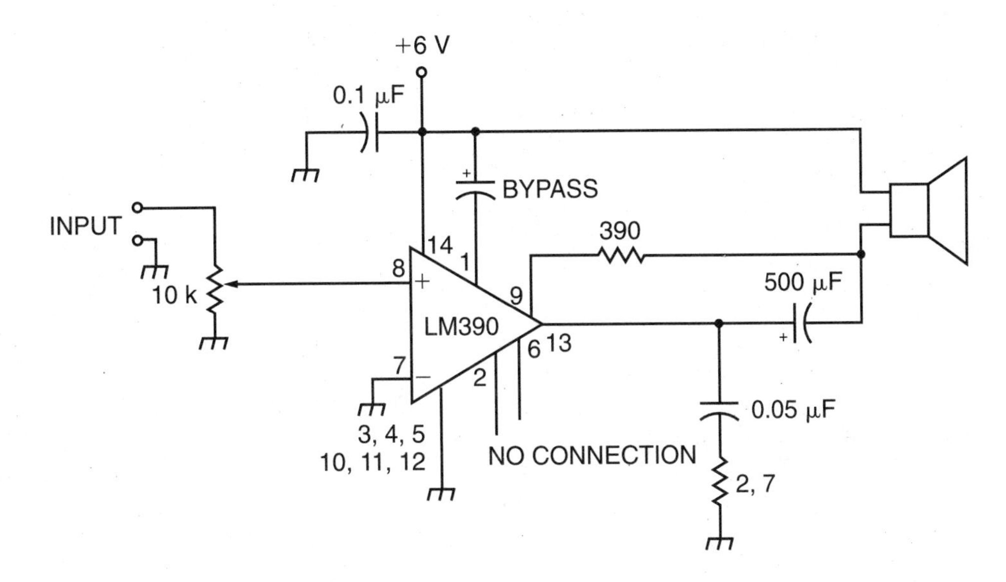

Amplifier with gain of 200 and minimum Cb

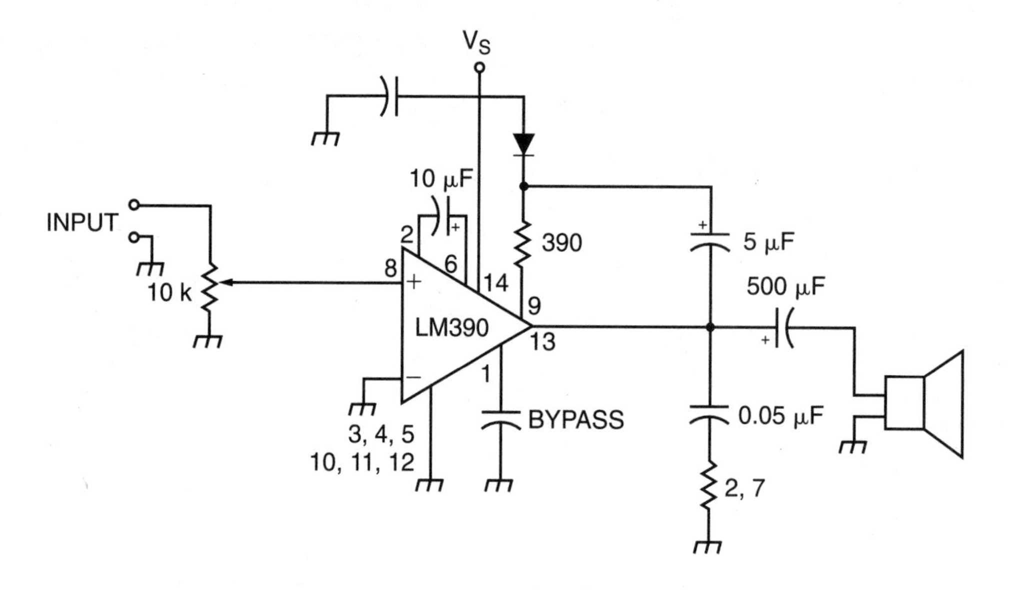

2.5-watt bridge amplifier

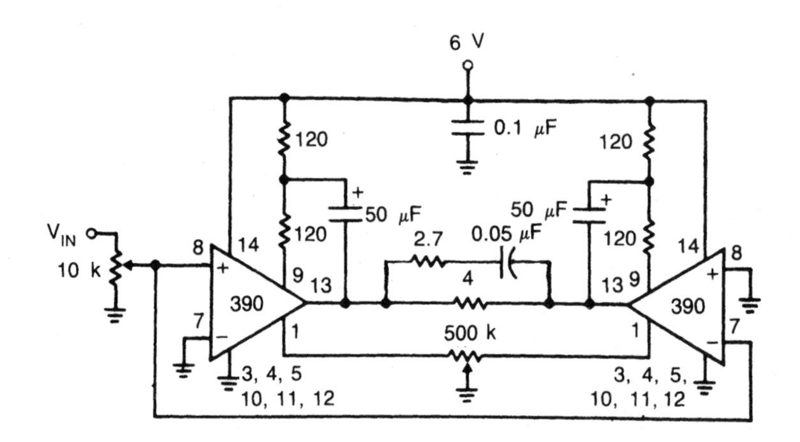

Amplifier with bass boost

Low audio frequencies are given a little extra "oompf" by this amplifier circuit.

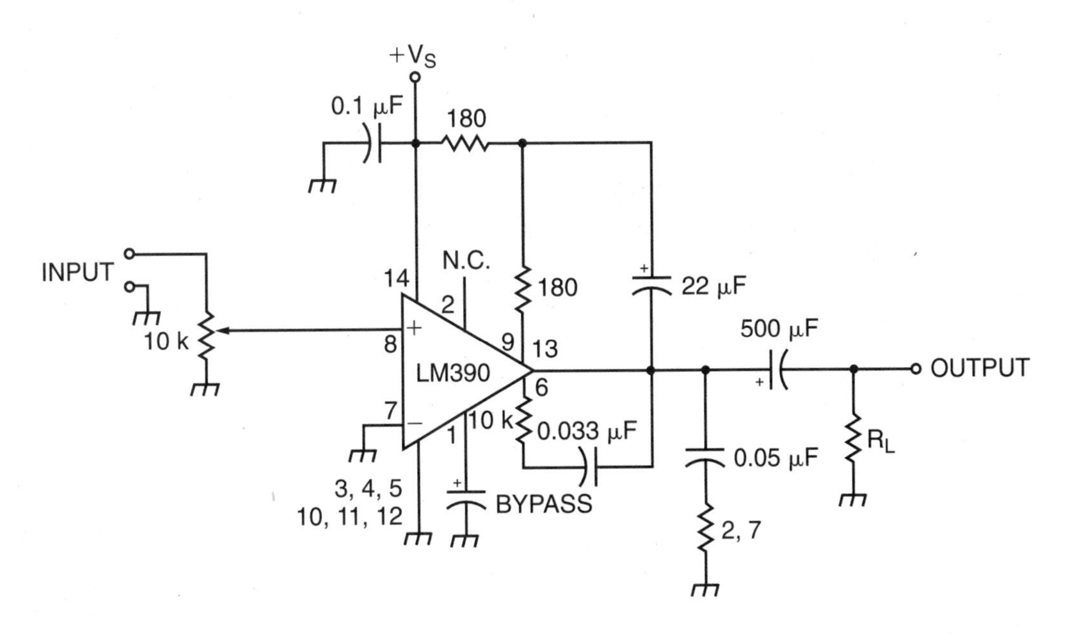

LM391 audio power driver

The 391 is an audio amplifier driver designed to drive external power transistors from 10 watts up to 100 watts. Notice that this level of output power does not (and should not) pass through the 391 itself, but through the external power transistors and their related circuitry. The 391 merely preconditions the signal for final amplification by the external transistors. The gain and bandwidth of this device are user selectable with appropriate external circuitry.

This chip is internally protected against thermal overloads and output faults, such as short circuits. An unusually wide range of supply voltages can be used with the 391, up to a maximum of ±50 V, or +100 V. The input signal voltage can be as high as 5 V under the actual supply voltage.

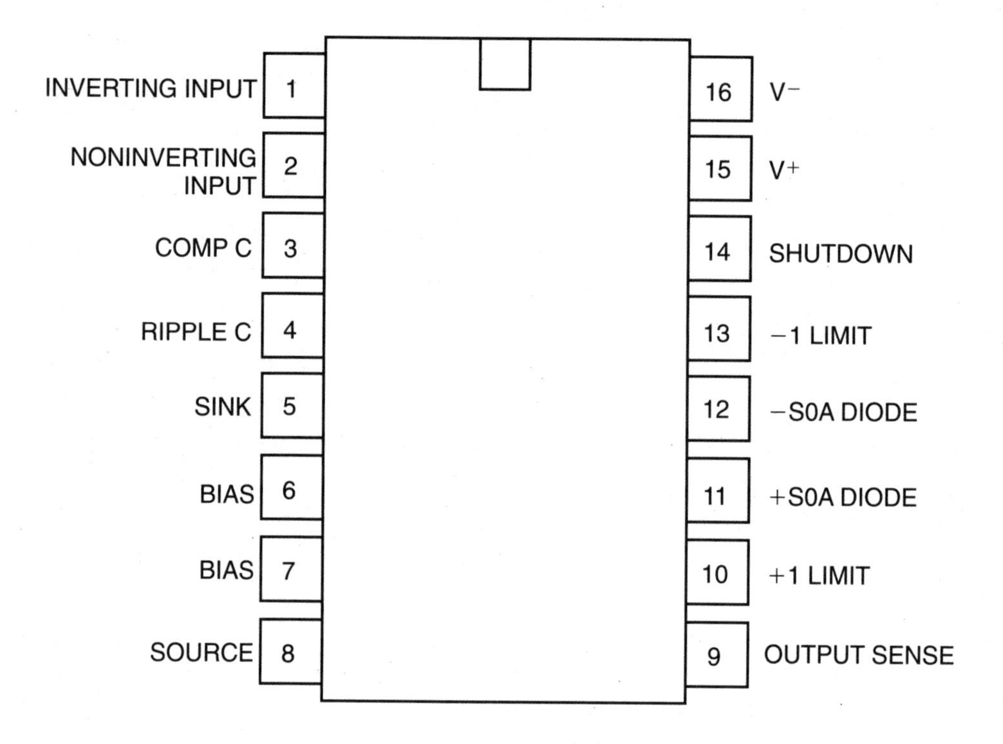

Basic audio amplifier

To avoid cluttering the diagram, the power supply connections are not shown. Of course they should always be assumed. An amplifier circuit will not work without a power source.

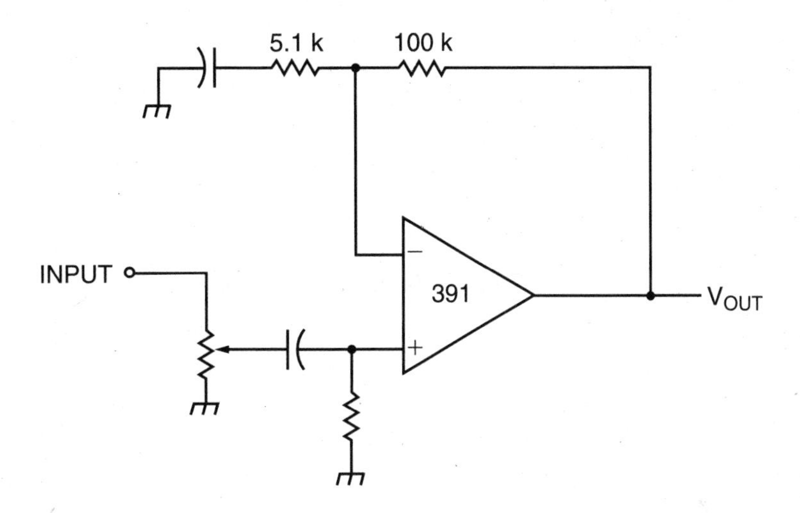

Bridged amplifier

To avoid cluttering the diagram, the power supply connections are not shown. Of course they should always be assumed. An amplifier circuit will not work without a power source.

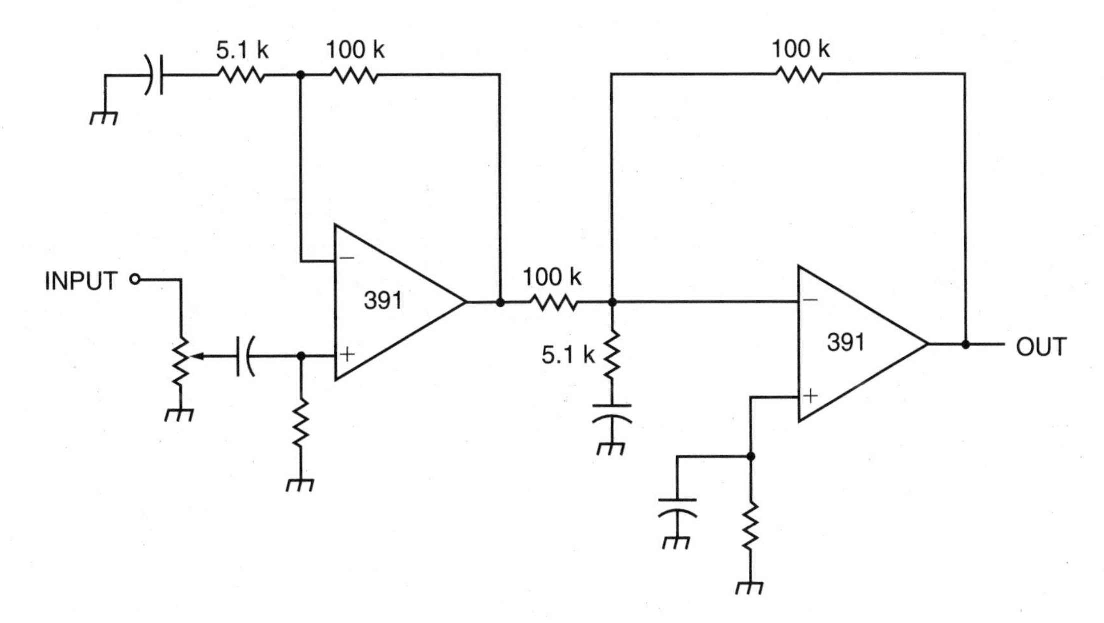

831 Low-voltage audio power amplifier

The 831 dual-audio power amplifier is designed to be operated from low voltages, ranging from 1.8 V to 6.0 V. It contains two independent audio amplifiers for stereo applications, or the two amplifiers can be combined for higher power bridged (BTL) operation. This chip can provide a power output of up to 440 mW into 8 ohms in the BTL mode. Each single amplifier stage is rated for 220 mW.

A patented compensation technique reduces high-frequency radiation, making the 831 very well suited for AM radio applications. Lower distortion and less wide-band noise are additional advantages of this internal compensation.

Most audio amplifier ICs require an input coupling capacitor, but the 831 is designed for direct coupling. The voltage gain can be adjusted with a single external resistor.

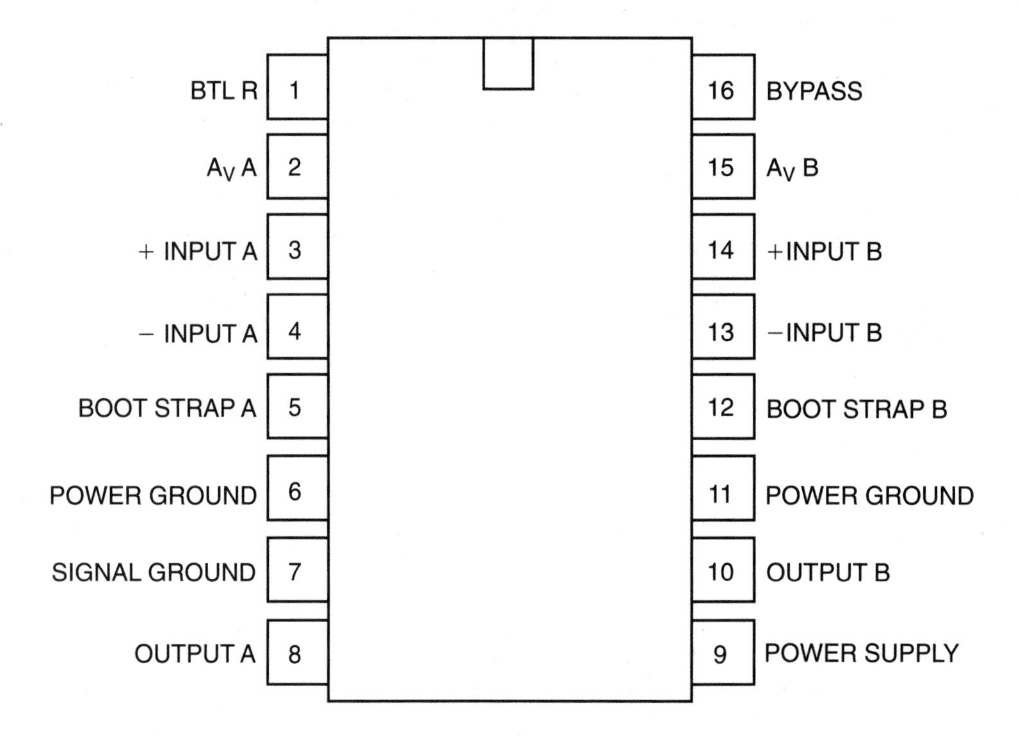

Basic stereo amplifier

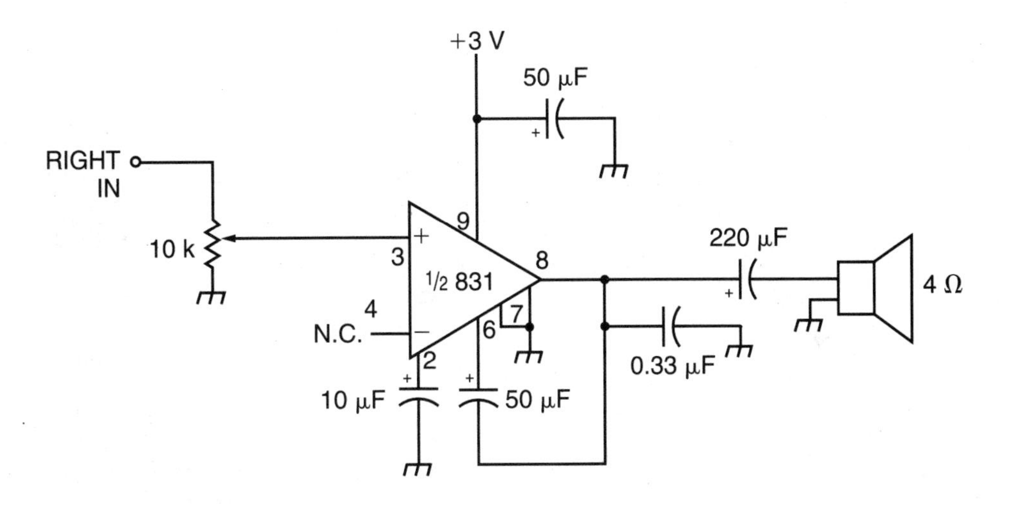

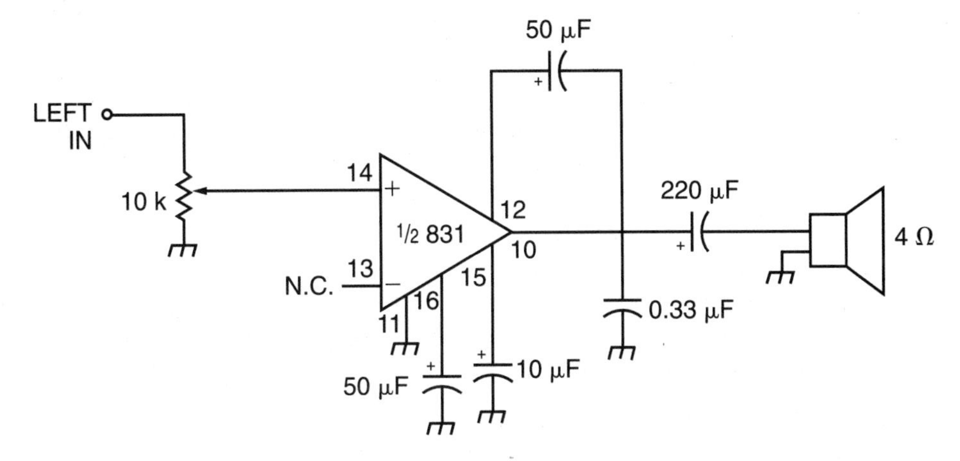

BTL amplifier with minimum parts

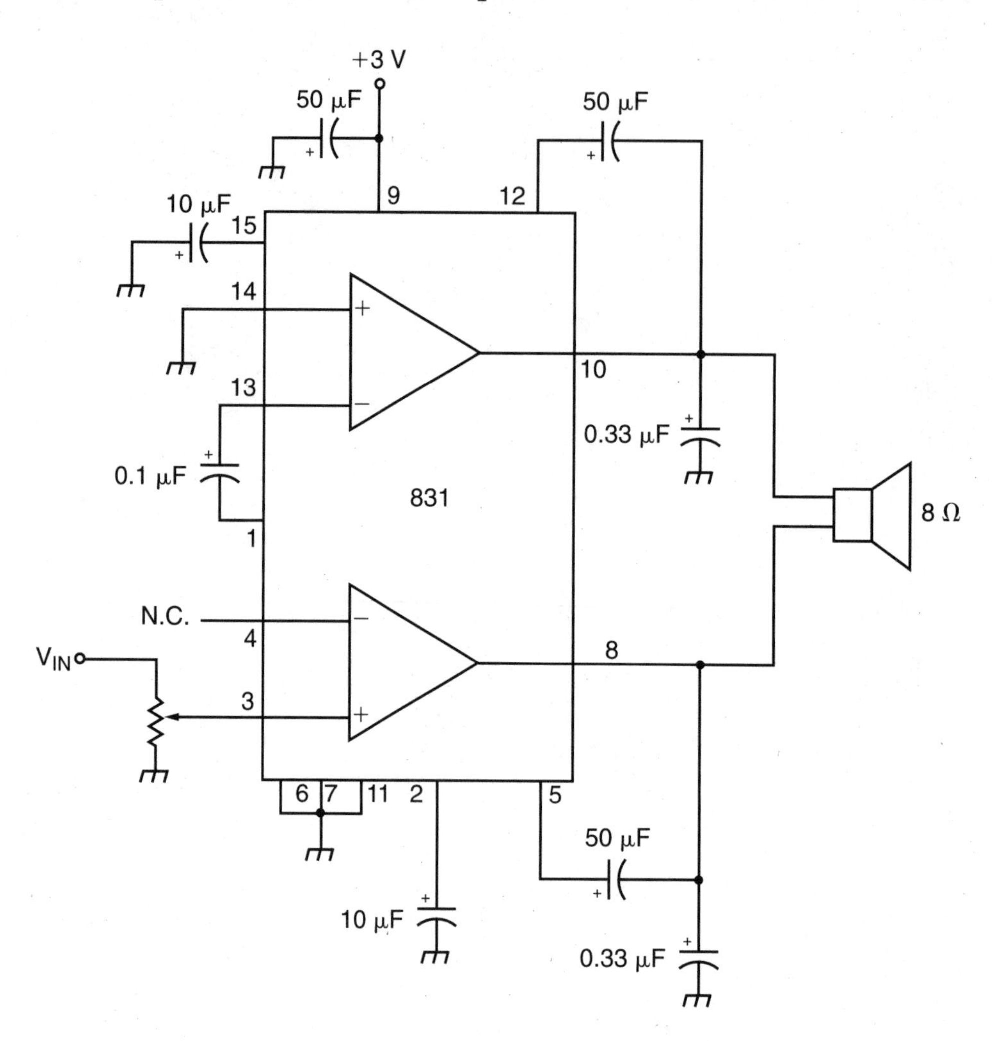

BTK amplifier for hi-fi quality

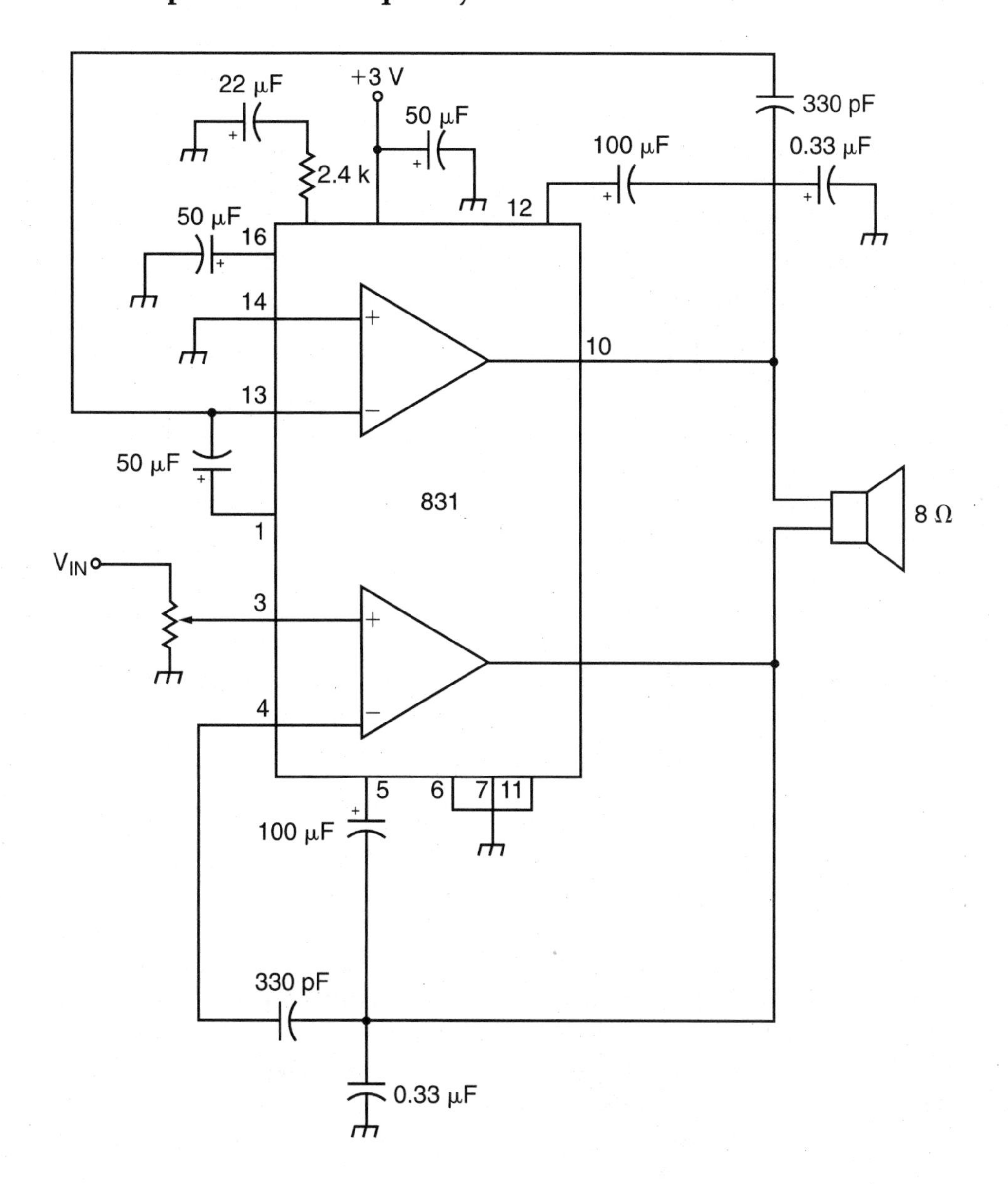

Dual amplifier for hi-fi quality

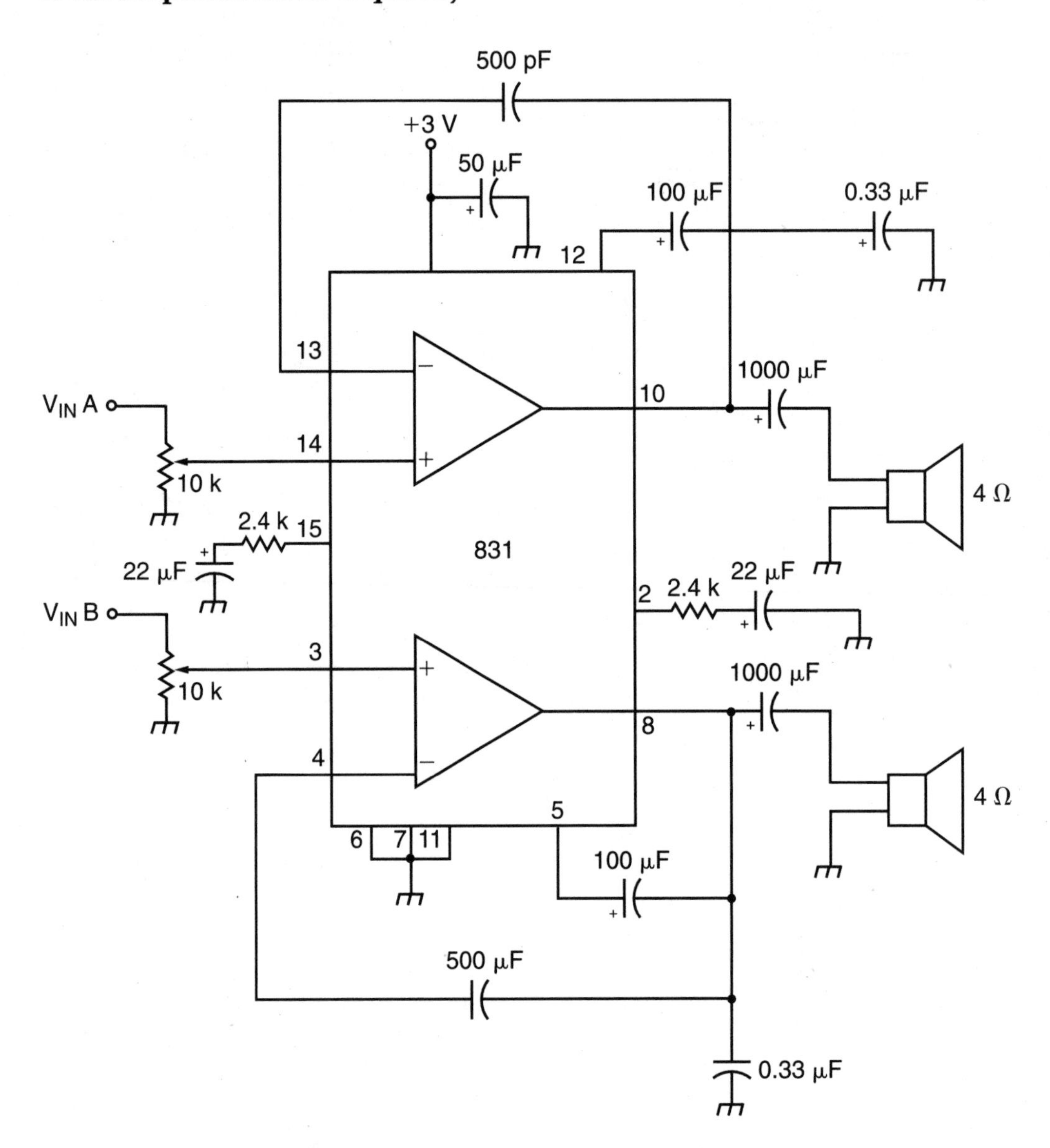

1875 20-watt audio power amplifier

The 1875 is a fairly high powered audio amplifier IC, capable of putting out 20 watts into 4-ohm or 8-ohm loads, with very low distortion. The 1875 can provide high gain, and a large output voltage swing. It is internally compensated and stable for gains of 10 or more. It also has a fast slew rate.

This chip features internal protection for ac or dc short circuits to ground, as well as internal current limiting and thermal shut-down protection.

The 1875 is designed for use with a minimum of external components.

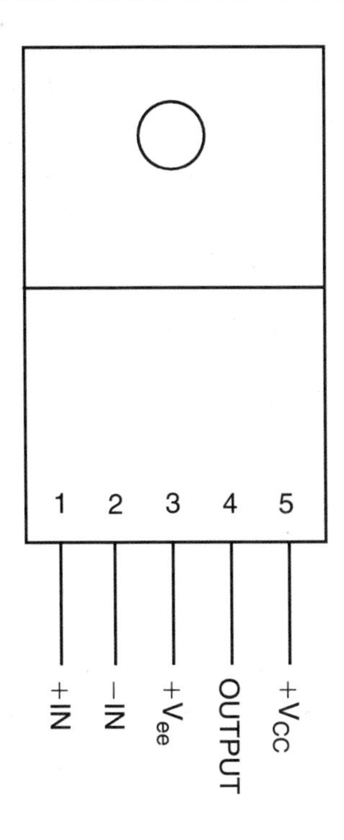

20-watt audio amplifier

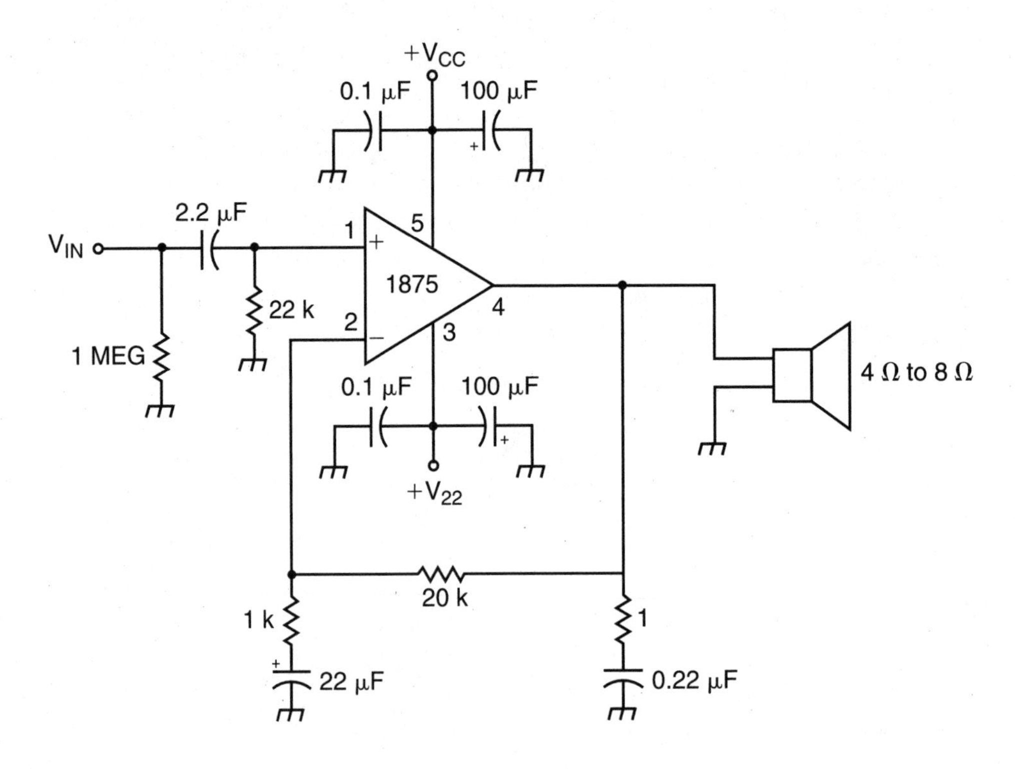

20-watt audio amplifier with single-ended power supply LM18877 dual-audio power amplifier

The 1877 contains a pair of audio amplifiers capable of continuously delivering 2 watts into 8-ohm loads, with very few external components. Each of the amplifiers on this chip is biased from a common internal voltage regulator, which results in high power supply rejection and output zero point centering.

The 1877 is internally compensated for all gains of more than ten. These amplifiers also offer very low cross-over distortion and low audio band noise. The internal circuitry is protected against ac short circuits, and includes thermal shut-down.

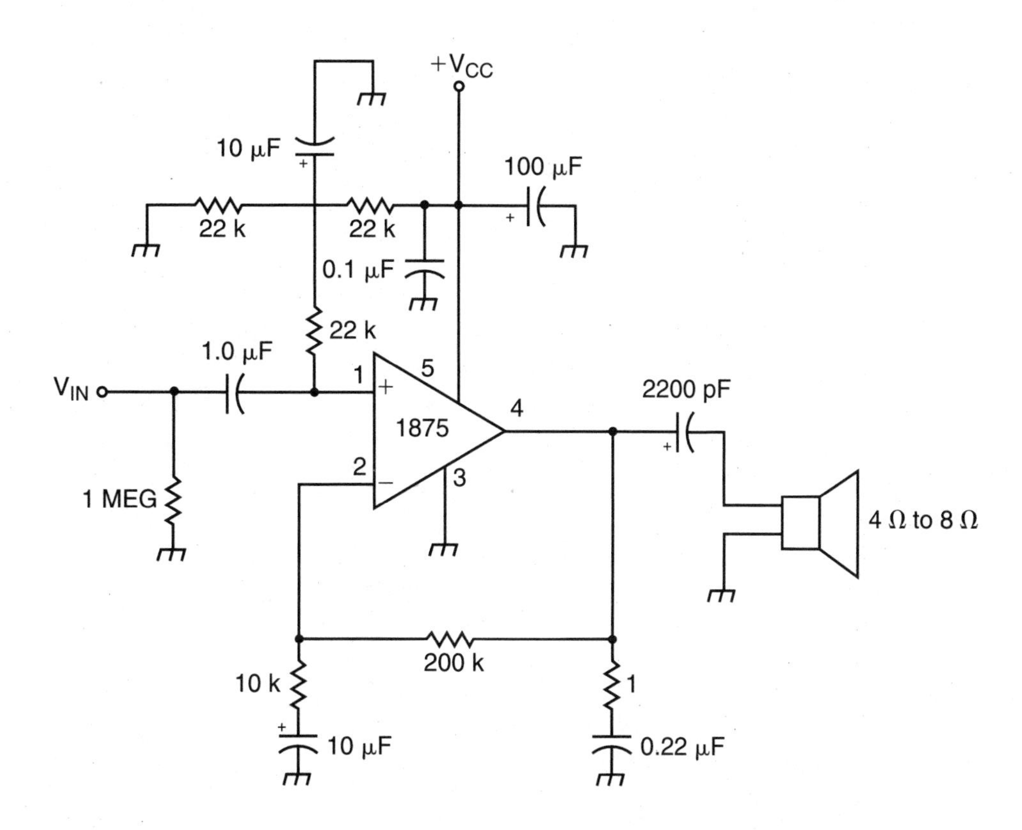

Stereo phonograph amplifier with bass tone control

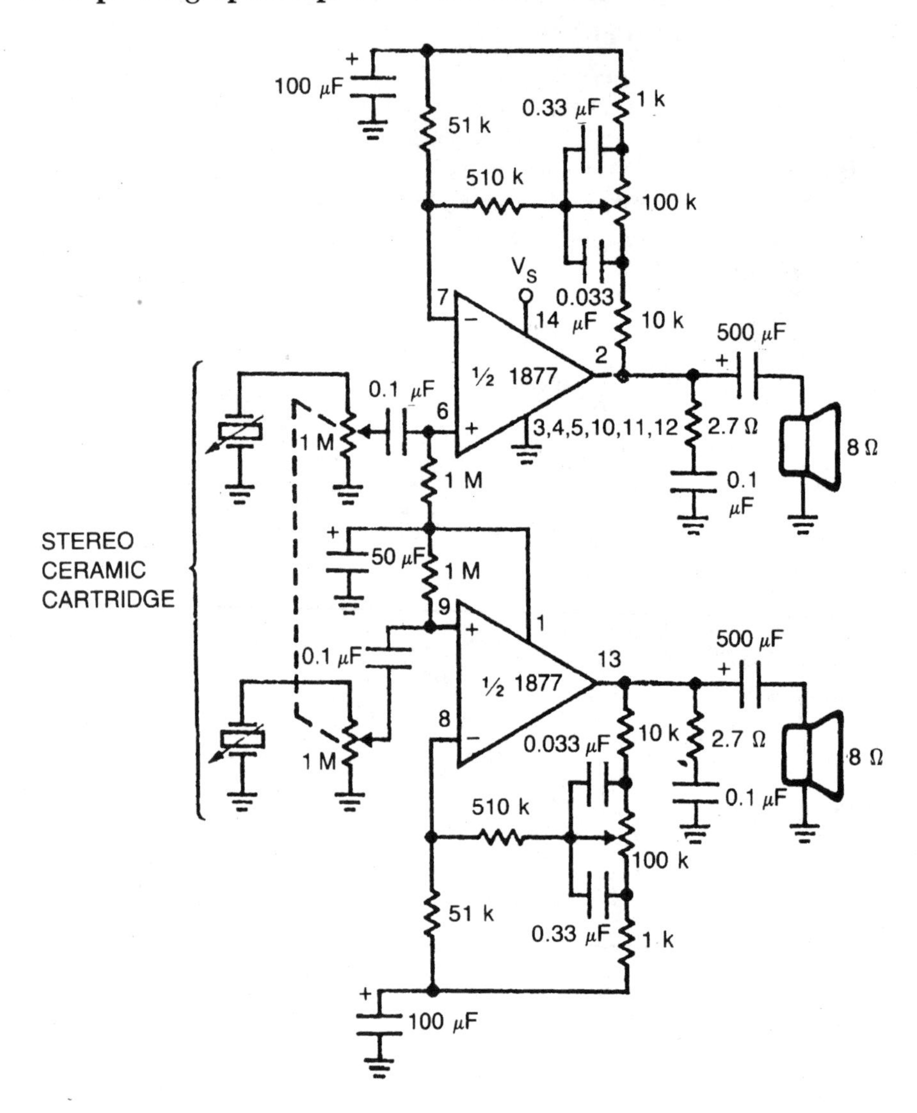

General-purpose stereo amplifier

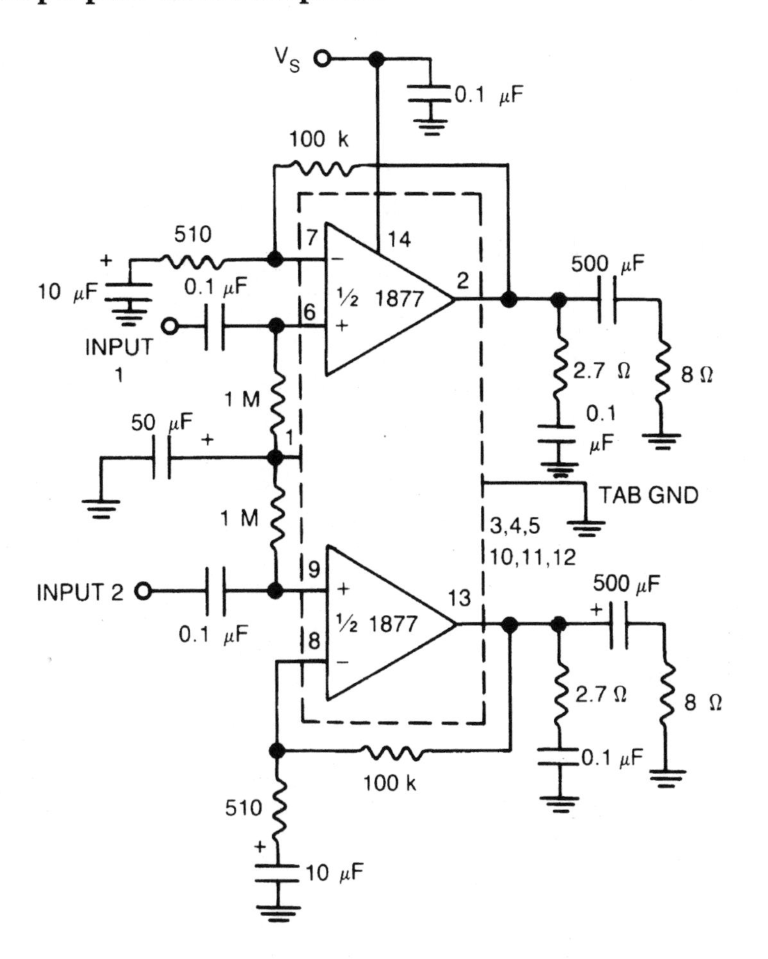

Inverting unity gain amplifier

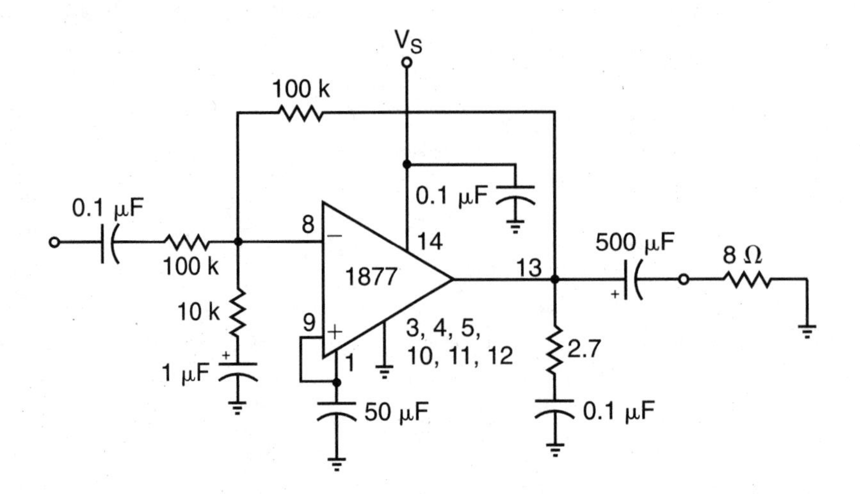

6
SECTION
Timers, oscillators, and signal generators

Generally, electronic circuits are designed to do something to some sort of signal from an outside source. In some cases, however, it is necessary to originate a signal. An oscillator produces an ac signal with a specific waveform and frequency.

In addition to oscillator and signal generating ICs, this chapter also features timers. Essentially, a timer is a linear multivibrator. A multivibrator, like a digital gate, has just two possible output states, HIGH or LOW. There are three basic types of multivibrators:

The *monostable multivibrator* has just one stable output state. (This is often called a *timer.*)
The *bistable multivibrator* has two stable output states. (Often called a *flip-flop.*)
The *astable multivibrator* has no stable output states. (Essentially a rectangle-wave generator.)

Most timer ICs can be operated as either monostable multivibrators or astable multivibrators. They usually cannot function as bistable multivibrators, but this type of operation usually is not appropriate to linear applications anyway.

322 precision timer

The 322 is one of a series of precision timers that offers great versatility with high accuracy. This chip can be operated from unregulated voltages ranging from 4.5 V to 40 V while maintaining constant timing periods from microseconds to hours.

An internal 3.15-V regulator is included in the 322 to reject any supply voltage fluctuations or changes, and to provide the user with a convenient reference voltage for other applications. External loads of up to 5 mA can be driven by this on-chip voltage regulator. An internal 2-V divider between the reference and ground sets the device's timing period to 1 RC. The timing period can be voltage-controlled by driving this divider with an external source through the V_{adj} pin. Timing ratios of up to 50:1 can be easily achieved.

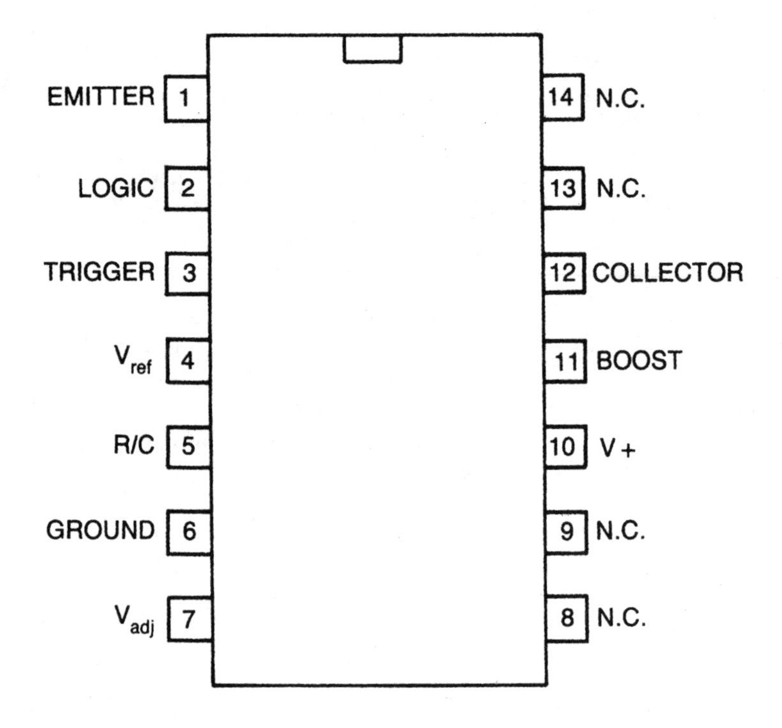

Monostable multivibrator

The timing period is equal to the product of the values of resistor R and capacitor C.

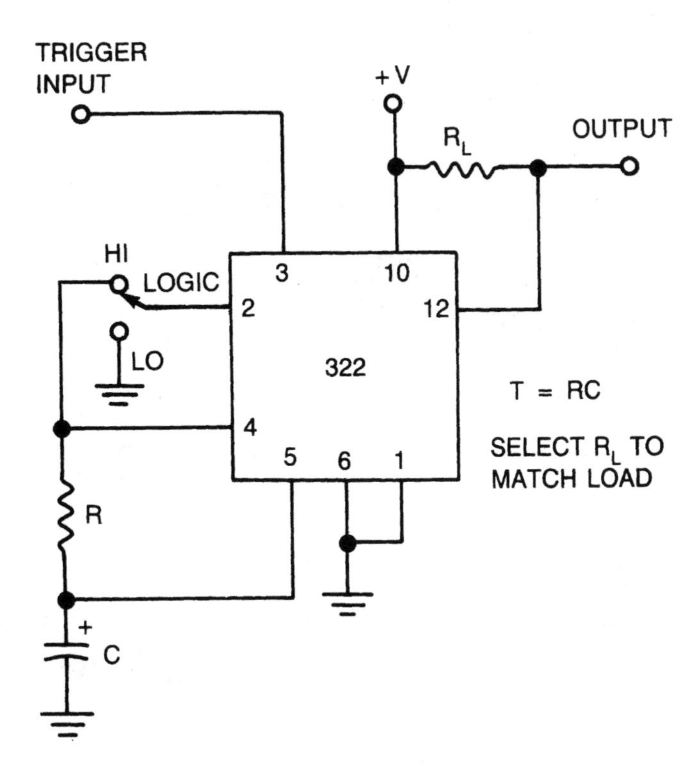

Astable multivibrator

The timing period (and therefore the output signal frequency) is determined by the values of resistor R and capacitor C.

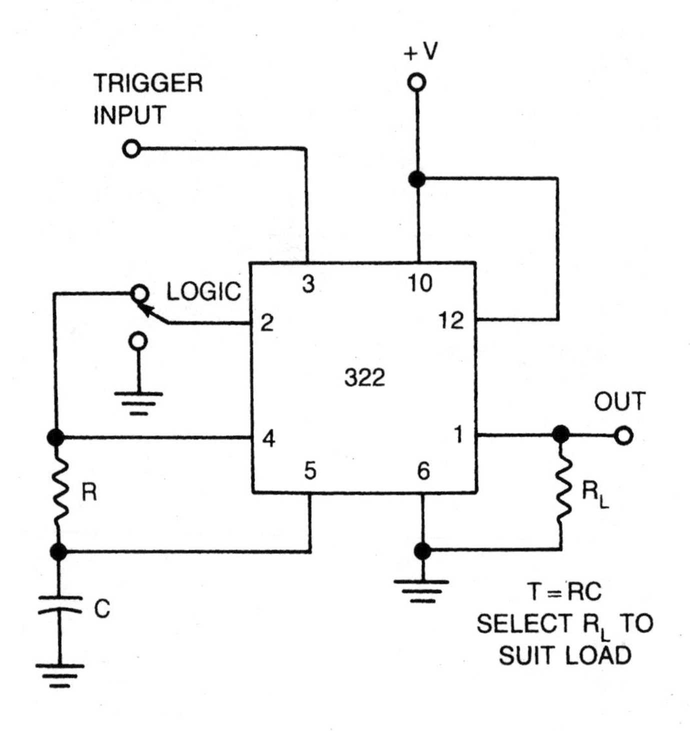

555 timer

The 555 timer is probably one of the most popular ICs ever. It has been around for years, and is unlikely to be discontinued in the foreseeable future, despite the appearance of a number of various improved and more specialized versions. Despite being a relatively simple device, nominally designed for just two basic circuits—a monostable multivibrator or an astable multivibrator—the 555 is incredibly versatile, and has been put to work in countless circuits and applications. It is very readily available, inexpensive, and quite easy to work with. For most basic applications, only a few external components are needed.

The timing period of the 555 is set by a straightforward external resistor/capacitor combination, according to a fairly simple formula.

$$T = 1.1RC$$

For reliable operation, the timing resistor should have a value between 10 KΩ and 14 MΩ, and the timing capacitor's value should be somewhere between 100 pF and 1000 µF. The larger either of these component values is, the longer the resulting timing period will be. The minimum and maximum component values given here produce a practical timing range from a low of 1.1 mS up to a high of 15,400 seconds (or 4 hours, 16 minutes, and 40 seconds). The timer certainly has quite an impressive range, especially for such an inexpensive and easy-to-use chip.

In astable multivibrator applications, two timing resistors are used, but the basic operating principles still apply. Two resistors are used in the astable circuit, because different timing periods are needed for the LOW and HIGH portions of the cycle.

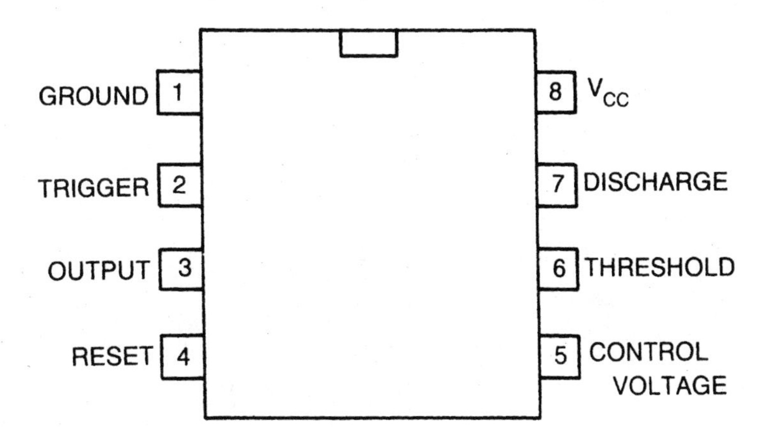

Monostable multivibrator

This circuit's timing period is a little greater than the product of the values of resistor R and capacitor C. The timing formula follows.

$$T = 1.1RC$$

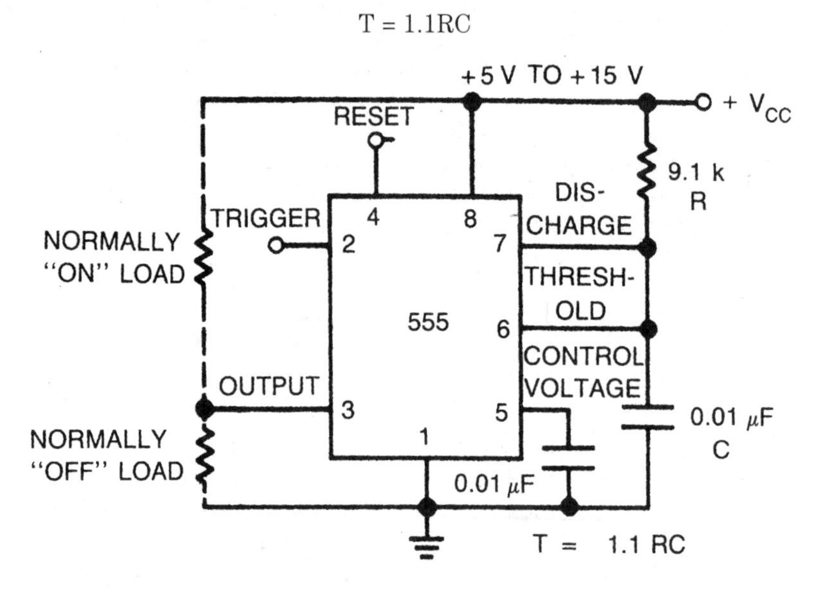

Astable multivibrator

The timing period (and therefore the output signal frequency) is determined by the values of resistors R_a and R_b, and capacitor C. The formula follows.

$$F = \frac{1}{[0.693C(R_a + R_b) + 0.693CR_b]}$$

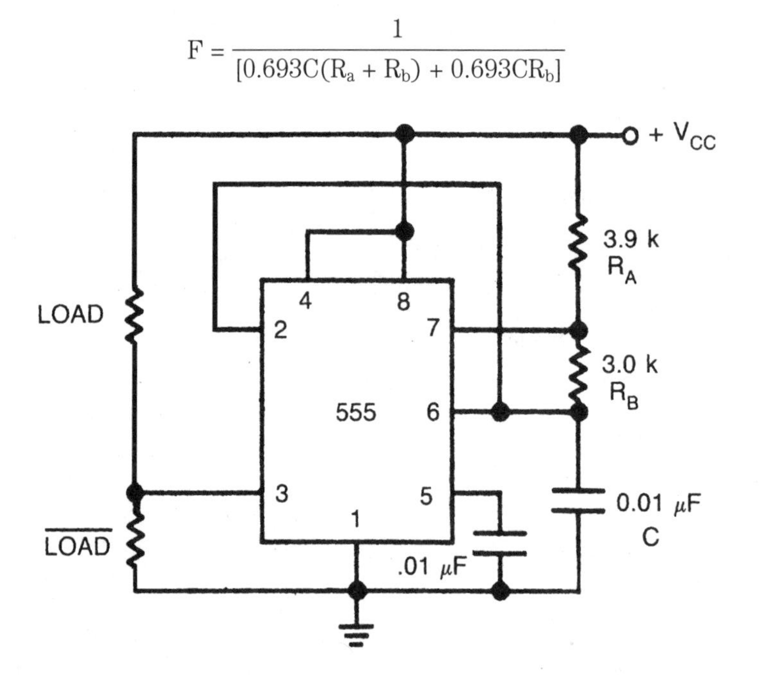

Pulse width modulator

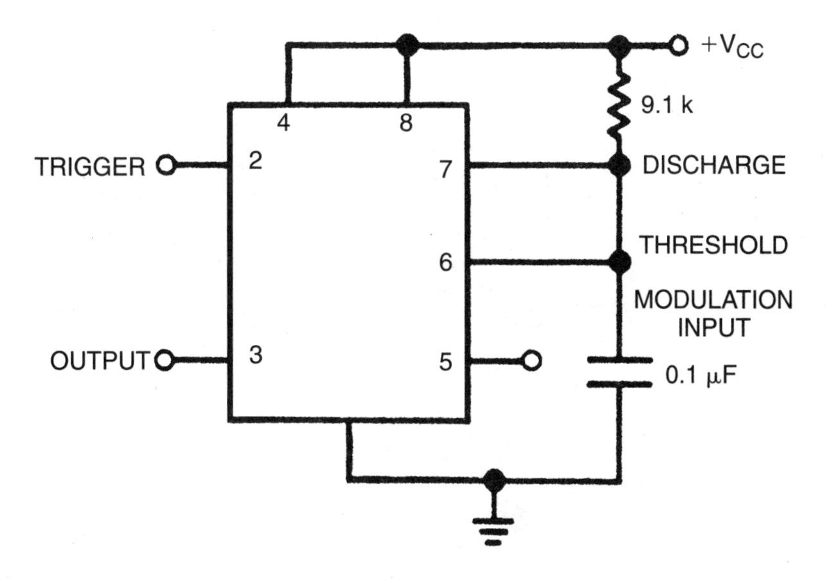

Pulse position modulator

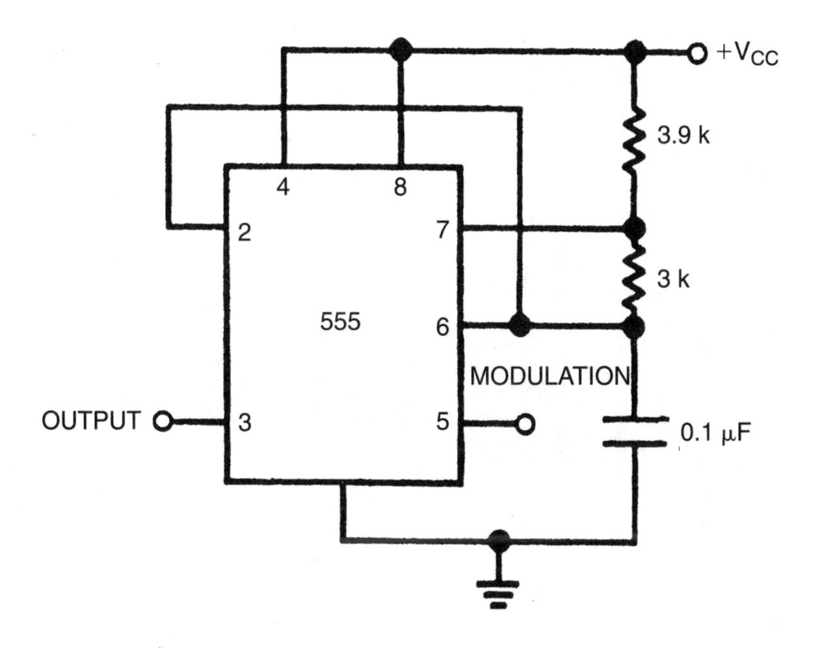

One-shot timer

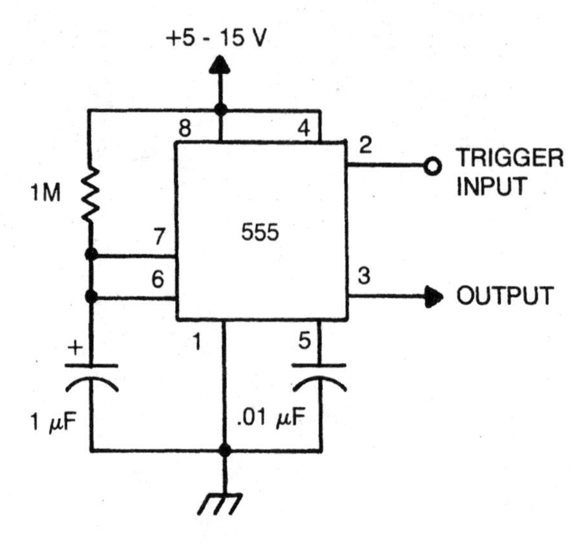

Pulse generator

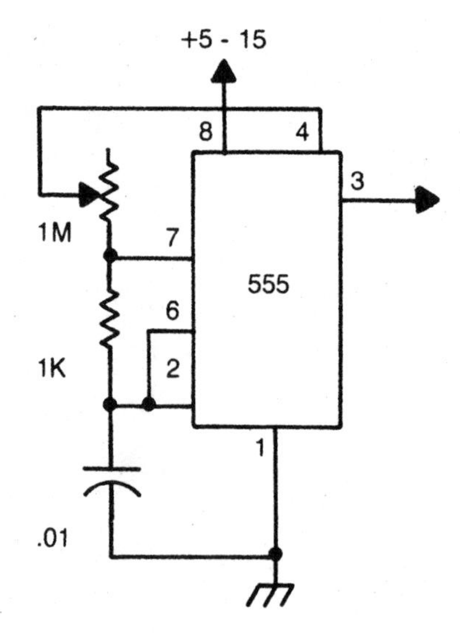

50 percent duty-cycle oscillator

The ordinary 555 astable multivibrator's circuit cannot generate a true square wave with a 50 percent duty cycle. This clever circuit forces the 555 to put out a good square-wave signal with equal HIGH and LOW times.

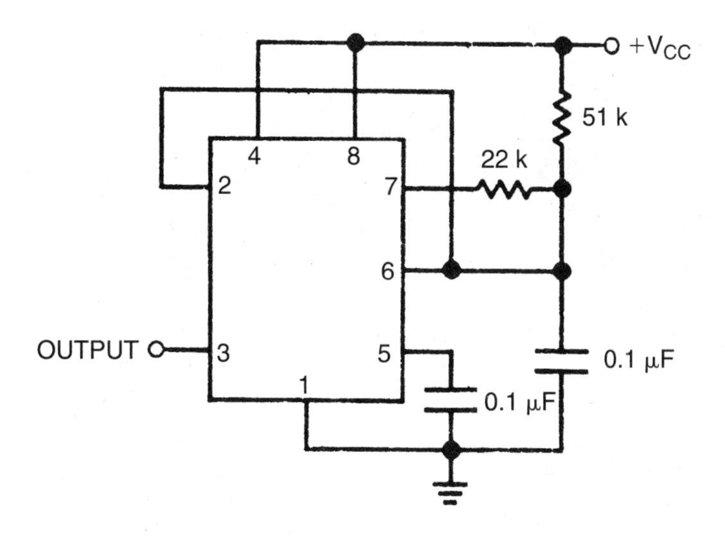

Timed relay

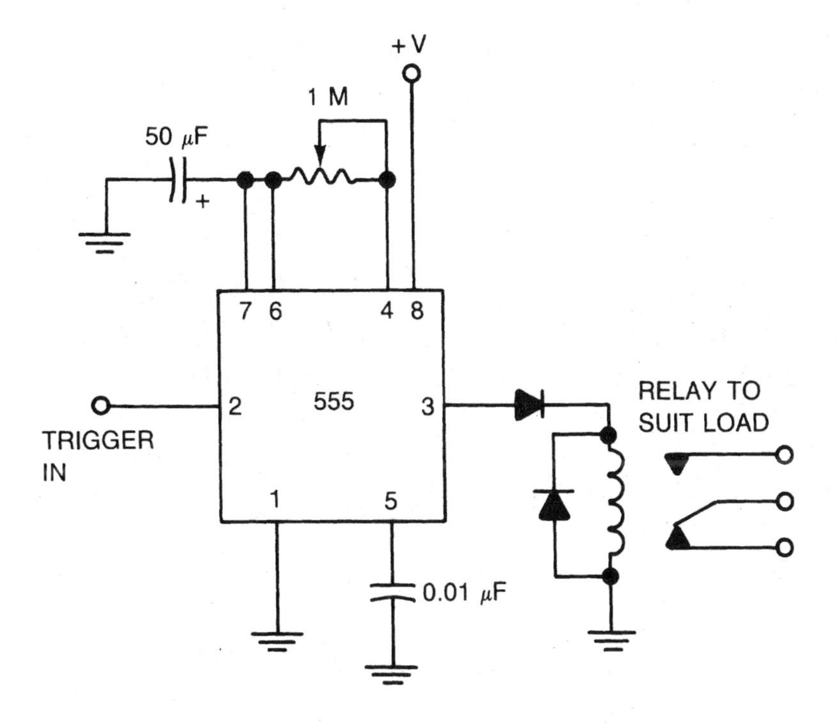

ON or OFF control

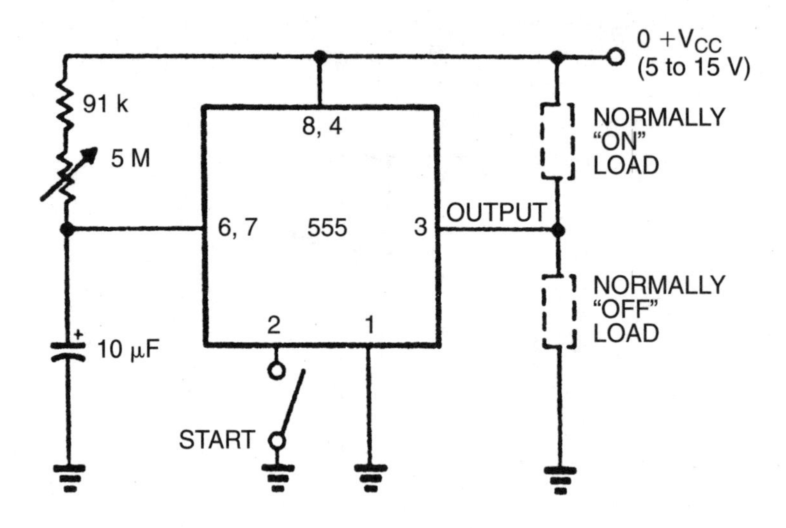

Switch debouncer

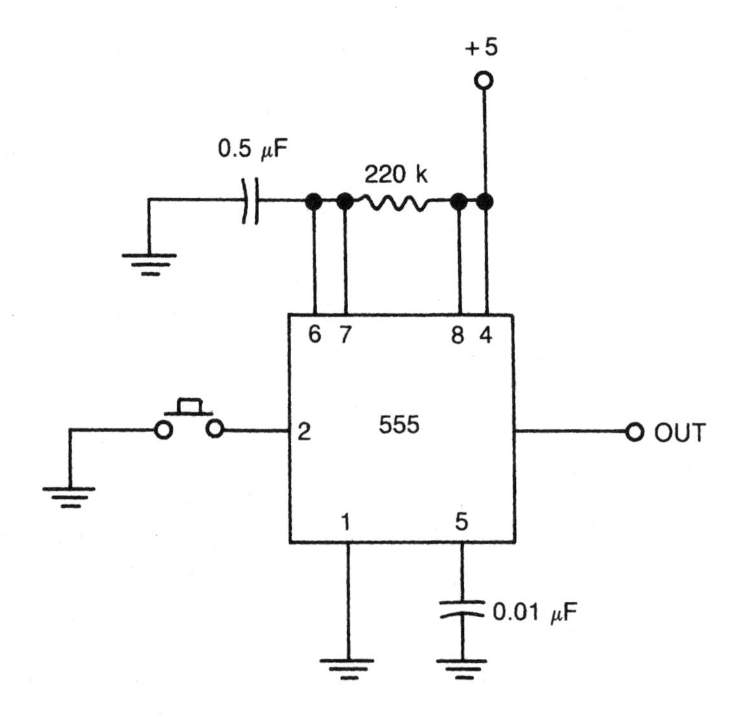

Timed touch switch

The timer is triggered by a touch of a finger to the conductive touch pad. It is very, very important to use *battery power only* for this type of circuit. No touch switch circuit should use ac power. If some unexpected short circuit or other fault causes the ac voltage to reach the touch plate, the results could be tragic. Don't take foolish chances, no matter how unlikely you think such a problem may be. There is always some finite chance of a harmful or even deadly accident. It's not worth taking the risk. Use battery power only.

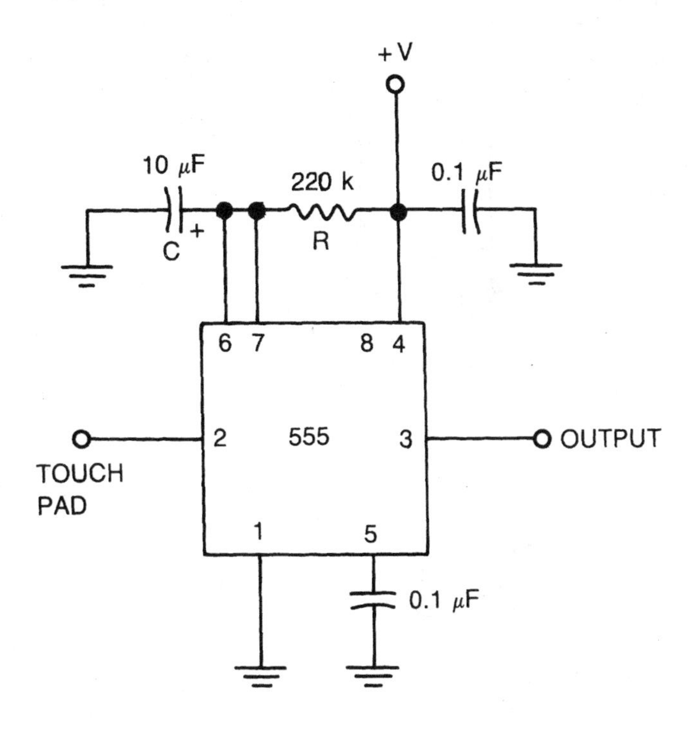

Frequency divider

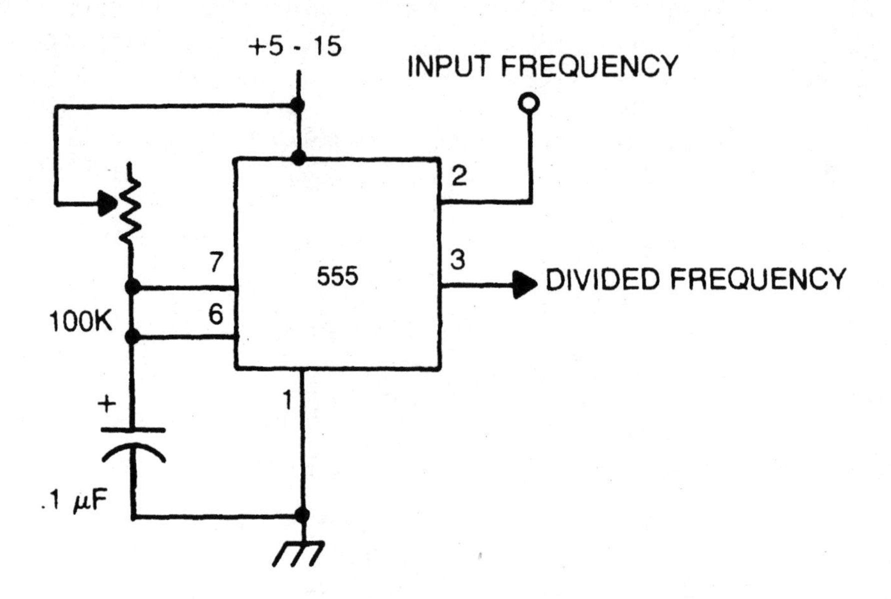

Very low-rate pulse generator

Basically, the 555 is wired as an ordinary astable multivibrator. Its output is fed through a CNOS counter (the CD4017), which puts out one output pulse for every ten input pulses. This means the actual output frequency is one-tenth the value calculated for the 555.

$$F = \frac{1}{[6.93(R_a+R_b) + 6.93CR_b]}$$

The counter can be wired for other division rates, if desired. See the circuit suggestions for this IC back in Sec. 2.

Longer timing periods (lower output frequencies) can be achieved by adding more counter stages in cascade. Again, see Sec. 2 for more suggestions on using the CD4017.

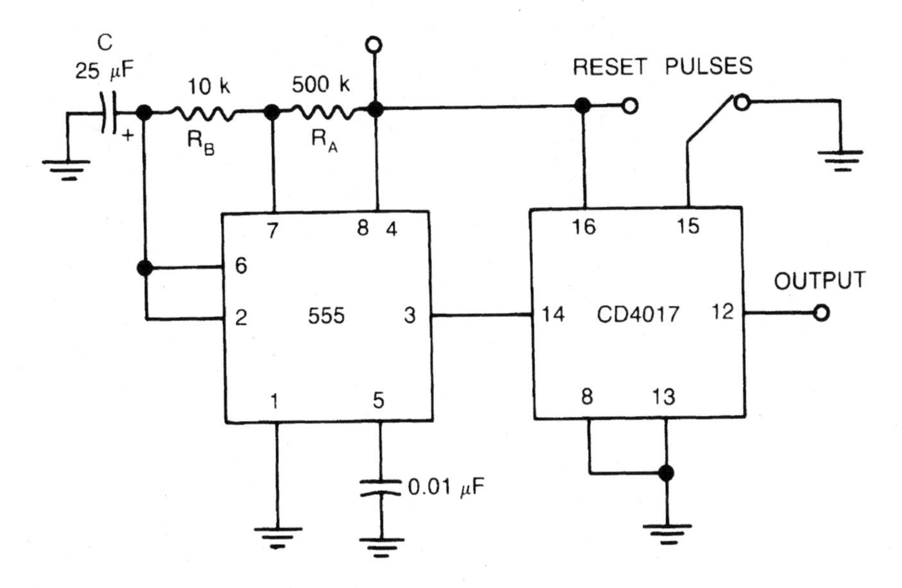

Triangle wave generator

The output resistor and capacitor filter down the strength of the harmonics, producing a fair approximation of a triangle wave. The triangle wave will be somewhat distorted and asymmetrical, because the 555 does not generate a true square wave with a 50 percent duty cycle. This will alter the actual harmonic content of the output signal somewhat.

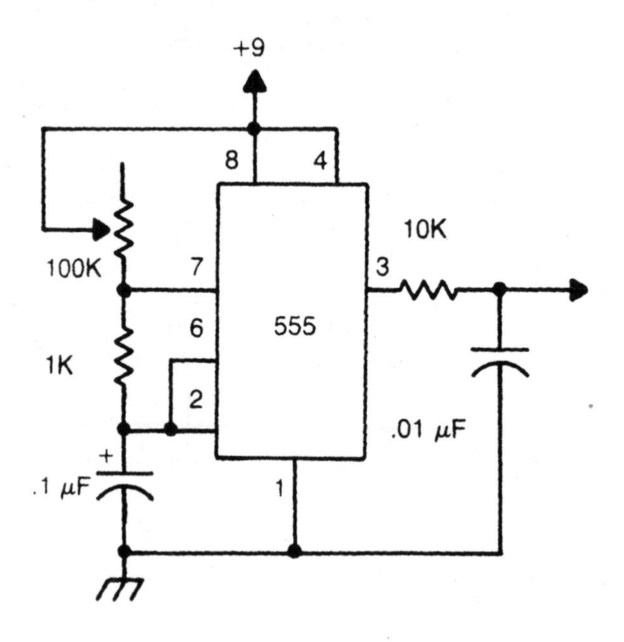

Tone burst generator

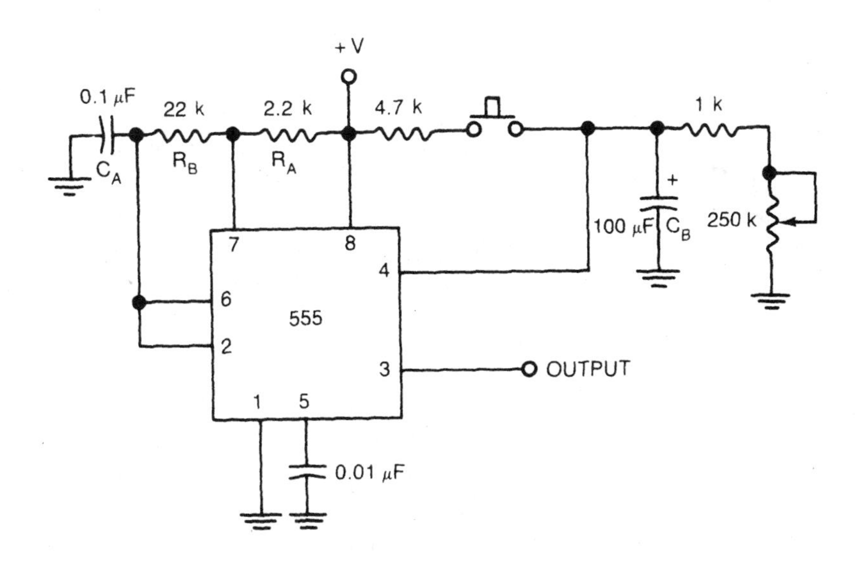

Toy organ

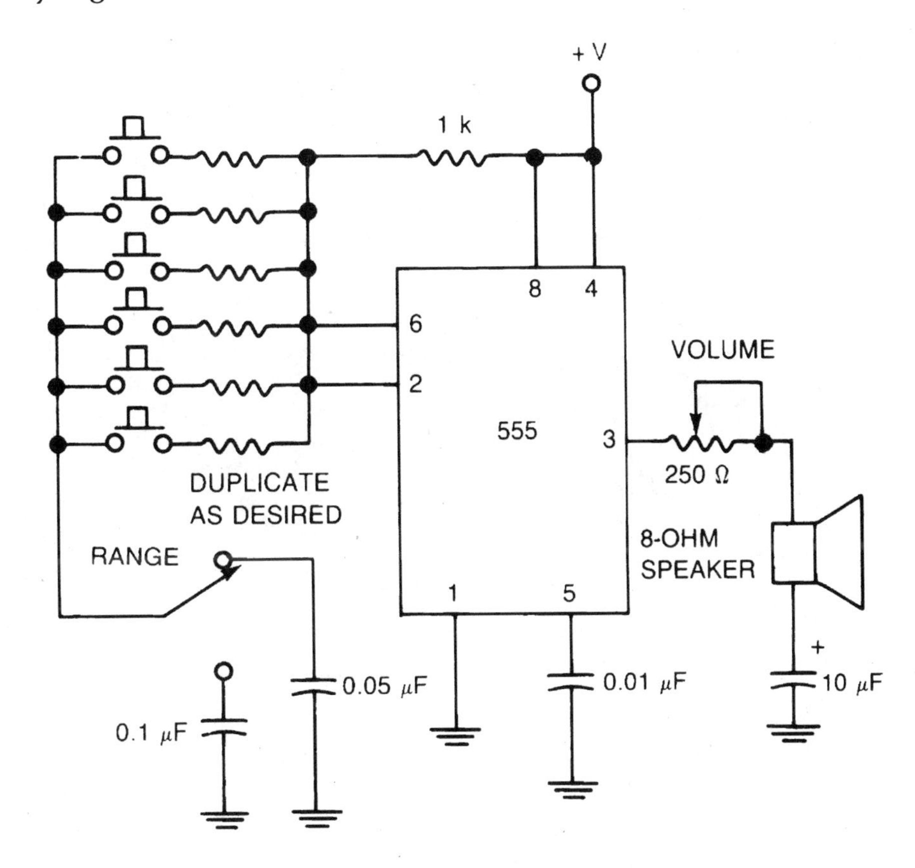

Telephone dialing tone encoder

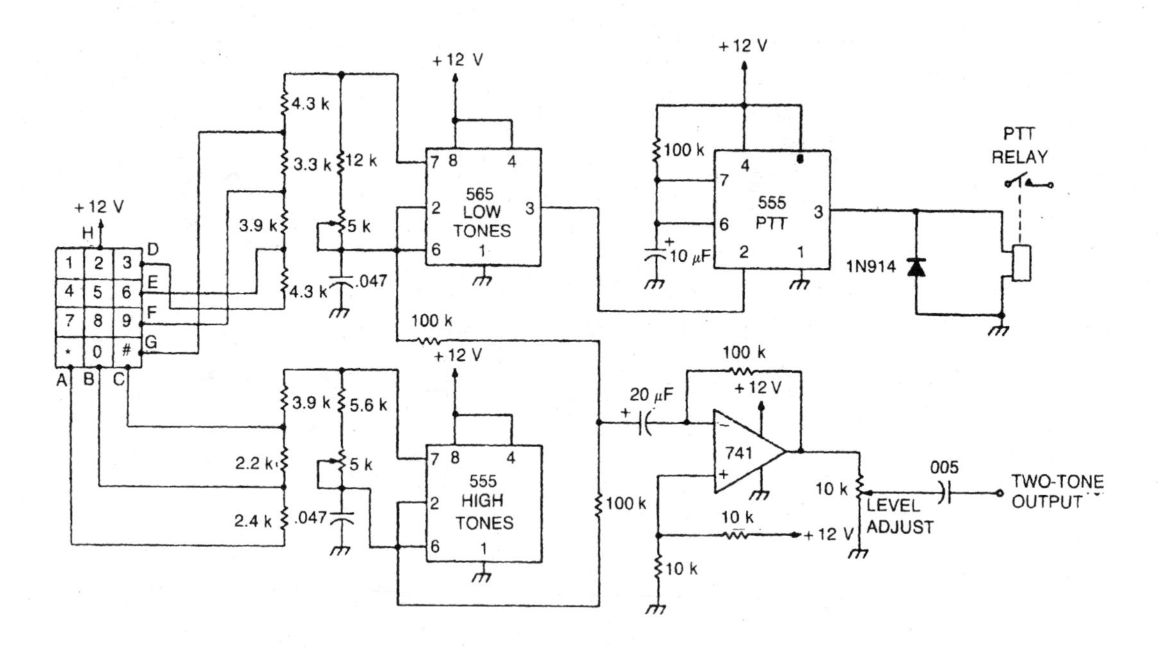

Missing pulse detector

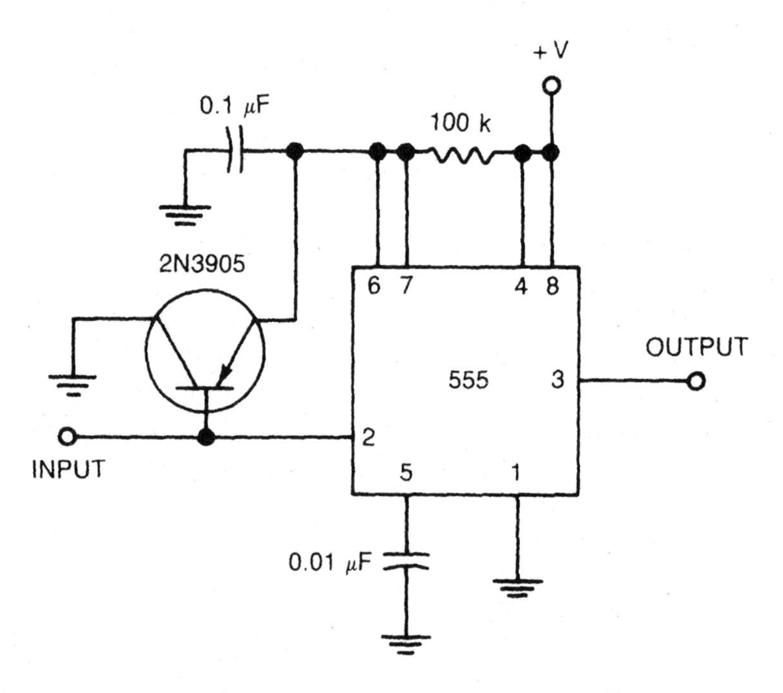

Light on alarm

The alarm sounds when the sensor (photoresistor) is sufficiently illuminated.

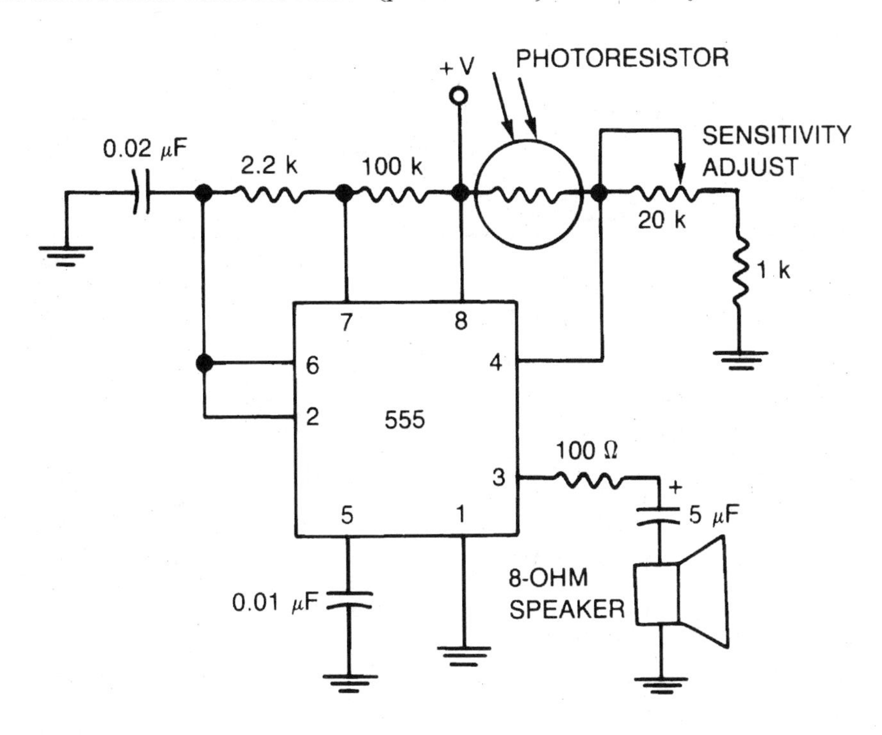

Light off alarm

The alarm sounds when the sensor (photoresistor) is not sufficiently illuminated.

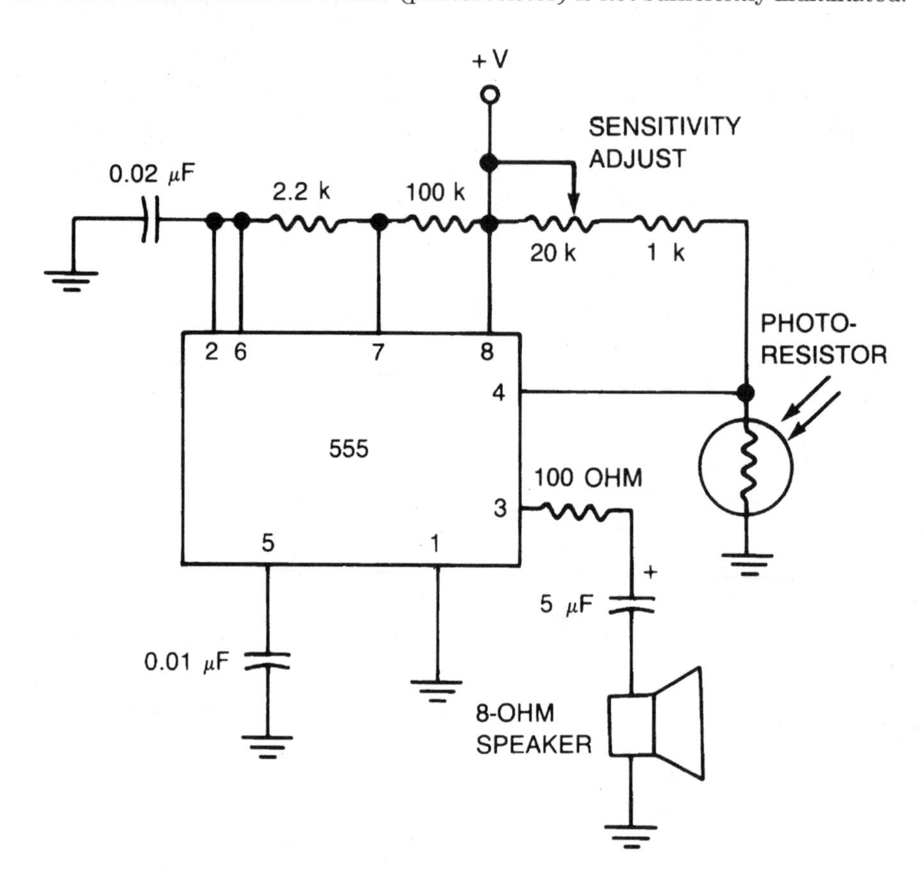

556 dual timer

The 556 contains two 555-type timer sections. These two sections are functionally independent of one another, except for the power-supply connections. One timer on the 556 can be used in a monostable multivibrator circuit, while the other is used in an astable multivibrator circuit, or they can be used in the same operating mode.

Each of the timer stages in the 556 is the same as the 555 timer earlier. They are directly interchangeable, except for the necessary changes in the pin-out configurations.

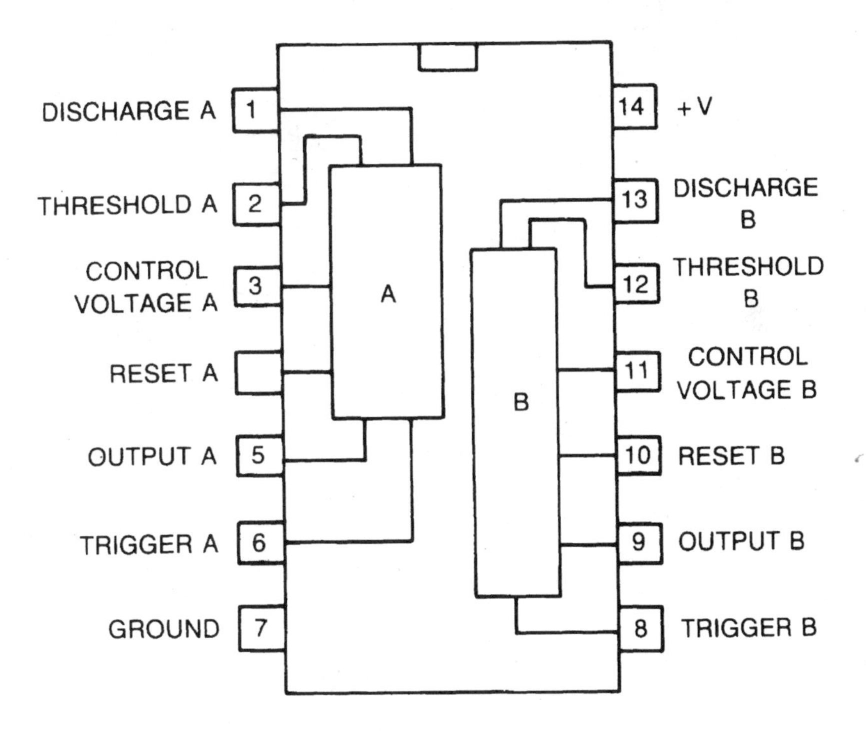

Gated tone generator

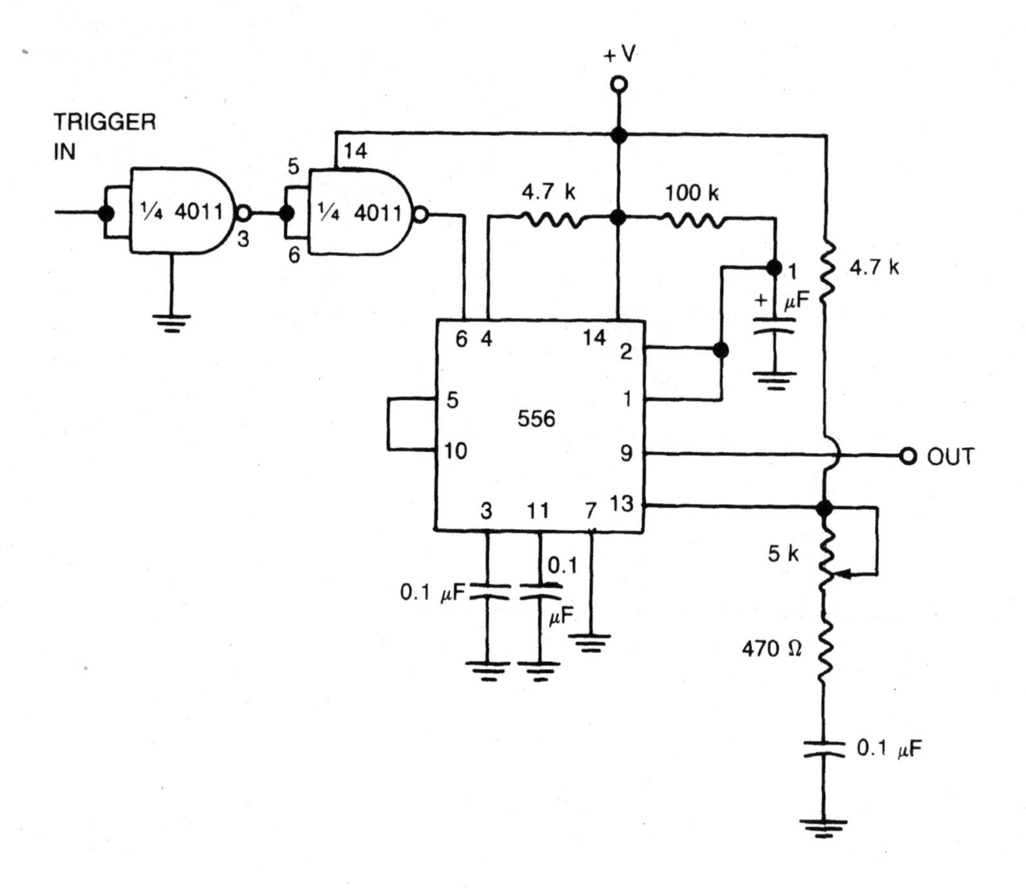

Tone burst generator

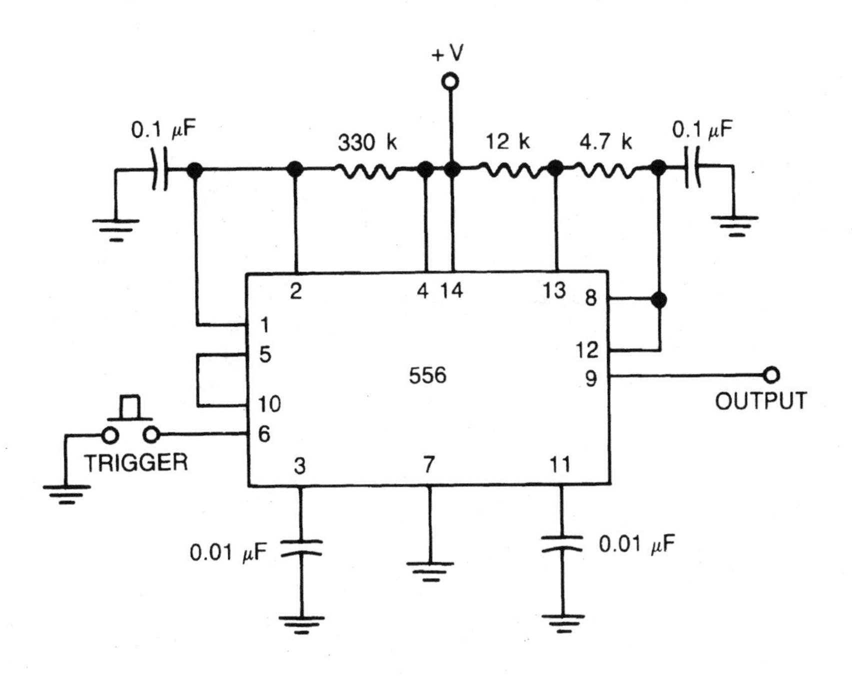

Extended range monostable multivibrator

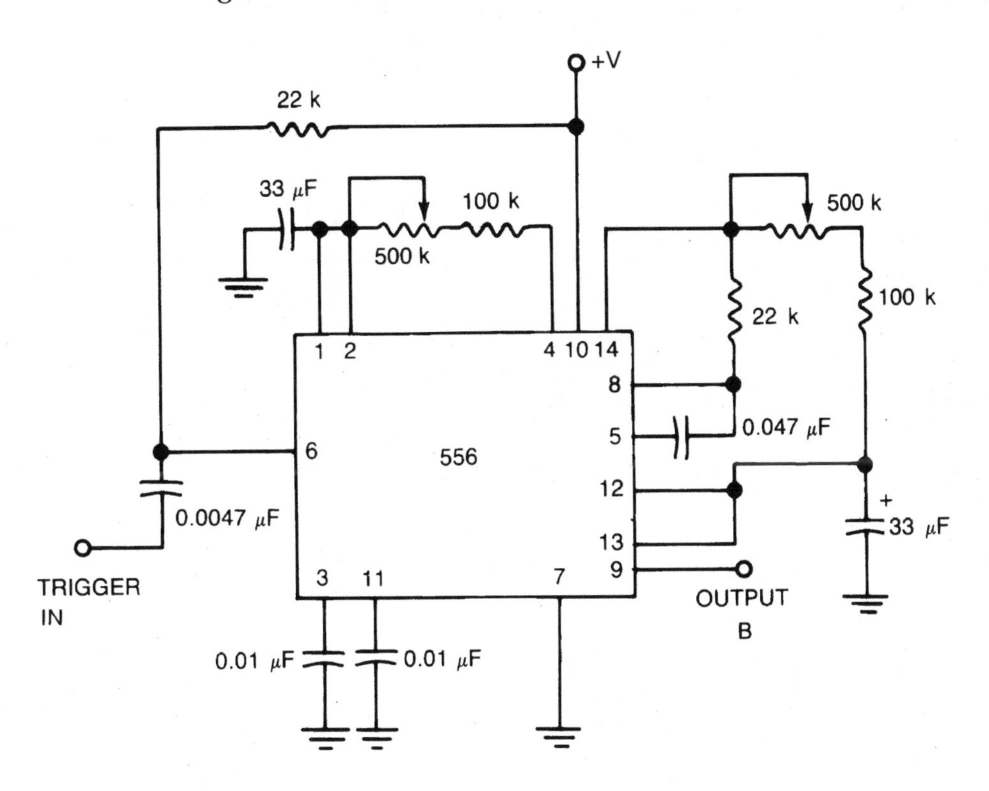

Pulsed relay driver

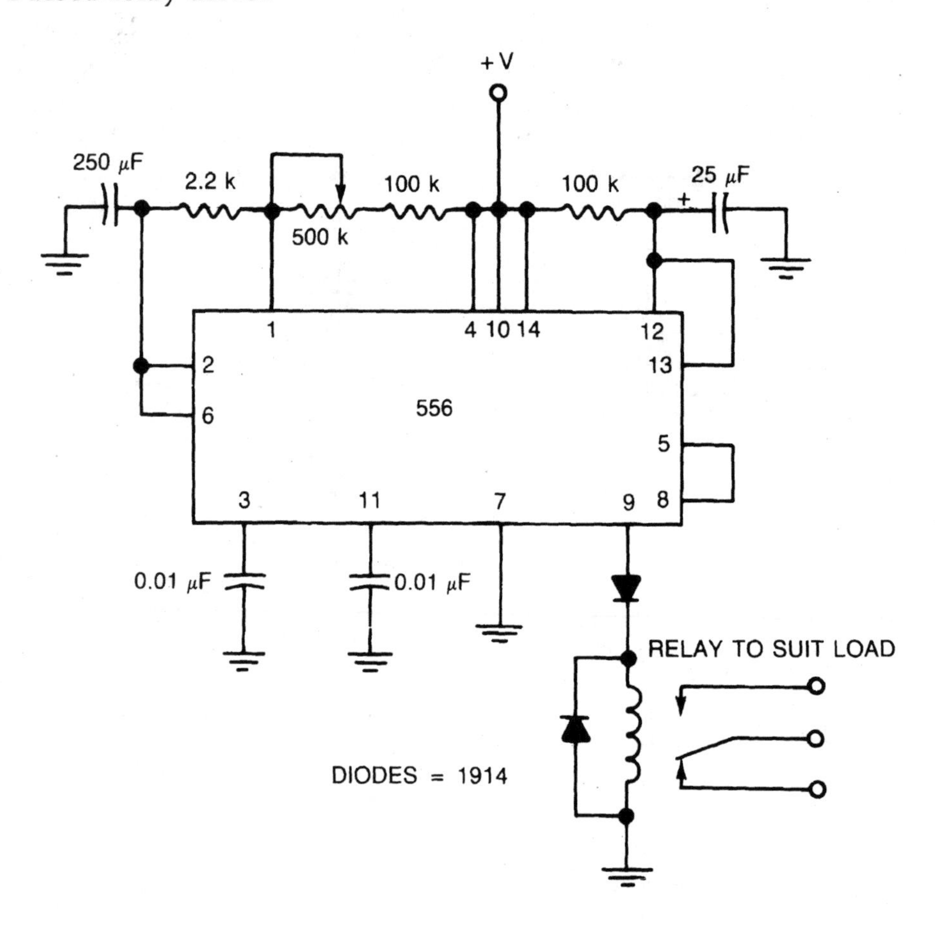

558 quad timer

The 558 contains four somewhat simplified 555-type timers in a single 16-pin package. Not all functions are brought out to the pins, so the 558 cannot be substituted in all 555 applications. This device is intended primarily for operation in the monostable multivibrator, but it can be tricked into functioning as an astable multivibrator.

Functionally, each 558 timer stage is equivalent to a 555.

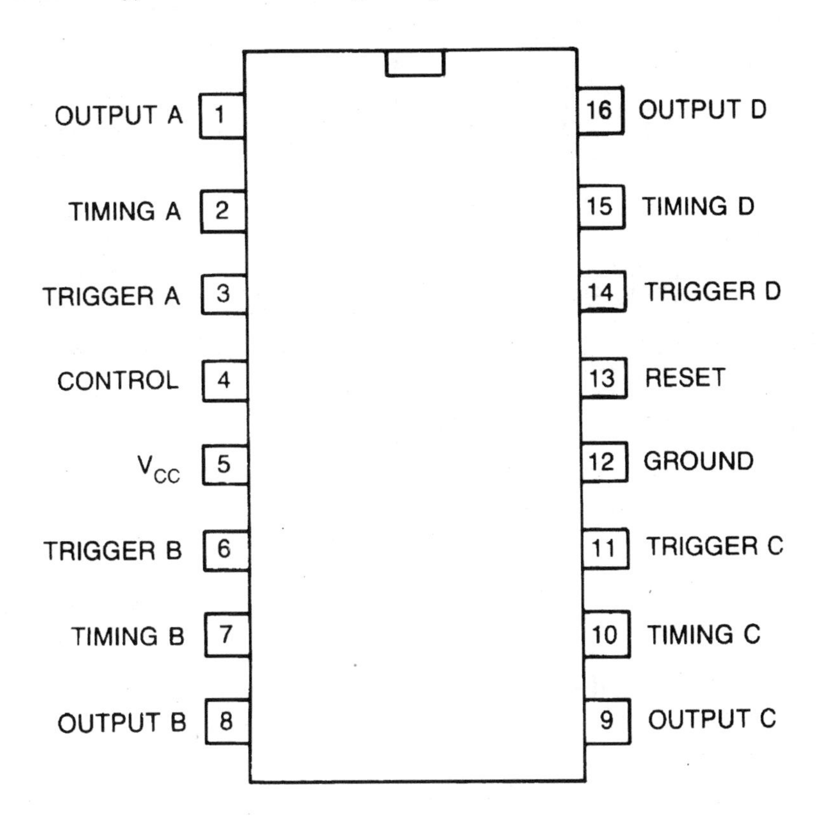

Rectangle-wave generator with independent pulse-width control

The output frequency can be adjusted without changing the pulse width, or duty cycle, or vice versa.

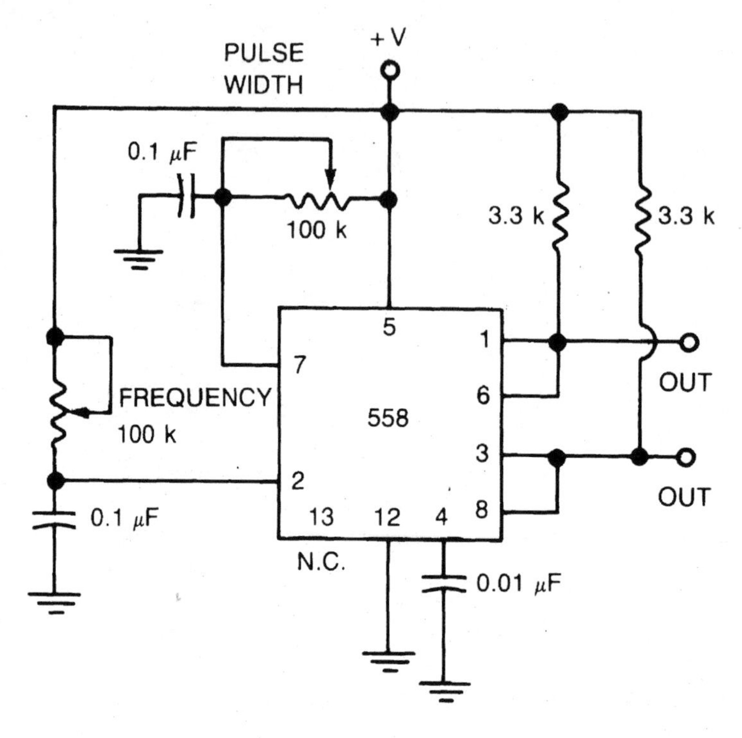

Monostable multivibrator

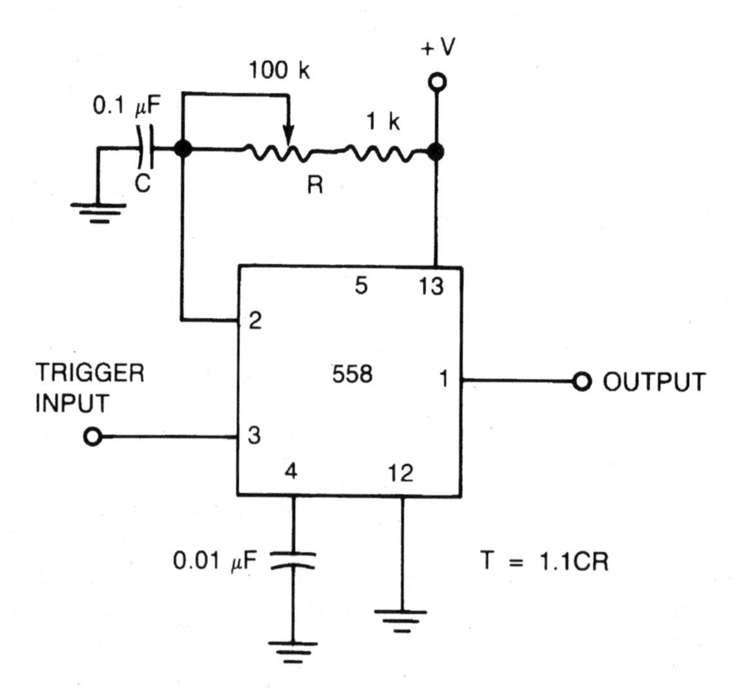

Astable multivibrator

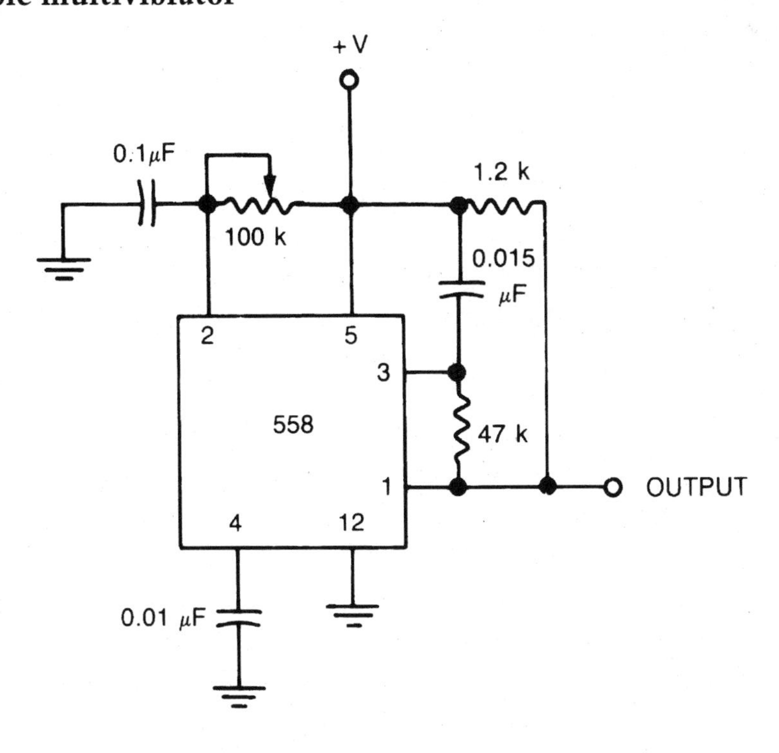

Extended range timer

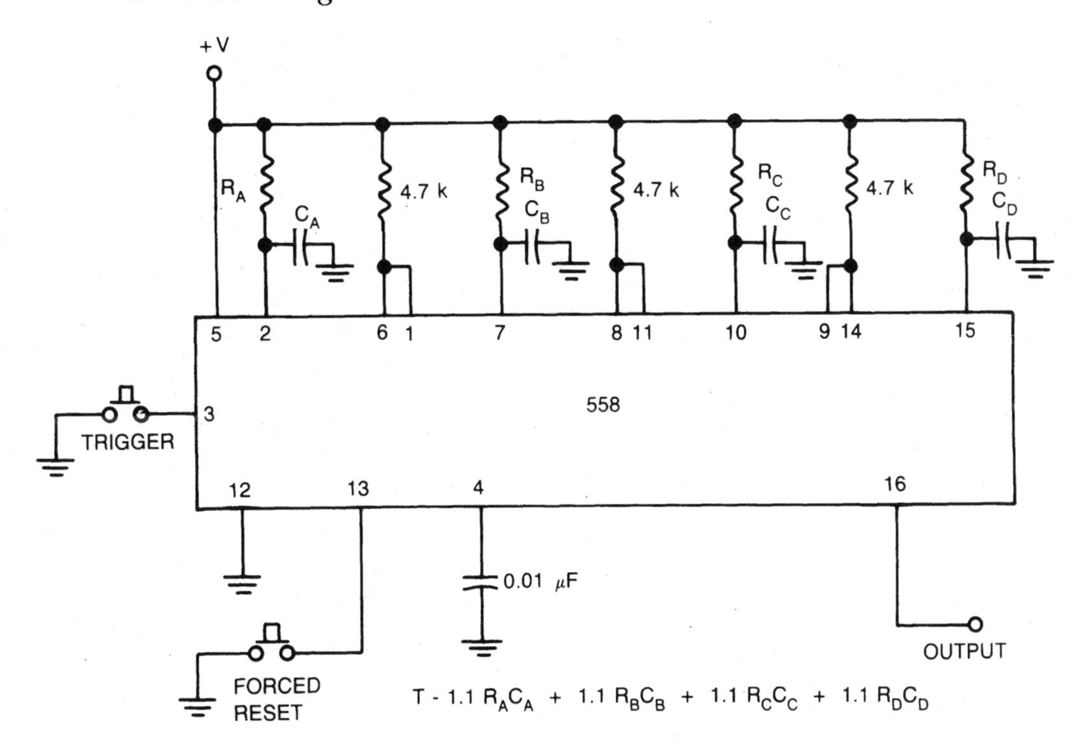

LM565 phase-locked loop

The 565 is an inexpensive general-purpose PLL (phase-locked loop). The VCO (voltage-controlled oscillator) frequency is set with an external resistor and capacitor. By varying the resistance, a 10:1 frequency tuning range can be achieved with a single capacitor. Similarly, an external resistor/capacitor pair sets the circuit's bandwidth, response speed, capture and pull-in range over a wide range.

The internal circuitry of the 565 includes a stable, highly linear VCO, and a double balanced phase detector with good carrier suppression.

The supply voltage for this device can be anything from ±5 to ±12 V.

Square-wave and linear triangle-wave outputs are available from this chip. The square-wave output is TTL and DTL compatible.

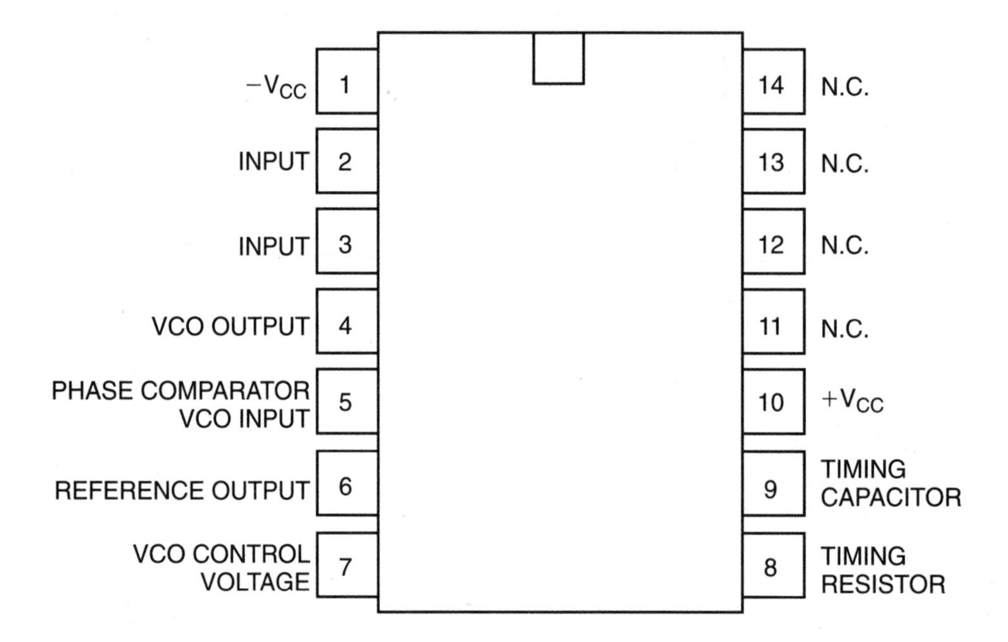

FSK demodulator

This circuit is designed to decode frequency-shift-keying signals switching between the standardized frequencies of 2025 Hz and 2225 Hz.

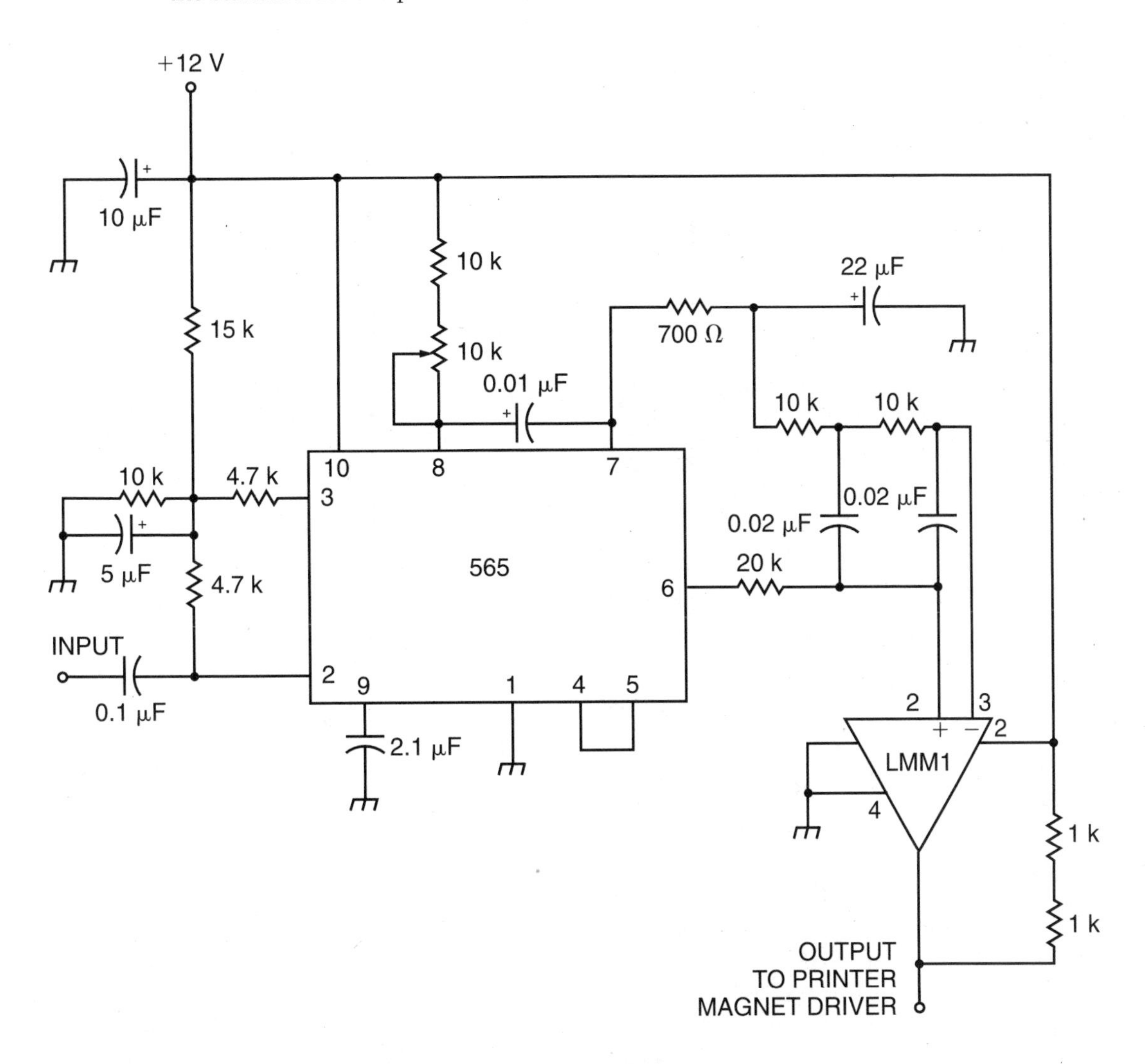

Frequency multiplier

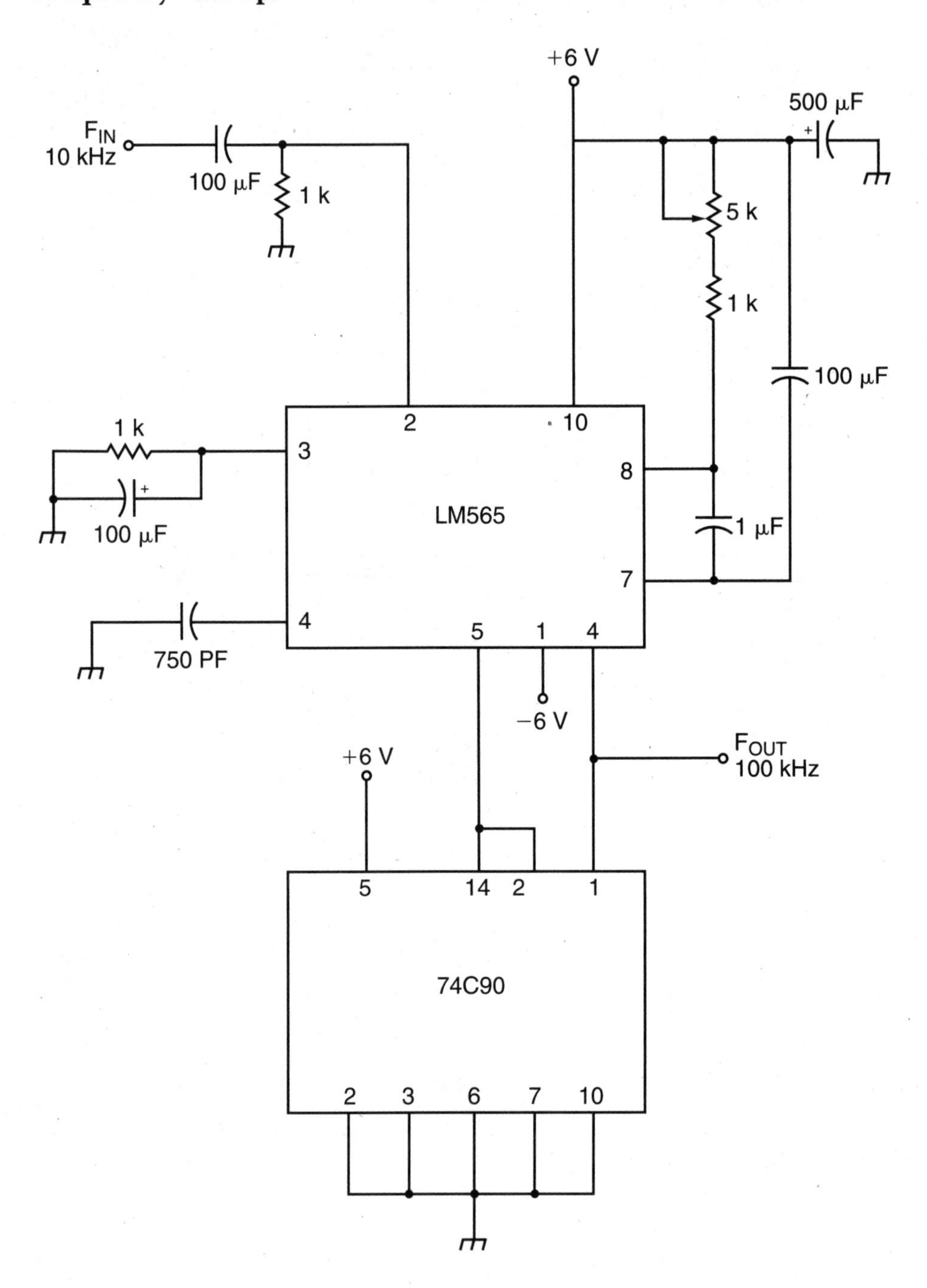

LM566 voltage-controlled oscillator

The 566 is a general-purpose voltage-controlled oscillator. It has two outputs, one generating a square wave and the other a triangle wave.

The control voltage's effect on the output signal's frequency is very linear, and a 10-to-1 frequency range can be achieved with a single fixed capacitor. The base signal frequency is set by an external resistor and capacitor. It can then be varied and controlled dynamically by an input voltage or current.

The 556 can be powered over a fairly wide range of supply voltages extending from 10 to 24 V.

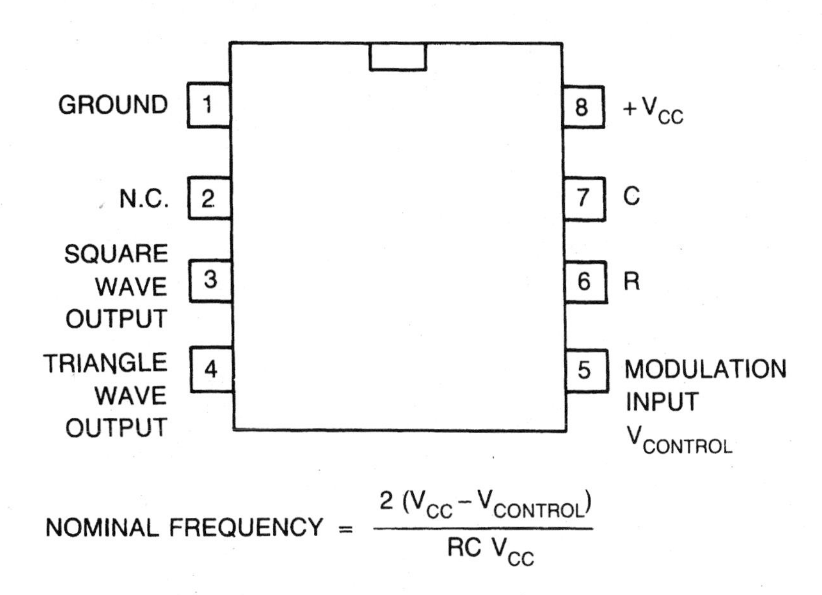

$$\text{NOMINAL FREQUENCY} = \frac{2\,(V_{CC} - V_{CONTROL})}{RC\,V_{CC}}$$

Basic VCO

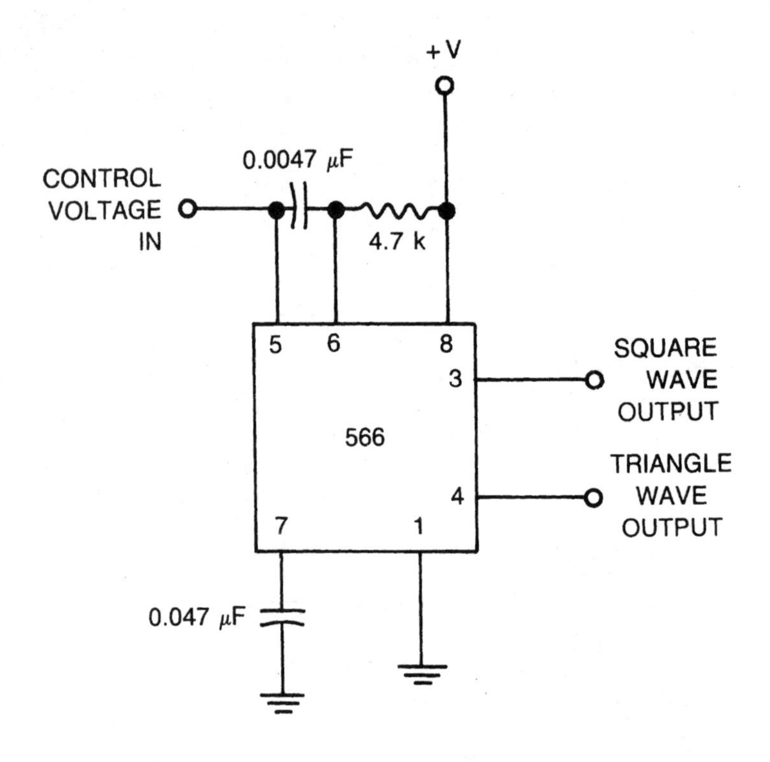

2-waveform oscillator

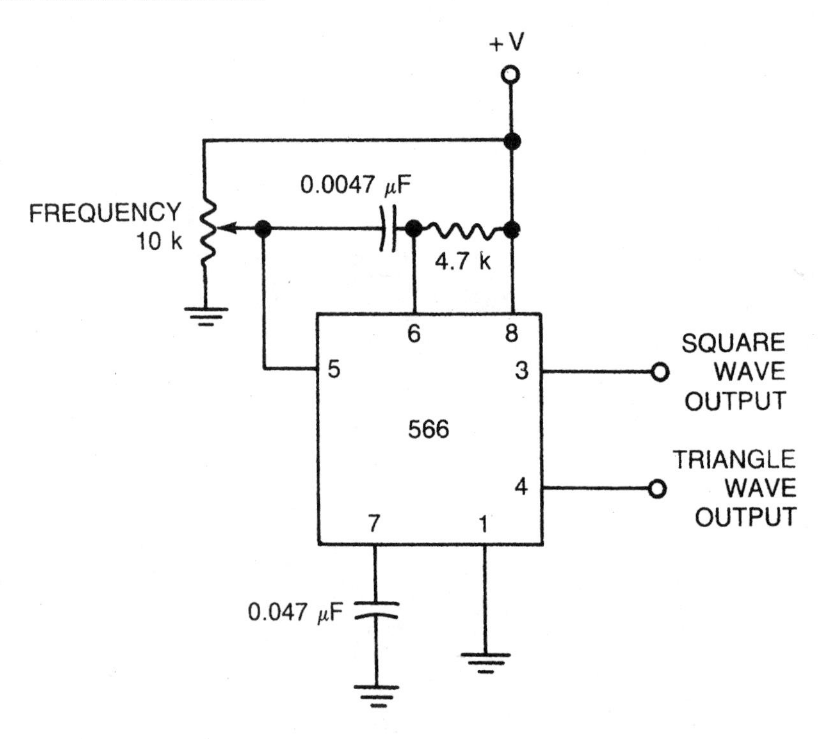

Vibrato tone generator

Vibrato is a rapid, but audible, repeating fluctuation in pitch. A little bit of vibrato can make electronic musical instruments sound more alive and expressive. However, too much or too exaggerated a vibrato effect can quickly become objectionable.

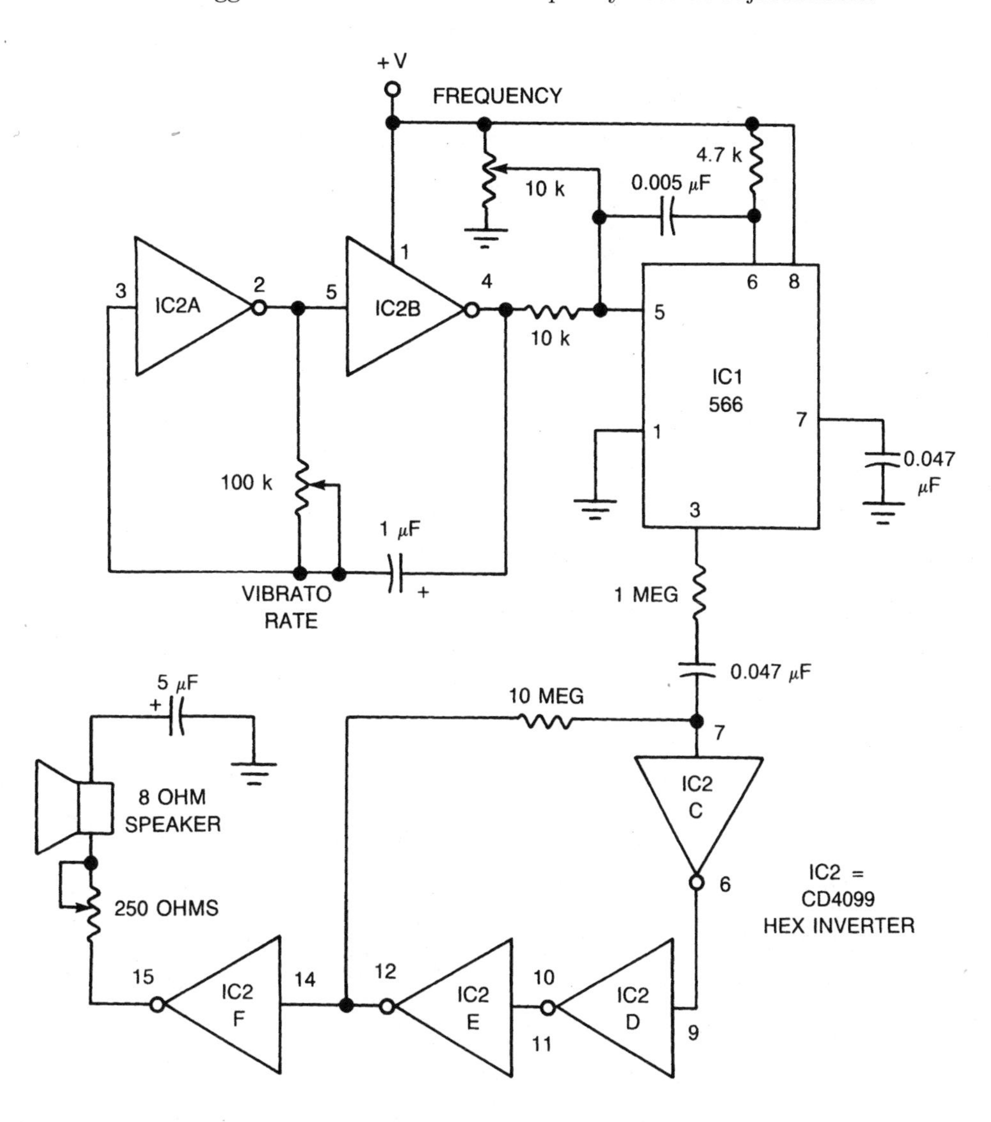

FSK generator

Frequency-shift-keying is used for analog transmission of digital data. See page 335 for an FSK decoder circuit.

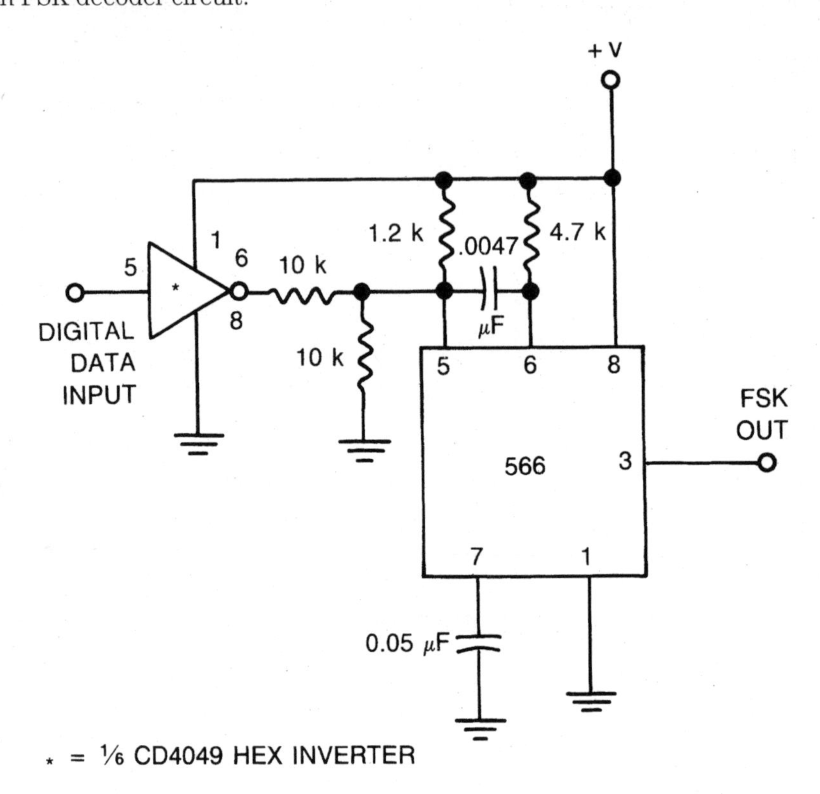

LM567 tone decoder

The 567 is a general-purpose tone decoder. When its input signal has a frequency within the circuit's preprogrammed passband, an on-chip transistor is saturated, switching to ground. This effect can be easily detected by external circuitry to control almost any desired function.

An on-chip VCO determines the center frequency of the decoder and drives an I and Q detector. A handful of external components are used to set up the key variables: center frequency, bandwidth, and output delay. An external resistor can cover a 20-to-1 frequency range. The center frequency, which is highly stable, can be adjusted over a very wide range, from 0.01 Hz up to 500 kHz. The bandwidth can be adjusted from 0 to 14 percent.

The 567 is quite good at rejecting noise and out-of-band signals, and offers good immunity to false signals.

This device's output can sink up to 100 mA. It is compatible with TTL or CMOS logic circuits, when appropriate supply voltages are used.

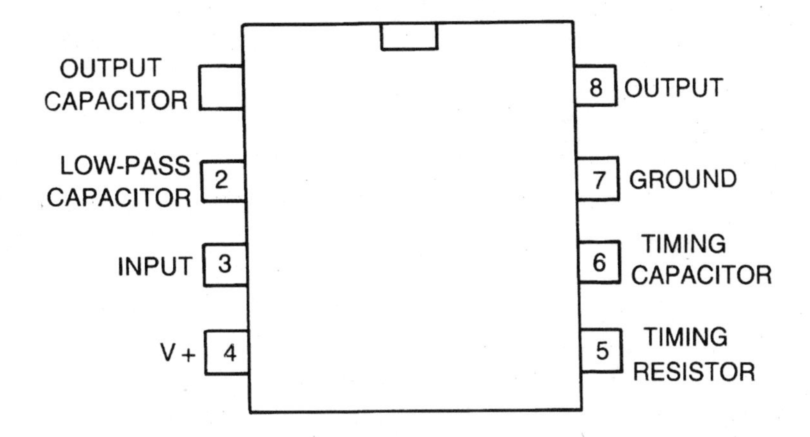

Oscillator with double-frequency output

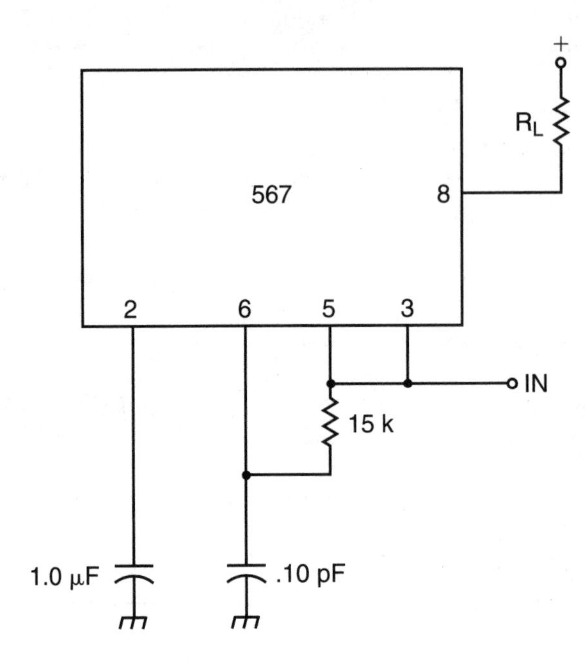

Dual-frequency output oscillator

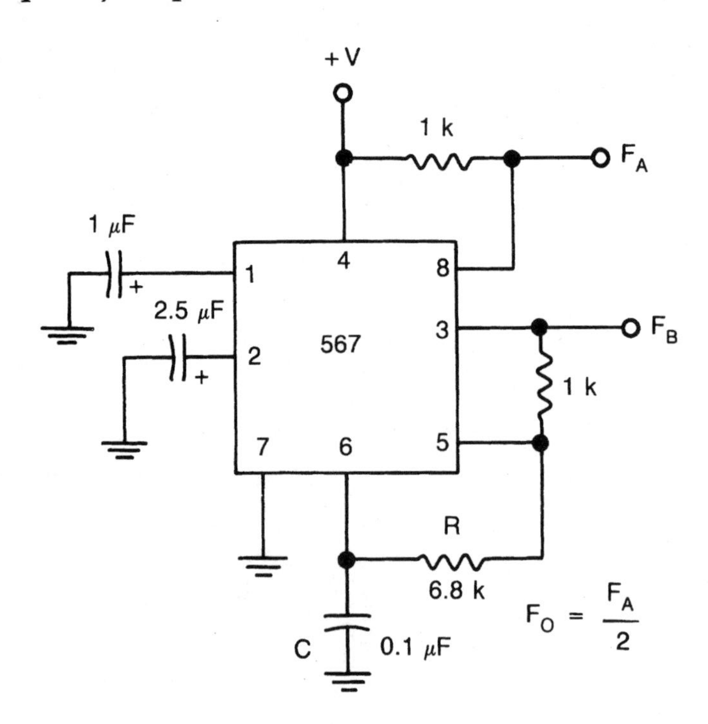

Precision oscillator to drive 100 mA loads

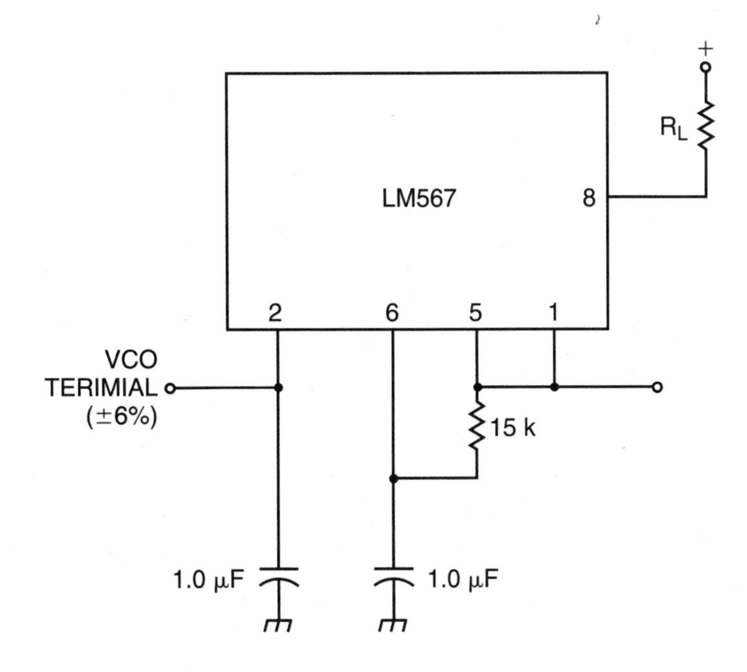

Oscillator with quadrature output

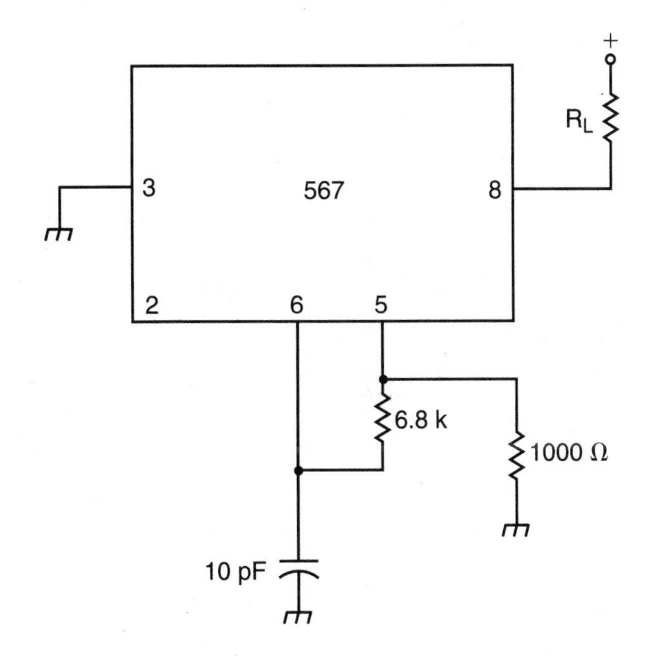

Oscillator with phase-shifter output

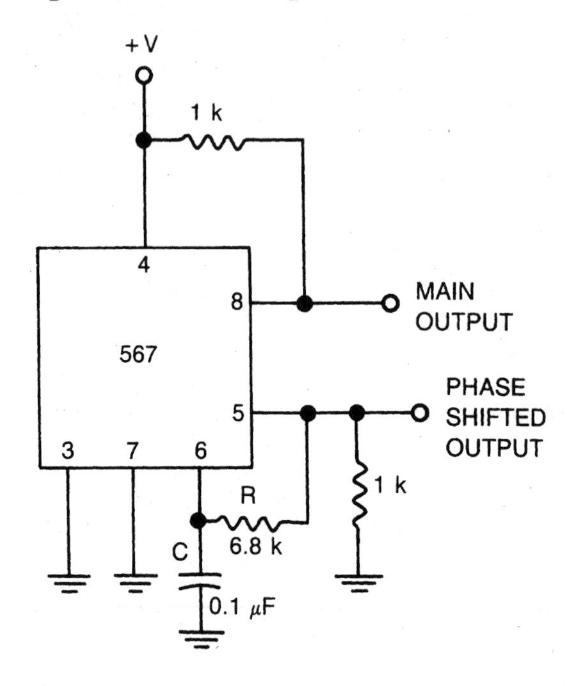

Tone decoder

The LED lights up when the input signal is within the preset frequency band. The center frequency is defined by the values of resistor R and capacitor C, using the following formula.

$$F_o = \frac{1.1}{RC}$$

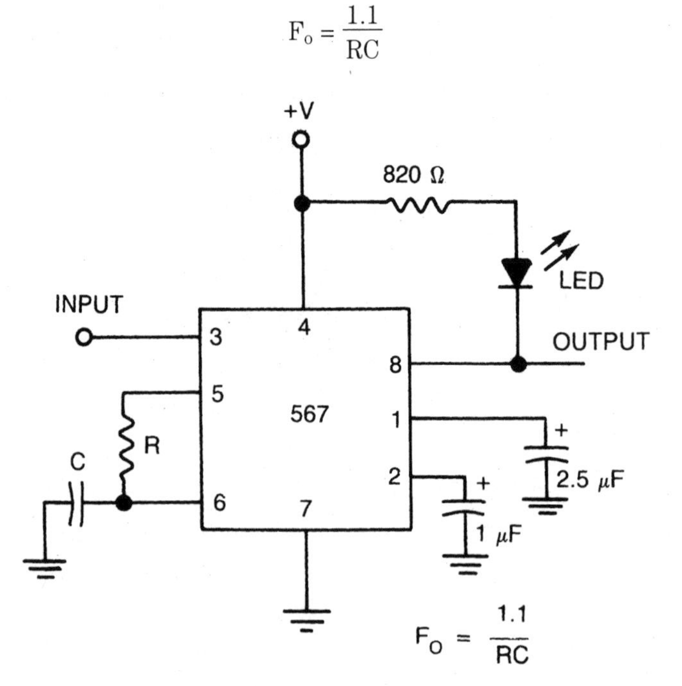

Narrow-range tone decoder

This tone decoder circuit has a more selective passband than the basic circuit on the previous page.

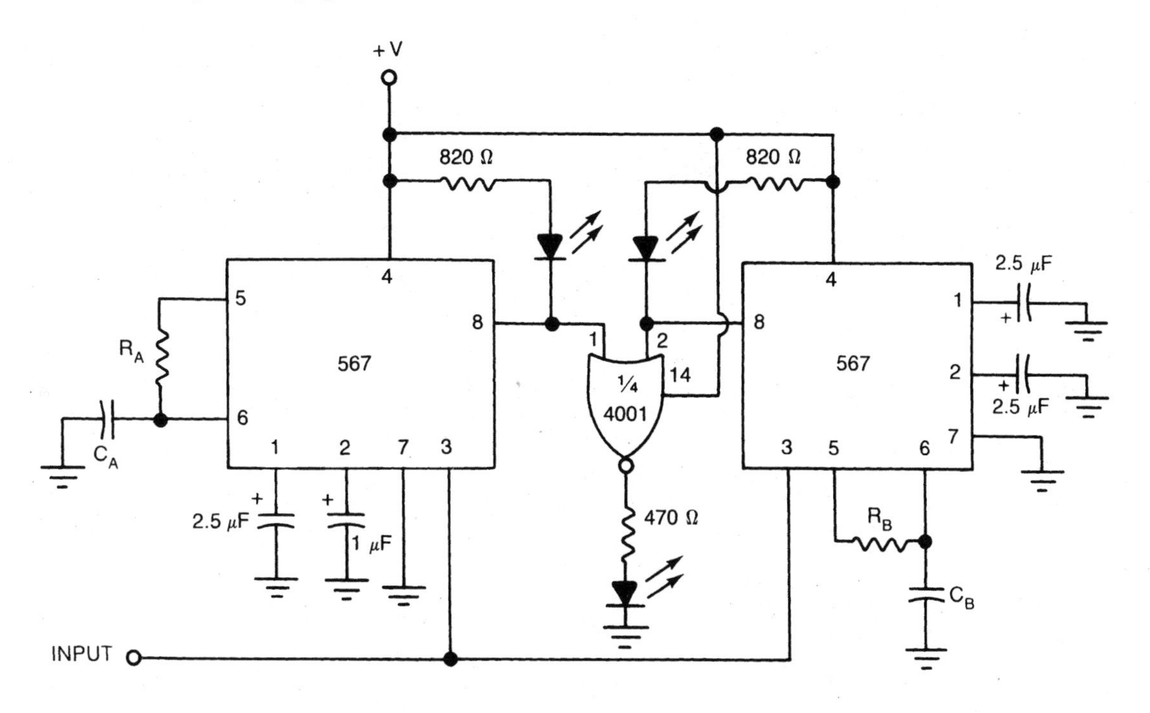

LM3909 LED flasher/oscillator

The 3909 is a simple 8-pin oscillator IC. It is intended primarily for LED flasher circuits, but it can also be used in many other low-frequency oscillator applications as well.

A simple external capacitor is used to set the frequency of the on-chip oscillator. The timing resistors are provided within the chip itself. This timing capacitor also provides a voltage boost, so pulses of over 2 V are delivered to the external LED (or other load device), even with supply voltages as low as 1.5 V.

Being inherently self-starting, the basic 3909 LED flasher circuit requires only the IC, a battery, the LED, and a timing capacitor.

A slow (pin 8) or fast (pin 1) flash rate can be set up, depending on where the external timing capacitor is connected to the chip.

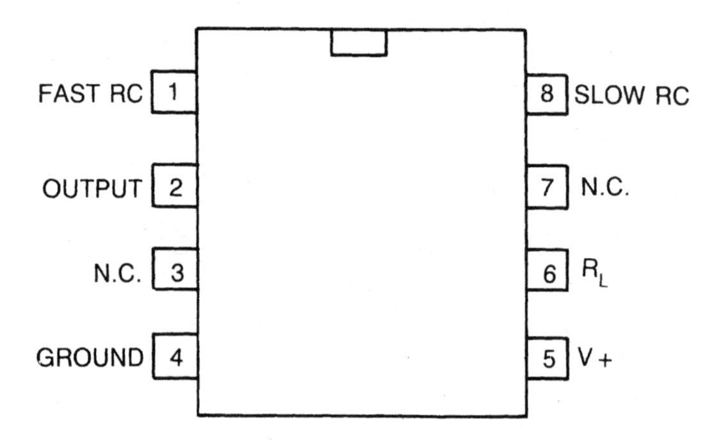

Basic LED flasher

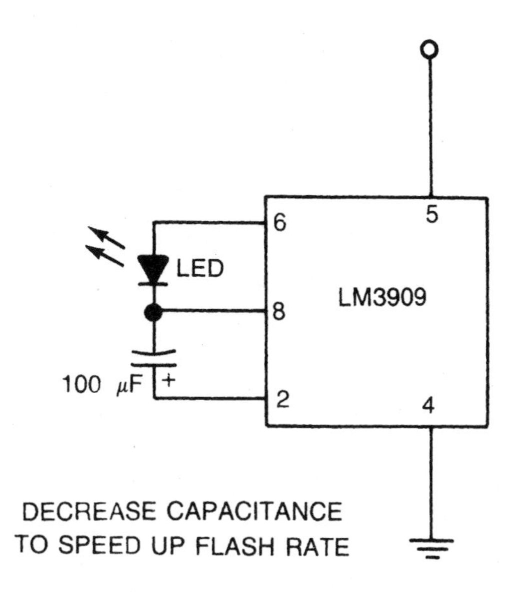

Minimum power 1.5 V LED flasher

Current drain in this circuit is extremely low. Even a common penlite battery will keep the LED flashing a long, long time. The battery powering this circuit should last almost as long as if it was just sitting on a shelf.

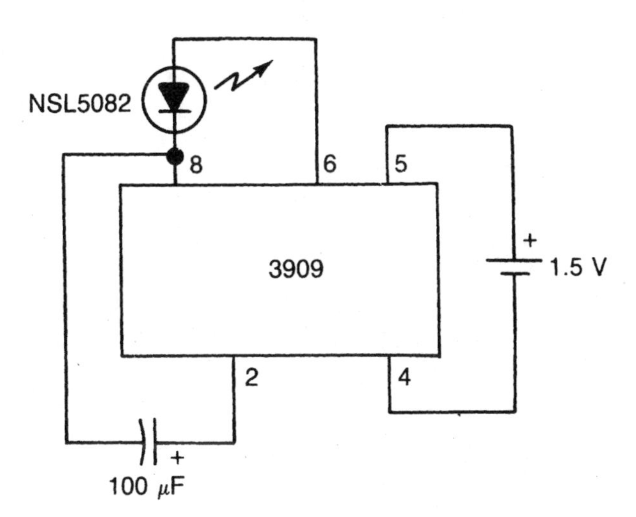

3-volt flasher

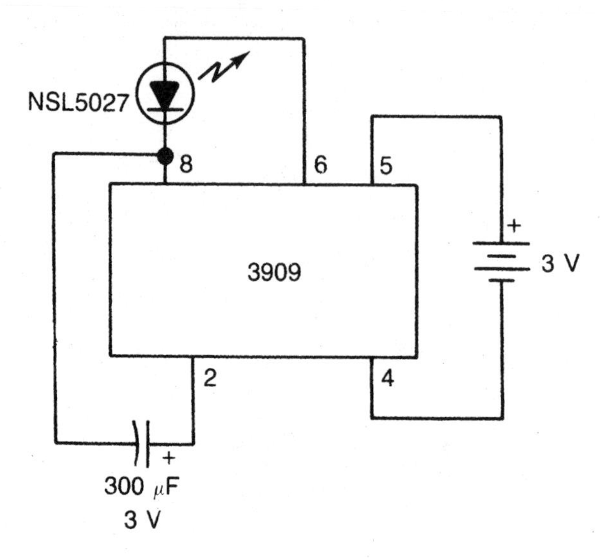

Fast blinker

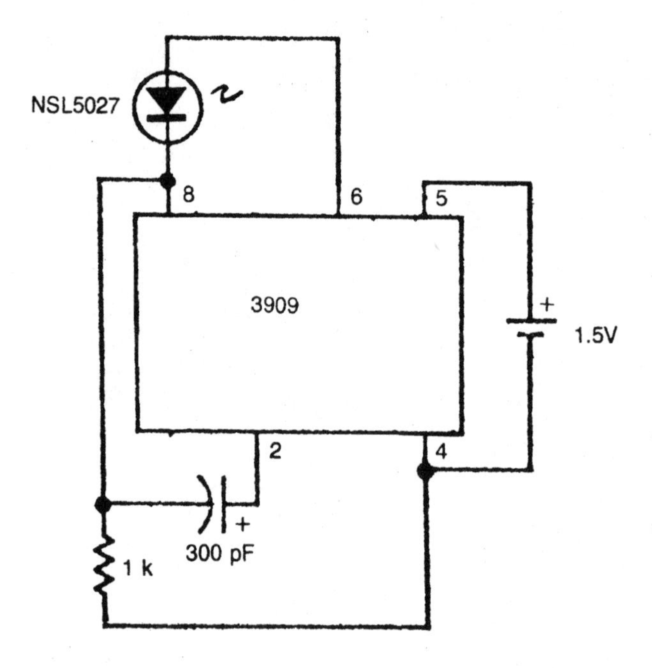

Photosensitive oscillator

The stronger the light striking the sensor surface of the photoresistor, the higher the pitch of the emitted tone.

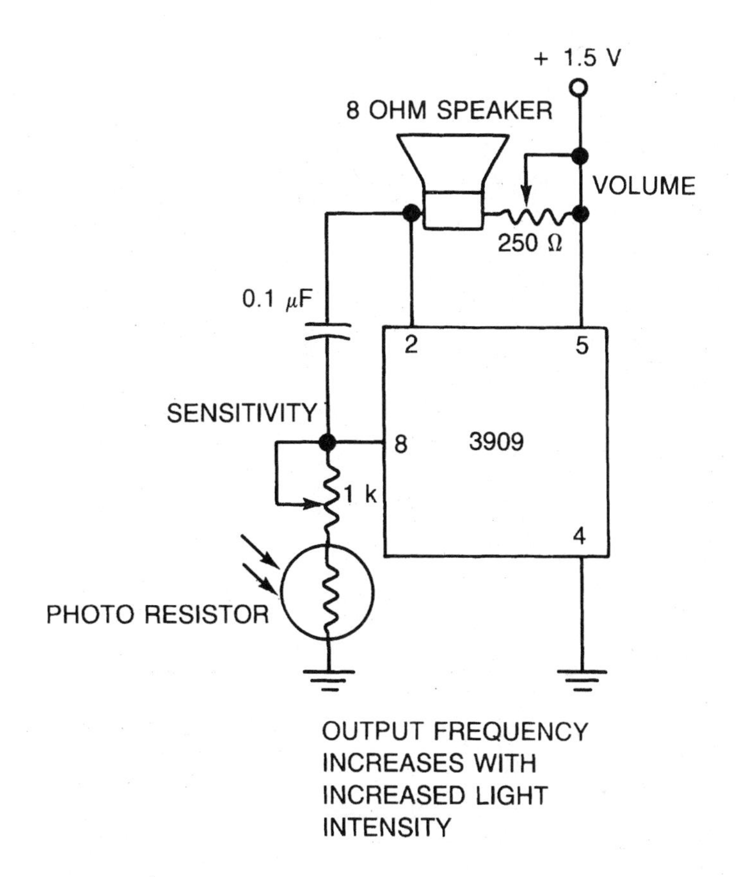

Bird chirp simulator

For some really strange effects, try experimenting with other values for the two capacitors.

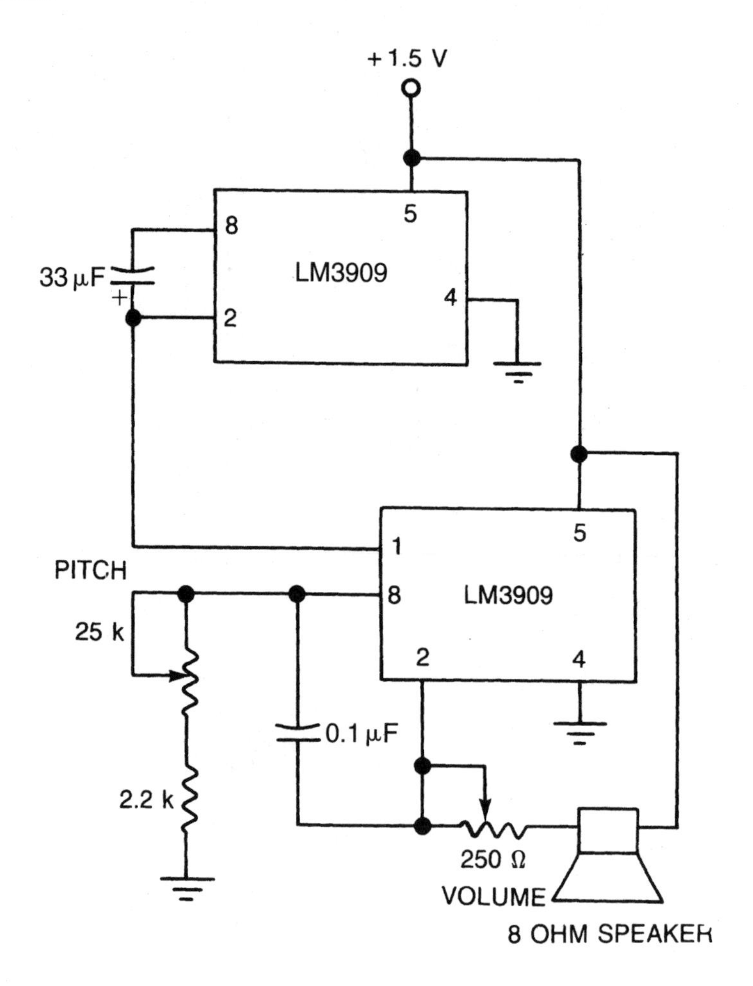

TTL controlled LED flasher

When a logic HIGH is presented to the input of this circuit, the LED starts flashing. A logic LOW input turns off the flasher, so the LED remains dark.

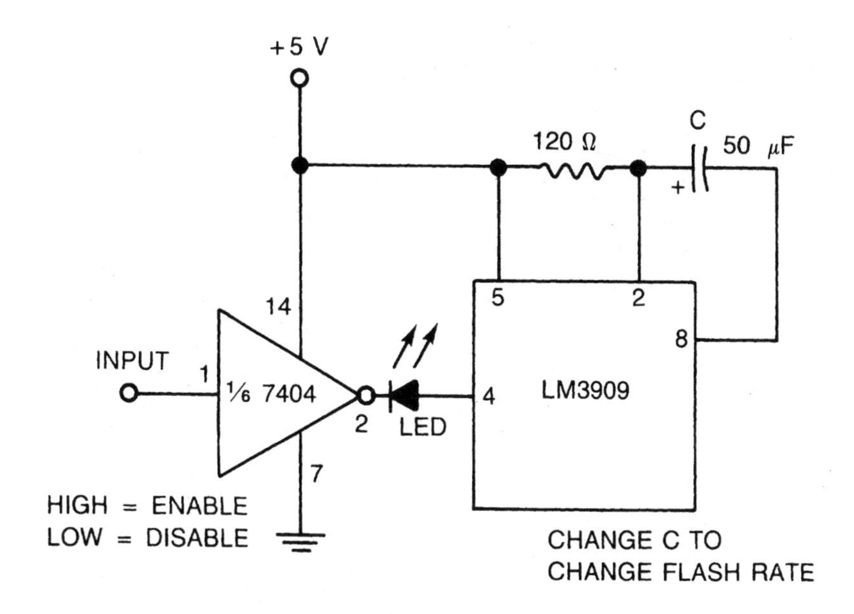

Two-tone siren

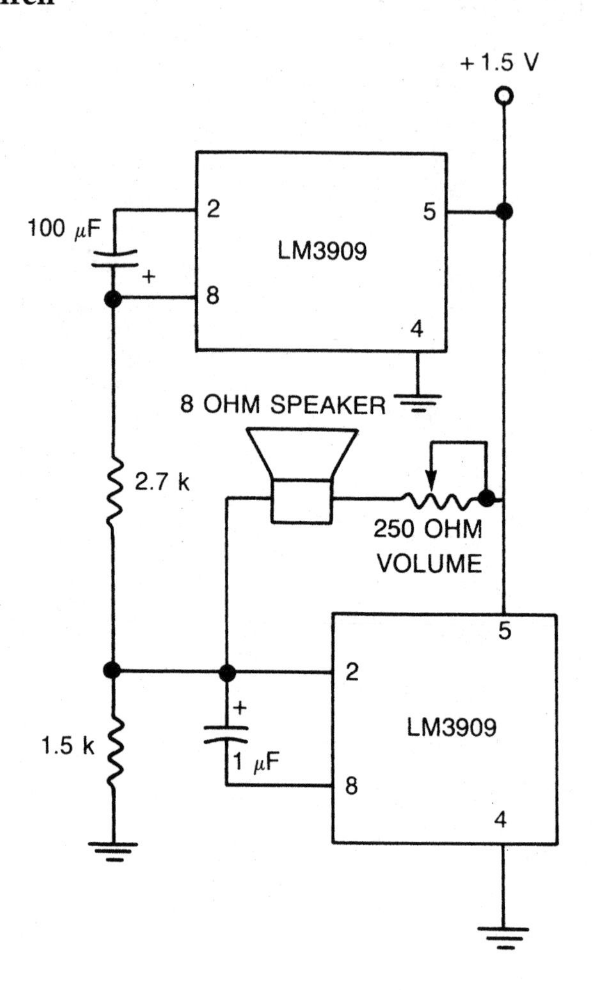

4-parallel LED flasher

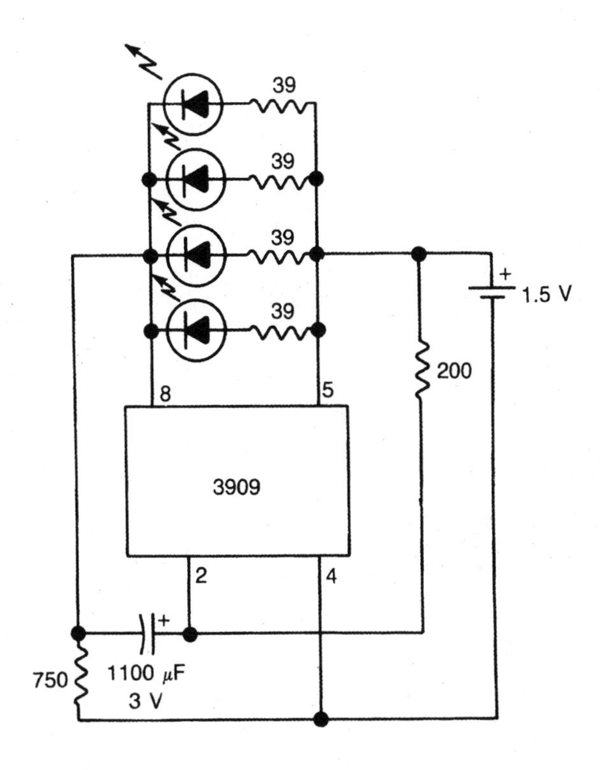

High-efficient 4-parallel LED flasher

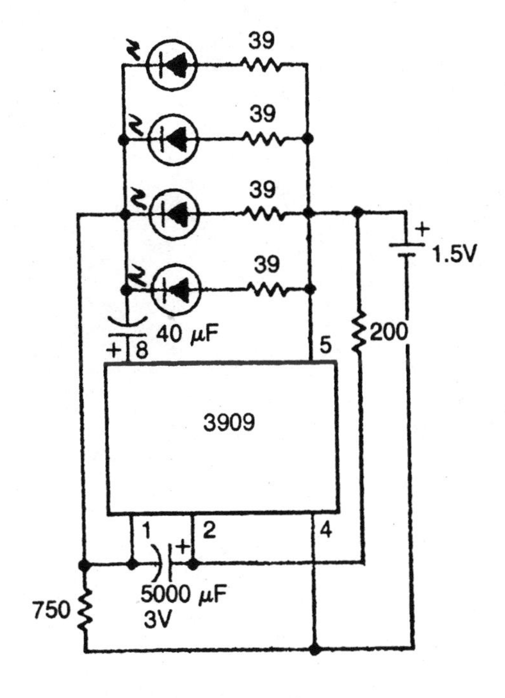

Variable rate flasher

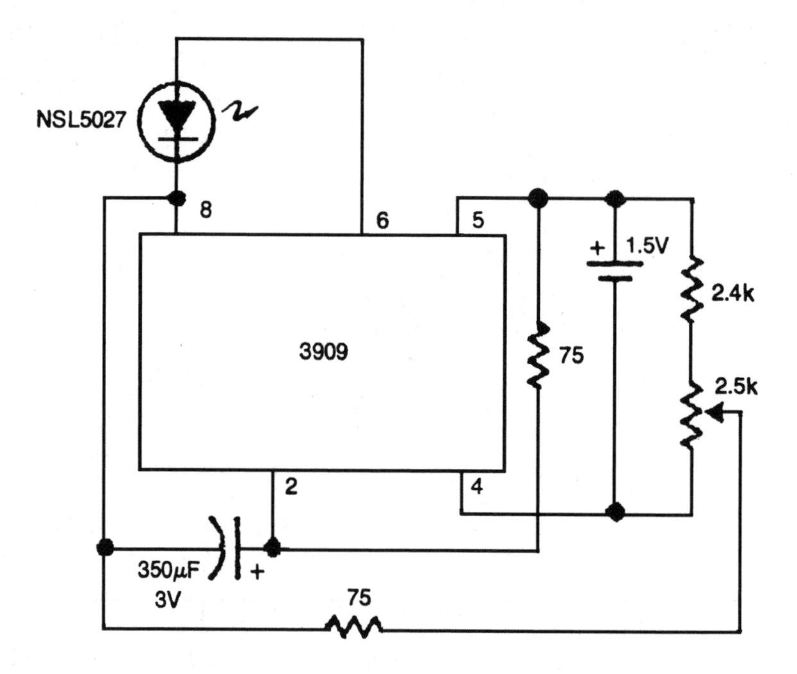

Incandescent bulb flasher

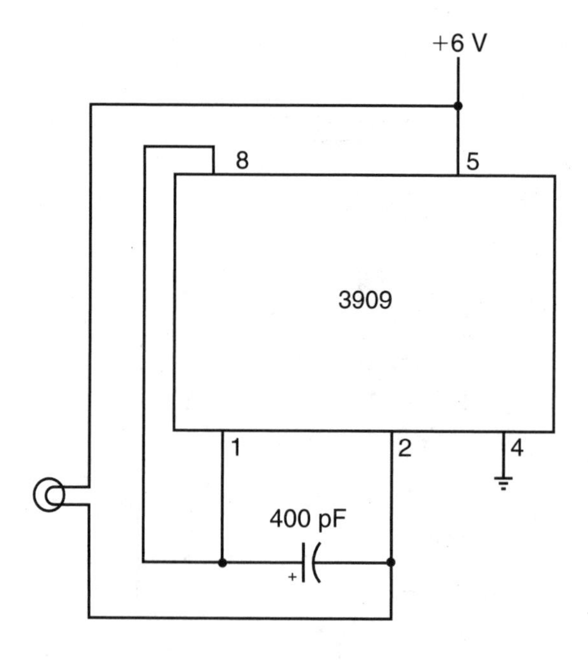

Light activated oscillator

A tone sounds when the light level striking the photoresistor is bright enough.

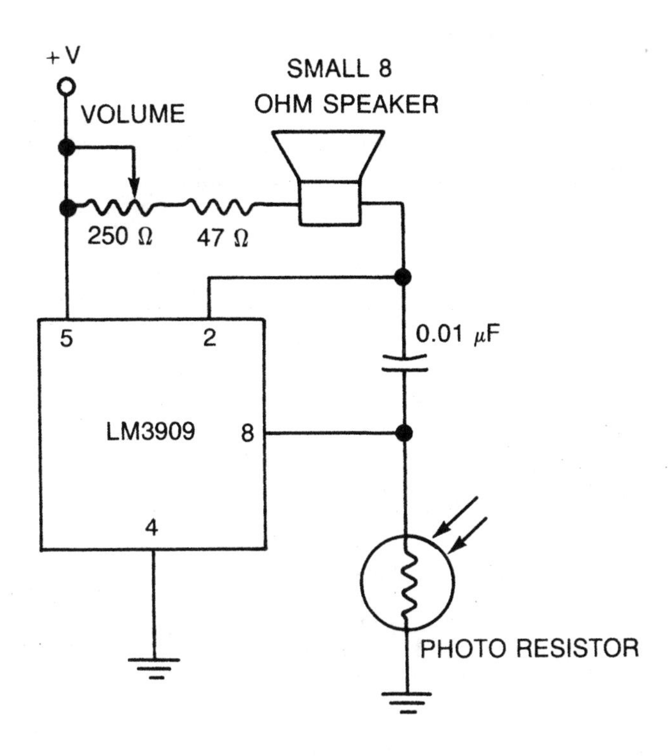

Liquid sensor/alarm

A tone sounds when the two probes are shorted together by being immersed in the same liquid. Use *battery power only* for this circuit. Don't risk the possibility of a short circuit. If by some fluke live ac was conducted into the liquid, the results could be tragic, or even fatal.

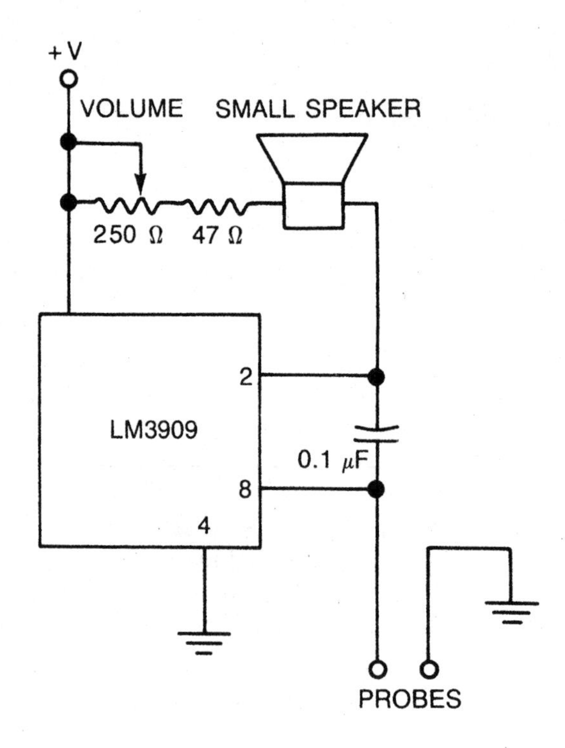

Relay driver oscillator

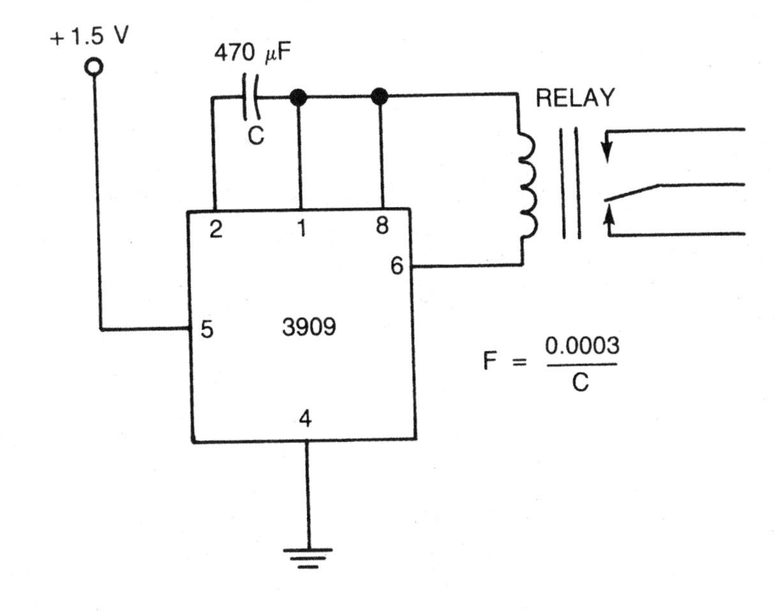

Variable frequency relay driver oscillator

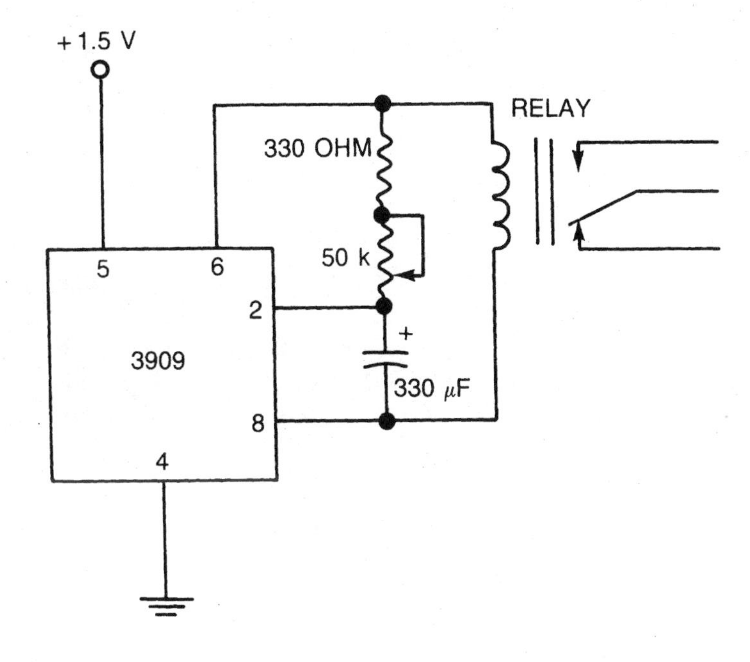

Emergency lantern/flasher

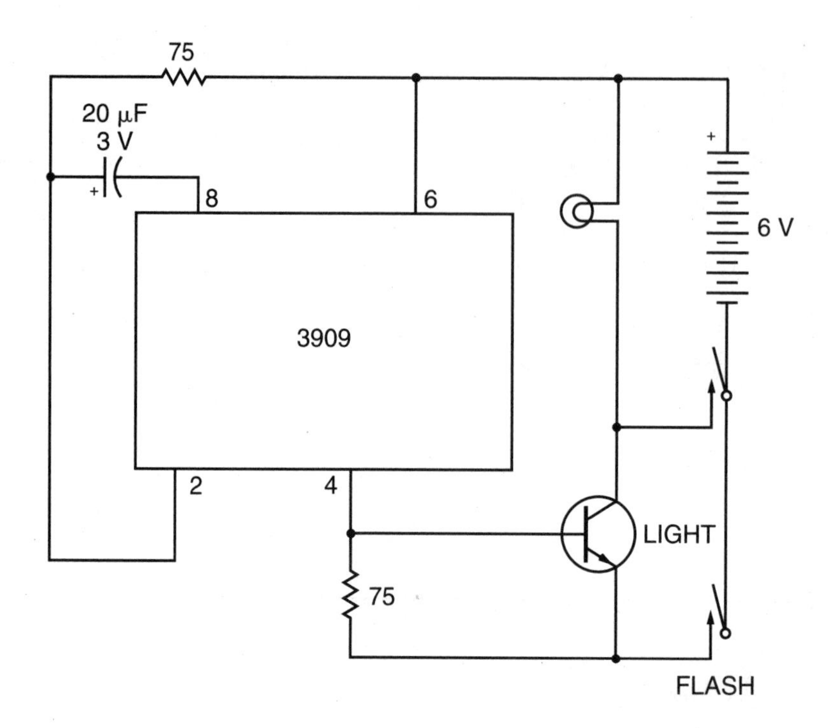

LED booster

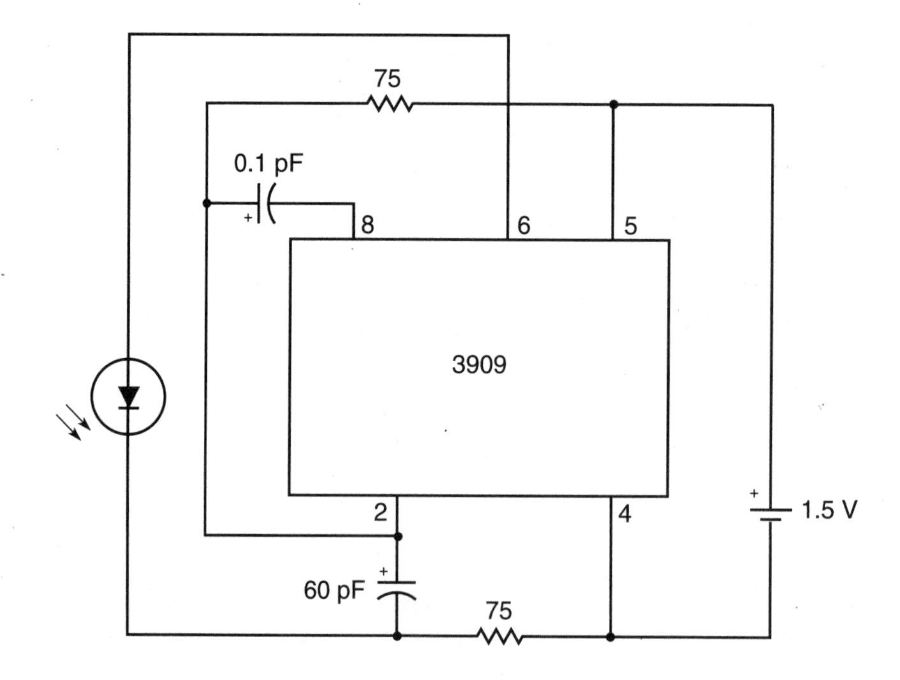

1 kHz square-wave oscillator

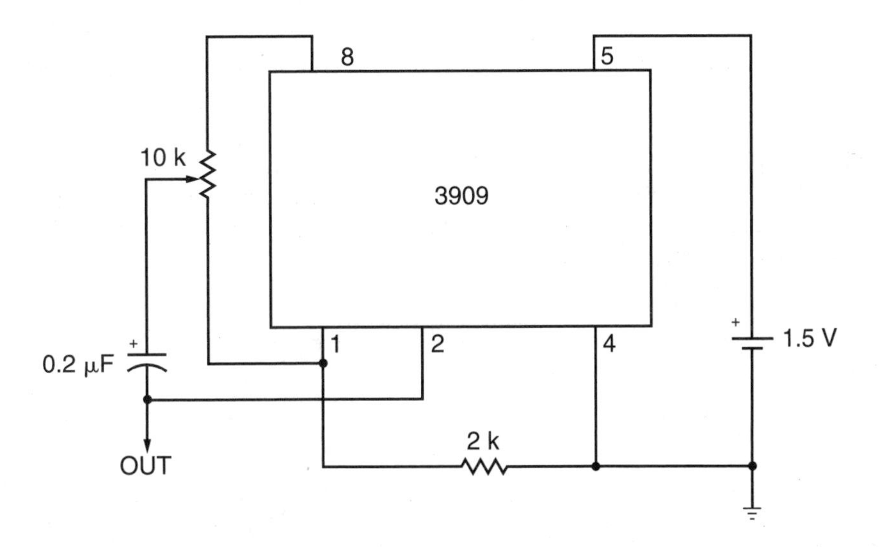

"Buzz Box" continuity and coil checker

This is a handy piece of simple test equipment for finding short circuits or opens.

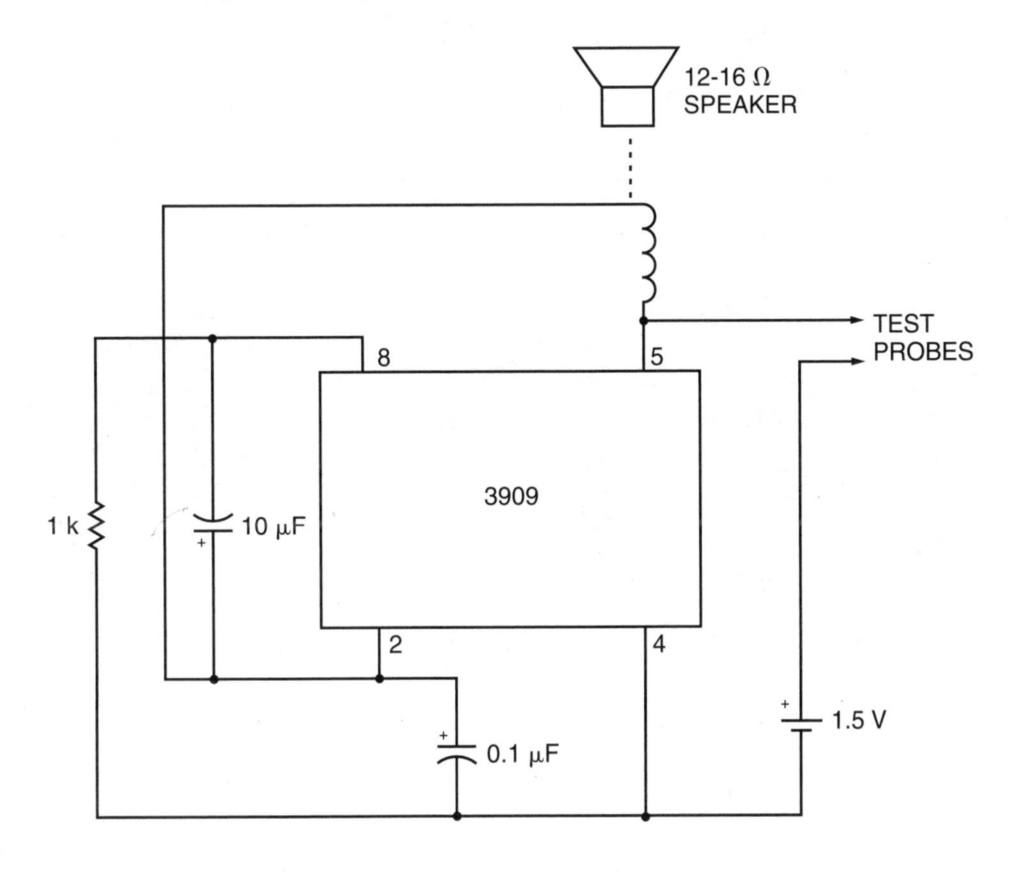

MM5369 17-stage oscillator/driver

The 5369 uses a quartz crystal to set its precise base frequency. Then a 17-stage binary divider drops the base frequency to a precise reference of a lower value. Both the original base frequency and the divider output can be tapped off from this chip.

CMOS circuitry is used in the 5369. This device can be used in either analog or digital circuits. Of course, the output signals are in the form of square waves.

The supply voltage can be anything from 3 V up to 15 V, and current consumption is low. Assuming the supply voltage is 10 V, the 5369 can handle signal frequencies up to 4 MHz.

This device is also called a 60 Hz timebase generator.

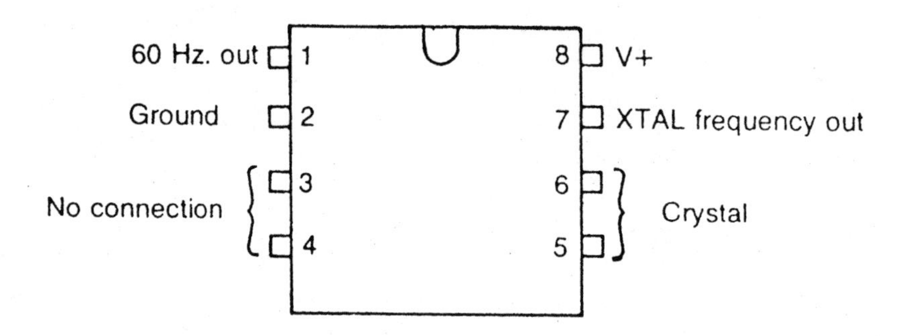

60 Hz timebase

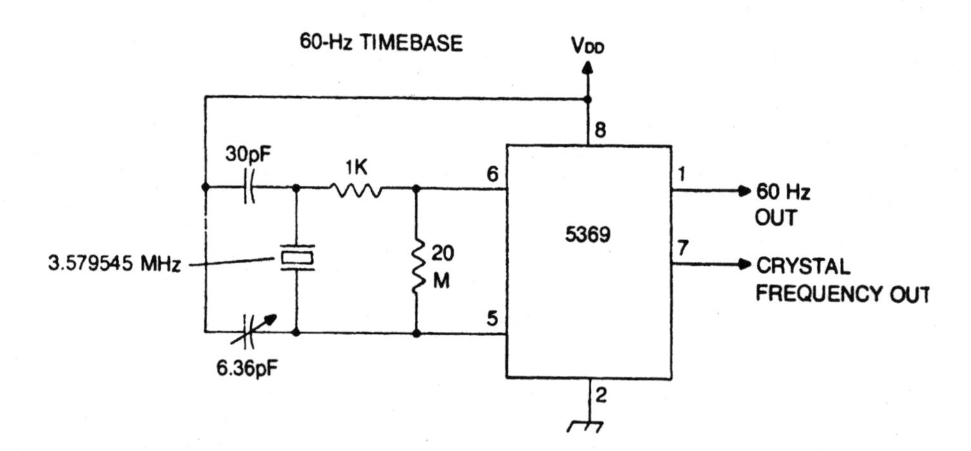

1 Hz timebase

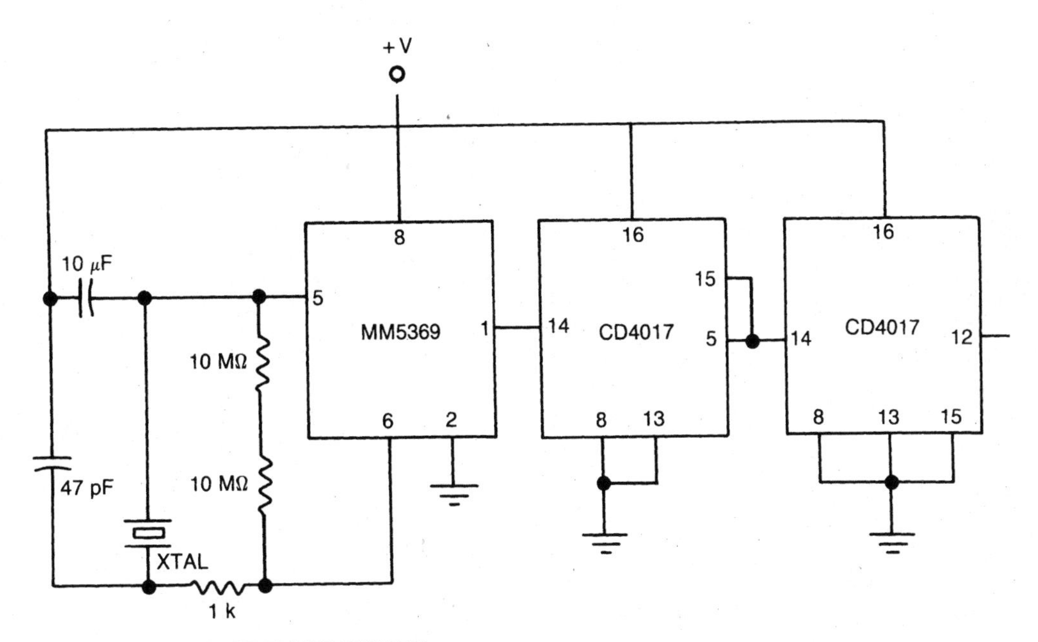

XTRAL = 3.58 COLOR BURST CRYSTAL

7

SECTION

RF amplifiers and related devices

Probably the most commonly used type of electronic circuit is the amplifier. Almost all electronic systems of any complexity include at least one amplifier stage. Audio amplifiers typically only need to handle frequencies up to about 20 kHz, and generally involve few inherent problems to the circuit designer. In radio frequency (RF) applications, however, the high-frequency signals involved introduce a number of special considerations and problems. RF signals can have frequencies of several hundred kilohertz, or even megahertz. Such high-frequency signals tend to behave somewhat differently than lower frequency signals. For example, two adjacent conductors can act as a small capacitance which is ignored by AF signals, but an RF signal might see it as a partial or complete short circuit.

In this section we will look at ICs designed for the amplification and other modification of signals within the radio frequency range. A number of related devices for RF applications are also covered here. This chapter is not restricted just to amplifiers. All of these ICs are intended for use in radio and television receivers and communications systems, as well as related applications.

Because RF circuits operate at much higher signal frequencies than audio or dc circuits, greater care must be taken when working on circuit layouts, lead lengths, and heat sinks.

1310 phase-locked loop FM stereo demodulator with indicating lamp

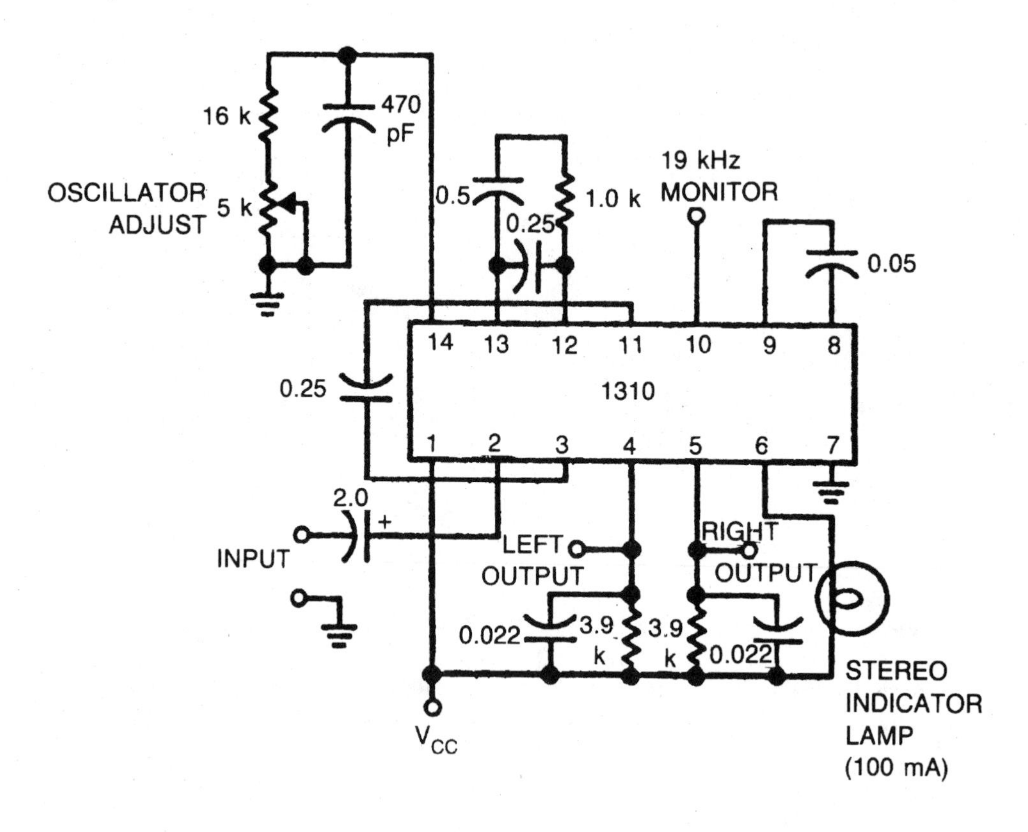

1350 integrated video amplifier

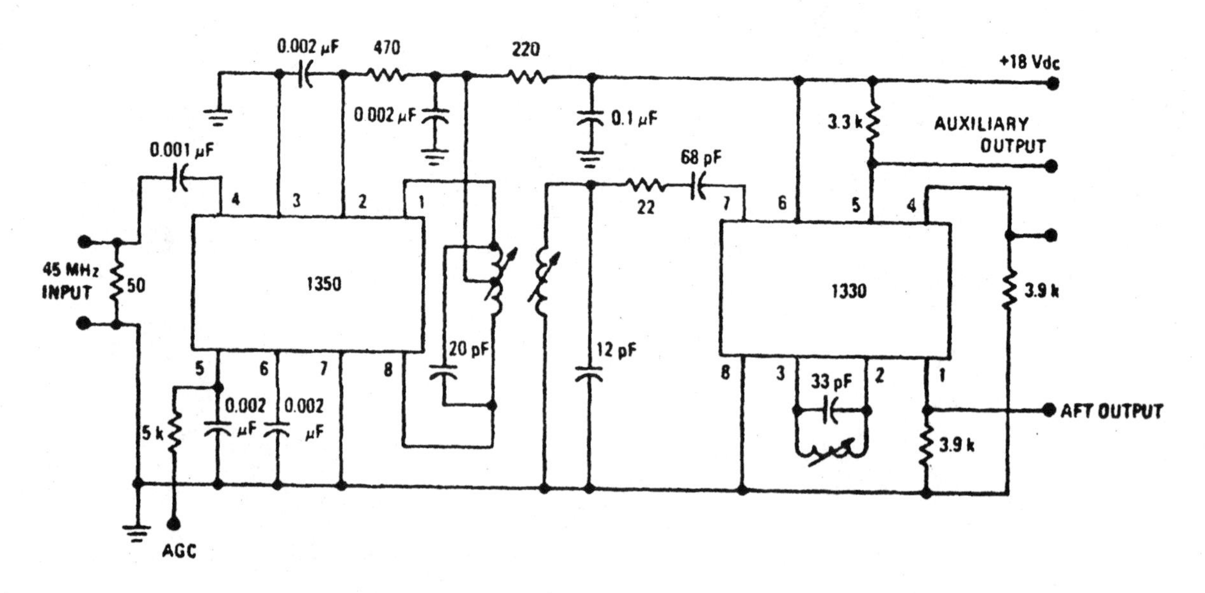

1496 balanced modulator/demodulator (+12 V single-ended supply)

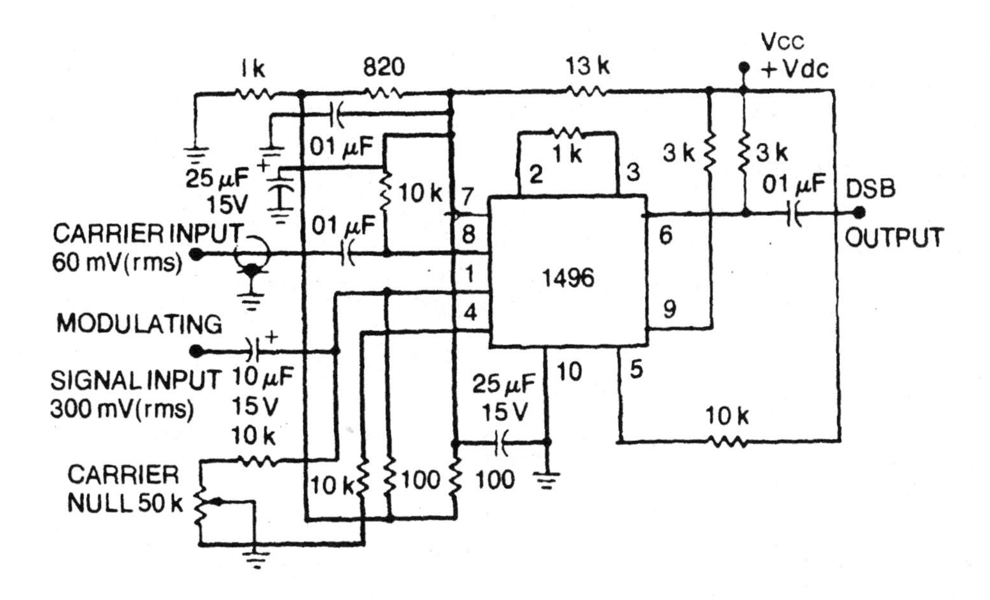

Balanced modulator/demodulator

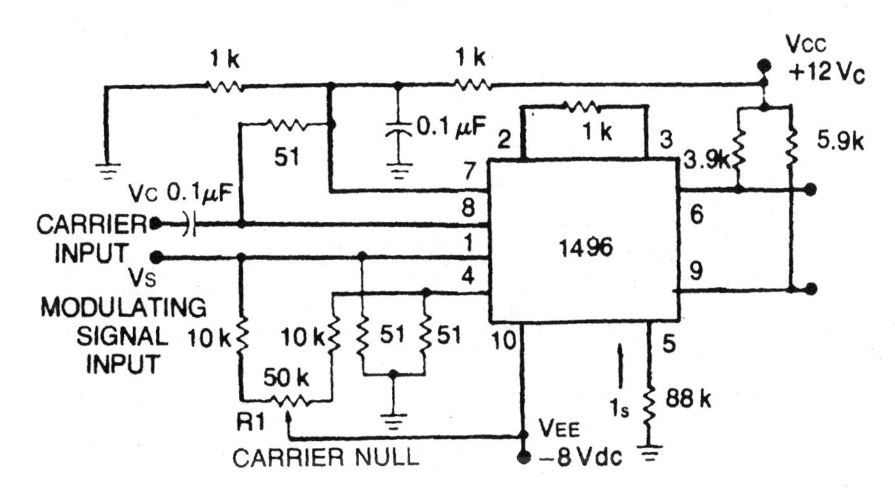

Low-frequency doubler

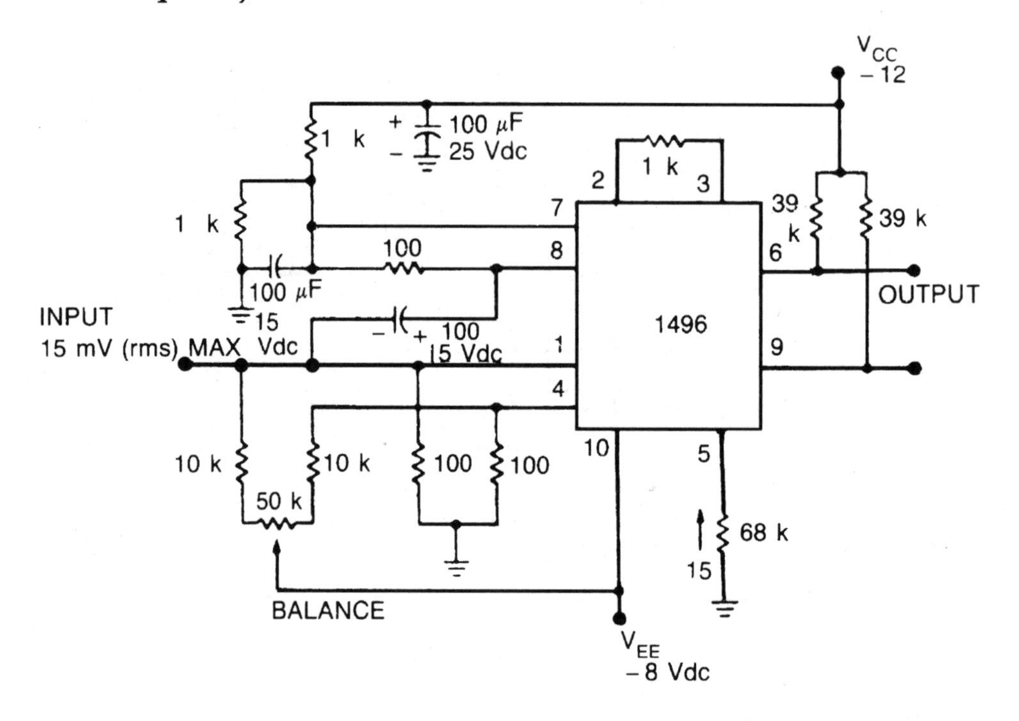

Broadband frequency doubler

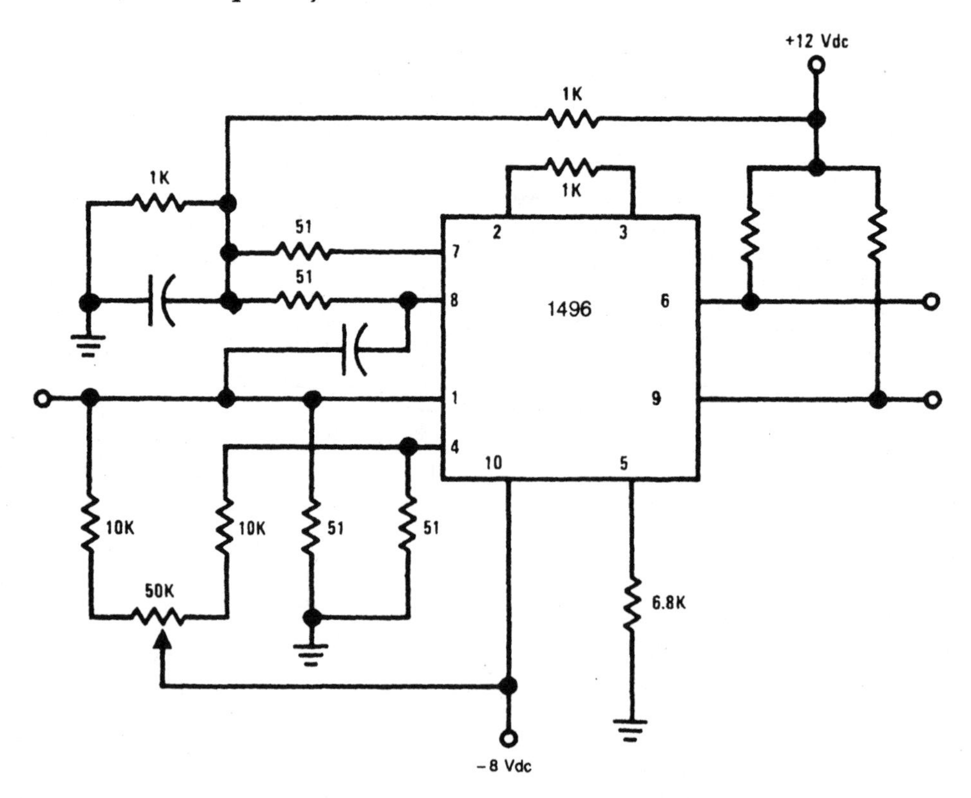

150 to 300 MHz doubler

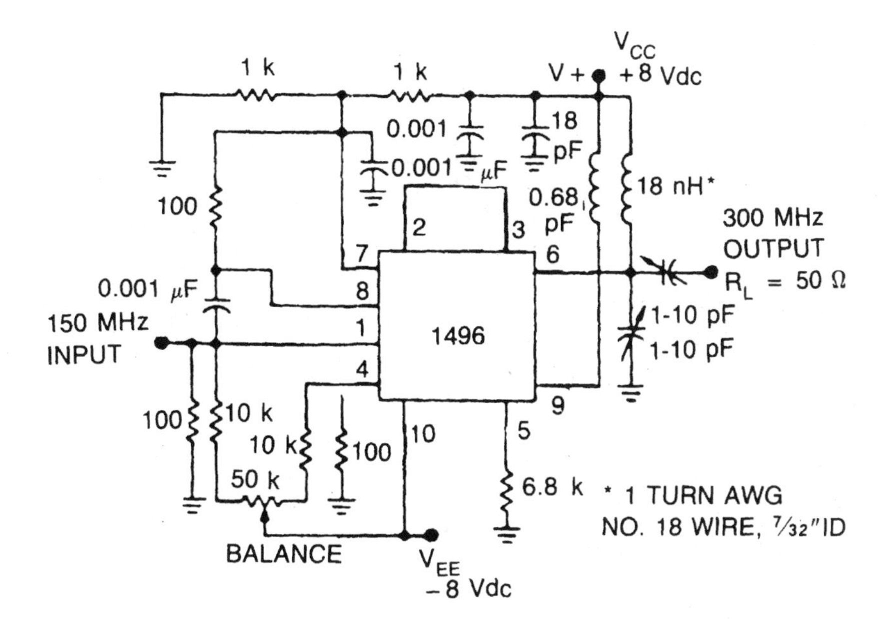

AM modulator circuit

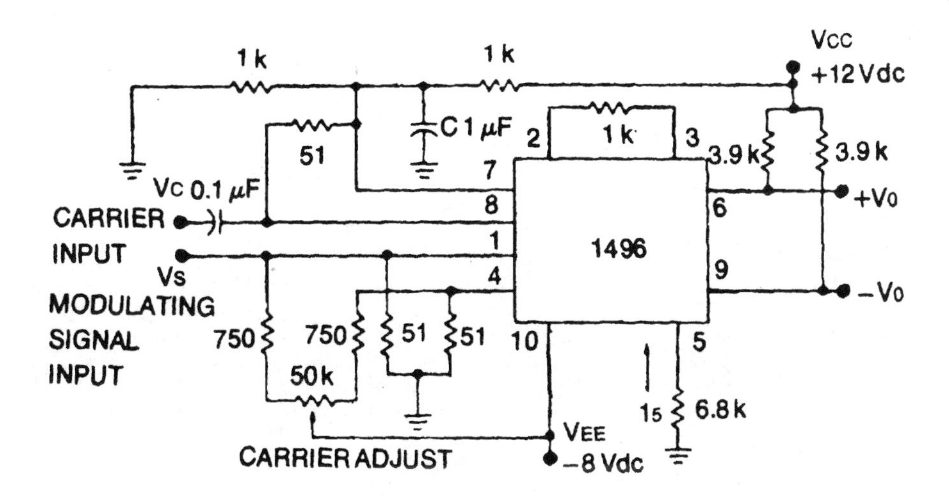

SSB product detector

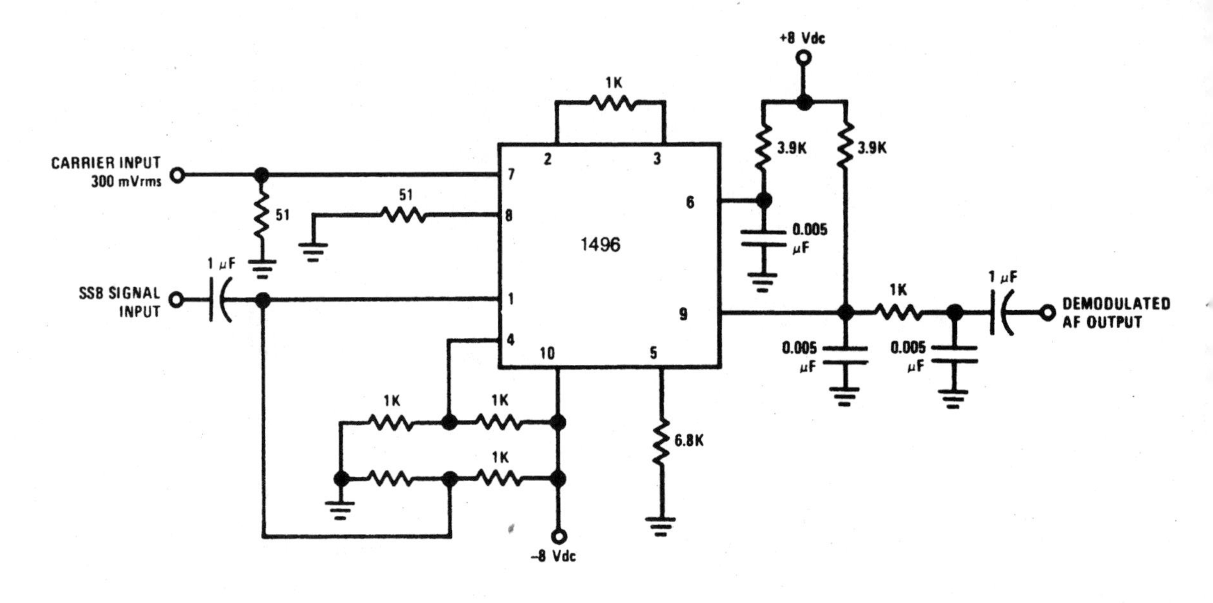

Doubly balanced mixer

This circuit has broadband inputs and a 9.0 MHz tuned output.

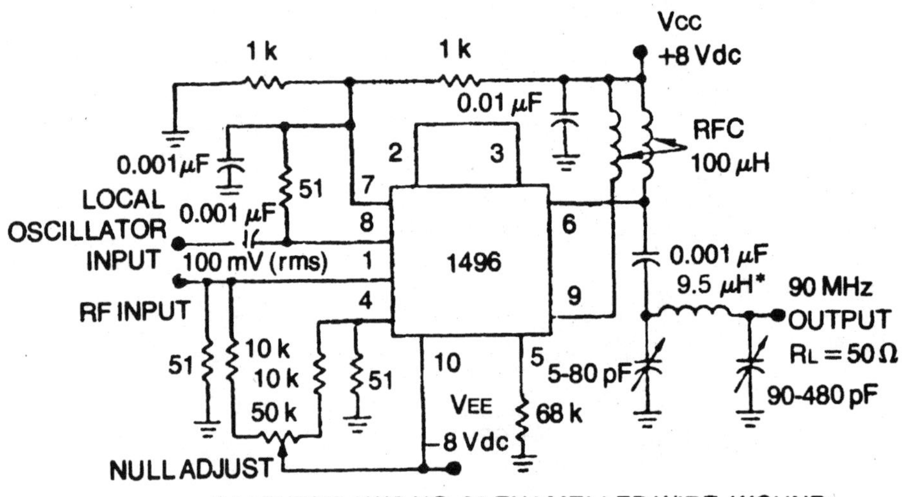

MC2833 low-power FM transmitter system

The 2833's internal circuitry includes a built-in microphone amplifier, VCO, and two auxiliary transistors to form the basis of a one-chip FM transmitter sub-system, particularly well-suited for such applications as cordless telephones, wireless microphones, and FM communication equipment. Only a very few external components are required for a functional FM transmitter circuit.

This device can be operated from a fairly wide range of supply voltages, extending from 2.8 V to 9.0 V. The drain current is quite low, with a typical value of about 2.9 mA.

Using the on-chip transistor amplifiers, the power output can be boosted to +10 dBm. Using direct RF output, the 2833 can achieve −30 dBm power output to 60 MHz.

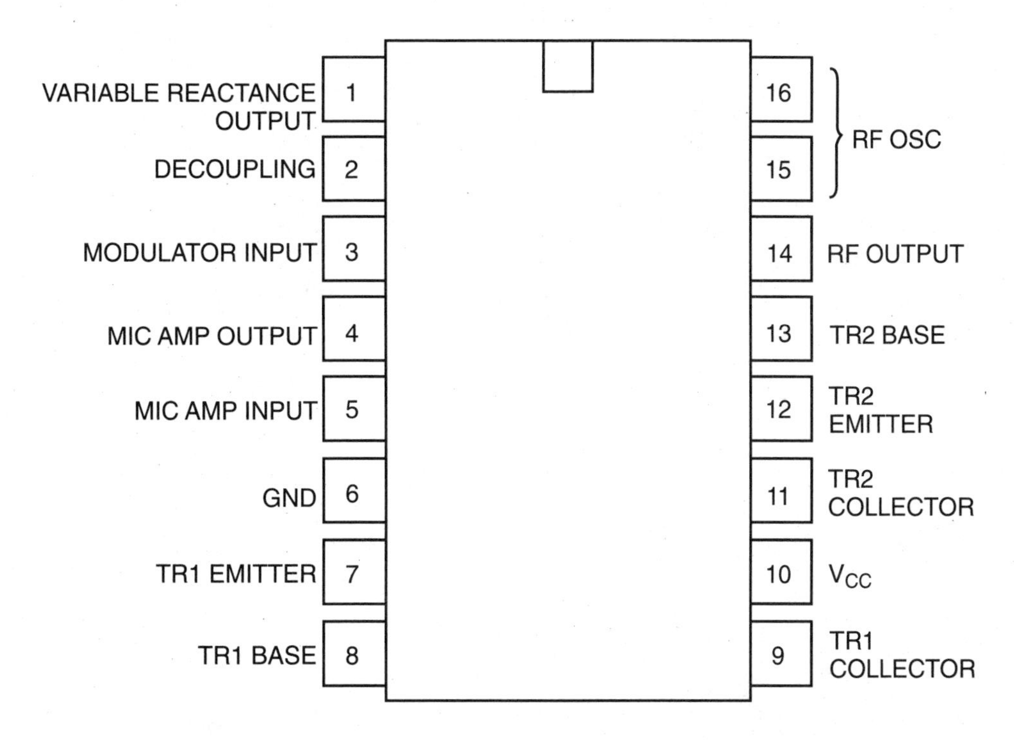

Test circuit

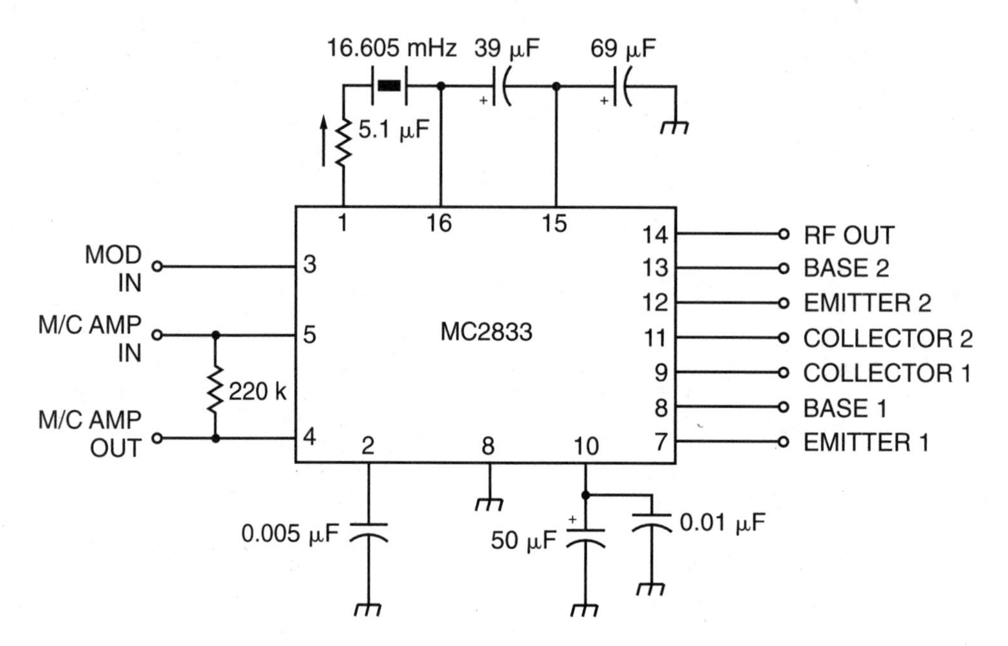

(*Courtesy of Motorola.*)

Single chip VHF narrowband FM transmitter

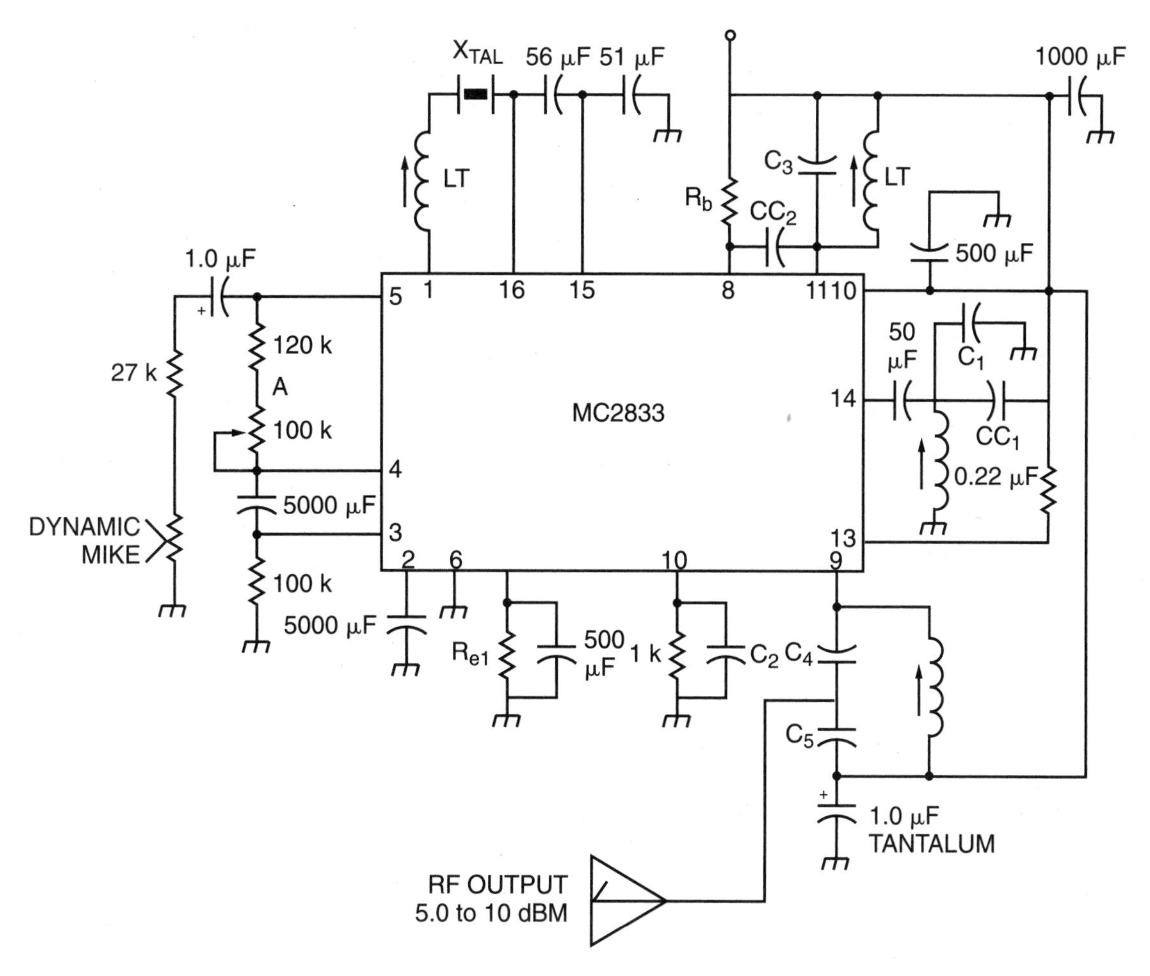

(*Courtesy of Motorola.*)

8

SECTION

Voltage regulators and power supply devices

Except for a very few simple, passive circuits, all electronics circuits require a power supply of some sort. Most digital circuitry and high-precision linear circuitry usually requires very exact and consistent supply voltages (and sometimes currents). To meet this sort of need, which is constantly increasing with new developments in the electronics field, numerous voltage regulator ICs have been created, along with a number of other devices for various power supply functions.

In itself, a power supply is rather mundane and not too interesting. Yet, it is essential for almost any other type of electronics circuit to function. Despite the apparent simplicity of the function, there are an incredible number of voltage regulator ICs on the market, with new devices appearing every month. This section looks at just a few representative examples.

117 integrated adjustable output voltage regulator (1.2- to 25-V)

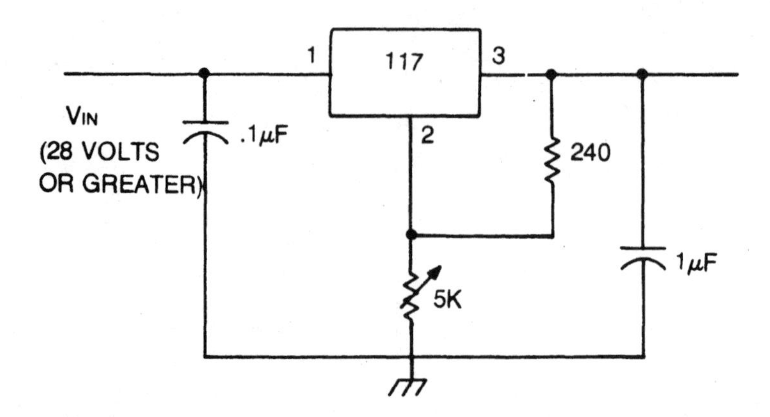

1-amp current regulator

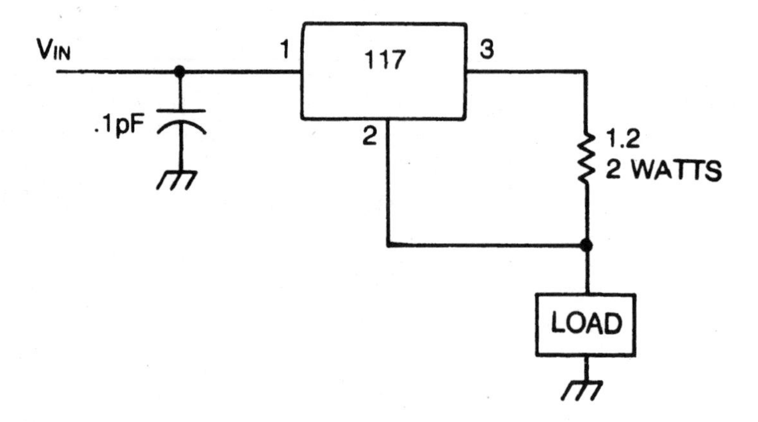

1.2 to 20 V adjustable regulator with 4 mA minimum load current

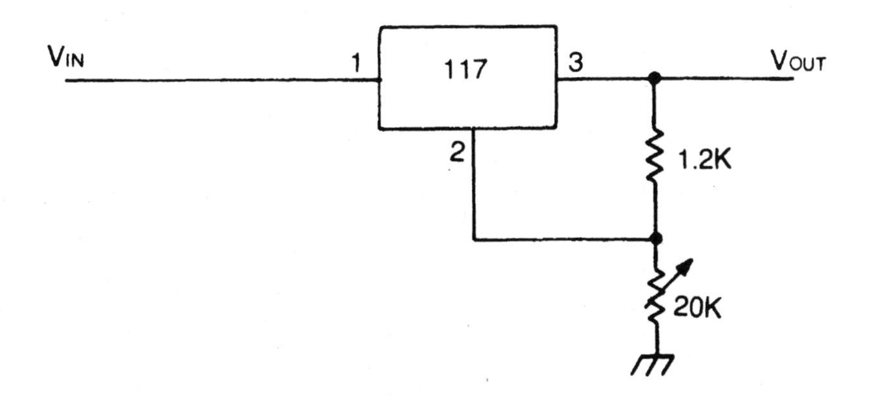

Slow turn-on 15-V regulator

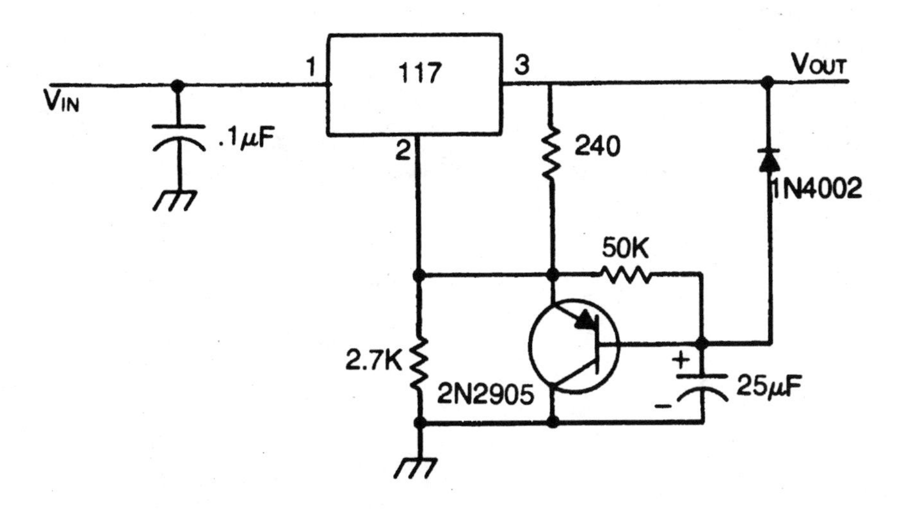

5-V logic regulator with electronic shutdown

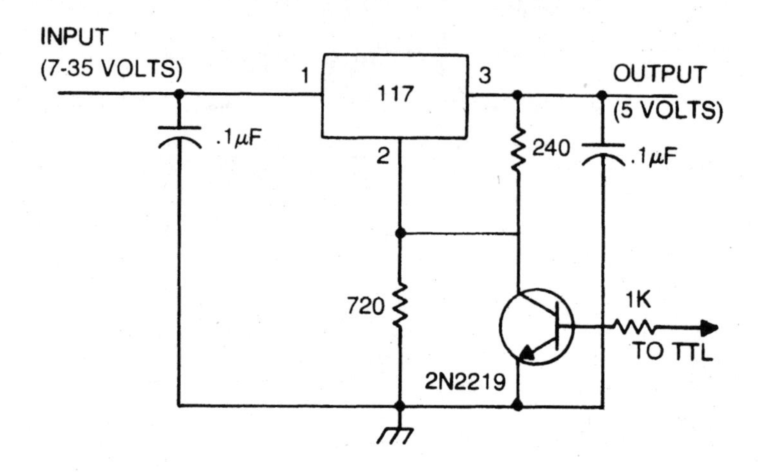

Precision current limiter

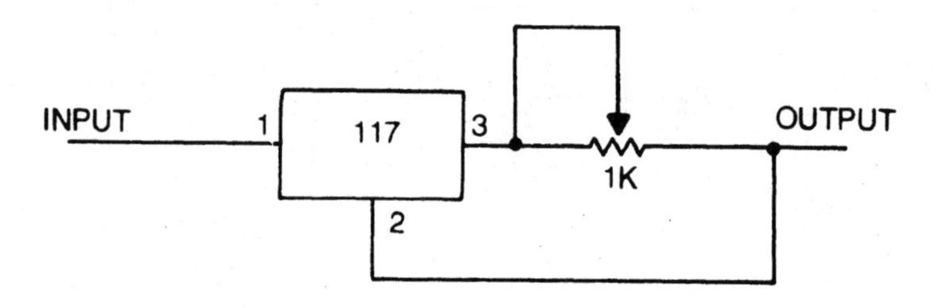

0- to 30-V regulator

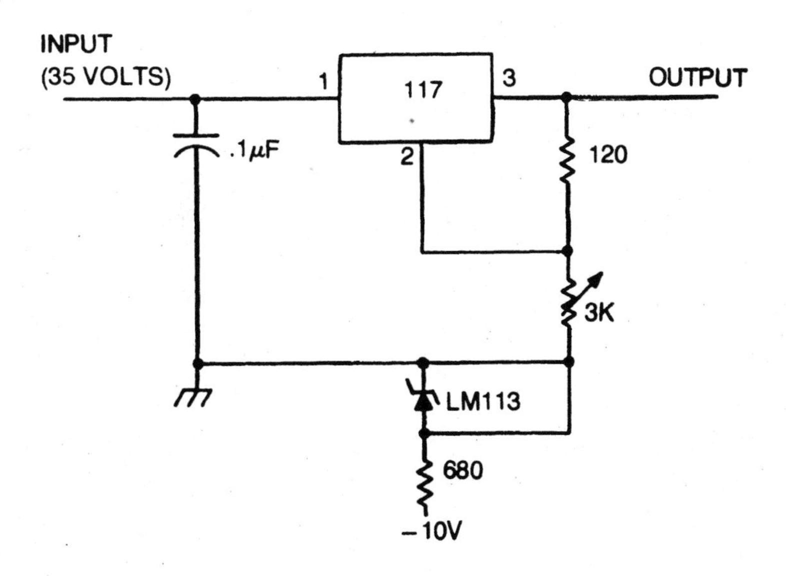

Highly stable 10-V regulator

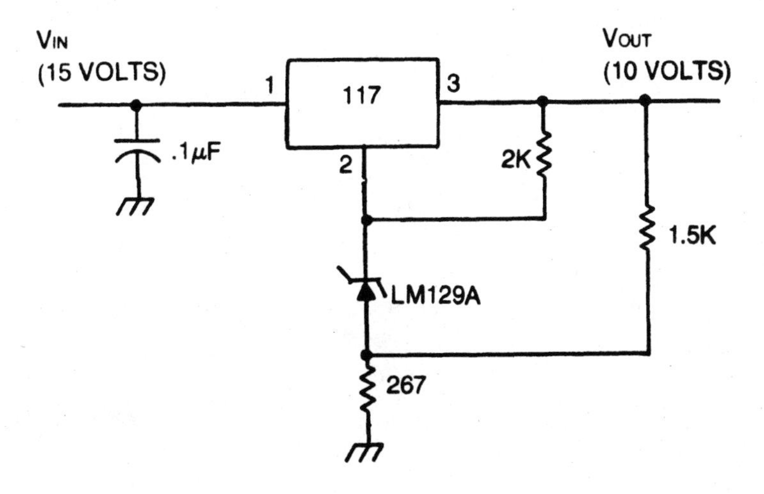

12-V battery charger

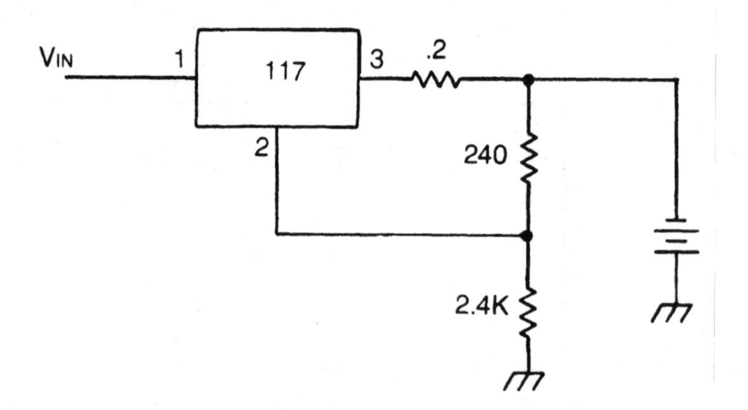

Adjustable regulator with improved ripple suppression

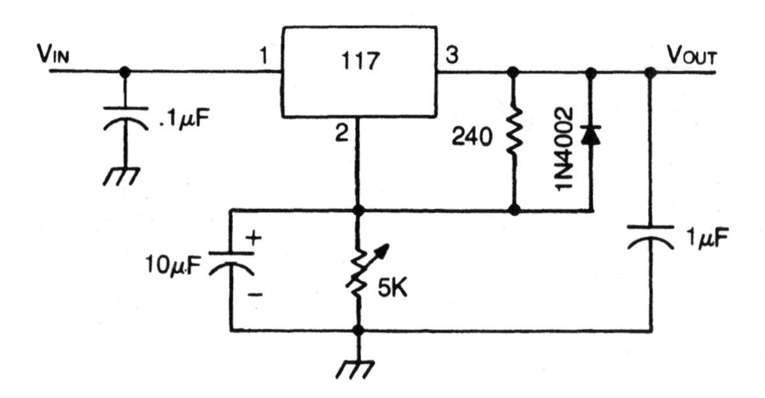

AC voltage regulator

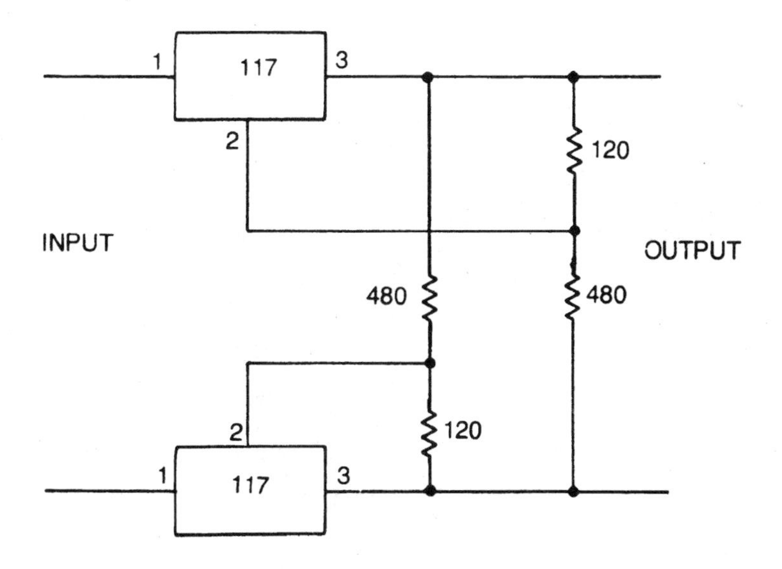

LM123 3-amp, 5-V positive voltage regulator

The 123 is a 3-terminal positive voltage regulator with a preset 5-V output and a load driving capability of a full 3 amperes. Special circuit design and processing techniques are used to provide the high output current without sacrificing the regulation characteristics of lower current devices.

This 3-amp regulator is virtually blow-out proof. Current limiting, power limiting, and thermal shutdown provide a high level of reliability.

No external components are required for basic operation of the 123. If the device is more than 4 inches from the filter capacitor, however, a 1 μF solid tantalum capacitor should be used on the input. It is also advisable to use a 0.1 μF or larger capacitor across the output to swamp out any stray load capacitance, or to reduce any load transient spikes created by fast-switching digital logic.

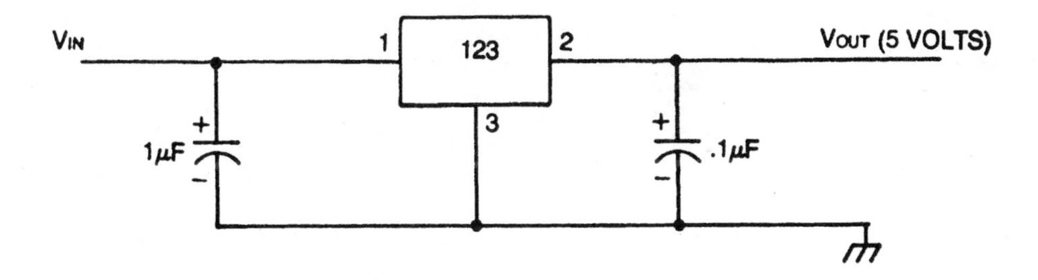

0- to 10-V adjustable regulator

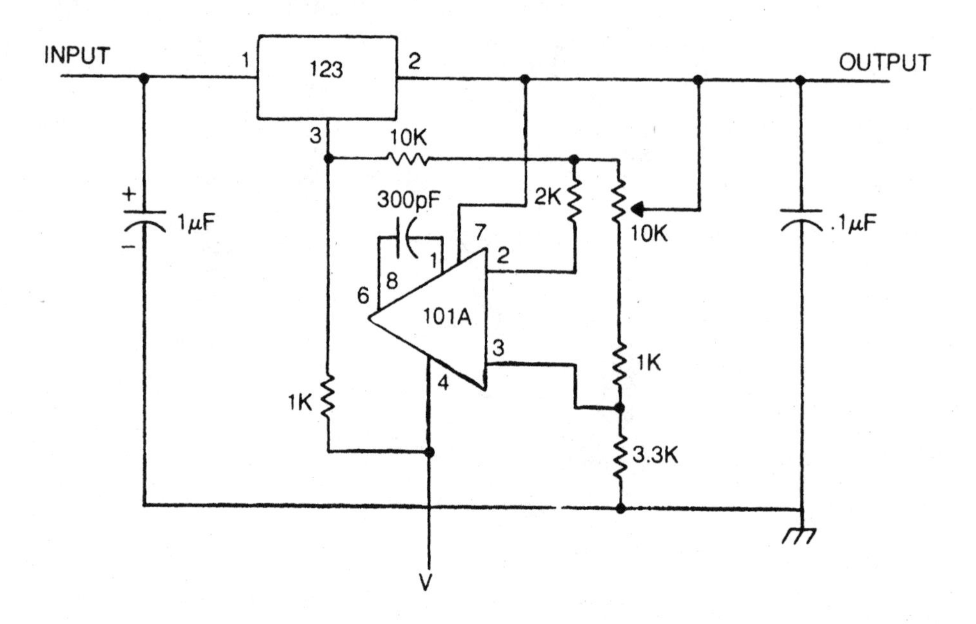

LM317 1.2 to 37 V regulator

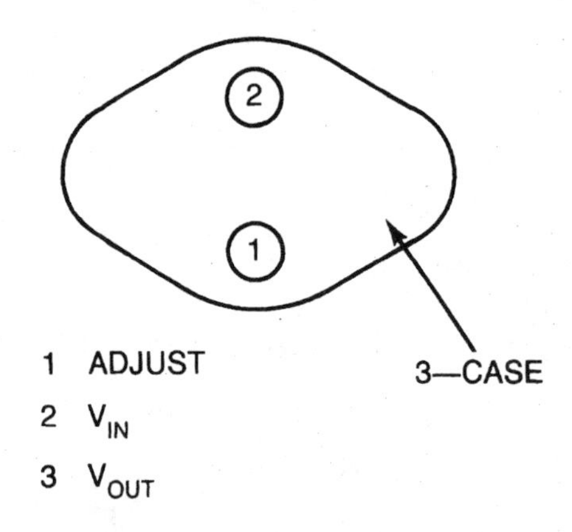

Nicad battery charger

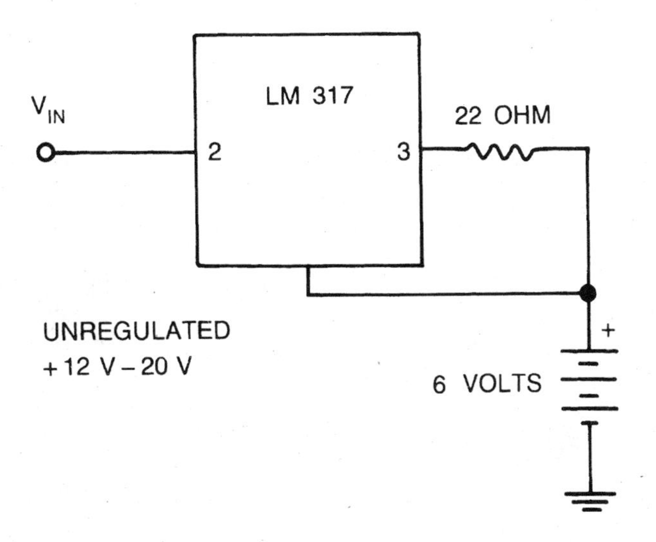

Variable voltage regulator

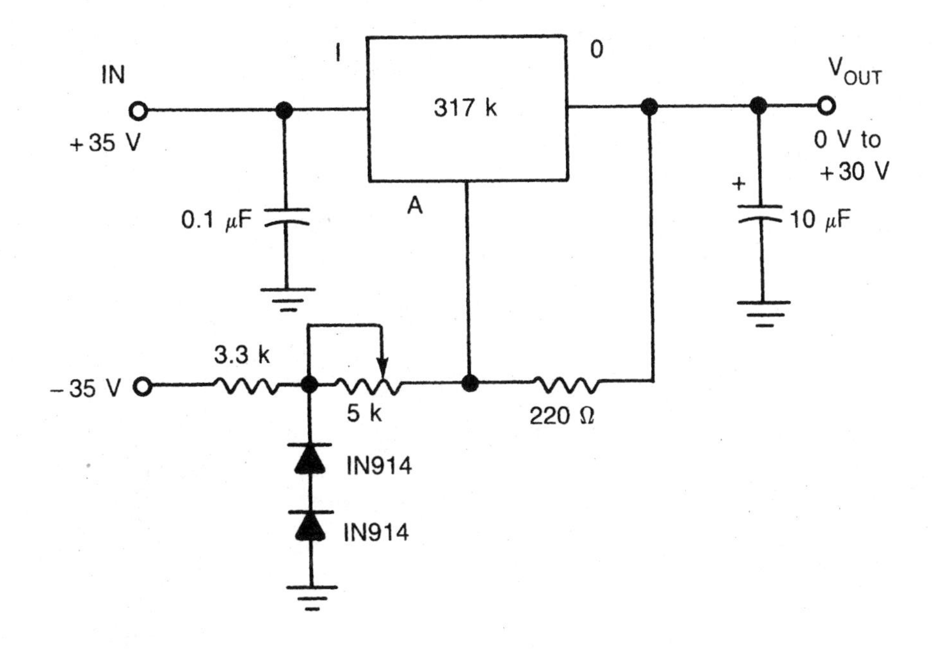

Alternate variable voltage regulator

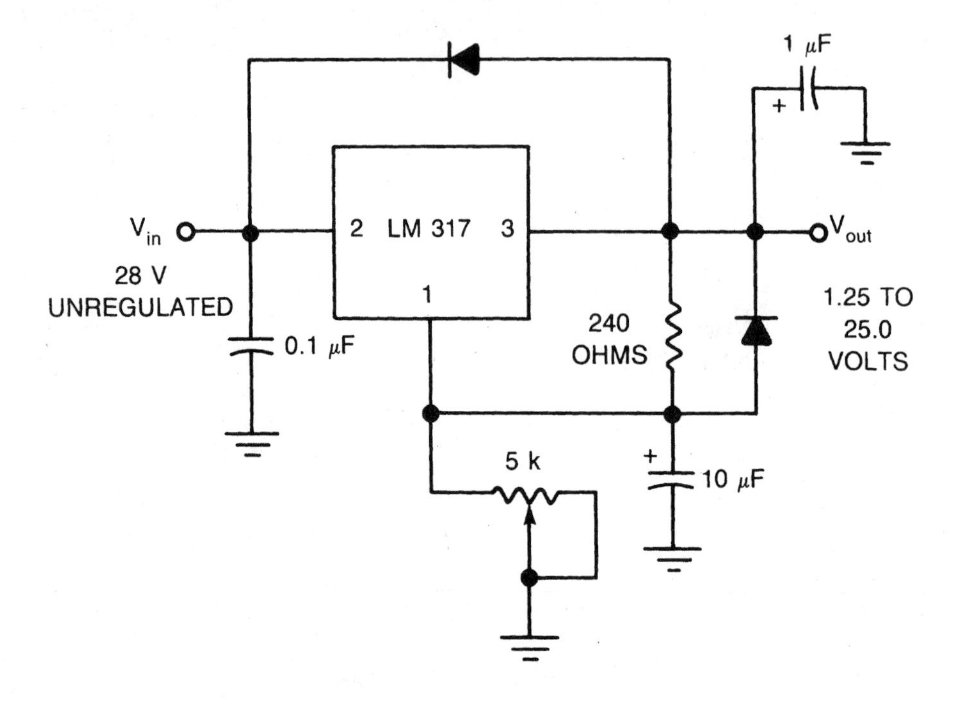

Precision current limiter

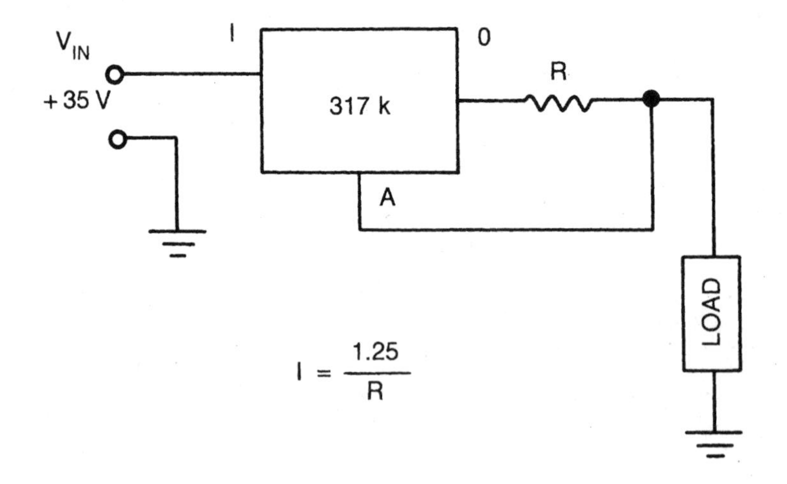

$$I = \frac{1.25}{R}$$

LM337 3-terminal adjustable negative voltage regulator

The 337 is a 3-terminal voltage regulator that can put out an externally adjustable negative voltage anywhere between −1.2 V to −37 V. Only two external resistors are required to set the desired output voltage of this device. It can supply a hefty amount of current as well. It is rated for output currents exceeding 1.5 A. The 337 is also designed for maximum reliability, with internal current limiting, thermal shutdown, and safe area compensation.

By connecting a resistor between the adjustment and output pins, this IC can also be used in precision current regulator applications.

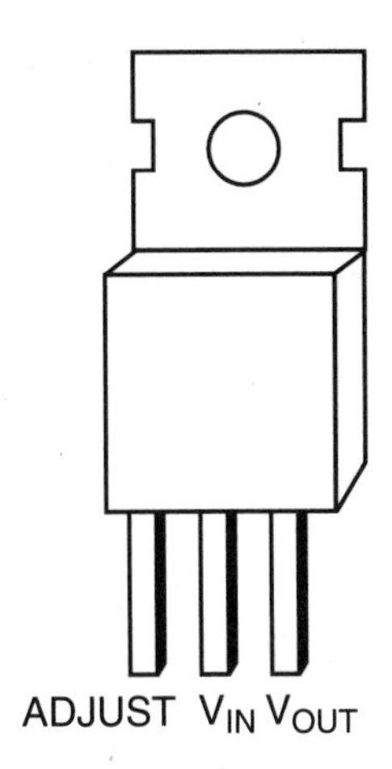

Standard negative voltage regulator

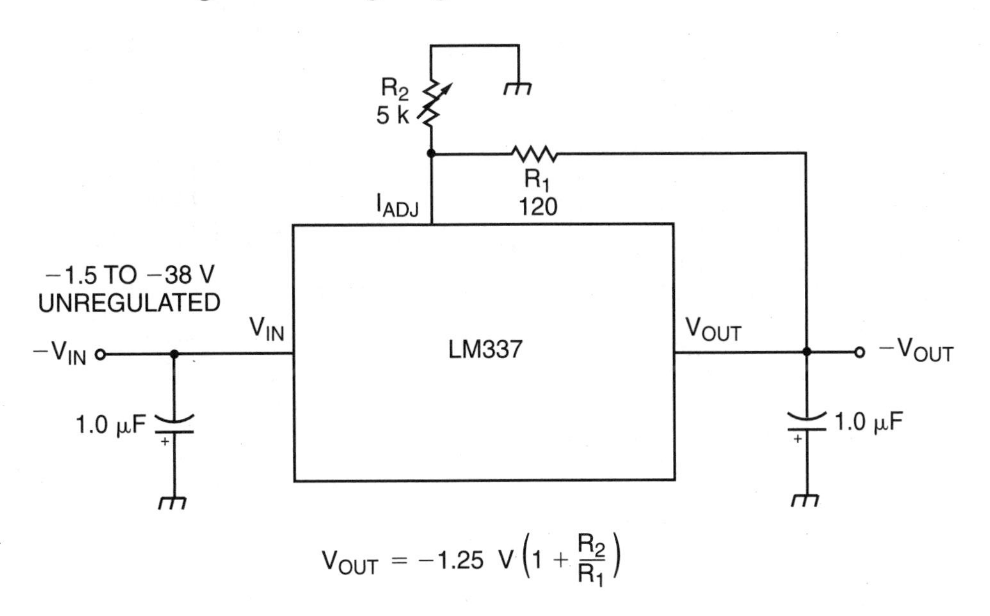

$$V_{OUT} = -1.25\ V\left(1 + \frac{R_2}{R_1}\right)$$

Constant current source

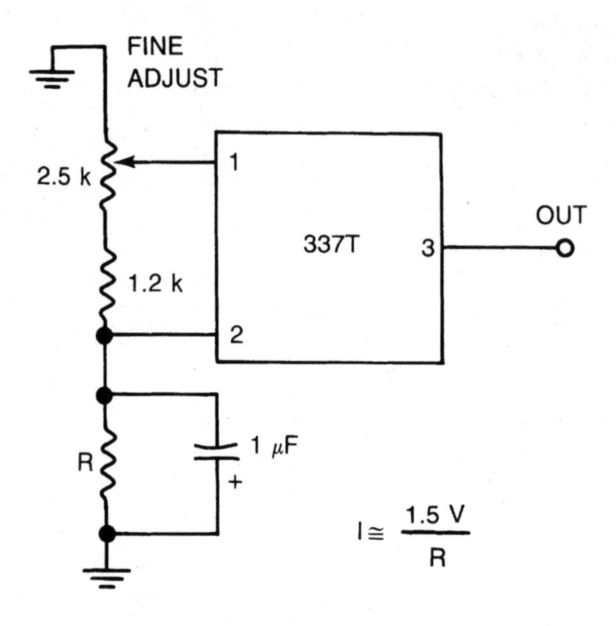

340 series voltage regulator

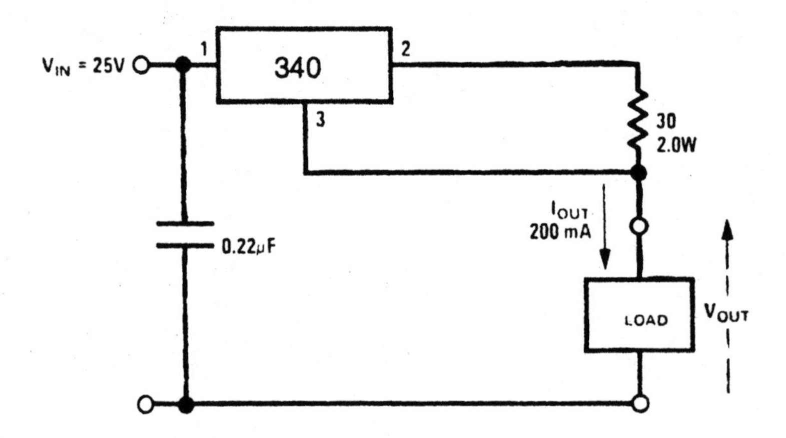

10-V regulator

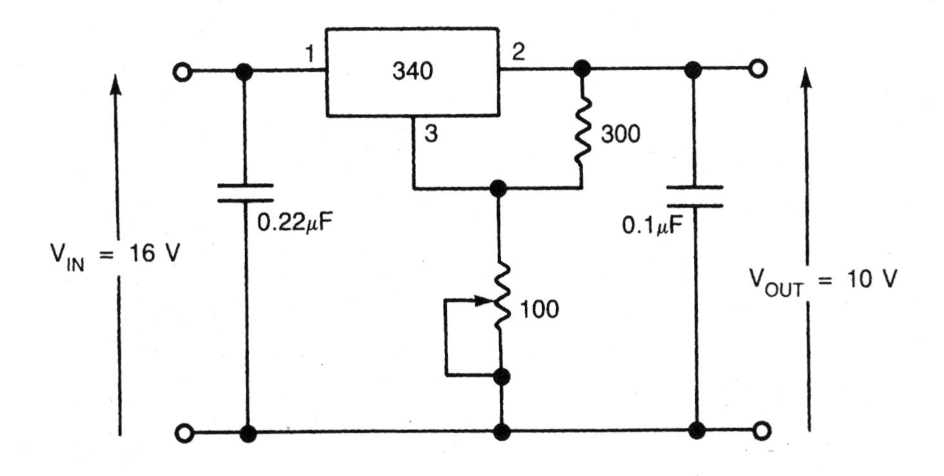

Variable high-voltage regulator with short-circuit and over-voltage protection

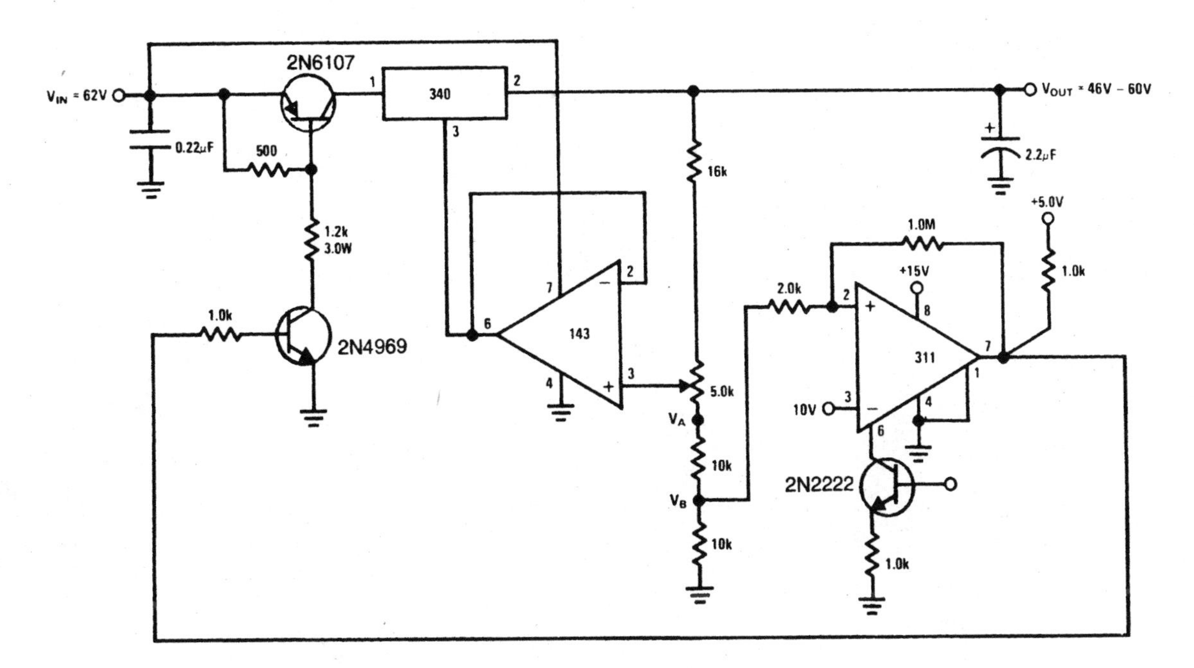

Variable output regulator

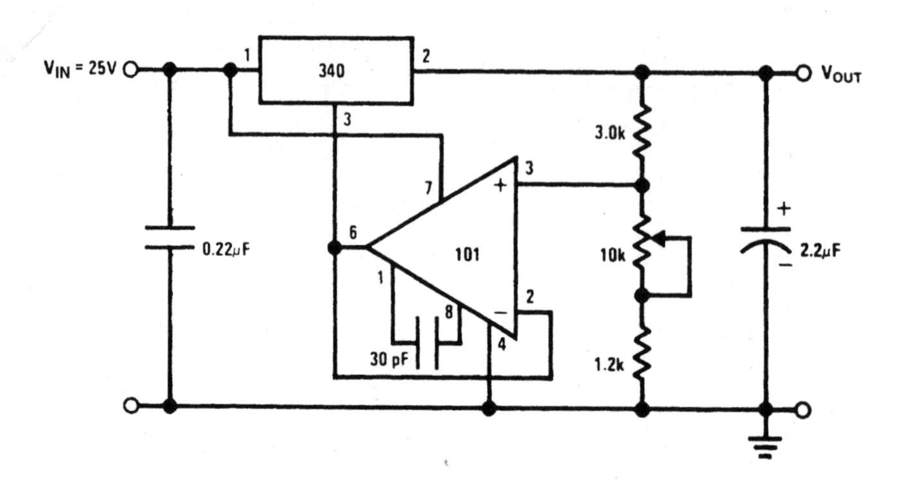

15-V 5.0-A regulator with short-circuit current limiting

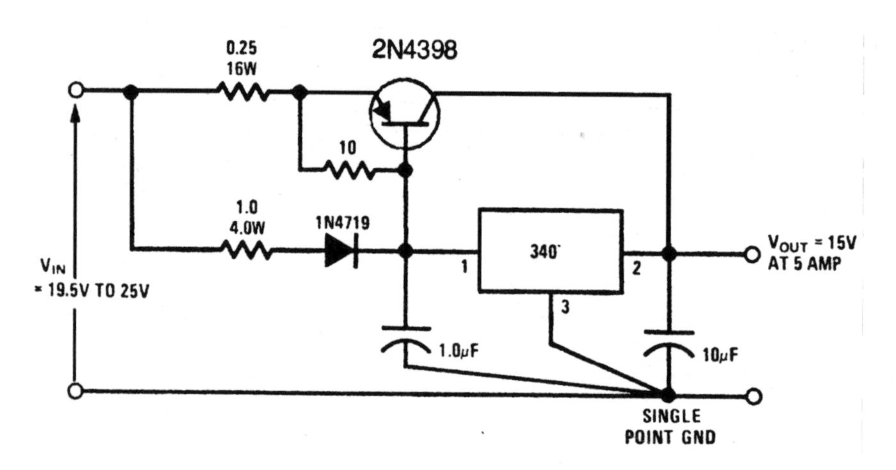

Electronic shutdown circuit

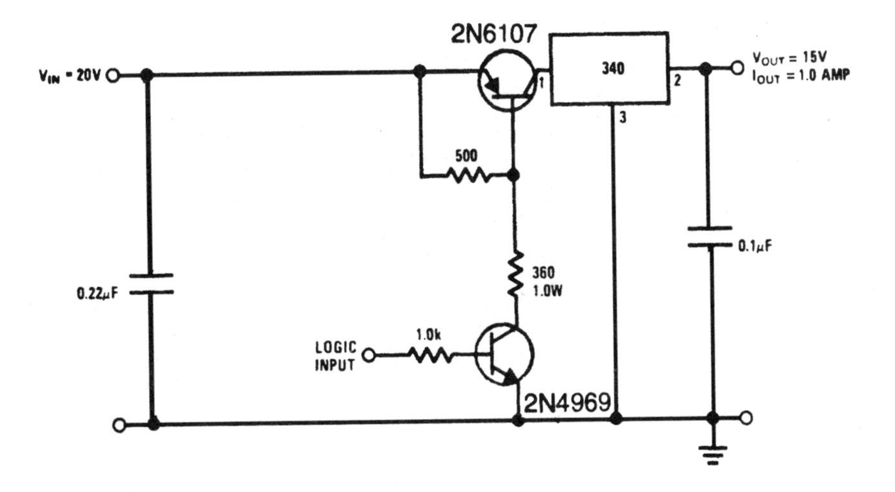

Tracking dual supply ±5.0 V to ±18 V

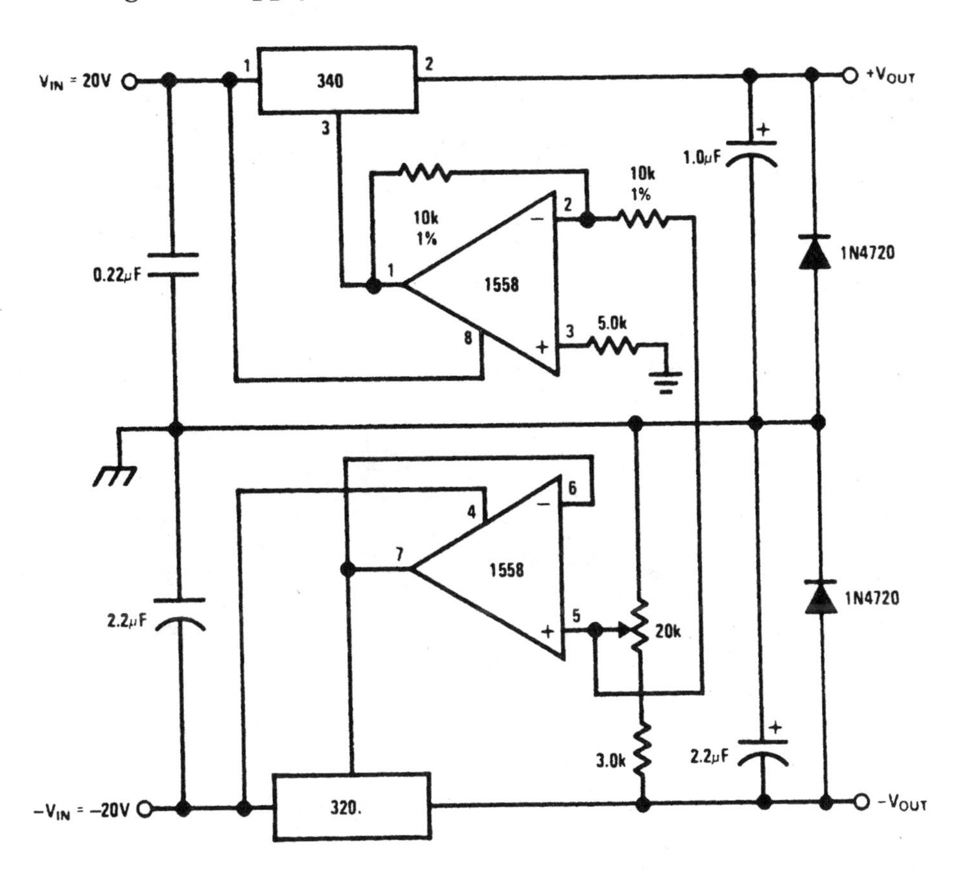

TL431 programmable precision reference

The 431 is a 3-terminal programmable shunt regulator diode. It is available in three different package styles, as illustrated. Notice that there are only three active connections, even in the 8-pin packages: anode, cathode, and reference. A stable 2.5-V reference voltage is available from this last pin.

The 431 can be user-programmed for any voltage ranging from Vref (2.5 V) up to 36 V. Just two external resistors are required to program the device's output voltage. It can operate with currents ranging from 1.0 mA to 100 mA, and has a typical dynamic impedance of a mere 0.22 ohm.

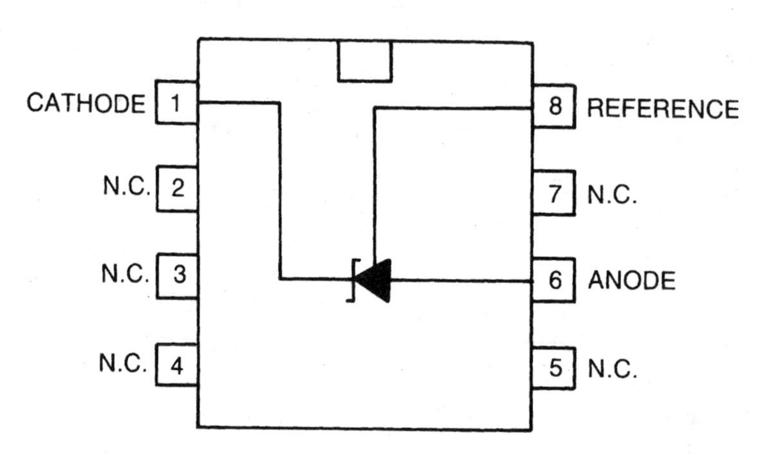

Variable voltage regulator

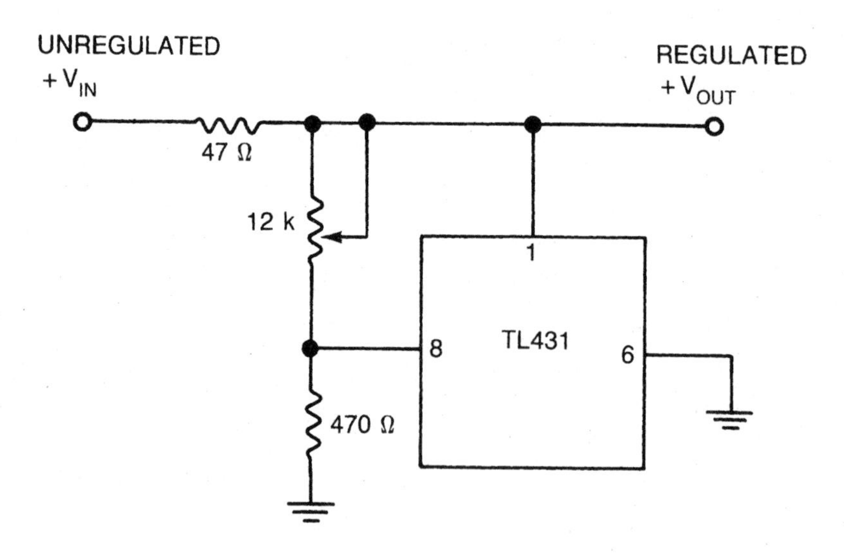

Alternator variable voltage regulator

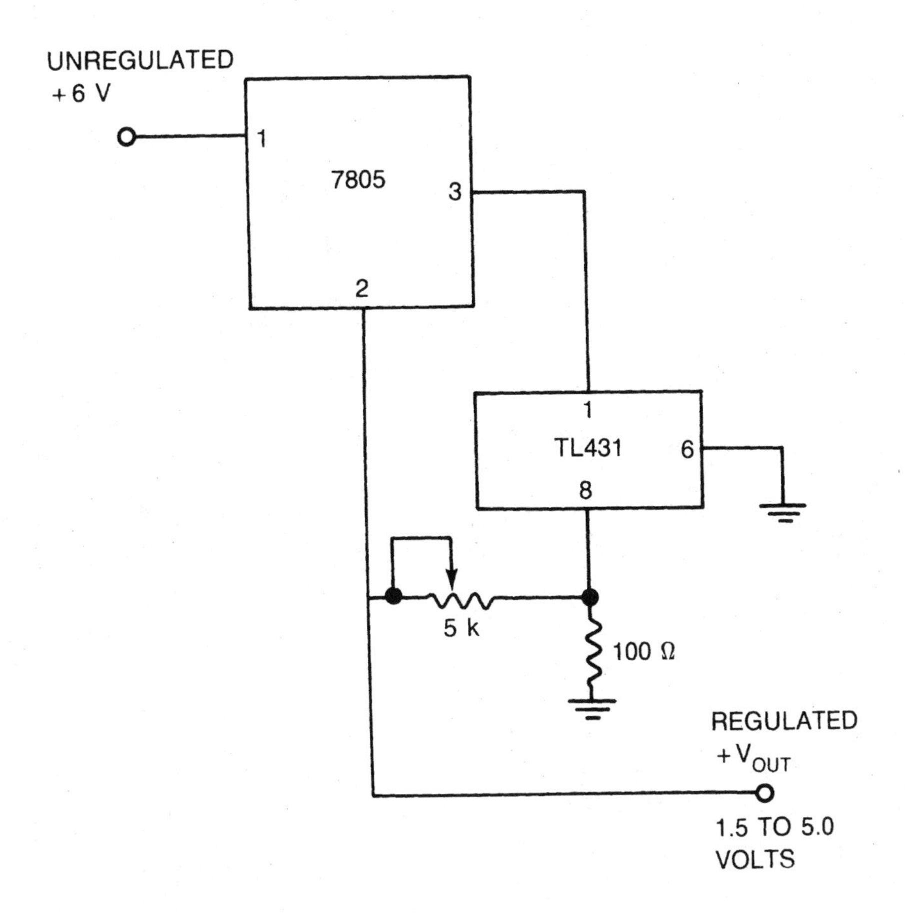

Timer

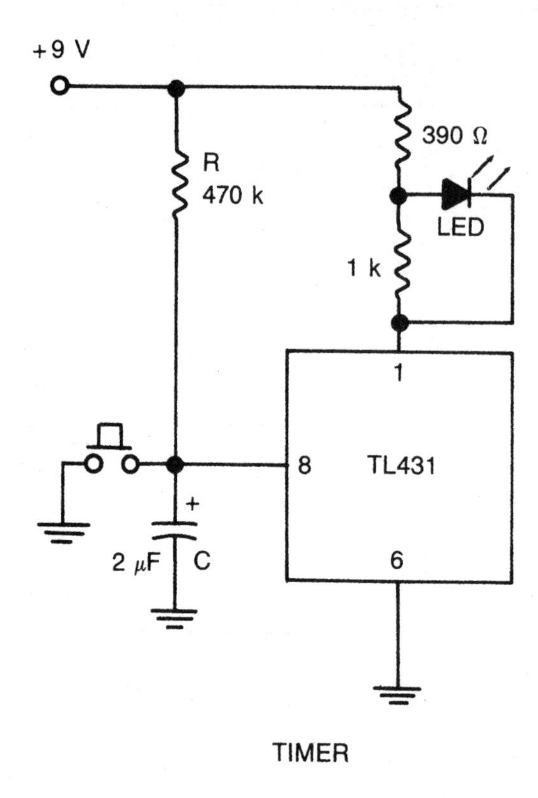

TIMER

TTL logic level detector

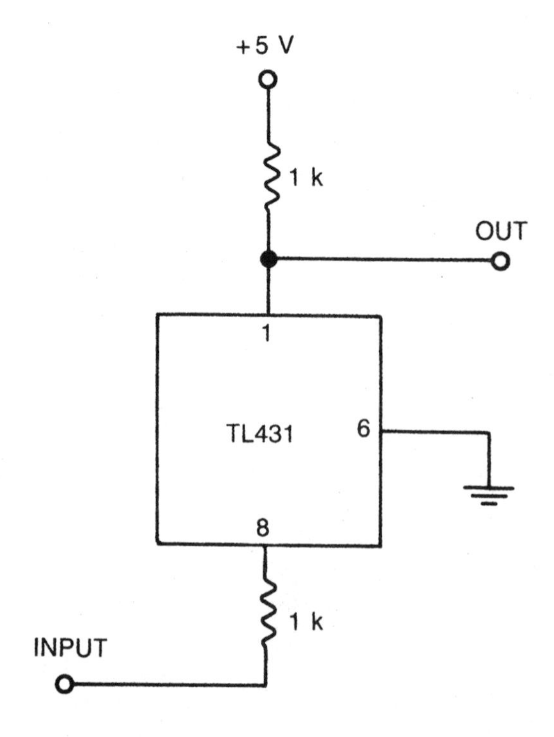

LM2931 series low-dropout voltage regulators

The 2931 is available either as a fixed 5.0 V or an adjustable output voltage regulator. It is offered in a variety of package styles. The fixed voltage version is available with either ±5.0 or ±3.8 percent tolerances. The adjustable output devices are rated for a ±5.0 percent tolerance.

The 2931 can handle output currents higher than 100 mV and input voltages as high as 40 V. The adjustable output voltage version can be set up for output voltages ranging from 3 to 24 V.

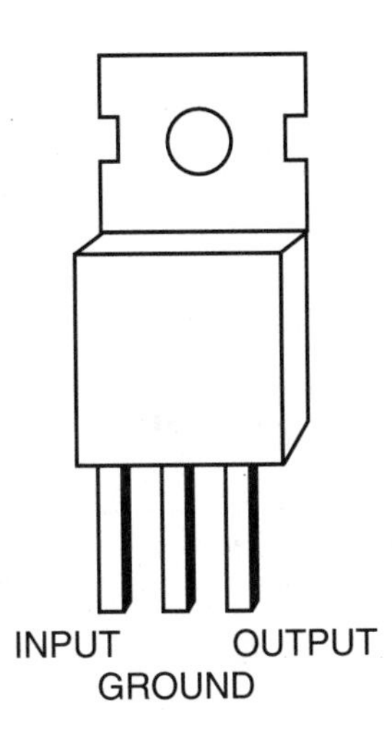

LM2931C low-dropout adjustable voltage regulator

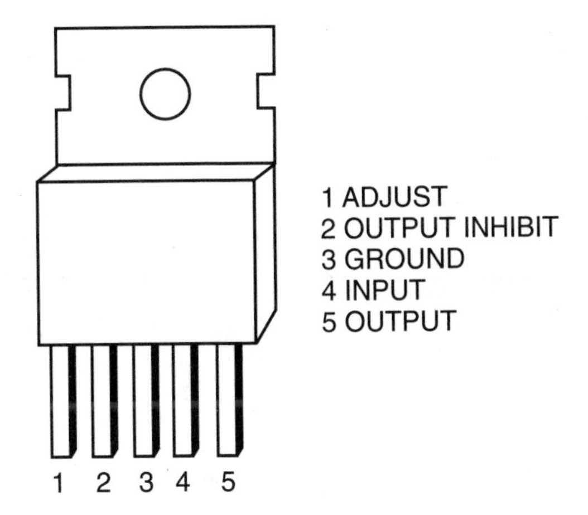

LM2931D low-dropout voltage regulator

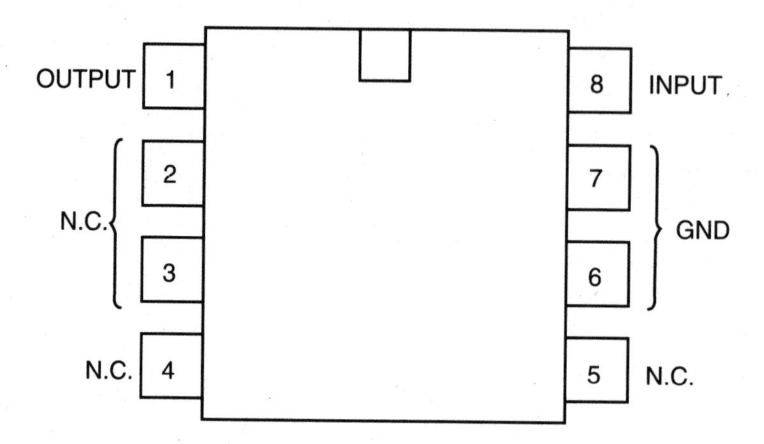

LM2931CD adjustable low-dropout voltage regulator

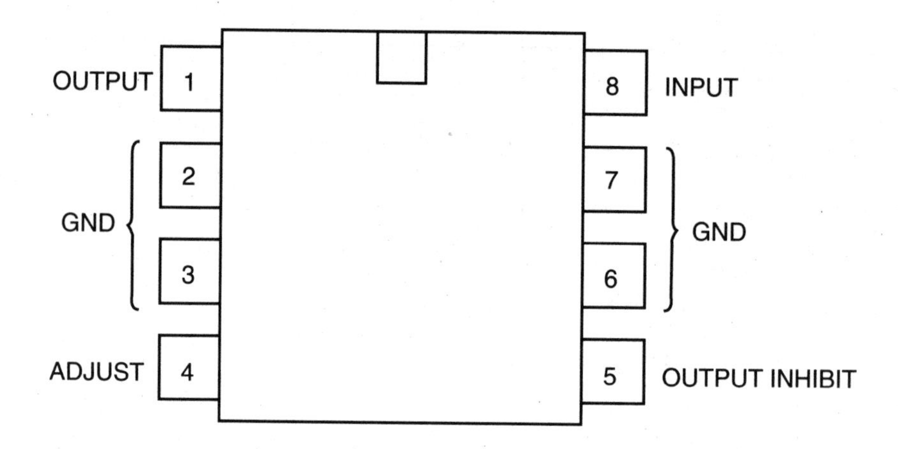

Fixed output regulator

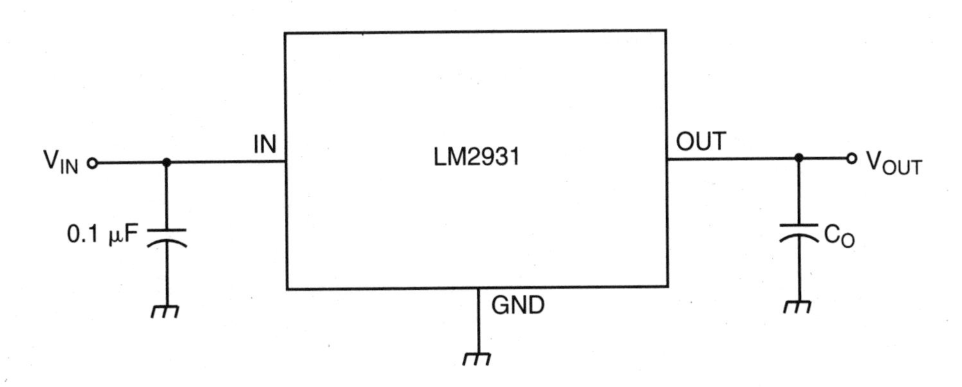

Adjustable output regulator

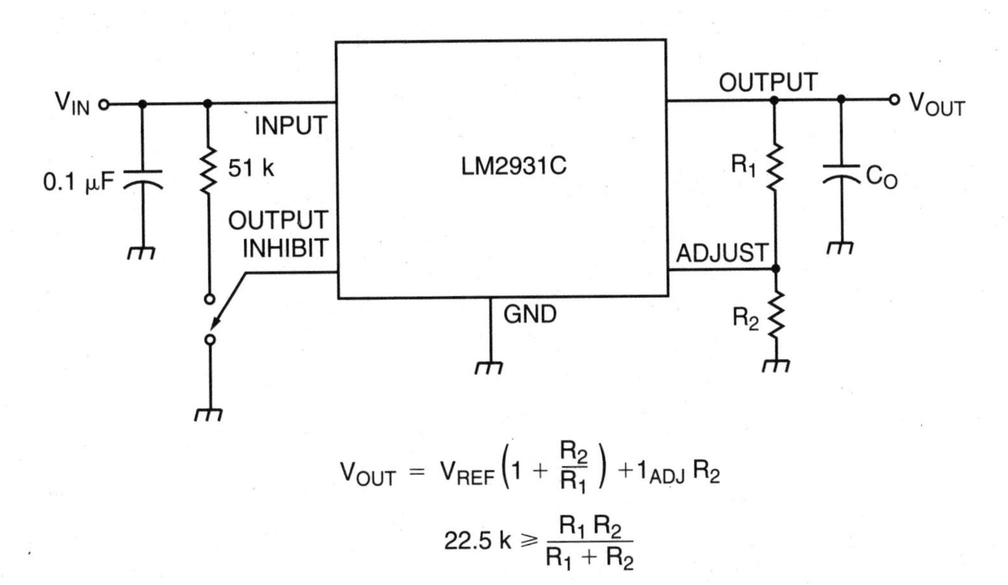

$$V_{OUT} = V_{REF}\left(1 + \frac{R_2}{R_1}\right) + 1_{ADJ} R_2$$

$$22.5\ k \geqslant \frac{R_1 R_2}{R_1 + R_2}$$

Low-differential voltage regulator

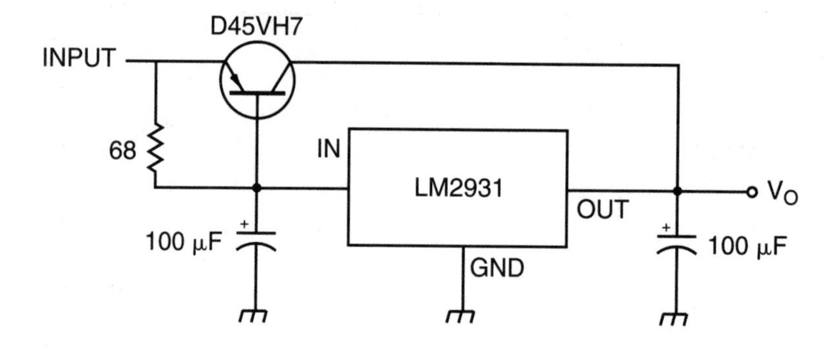

Current boost regulator with short-circuit protection

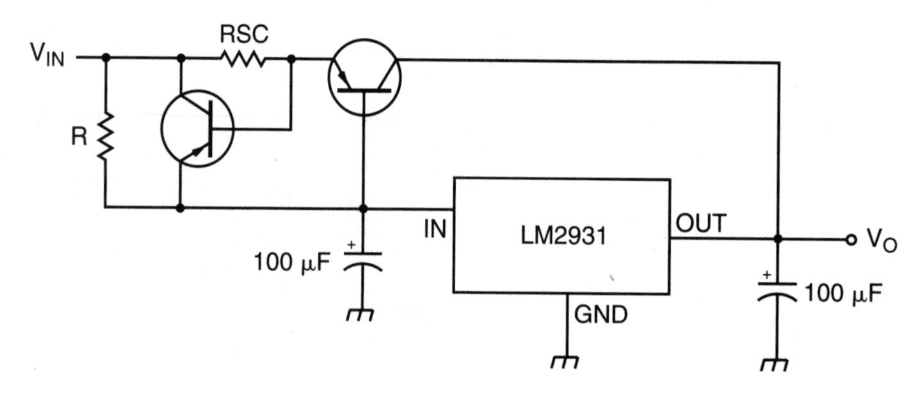

(*Courtesy of Motorola.*)

Constant intensity lamp flasher

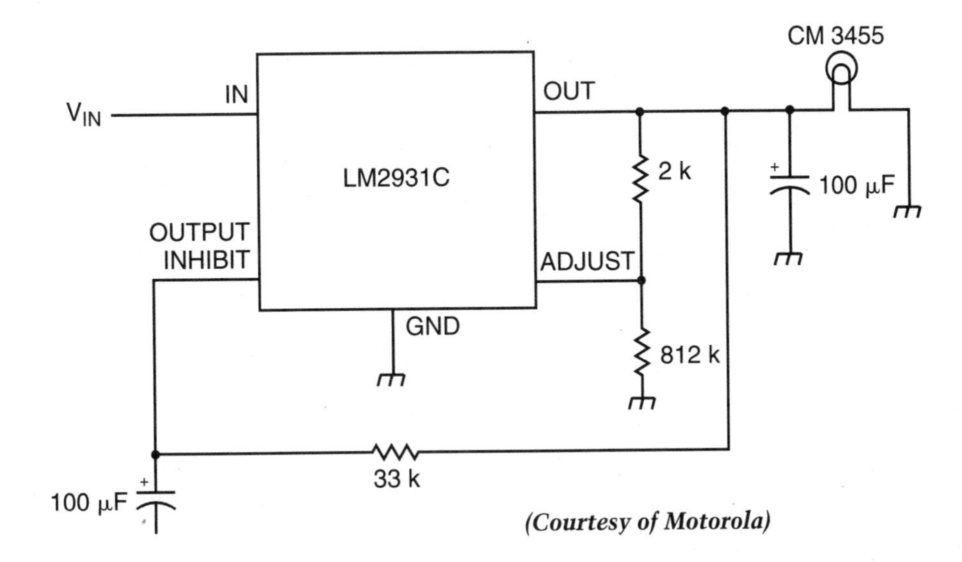

(Courtesy of Motorola)

LM2935 low-dropout dual voltage regulator

The 2935 is a dual positive 5.0 V low-dropout voltage regulator intended primarily for use in standby power systems. The main output of this device can supply up to 750 mA of current. This output can be turned on and off via the "Switch/Reset" pin (pin No. 4). The second output is not switchable. It is intended to supply power to volatile memory. It can drive loads up to 10 mA. When supplying 10 mA from the standby output, the quiescent current of the 2935 is no greater than 3.0 mA.

The 2935 was designed for use in harsh environments, such as in automotive applications. It is relatively immune to many typical input supply voltage problems, such as reversed battery (−12 V), doubled battery (+24 V), and load dump transients (up to +60 V).

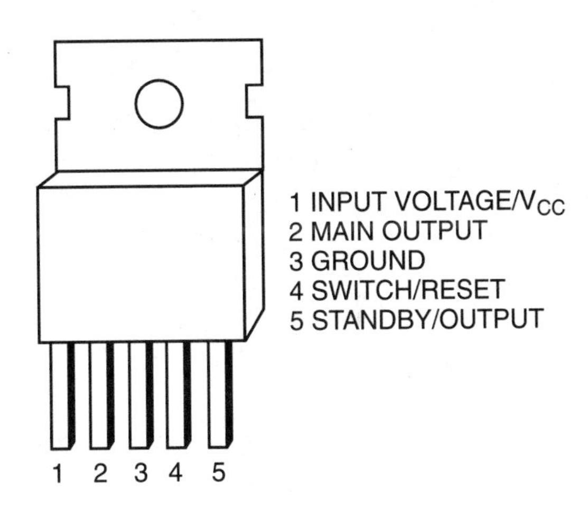

Low-dropout dual voltage regulator

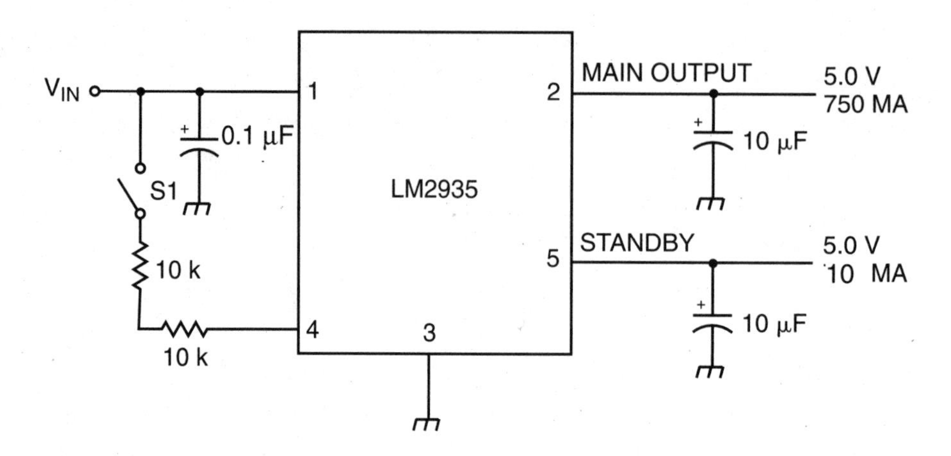

7805 +5-V regulator

The 7805 is a popular and easy-to-use voltage regulator IC. It has just three pins: input (pin No. 1), output (pin No. 2), and common, or ground (pin No. 3).

The 7805 is designed for a well-regulated output of 5 V. This voltage regulator chip can handle currents over 1.5 ampere, if adequate heat-sinking is provided.

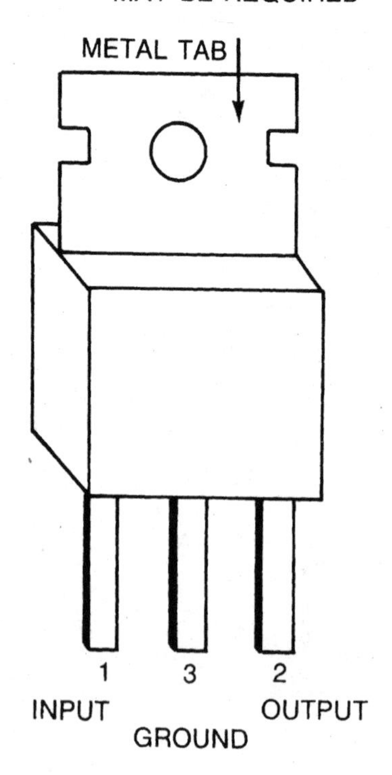

+5-V regulator

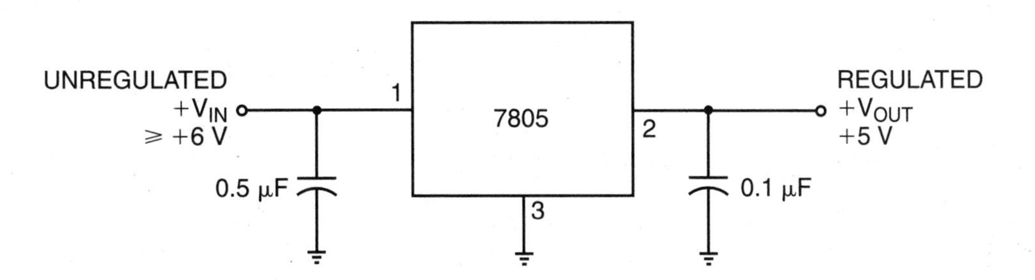

Current regulator

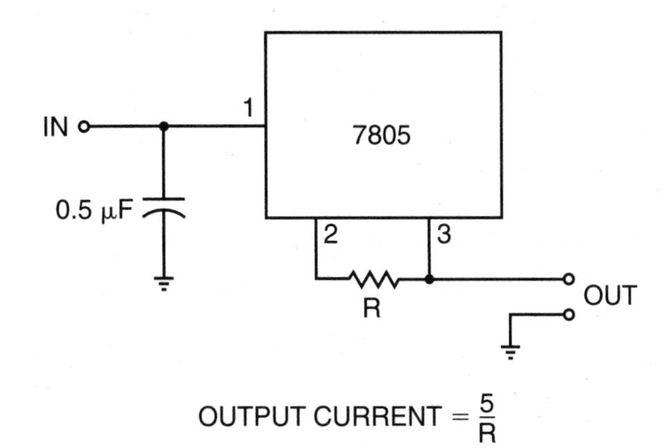

OUTPUT CURRENT $= \frac{5}{R}$

Very high-current voltage regulator with short-circuit protection

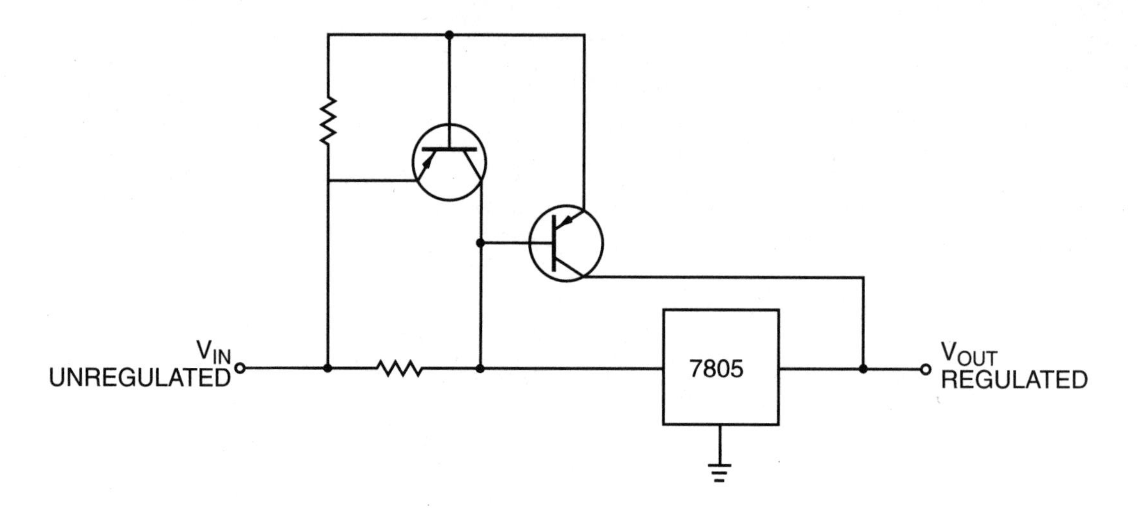

Variable output voltage regulator

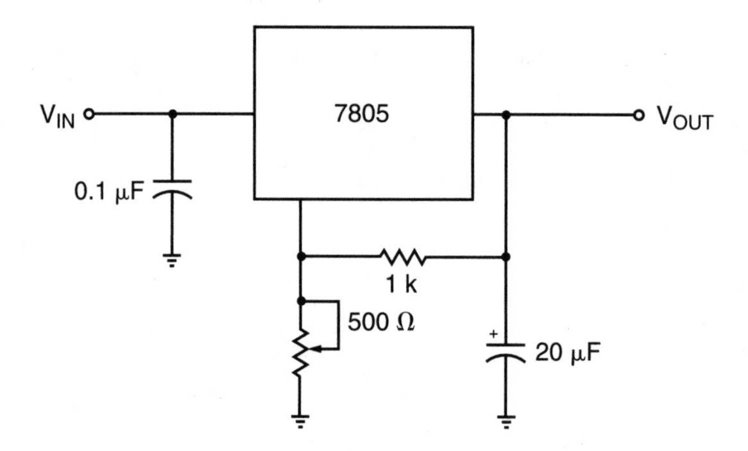

Complete 5-V regulated power supply

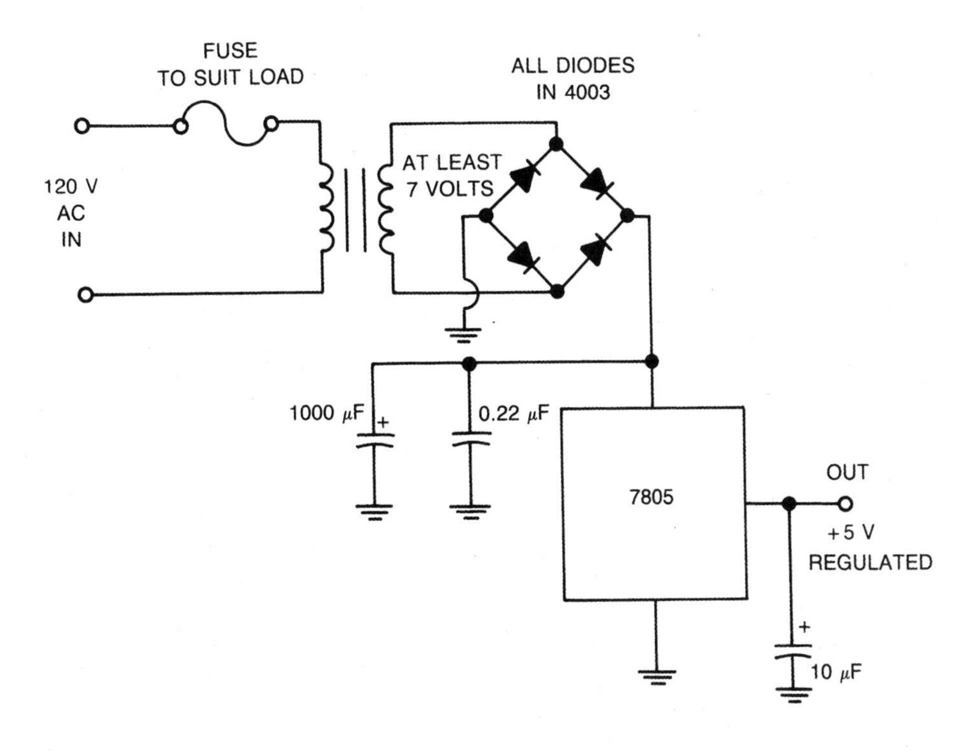

Increased current voltage regulator

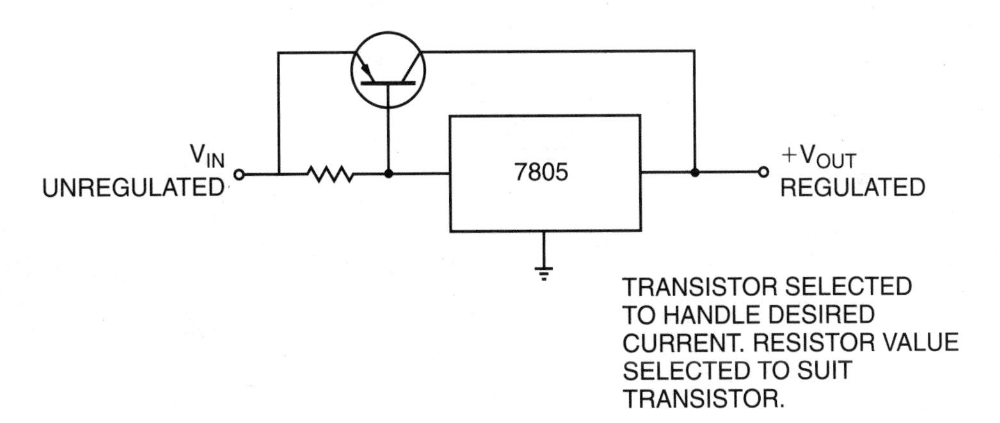

78L05 5-V integrated voltage output regulator

The 78L05 is a low-power version of the 7805.

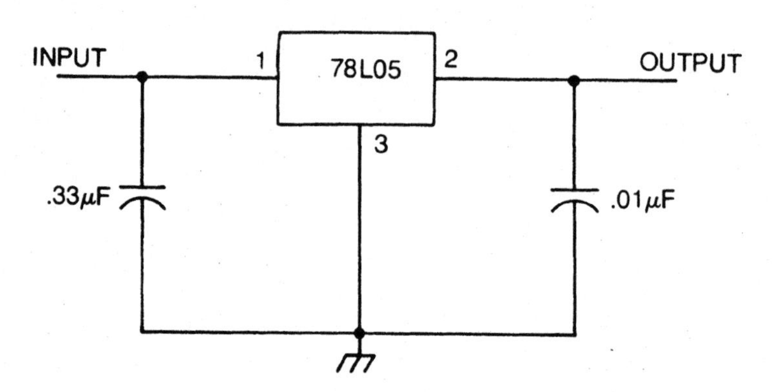

Variable output regulator

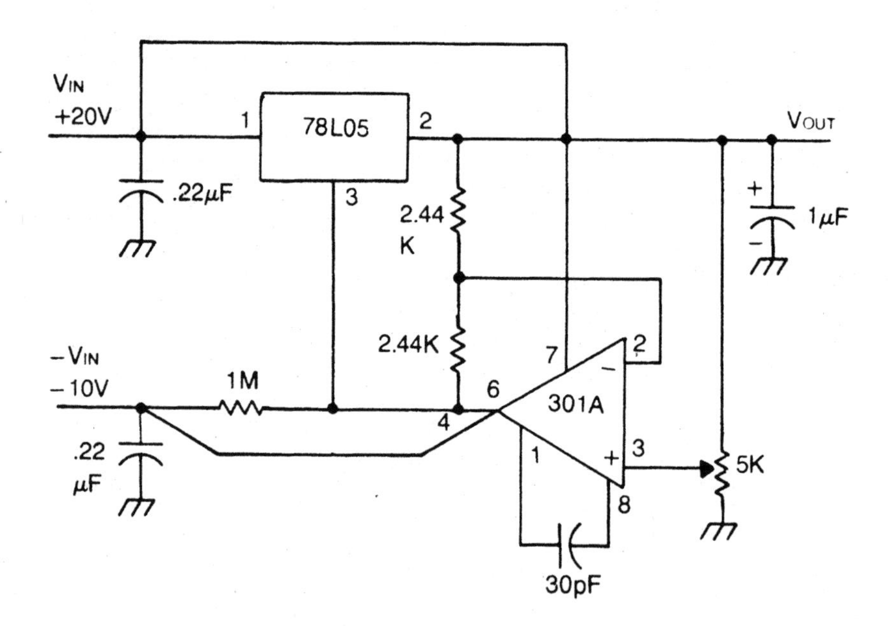

7812 +12-V regulator

The 7812 is a popular and easy-to-use voltage regulator IC. It has just three pins: input (pin No. 1), output (pin No. 2), and common, or ground (pin No. 3).

The 7812 is designed for a well-regulated output of 12 V. This voltage regulator chip can handle currents over 1.5 amperes, if adequate heat-sinking is provided.

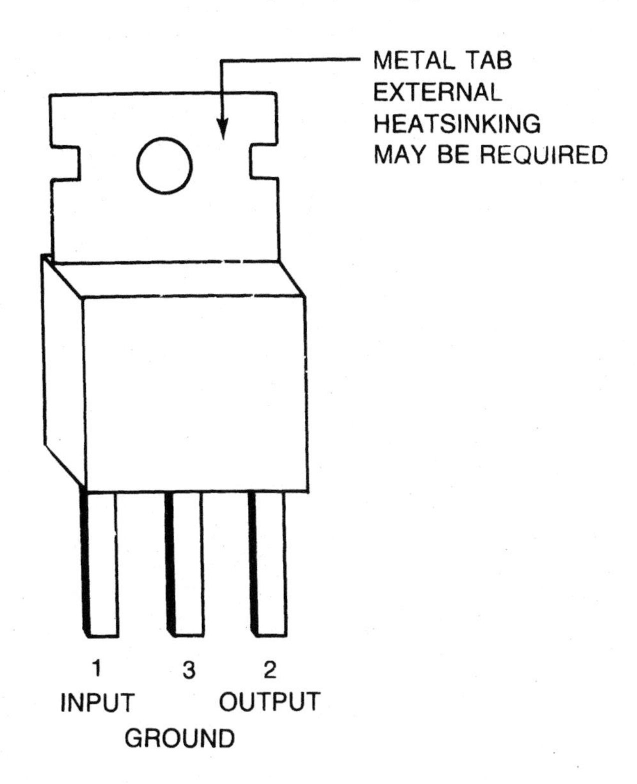

+12-V regulator

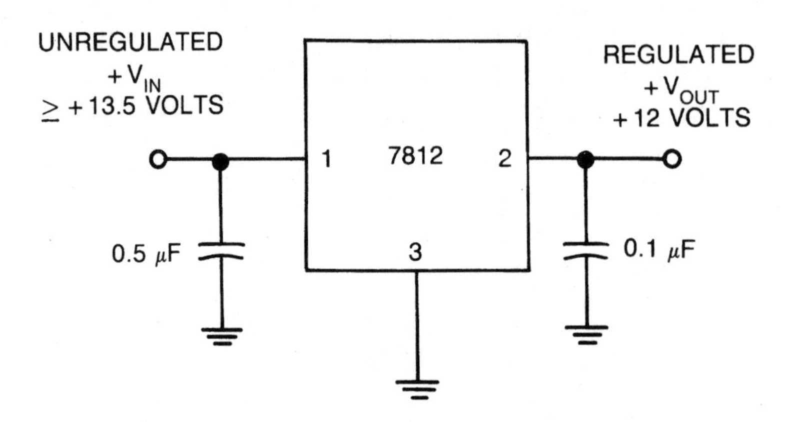

Current regulator

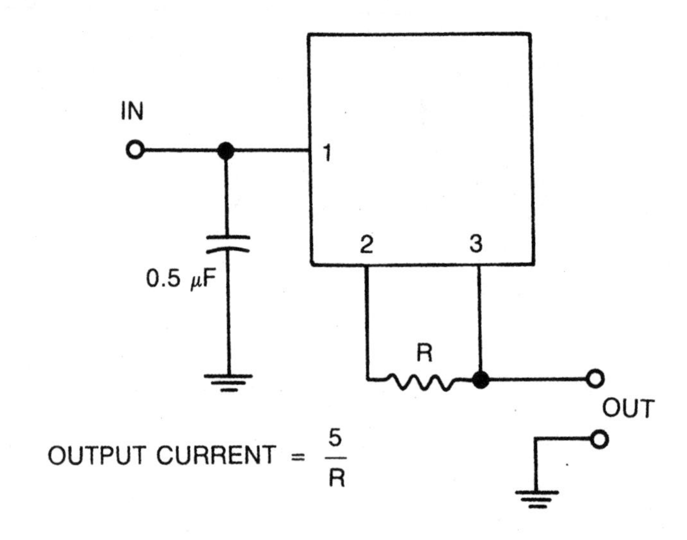

Very high-current voltage regulator with short-circuit protection

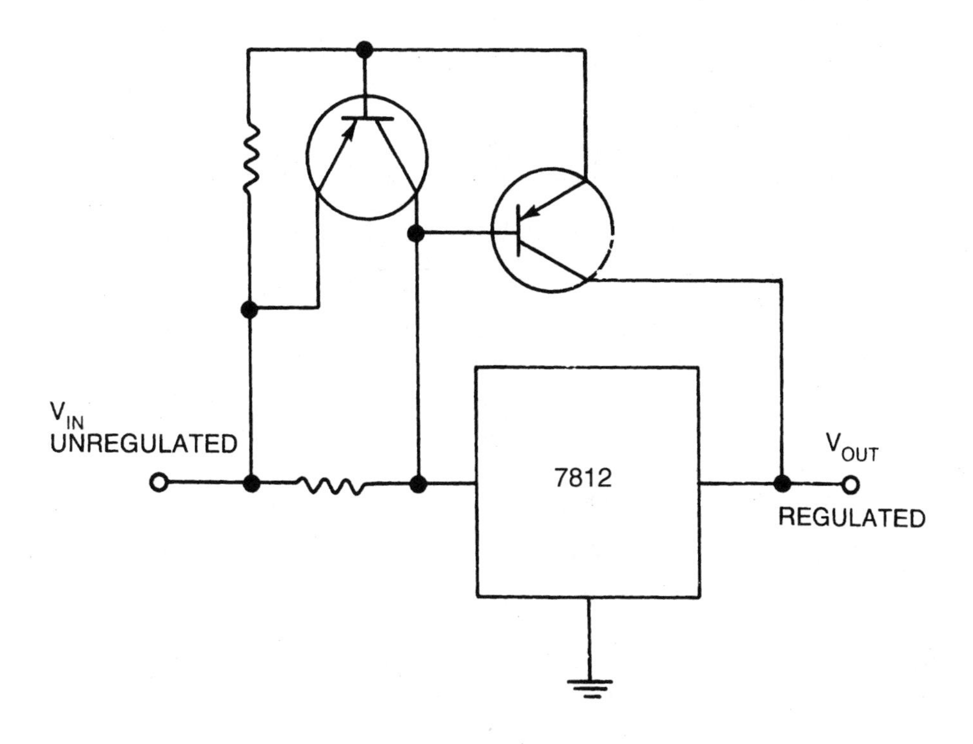

Variable output voltage regulator

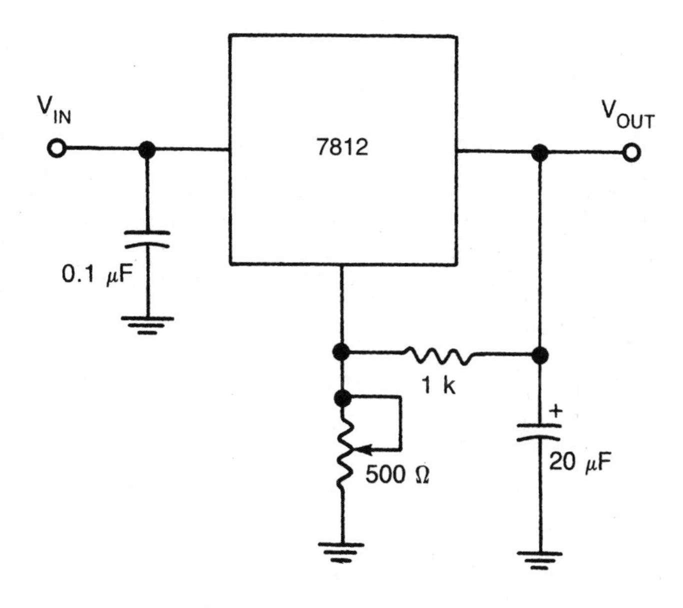

Increased voltage output

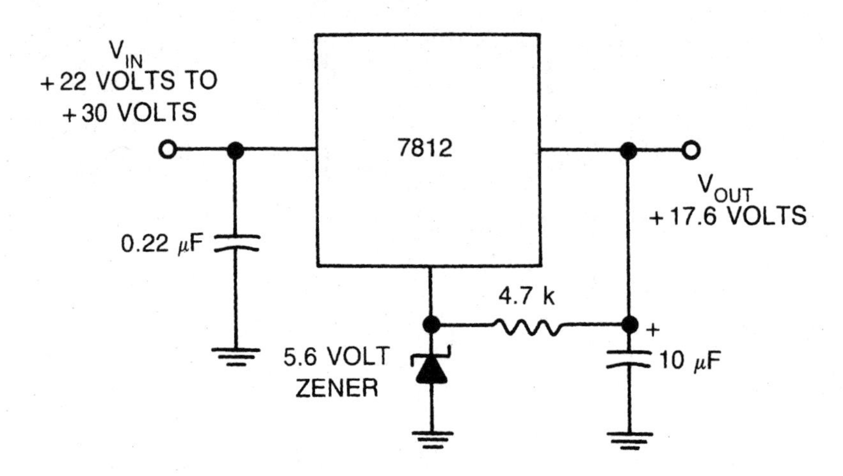

7815 15-V voltage regulator

The 7815 is a popular and easy-to-use voltage regulator IC. It has just three pins: input (pin No. 1), output (pin No. 2), and common, or ground (pin No. 3).

The 7815 is designed for a well-regulated output of 15 V. This voltage regulator chip can handle currents over 1.5 amperes, if adequate heat-sinking is provided.

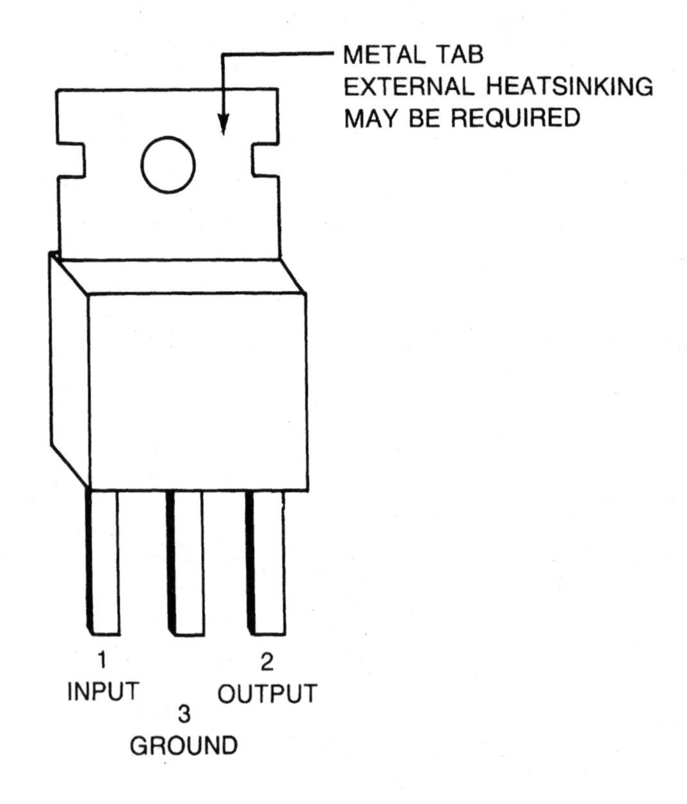

+15-V regulator

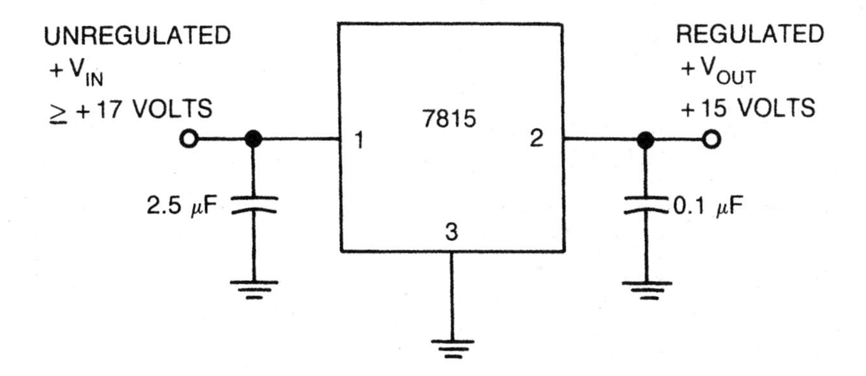

Current regulator

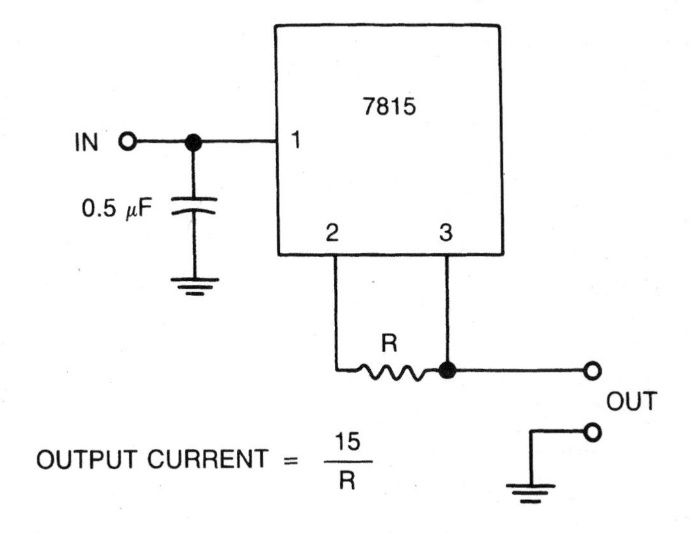

MC7900 series negative fixed voltage regulators

The 7900 series of fixed voltage regulators are essentially negative voltage versions of the popular 7800 series. These devices are simple to use, with three pins (input, output, and common). They are available in a variety of standard output voltages, and can deliver output currents greater than 1.0 ampere.

The 7900 series voltage regulators feature internal short-circuit protection, and thermal overload protection.

They are so complete that no external components at all are required for basic operation.

7905 negative 5-V series voltage regulator

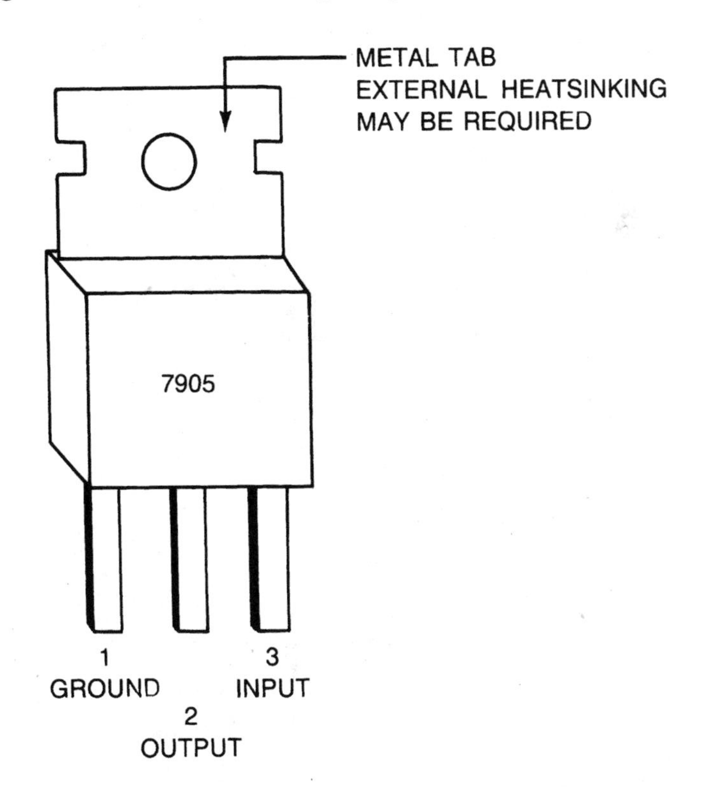

Dual-trimmed supply

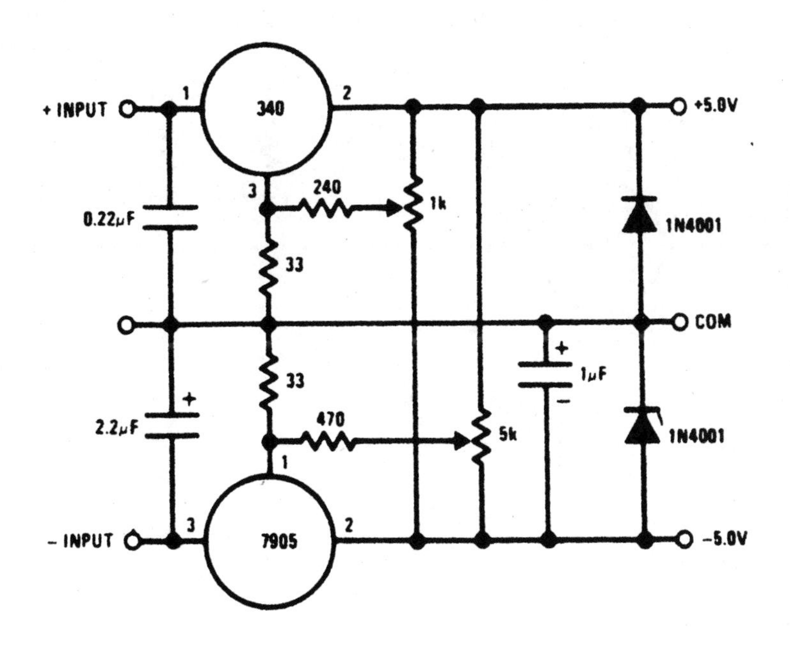

Current source

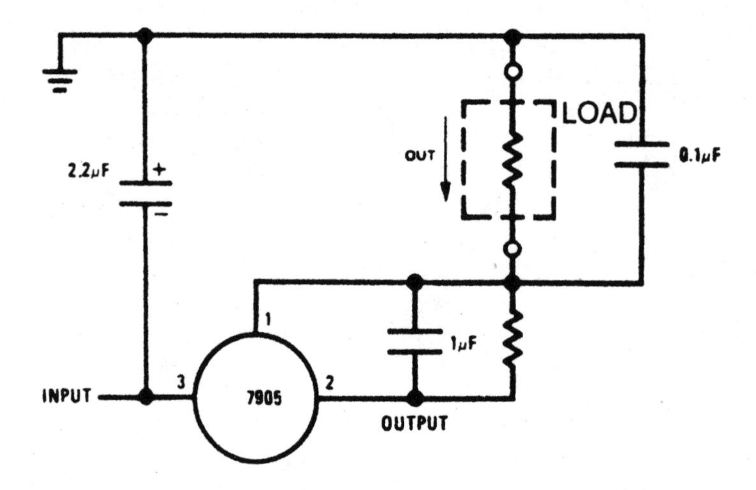

Highly stable 1-amp regulator

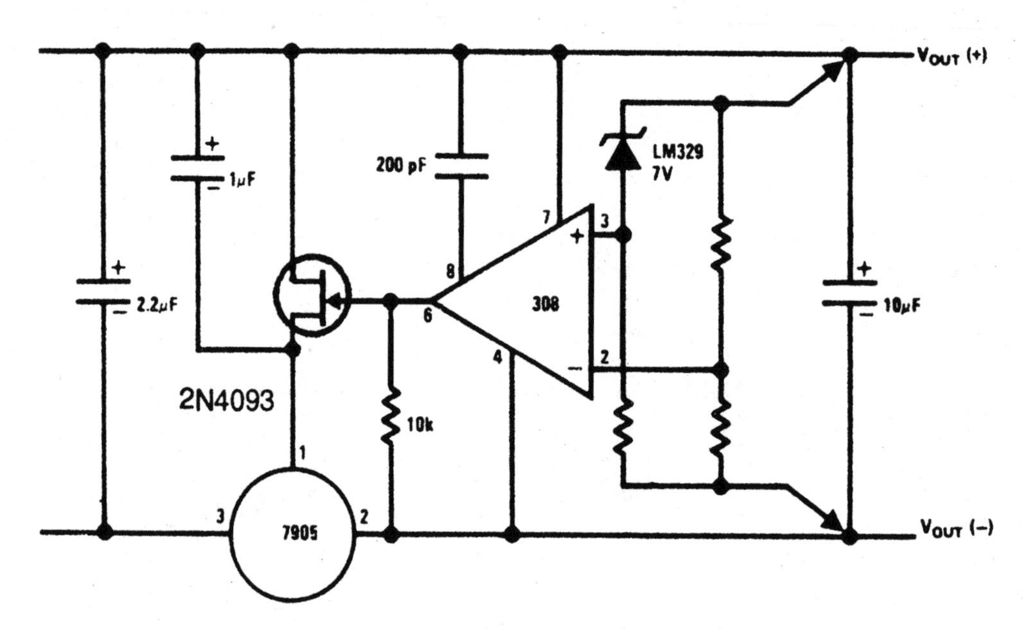

9

SECTION

Other linear devices

The last few sections have examined various common types of linear ICs. But there are many other, less common linear functions as well. This section contains a miscellaneous selection of linear devices that don't quite fit into any of the other section headings, including, but not limited to, voltage comparators, multipliers, and arrays.

LM339/LM339A low-power, low-offset quad voltage comparator

The LM339 series consists of four independent precision voltage comparators, with an offset voltage specification as low as 2 mV maximum for all four comparator stages.

This device was designed specifically to operate from a single-ended power supply over a wide range of voltages. Operation from split power supplies is also possible, and is preferred for the A suffix versions. The low-power supply current drain of the 339 is independent of the magnitude of the power supply voltage. These comparators also have a unique characteristic in that the input common-mode voltage range includes ground, even when the chip is operated from a single-ended supply voltage.

Application areas include limit comparators, simple analog-to-digital converters, pulse generators, square-wave generators, time-delay generators, wide-range VCOs, window comparators, MOS clock timers, multivibrators, and high-voltage digital logic gates.

The 339 series is designed to directly interface with TTL and CMOS circuitry. When operated from a split power supply, this device will directly interface with MOS logic, where the low-power drain of the LM339 is a distinct advantage over most standard voltage comparator devices.

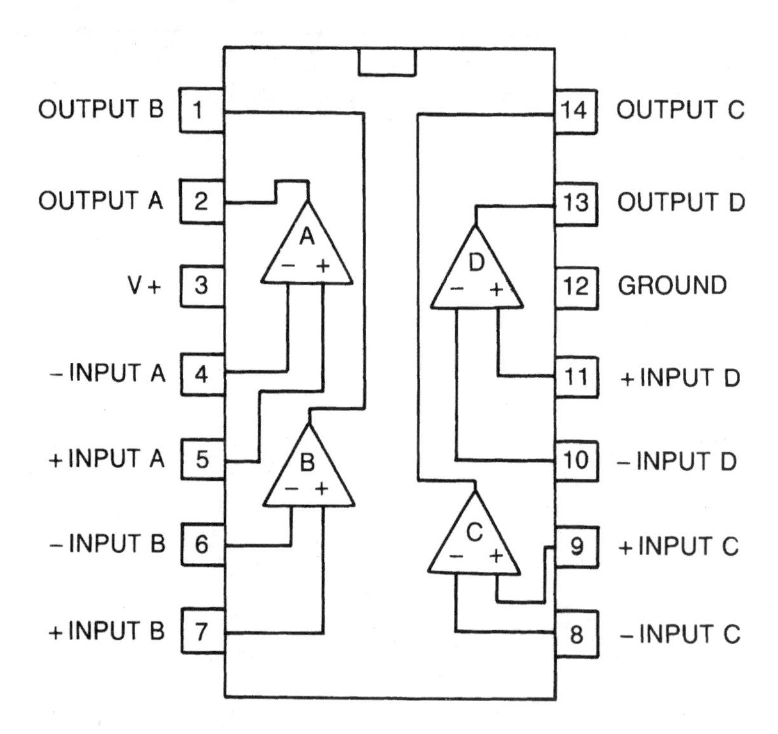

Basic comparator

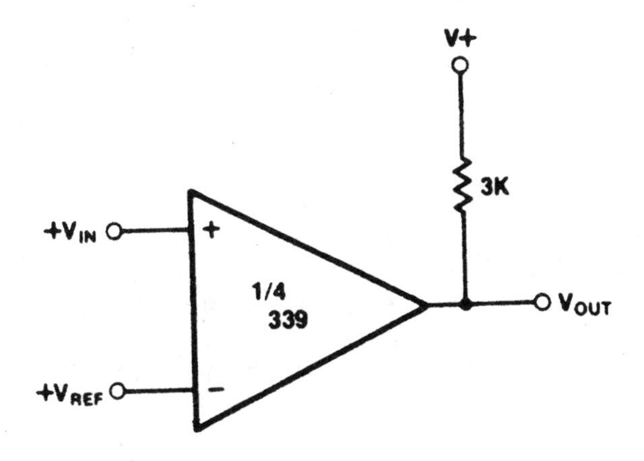

Comparator with hysteresis

The addition of hysteresis limits output chatter and false signals when the input voltage is very close to the reference voltage, and/or there is the possibility of noise or interference on the signal line.

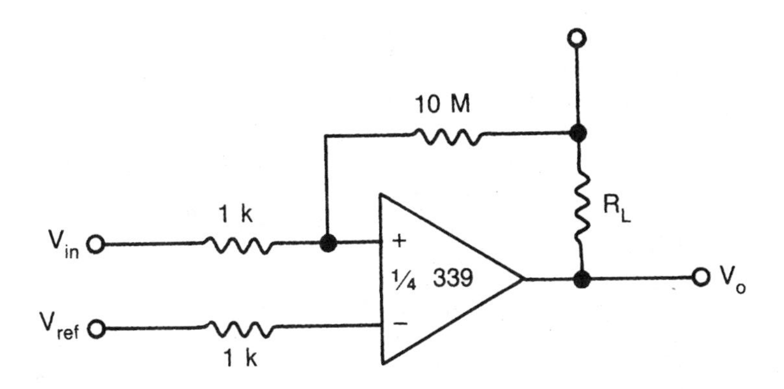

Limit comparator

The output (taken from the transistor's collector) indicates whether or not the input voltage is within its preset limits, defined by the three resistors in the voltage divider string (R_1, R_2, and R_3). These resistance values interact, but, roughly speaking, R_1 sets the upper voltage limit, R_3 sets the lower voltage limit, and R_2 controls the distance between the upper and lower limits of the reference band.

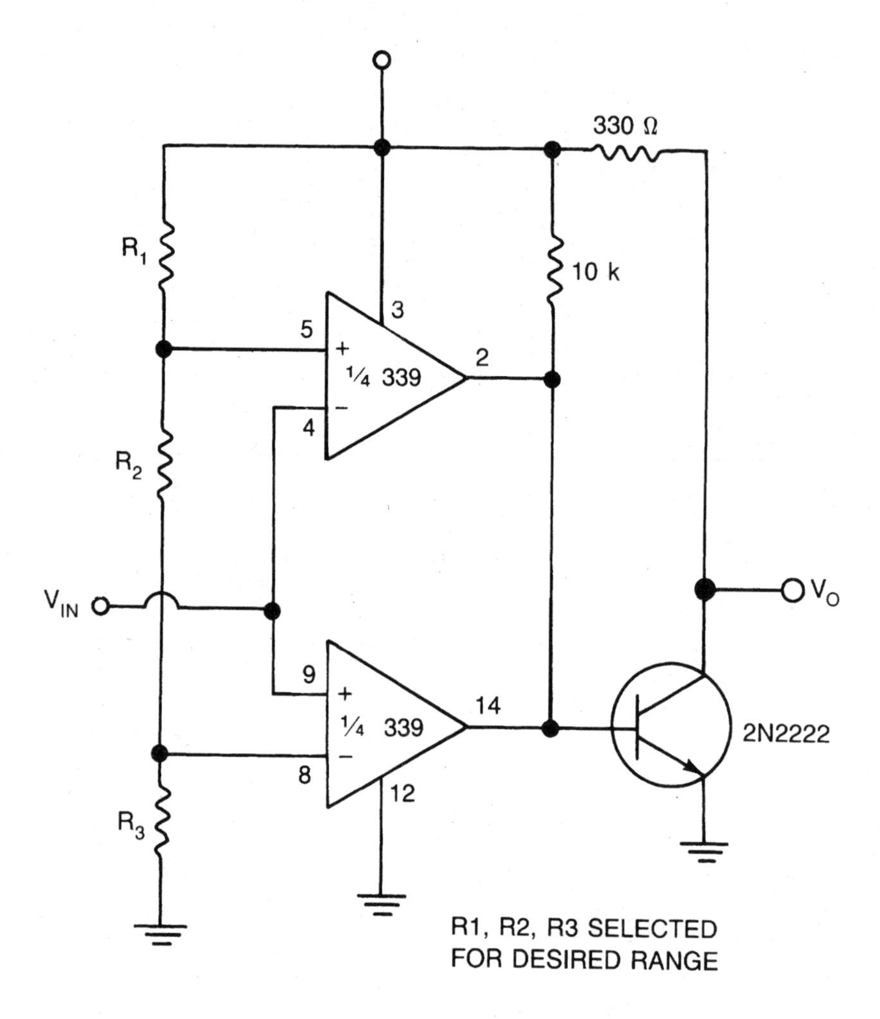

One-shot multivibrator

This circuit is also known as a *monostable multivibrator* or a *timer.*

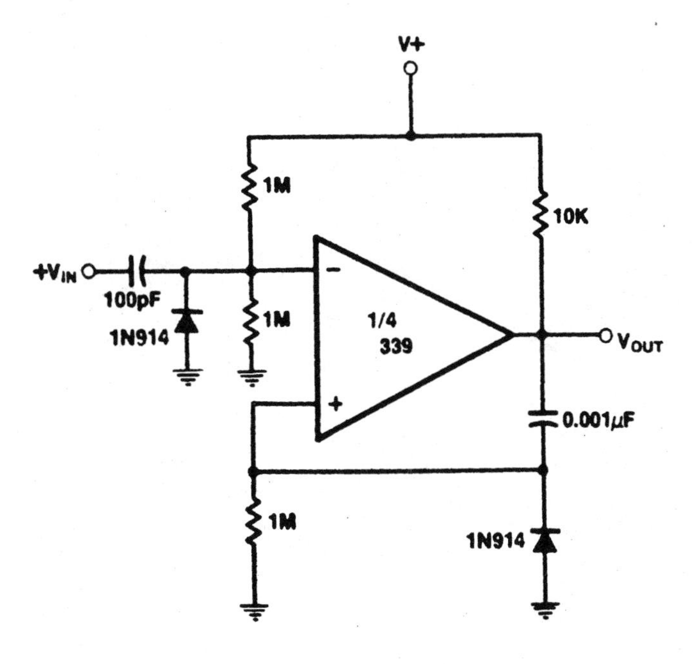

Bistable multivibrator

This type of circuit is also known as an *RS flip-flop.* The output reverse states (from HIGH to LOW or vice versa) each time it is triggered. The S input "sets" the output to *high.* The R input "resets" the output to *low.*

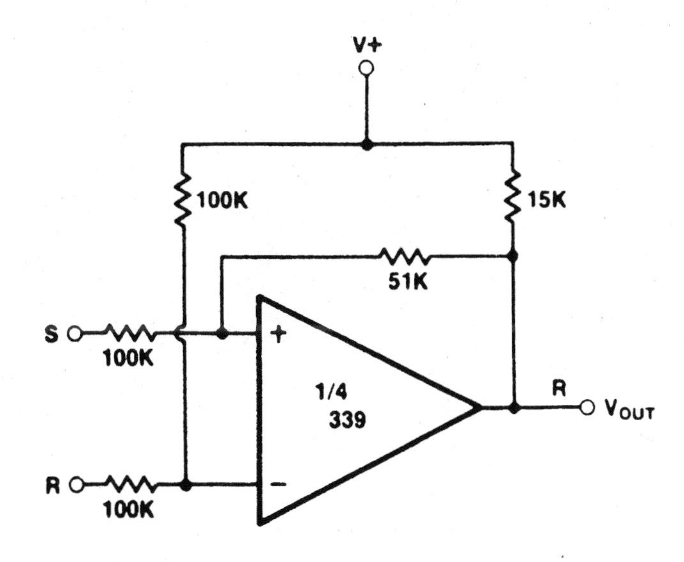

Driving TTL

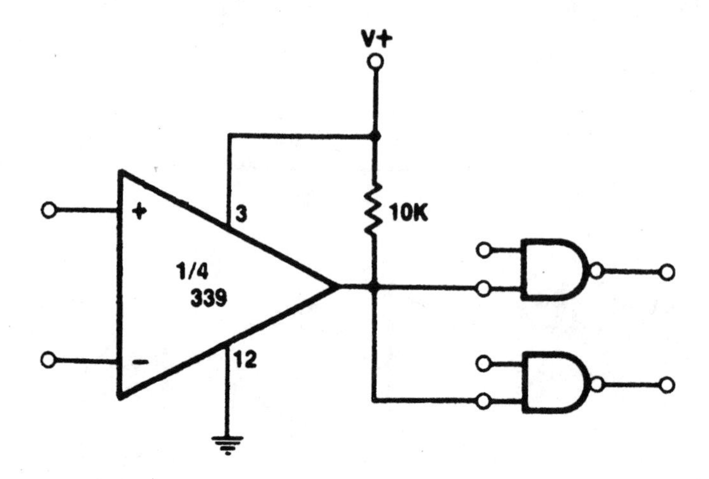

Driving CMOS

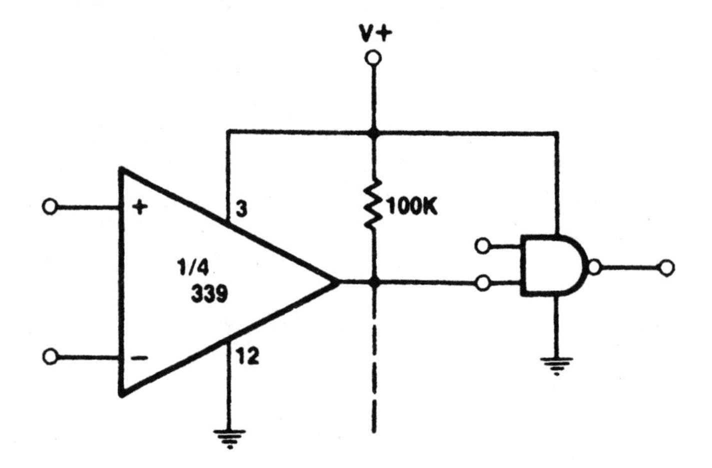

AND gate

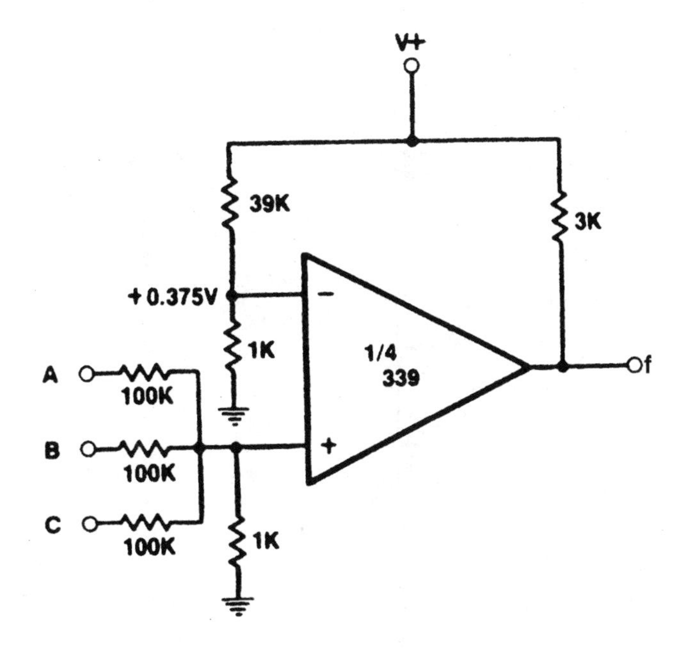

OR gate

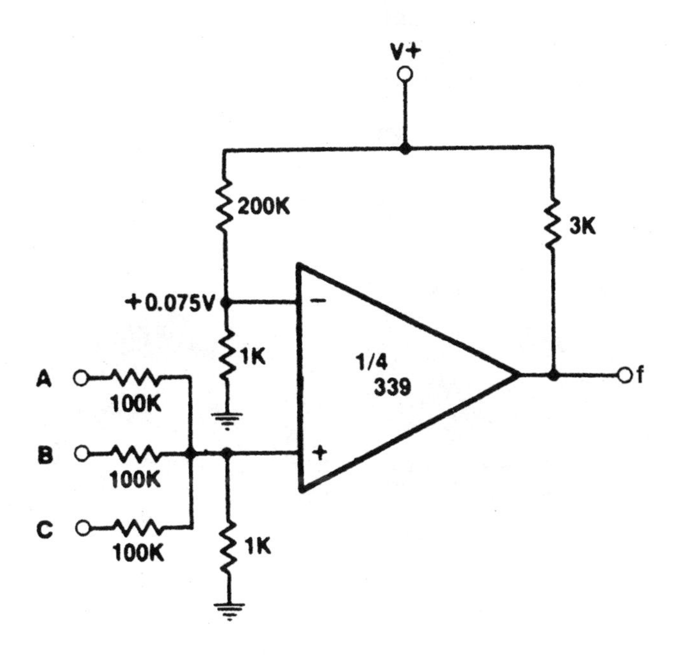

LM1036 dual dc-operated tone/volume/balance controller

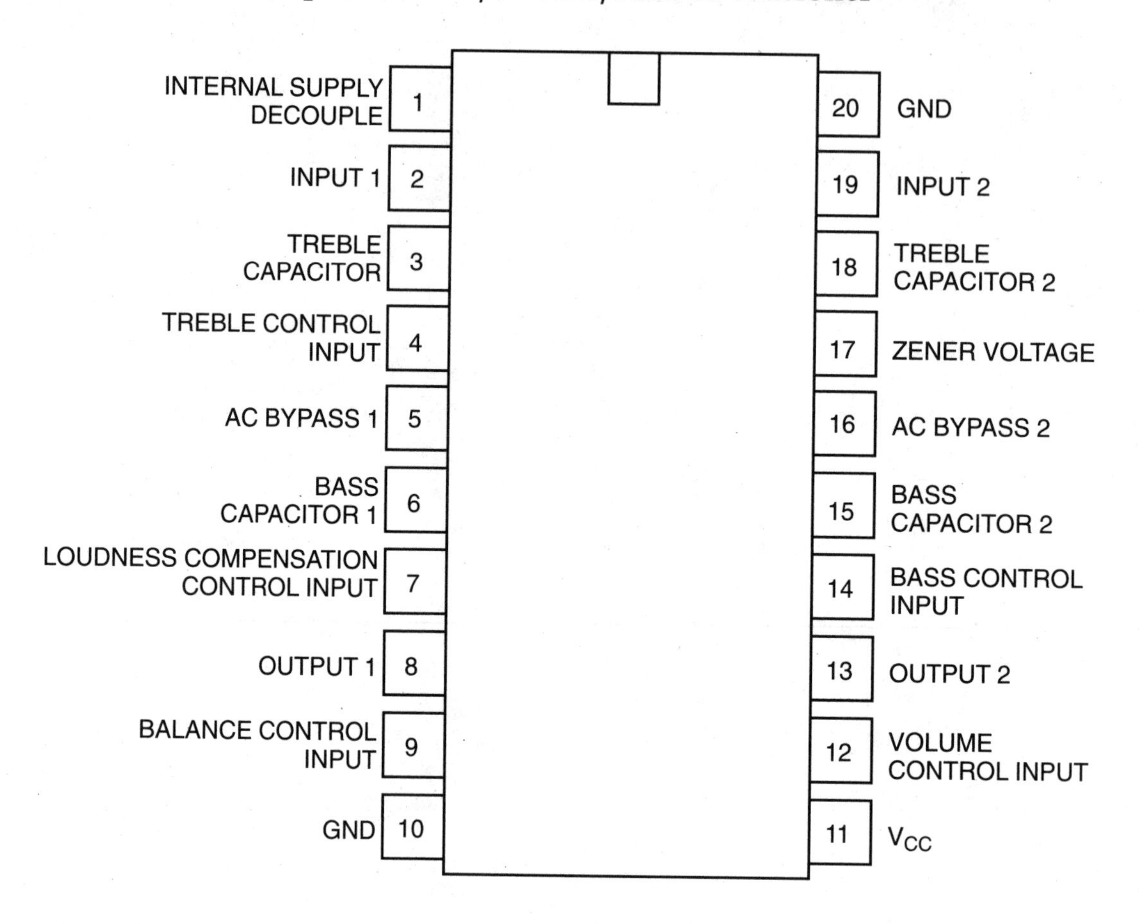

Basic dual dc-operated tone/volume/balance circuit

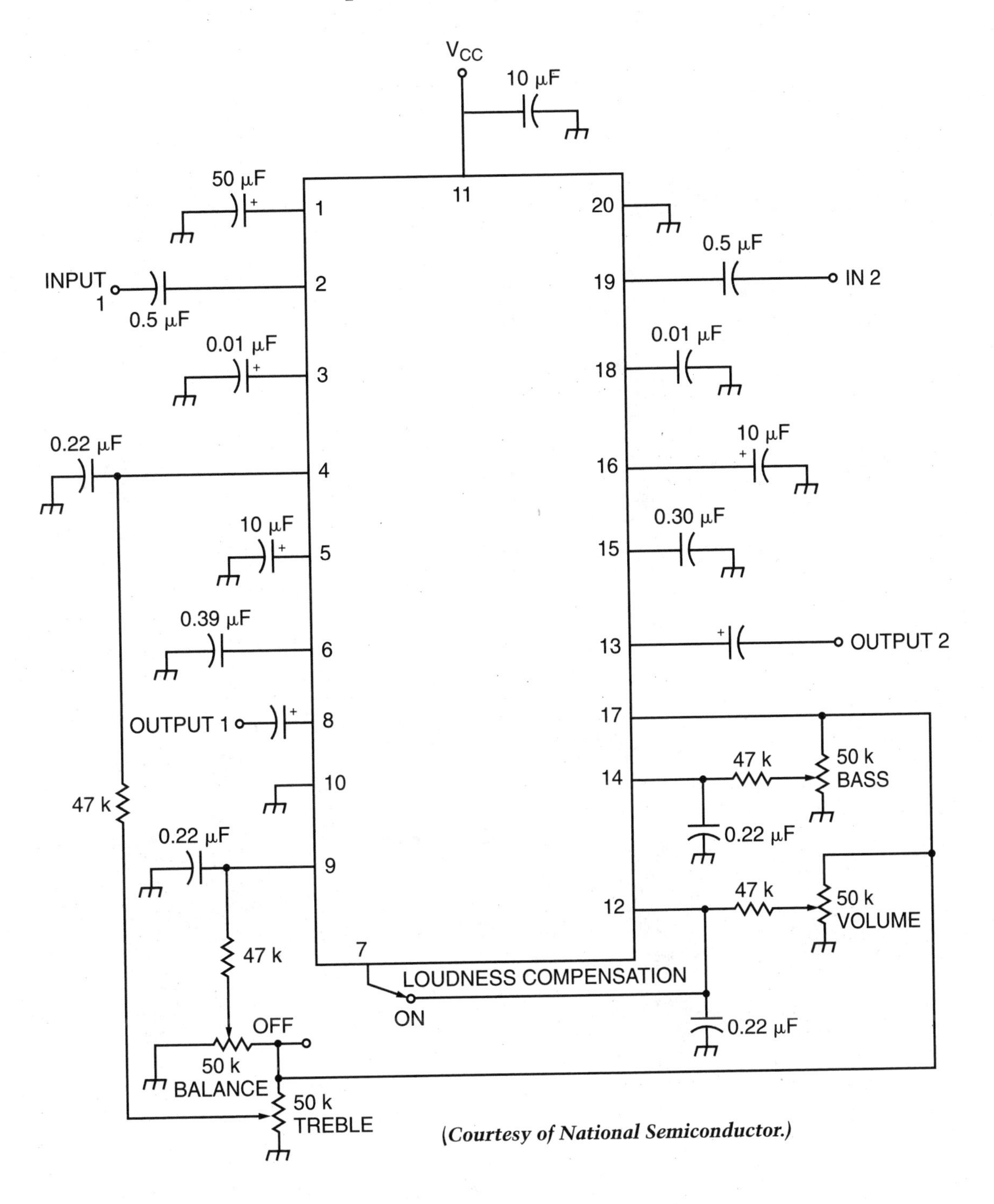

(*Courtesy of National Semiconductor.*)

Modification with additional bass-boost and loudness control

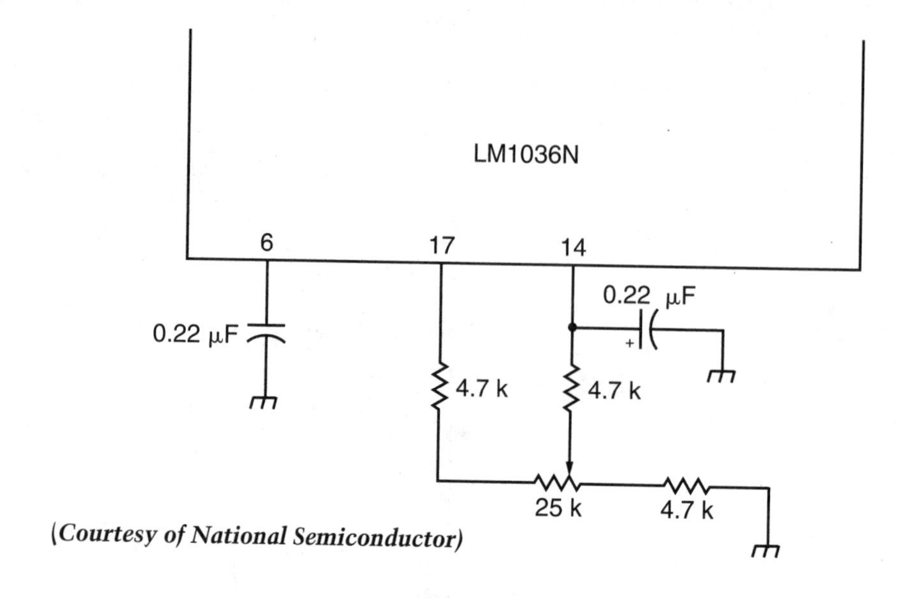

(*Courtesy of National Semiconductor*)

1494 linear 4-quadrant integrated multiplier with square root circuit

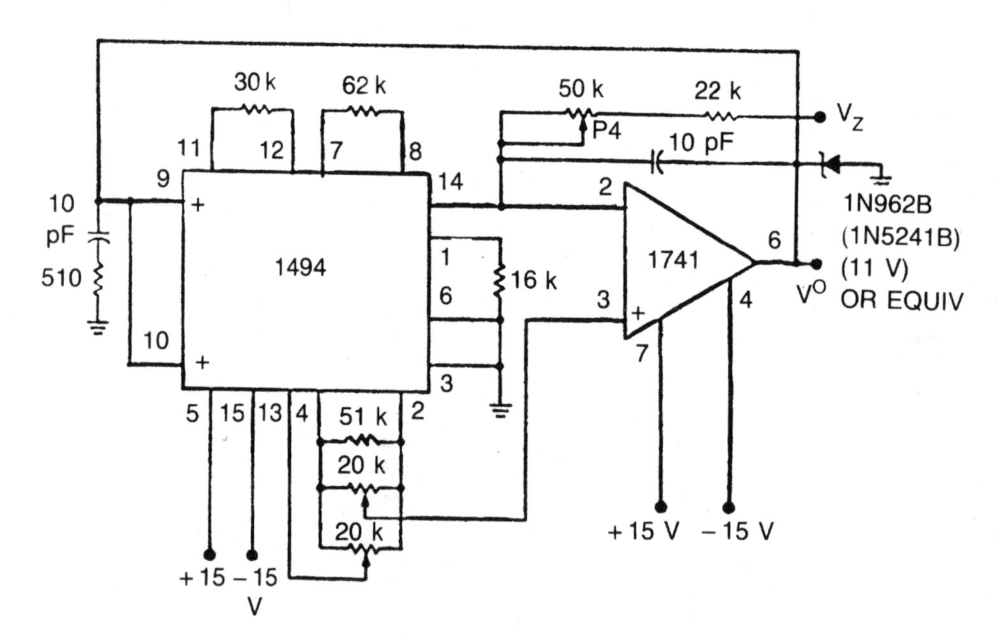

Squaring circuit

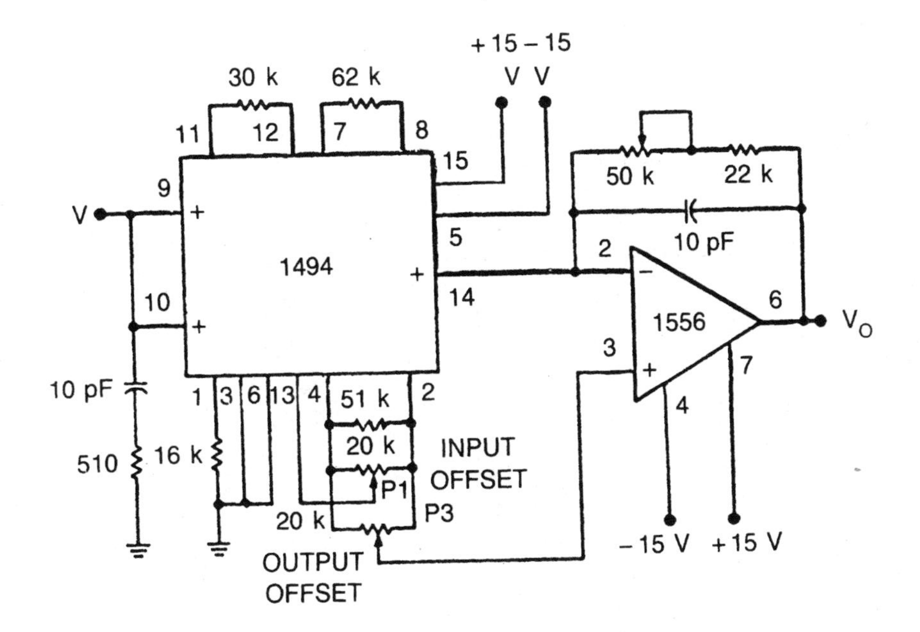

10

SECTION

Digital/analog hybrids

Ordinarily digital and linear (or analog) circuitry are two separate things. A digital circuitry can recognize only two voltage levels—LOW and HIGH, while an analog circuit responds to a continuous range of intermediate voltage levels. In some applications, however, it may be desirable to combine digital and analog functions in a single circuit or system.

Some circuits, by their nature, can function either way. In this final section, we will look at one example of such a hybrid analog/digital device.

7555 CMOS timer

The 7555 is a CMOS version of the popular 555 timer. It can be used in either analog or digital applications. Naturally, it is directly compatible with CMOS gates. Otherwise, it is functionally identical to the 555, and is pin-for-pin compatible.

Despite being a relatively simple device, nominally designed for just two basic circuits—a monostable multivibrator or an astable multivibrator—the 7555, like the 555, is incredibly versatile, and has been put to work in countless circuits and applications. It is very readily available, inexpensive, and quite easy to work with. For most basic applications, only a few external components are needed.

The timing period of the 7555 is set by a straightforward external resistor/capacitor combination, according to a fairly simple formula.

$$T = 1.1RC$$

For reliable operation, the timing resistor should have a value between 10 KΩ and 14 MΩ, and the timing capacitor's value should be somewhere between 100 pF and 1000 µF. The larger either of these component values is, the longer the resulting timing period will be.

In astable multivibrator applications, two timing resistors are used, but the basic operating principles still apply. Two resistors are used in the astable circuit, because different timing periods are needed for the LOW and HIGH portions of the cycle.

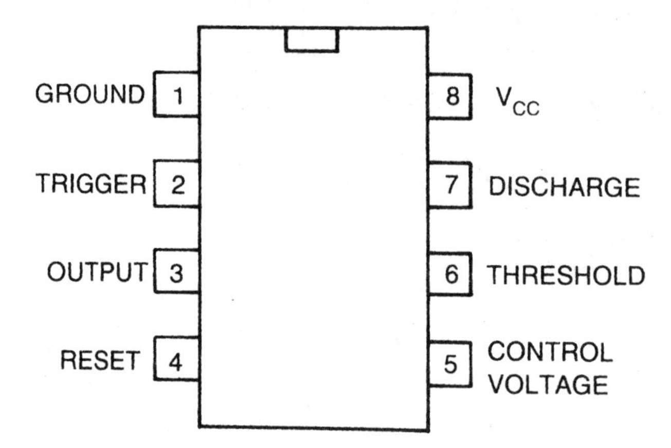

Basic monostable multivibrator

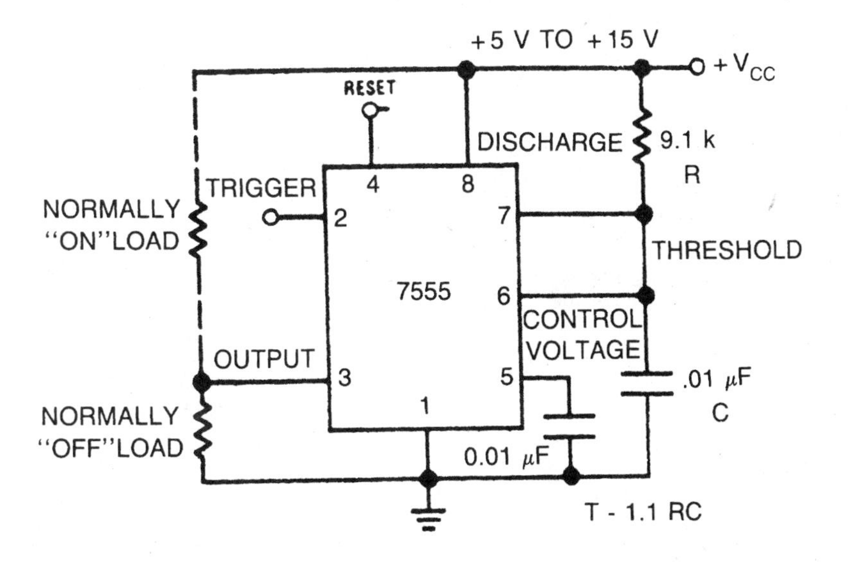

Basic astable multivibrator

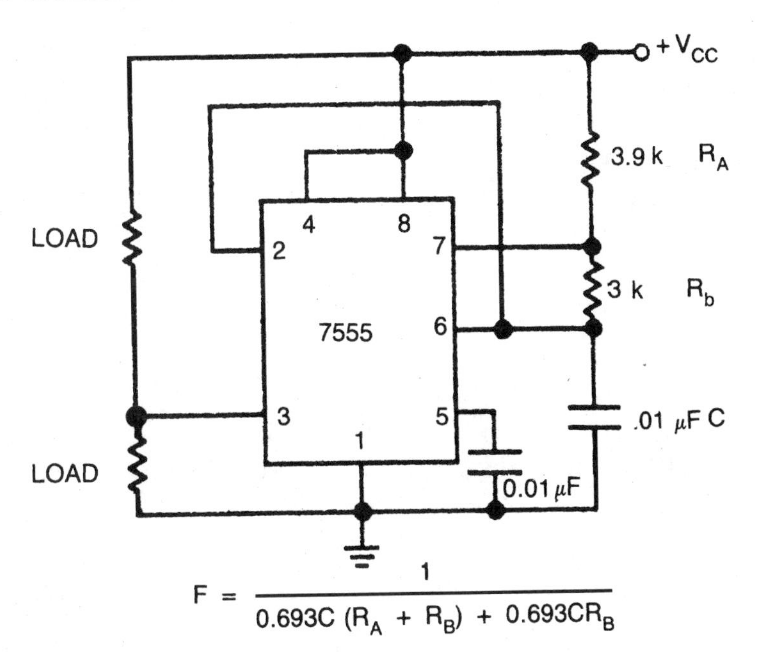

$$F = \frac{1}{0.693C\,(R_A + R_B) + 0.693CR_B}$$

A

APPENDIX

Symbols and definitions

DC voltages

All voltages are referenced to ground. Negative voltage limits are specified as absolute values (i.e., 10 V is greater than −1.0 V).

V_{cc} — *Supply voltage:* The range of power supply voltage over which the device is guaranteed to operate within the specified limits.

V_{cd} (Max) — *Input clamp diode voltage:* The most negative voltage at an input when the specified current is forced out of that input terminal. This parameter guaranteed the integrity of the input diode intended to clamp negative ringing at the input terminal.

V_{ih} — *Input HIGH voltage:* The range of input voltages recognized by the device as a logic HIGH.

V_{ih} (Min) — *Minimum Input HIGH voltage:* This value is the guaranteed input HIGH threshold for the device. The minimum allowed input HIGH in a logic system.

V_{il} — *Input LOW voltage:* The range of input voltages recognized by the device as a logic LOW.

V_{il} (Max) — *Maximum Input LOW voltage:* This value is the guaranteed input LOW threshold for the device. The maximum allowed input LOW in a logic system.

V_m — *Measurement voltage:* The reference voltage level on ac waveforms for determining ac performance. Usually specified as 1.5 V for most TTL families, but 1.3 V for the low-power Schottky 74LS family.

V_{oh} (Min) — *Output HIGH voltage:* The minimum guaranteed HIGH voltage at an output terminal for the specified output current I_{oh} and at the minimum V_{cc} value.

V_{ol} (Max) *Output* LOW *voltage:* The minimum guaranteed LOW voltage at an output terminal sinking the specified load current I_{ol}.

V_{t+} (Min) *Positive-going threshold voltage:* The input voltage of a variable threshold device which causes operation according to the specification as the input transition rises from below V_{t-}.

B
APPENDIX

Index of devices

Section	Device Number	Function
4	TL062	Dual low-power JFET input operational amplifier
8	117	Integrated adjustable voltage regulator
8	LM123	3-amp, 5-V positive voltage regulator
4	124	Quad operational amplifier
6	322	Precision timer
4	324	Quad operational amplifier
8	LM337	3-terminal adjustable negative voltage regulator
9	LM339/LM339A	Low-power, low-offset quad voltage comparator
8	340	Series voltage regulator
4	353	Wide-bandwidth, dual JFET input operational amplifier
5	LM380N	Audio power amplifier
5	LM384	5-watt audio power amplifier
5	LM386N	Low-voltage audio power amplifier
5	LM387	Low-noise dual preamplifier
5	387A	Low-noise preamplifier
5	LM389	Low-voltage audio power amplifier with NPN transistor array
5	LM390	1-watt battery-operated audio power amplifier
5	LM391	Audio power driver
8	TL431	Programmable precision reference
6	555	Timer
6	556	Dual timer
6	558	Quad timer
6	LM565	Phase-locked loop

Section	Device Number	Function
6	LM566	Voltage-controlled oscillator
6	LM567	Tone decoder
4	709	Monolithic operational amplifier
4	741	Internally compensated operational amplifier
4	747	Dual operational amplifier
4	748	High-performance operational amplifier
5	831	Low-voltage audio power amplifier
9	LM1036	Dual dc operated tone/volume/balance controller
7	1310	Phase-locked loop FM stereo demodulator
7	1350	Integrated amplifier
3	DS1386/DS1486	Watchdog Time-Keeper
9	1494	Linear 4-quad integrated multiplier
7	1496	Balanced modulator/demodulator
3	DS1620	Digital thermometer/thermostat
5	1875	20-watt audio power amplifier
5	LM18877	Dual audio power amplifier
7	MC2833	Low-power FM transmitter system
8	LM2931	Series low-dropout voltage regulators
8	LM2935	Low-dropout dual voltage regulator
4	3301	Quad single supply operational amplifier
4	LM3900	Quad Norton amplifier
6	LM3909	LED flasher/oscillator
2	4001	Quad 2-input NOR gate
2	4011	Quad 2-input NAND gate
2	4012	Dual 4-input NAND gate
2	4017	Decade counter/decoder
2	4023	Triple 3-input NAND gate
2	4028	BCD-to-decimal decoder
2	4046	Micropower phase-locked loop
2	4049	HEX inverting buffer
2	4051	8-channel analog multiplexer
2	4066	Quad bilateral switch
2	4070	Quad 2-input EXCLUSIVE-OR gate
2	4081	Quad 2-input and buffered B-series gate
2	4511	BCD-to-7-segment display latch/decoder/driver
2	4518	Dual BCD counter
6	MM5369	17-stage oscillator/driver
1	7400	Quad 2-input NAND gate
1	7402	Quad 2-input NOR gate
1	7404	HEX inverter
1	7408	Quad 2-input AND gate
1	7432	Quad 2-input OR gate
1	7447	BCD-to-7-segment decoder/driver
1	7473	Dual JK flip-flop with clear
1	7474	Dual D positive edge triggered flip-flop with preset and clear

Section	Device Number	Function
1	7476	Dual JK flip-flop
1	7490	Decade counter
8	7805	+5-V regulator
8	78105	5-V integrated voltage regulator
8	7812	+12-V regulator
8	7815	+15-V voltage regulator
8	MC7900	Series negative fixed voltage regulators
3	MC14490	HEX contact bounce eliminator
1	74123	Dual retriggerable one-shot with clear
1	74132	Quad 2-input NAND Schmitt trigger
1	74138	1-of-8 decoder/demultiplexer
1	74192	BCD up/down counter
1	74193	4-bit up/down counter

Index

Illustrations are in **boldface.**

A

B

E

F

H

I

J

L

M

N

O

T

U

V

W

X

Z